Challenges and Strategies of Dryland Agriculture

Challenges and Strategies of Dryland Agriculture

Srinivas C. Rao and John Ryan, *Co-editors*
Managing Editors: Lisa K. Al-Amoodi
Editor-in-Chief ASA Publication: Kenneth A. Barbarick
Editor-in-Chief CSSA Publication: Craig A. Roberts

CSSA Special Publication 32

Crop Science society of America, Inc.
American Society, Wisconsin, USA
Madison, Wisconsin, USA

SCIENTIFIC
PUBLISHERS

Published by:
Scientific Publishers
5A, New Pali Road, P.O. Box 91
Jodhpur 342 001, India
E-mail: info@scientificpub.com
Website: www.scientificpub.com

Print : 2024

Disclaimer: Whereas every effort has been made to avoid errors and omissions, this publication is being sold on the understanding that neither the editors (or authors of chapters in edited volumes) nor the publishers nor the printers would be liable in any manner to any person either for an error or for an omission in this publication, or for any action to be taken on the basis of this work. Any inadvertent discrepancy noted may be brought to the attention of the publishers, for rectifying it in future editions, if published.

ISBN: 978-81-72-334-55-0 (Hardbound)
ISBN: 978-93-87991-09-5 (E-book)

Crop Science society of America, Inc.

American Society, Wisconsin, USA

677 South Segoe Road, Madison, Wisconsin, USA

Printed in India

CONTENTS

FOREWORD

The world has made remarkable progress in maintaining adequate food supplies during the past quarter century by introducing yield-increasing technologies such as better genetics, crop protection products, and more efficient use of fertilizers and irrigation. Far more people depend on irrigation in the modern world than during the times of ancient Sumeria. The spread of irrigation has been the key factor in increasing global crop yields. But future water scarcity presents the single biggest threat to future food production. The shift of water from agriculture to the growing cities and industry almost certainly will impact global food production. This means that dryland agriculture will be increasingly important in meeting food requirement for the growing population. Advances in plant genetics and agronomic conservation technologies, when considered in concert, continue to provide the greatest opportunities to achieve sustainability and profitability in dryland agriculture and will continue to be the focus of the ARS research program.

The ARS is pleased to join the Crop Science Society of America and International Center for Agriculture Research in Dry Areas (ICARDA) in sponsoring a symposium "Challenges and Strategies for Dryland Agriculture" at the Tri-Societies Annual Meeting in November 2002 at Indianapolis, IN.

This special publication contains an impressive series of papers by an international group of experts on dryland agricultural production, conservation, and policy. The principles, philosophies, and technologies presented in this publication have the potential to contribute to improve food security and livelihoods for the people in dryland regions of the world.

Edward B. Knipling
Acting Administrater
USDA-ARS
Washington, DC

PREFACE

The world's population has more than doubled in the last half century, reaching 6 billion in 1999, and is projected to grow to 9.3 billion by the Year 2050. As a consequence of more people on this earth that need to be fed, in addition to rising incomes in some countries, the demand for food is expected to increase by 50% by 2015 and to more than double by 2050. It is this grim reality that agriculture as a profession faces—the challenge has to be met by everyone involved in the food production chain, from researchers to farmers, and all in between. Today, the world's population is better nourished than any time in history. The advances in global food production in the 20th century have dispelled, at least temporarily, the dire predictions of Malthus. Yet, despite such achievements, poverty and malnutrition, and their associated societal consequences, are the lot of numerous people in developing countries, especially in Africa and Asia.

In assessing modern agricultural practices and technology, it is important to distill both the limits and potential for sustaining global food supplies without degrading the resource base. Over the past three decades, expansion of irrigation, high-yielding varieties, and fertilizer input have been the major factors in achieving self-sufficiency in food grain production. On a worldwide basis, agriculture accounts for about 70% of all annual water withdrawals, and significant areas of irrigated lands are degraded to some extent by waterlogging and salinization. Another major pressure that leads to declining irrigation is high energy costs associated with delivering water to crops. With burgeoning populations, renewable fresh water resources are subject to severe competition between agriculture, industrial, and residential uses. Demand is increasing for all these uses. The shift of water from agriculture to urbanized societies and industry may hinder future global food production.

Given the pressure on the world's ecosystems, dryland agriculture, a sector that has been neglected in the past, will be increasingly important in meeting food requirements in the future. Globally, 90% of cropland is classified as dryland, and these lands provide 67% of all crop production and about half of the economic value of all crops. However, it is estimated that more than two-thirds of the potentially productive drylands are threatened by various forms of degradation. Therefore, it is essential that appropriate production technology be developed in the future to protect the fragile drylands in the process of meeting the needs of world's future population.

This timely symposium—the first major international meeting on the subject for many years—has brought dryland agriculture into the forefront of international agricultural research, as well as highlight the role of the world's major national and international research centers in addressing the problems of drylands. It was with this background, the Crop Science Society of America, the American Society of Agronomy, USDA-ARS, and the International Center for Agricultural Research in the Dry Area (ICARDA) sponsored a symposium addressing the wide array of issues associated with "Challenges and Strategies of Dryland Agriculture into the New Millennium" at the 2002 Annual Meetings of the Tri-Societies in Indianapolis, IN.

As Editors of this special publication, we express our gratitude to the authors who submitted their manuscripts in a timely manner to us. A particular thanks is due to those authors who, though not present or participating in the meeting itself, responded to the call for additional papers. Their contributions considerably complemented the subject-matter issues and undoubtedly widen the technical and geographical appeal of this volume. We also thank many scientists who served as reviewers for the chapters. In addition, we express our sincere appreciation to USDA-ARS Office of Technology Transfer, USDA-ARS Office of International Research Programs, USDA-ARS Grazing Lands Research Laboratory, and ICARDA for providing financial assistance for publication of this book.

Srinivas C. Rao, co-editor
USDA-ARS
Grazinglands Research Laboratory
El Reno, Oklahoma

John Ryan, co-editor
International Center for Agricultural
Research in the Dry Areas
Aleppo, Syria

ACKNOWLEDGMENTS

We would like to thank the following persons at ICARDA's Natural Resources Management Program who assisted in the typesetting, formatting, and graphics for the papers in this volume: Miss Rima El-Khatib, Administrative Assistant; Miss Zuka Moussatat, Secretary; and Mrs. Zuka Istanbouli, Secretary.

CONTRIBUTORS

Mohammad Abuzar	Institute of Sustainable Irrigated Agriculture, Department of Primary Industries, Tatura, Victoria 3616, Australia
Kanat Akshalov	Grain Research Institute, 474070 Institute, Shortandy District, Akmola Region, Kazakhstan
Darwin W. Anderson	Dept. Soil Science, University of Saskatchewan, Saskatoon, Saskatchewan, Canada S7N 5A8
John F. Angus	Commonwealth Scientific and Industrial Research Organization (CSIRO), Plant Industry, GPO Box 1600, Canberra 2601, Australia
Roger Armstrong	Victoria Institute for Dryland Agriculture, Department of Primary Industries, Horsham, Victoria 3401, Australia
Gunther Backes	Plant Research Department, Riso National Laboratory, P.O. Box 49, DK-4000, Roskid, Denmark
Michael Baum	International Center for Agricultural Research in the Dry Areas (ICARDA), P. O. Box 5466, Aleppo, Syria
Robert Belford	Victoria Institute for Dryland Agriculture, Department of Primary Industries, Horsham, Victoria 3401, Australia
Alan T. P. Bennie	Department of Soil Science, University of the Orange Free State, Bloemfontein, South Africa
Danie J. Beukes	Agricultural Research Center, Institute for Soil, Climate, and Water, Pretoris, South Africa
Mustapha Bounejmate	International Center for Agricultural Research in the Dry Areas (ICARDA), P. O. Box 5466, Aleppo, Syria
Salvatore Ceccarelli	International Center for Agricultural Research in the Dry Areas (ICARDA), P. O. Box 5466, Aleppo, Syria
Mushtaq A. Chaudhry	University of Arid Agriculture, Rawalpindi, Pakistan
William D. Dar	International Crop Research Institute for Semi-Arid Tropics (ICRISAT), Patancheru 502324, AP, India
Allen R. Dedrick	National Program Staff, 5601 Sunnyside Avenue, Beltsville, MD 20705
Eddy De Pauw	International Center for Agricultural Research in the Dry Areas (ICARDA), P. O. Box 5466, Aleppo, Syria
Adel El-Beltagy	International Center for Agricultural Research in the Dry Areas (ICARDA), P. O. Box 5466, Aleppo, Syria
Mohamed El-Mourid	International Center for Agricultural research in the Dry Areas, Tunis, BP 435 El Menzah, 1004 Tunis, Tunisia
Ismahane Elouafi	International Center for Agricultural Research in the Dry Areas (ICARDA), P. O. Box 5466, Aleppo, Syria

William Erskine International Center for Agricultural Research in the Dry Areas (ICARDA), P. O. Box 5466, Aleppo, Syria

Jurgan D. Garbacht USDA-ARS, Grazinglands Research Laboratory, 7202 W. Cheyenne St. El Reno, OK 73036

Berhanu Gebremedhin International Livestock Research Institute, P.O.Box 5689, Addis Ababa, Ethiopia

Anthony J. Good Conimbla Road, Cowra 2794, Australia

Stefania Grando International Center for Agricultural Research in the Dry Areas (ICARDA), P. O. Box 5466, Aleppo, Syria.

Keith R. Helyar Wagga Wagga Agricultural Institute, NSW Agriculture, PMB, Wagga Wagga NSW 2650, Australia

Malcom Hensley Department of Soil Science, University of the Orange Free State, Bloemfontein, South Africa

John Howieson Center for Rhizobium Studies, Division of Science, Murdoch University, South St, Murdoch, Western Australia 6150, Australia

Mark Imhof State Chemistry Laboratory, Department of Primary Industries, Werribee, Victoria 3030, Australia

H. Henry Janzen Agriculture and Agri-Food Canada, Research Center, 5403 1st Ave. S., Lethbridge, Alberta, Canada TIJ 4BI.

Ahmed Jahoor Plant Research Department, Riso National laboratory, P.O. Box 49, DK-4000, Roskid, Denmark

Aitkalym Kireyev Crop Husbandry Research Institute, KIZ, 37, Erlepesov Street, Almalybak, Kaskelen District, Almaty Region, Kazakhstan

Parviz Koohafkan Land and Water Development Division, B-749, Food and Agriculture Organization (FAO), Viale delle Terme di Caracalla 00100, Rome

Rattan Lal Carbon Management and Sequestration Center, 2021 Coffey Road, The Ohio State University, Columbus, Ohio 43210

Francis J. Larney Agriculture and Agri-Food Canada, Research Center, 5403 1st Ave. S., Lethbridge, Alberta, Canada TIJ 4BI

Eric M. McGaw International Crop Research Institute for Semi-Arid Tropics (ICRISAT), Patancheru 502324, AP, India

Rajinder S. Malhotra International Center for Agricultural Research in the Dry Areas (ICARDA), P. O. Box 5466, Aleppo, Syria

Lyudmila Martynova Cropping Research Institute, 720027 KIZ 91, T. Frunze Street, Bishkek, Kyrgyzstan

Herman S. Mayeux USDA-ARS, Grazinglands Research Laboratory, 7207 W. Cheyenne St., El Reno, OK 73036

Rakhim Medeubayev 487210 Krasniy vodopad, Saryagash District, South Kazakhstan Region, Kazakhstan

Ali M. Abd El Moneim International Center for Agricultural Research in the Dry Areas (ICARDA), P. O. Box 5466, Aleppo, Syria

Miloudi M. Nachit	International Center for Agricultural Research in the Dry Areas (ICARDA), P. O. Box 5466, Aleppo, Syria
Adel Nassar	Terbol Research Station, ICARDA, Terbol, Bekaa, Lebanon
Tidiane Ngaido	International Food Policy Research Institute, 2033 K Street, N.W. Washington, DC 20006
James Nuttall	Victoria Institute for Dryland Agriculture, Department of Primary Industries, Horsham, Victoria 3401, Australia
Rajendra S. Paroda	ICARDA Tashkent Office, Central Asia and Caucasus, Uzbekistan, P.O. Box 4564 6-106, Murtazaeva Street, Tashkent 700000
John Pender	International Food Research Institute, 2033 K-Street, Washington, DC 20006
Srinivas C. Rao	USDA-ARS, Grazinglands Research Laboratory, 7207 W. Cheyenne St., El Reno, OK 73036
Abdul Rashid	Land Resources Research Program, National Agricultural Research Center, Park Road, Islamabad 4500, Pakistan
D. V. R. Reddy	Donald Danforth Plant Sciences Center, 975 North Warson Road, St. Louis, MO 63132
Gangireddy Subba Reddy	Central Research Institute for DrylandAgriculture, Santoshnagar, Saidabad, P.O., Hyderabad 500059, A. P., India.
John Ryan	International Center for Agricultural Research in the Dry Areas (ICARDA), P. O. Box 5466, Aleppo, Syria.
Ashutosh Sarker	International Center for Agricultural Research in the Dry Areas (ICARDA), P. O. Box 5466, Aleppo, Syria.
Mohan C. Saxena	International Center for Agricultural Research in the Dry Areas (ICARDA), P. O. Box 5466, Aleppo, Syria
Jeanne M. Schneider	USDA-ARS, Grazinglands Research Laboratory, 7202 W. Cheyenne St. El Reno, OK 73036
Kapil D. Sharma	Central Research Institute for Dryland Agriculture, Santoshnagar, Saidabad, P.O., Hyderabad 500059, A. P., India
Kishori L. Sharma	Central Research Institute for Dryland Agriculture, Santoshnagar, Saidabad, P.O., Hyderabad 500059, A. P., India
Kadambot H. M. Siddique	Center for Legumes in Mediterranean Agriculture (CLIMA), Faculty of Natural and Agricultural Sciences, The University of Western Australia, 35 Stirling Highway, Crawley, Western Australia 6009, Australia
Harish P. Singh	Central Research Institute for Dryland Agriculture, Santoshnagar, Saidabad, P.O., Hyderabad 500059, A. P., India
Jo Slattery	Rutherglen Research Institute, Department of Primary Industries, RMB1145, Chiltern Valley Rd, Rutherglen, Victoria 3685, DX 218569, Australia

William J. Slattery	Australian Greenhouse Office, Department of Environment and Heritage, Australian Government, John Gorton Building, P.O. Box 621, Canberra, ACT 2061
Elwin G. Smith	Agriculture and Agri-Food Canada, Research Center, 5403 1st Ave. S., Lethbridge, Alberta, Canada TIJ 4BI
Jean L. Steiner	USDA-ARS, Grazinglands Research Laboratory, 7207 W. Cheyenne St., El Reno, OK 73036
Bobby A. Stewart	Dryland Agriculture Institute, West Texas A&M University, WTAMU Box 60278, Canyon, TX 79016-0001
Mekhlis Suleimenov	ICARDA Tashkent Office, Central Asia and Caucasus, Uzbekistan, P.O. Box 4564 6-106, Murtazaeva Street, Tashkent 700000
Sripada M. Udupa	International Center for Agricultural Research in the Dry Areas (ICARDA), P. O. Box 5466, Aleppo, Syria
Sui-Kwong Yau	Faculty of Agriculture and Food Science, American University of Beirut, P.O. Box 11-0236, Beirut, Lebanon
Khasan Yusupov	Galla-Aral Grain Research Institute, 704620 Institute, Olimlar Makhallasi, 1, Omonova Street, Jizzah Viloyat, Uzbekistan
Xunchang J. Zhang	USDA-ARS, Grazinglands Research Laboratory, 7202 W. Cheyenne St. El Reno, OK 73036

Conversion Factors for SI and non-SI Units

Conversion Factors for SI and non-SI Units

To convert Column 1 into Column 2, multiply by	Column 1 SI Unit	Column 2 non-SI Units	To convert Column 2 into Column 1, multiply by
		Length	
0.621	kilometer, km (10^3 m)	mile, mi	1.609
1.094	meter, m	yard, yd	0.914
3.28	meter, m	foot, ft	0.304
1.0	micrometer, μm (10^{-6} m)	micron, μ	1.0
3.94×10^{-2}	millimeter, mm (10^{-3} m)	inch, in	25.4
10	nanometer, nm (10^{-9} m)	Angstrom, Å	0.1
		Area	
2.47	hectare, ha	acre	0.405
247	square kilometer, km^2 (10^3 m)^2	acre	4.05×10^{-3}
0.386	square kilometer, km^2 (10^3 m)^2	square mile, mi^2	2.590
2.47×10^{-4}	square meter, m^2	acre	4.05×10^{3}
10.76	square meter, m^2	square foot, ft^2	9.29×10^{-2}
1.55×10^{-3}	square millimeter, mm^2 (10^{-3} m)^2	square inch, in^2	645
		Volume	
9.73×10^{-3}	cubic meter, m^3	acre-inch	102.8
35.3	cubic meter, m^3	cubic foot, ft^3	2.83×10^{-2}
6.10×10^{4}	cubic meter, m^3	cubic inch, in^3	1.64×10^{-5}
2.84×10^{-2}	liter, L (10^{-3} m^3)	bushel, bu	35.24
1.057	liter, L (10^{-3} m^3)	quart (liquid), qt	0.946
3.53×10^{-2}	liter, L (10^{-3} m^3)	cubic foot, ft^3	28.3
0.265	liter, L (10^{-3} m^3)	gallon	3.78
33.78	liter, L (10^{-3} m^3)	ounce (fluid), oz	2.96×10^{-2}
2.11	liter, L (10^{-3} m^3)	pint (fluid), pt	0.473

		Mass	
2.20×10^{-3}	gram, g (10^{-3} kg)	pound, lb	454
3.52×10^{-2}	gram, g (10^{-3} kg)	ounce (avdp), oz	28.4
2.205	kilogram, kg	pound, lb	0.454
0.01	kilogram, kg	quintal (metric), q	100
1.10×10^{-3}	kilogram, kg	ton (2000 lb), ton	907
1.102	megagram, Mg (tonne)	ton (U.S.), ton	0.907
1.102	tonne, t	ton (U.S.), ton	0.907
		Yield and Rate	
0.893	kilogram per hectare, kg ha^{-1}	pound per acre, lb $acre^{-1}$	1.12
7.77×10^{-2}	kilogram per cubic meter, kg m^{-3}	pound per bushel, lb bu^{-1}	12.87
1.49×10^{-2}	kilogram per hectare, kg ha^{-1}	bushel per acre, 60 lb	67.19
1.59×10^{-2}	kilogram per hectare, kg ha^{-1}	bushel per acre, 56 lb	62.71
1.86×10^{-2}	kilogram per hectare, kg ha^{-1}	bushel per acre, 48 lb	53.75
0.107	liter per hectare, L ha^{-1}	gallon per acre	9.35
893	tonne per hectare, t ha^{-1}	pound per acre, lb $acre^{-1}$	1.12×10^{-3}
893	megagram per hectare, Mg ha^{-1}	pound per acre, lb $acre^{-1}$	1.12×10^{-3}
0.446	megagram per hectare, Mg ha^{-1}	ton (2000 lb) per acre, ton $acre^{-1}$	2.24
2.24	meter per second, m s^{-1}	mile per hour	0.447
		Specific Surface	
10	square meter per kilogram, m^2 kg^{-1}	square centimeter per gram, cm^2 g^{-1}	0.1
1000	square meter per kilogram, m^2 kg^{-1}	square millimeter per gram, mm^2 g^{-1}	0.001
		Density	
1.00	megagram per cubic meter, Mg m^{-3}	gram per cubic centimeter, g cm^{-3}	1.00
		Pressure	
9.90	megapascal, MPa (10^6 Pa)	atmosphere	0.101
10	megapascal, MPa (10^6 Pa)	bar	0.1
2.09×10^{-2}	pascal, Pa	pound per square foot, lb ft^{-2}	47.9
1.45×10^{-4}	pascal, Pa	pound per square inch, lb in^{-2}	6.90×10^3

(continued on next page)

Conversion Factors for SI and non-SI Units

To convert Column 1 into Column 2, multiply by	Column 1 SI Unit	Column 2 non-SI Units	To convert Column 2 into Column 1, multiply by
		Temperature	
1.00 (K – 273)	kelvin, K	Celsius, °C	1.00 (°C + 273)
(9/5 °C) + 32	Celsius, °C	Fahrenheit, °F	5/9 (°F – 32)
		Energy, Work, Quantity of Heat	
9.52×10^{-4}	joule, J	British thermal unit, Btu	1.05×10^{3}
0.239	joule, J	calorie, cal	4.19
10^{7}	joule, J	erg	10^{-7}
0.735	joule, J	foot-pound	1.36
2.387×10^{-5}	joule per square meter, J m^{-2}	calorie per square centimeter (langley)	4.19×10^{4}
10^{5}	newton, N	dyne	10^{-5}
1.43×10^{-3}	watt per square meter, W m^{-2}	calorie per square centimeter minute (irradiance), cal cm^{-2} min^{-1}	698
		Transpiration and Photosynthesis	
3.60×10^{-2}	milligram per square meter second, mg m^{-2} s^{-1}	gram per square decimeter hour, g dm^{-2} h^{-1}	27.8
5.56×10^{-3}	milligram (H_2O) per square meter second, mg m^{-2} s^{-1}	micromole (H_2O) per square centimeter second, µmol cm^{-2} s^{-1}	180
10^{-4}	milligram per square meter second, mg m^{-2} s^{-1}	milligram per square centimeter second, mg cm^{-2} s^{-1}	10^{4}
35.97	milligram per square meter second, mg m^{-2} s^{-1}	milligram per square decimeter hour, mg dm^{-2} h^{-1}	2.78×10^{-2}
		Plane Angle	
57.3	radian, rad	degrees (angle), °	1.75×10^{-2}

	Electrical Conductivity, Electricity, and Magnetism		
10	siemen per meter, $S\ m^{-1}$	millimho per centimeter, $mmho\ cm^{-1}$	0.1
10^4	tesla, T	gauss, G	10^{-4}
	Water Measurement		
9.73×10^{-3}	cubic meter, m^3	acre-inch, acre-in	102.8
9.81×10^{-3}	cubic meter per hour, $m^3\ h^{-1}$	cubic foot per second, $ft^3\ s^{-1}$	101.9
4.40	cubic meter per hour, m^3 h–1	U.S. gallon per minute, $gal\ min^{-1}$	0.227
8.11	hectare meter, ha m	acre-foot, acre-ft	0.123
97.28	hectare meter, ha m	acre-inch, acre-in	1.03×10^{-2}
8.1×10^{-2}	hectare centimeter, ha cm	acre-foot, acre-ft	12.33
	Concentrations		
1	centimole per kilogram, $cmol\ kg^{-1}$	milliequivalent per 100 grams, meq $100\ g^{-1}$	1
0.1	gram per kilogram, $g\ kg^{-1}$	percent, %	10
1	milligram per kilogram, $mg\ kg^{-1}$	parts per million, ppm	1
	Radioactivity		
2.7×10^{-11}	becquerel, Bq	curie, Ci	3.7×10^{10}
2.7×10^{-2}	becquerel per kilogram, $Bq\ kg^{-1}$	picocurie per gram, $pCi\ g^{-1}$	37
100	gray, Gy (absorbed dose)	rad, rd	0.01
100	sievert, Sv (equivalent dose)	rem (roentgen equivalent man)	0.01
	Plant Nutrient Conversion		
	Elemental	***Oxide***	
2.29	P	P_2O_5	0.437
1.20	K	K_2O	0.830
1.39	Ca	CaO	0.715
1.66	Mg	MgO	0.602

1 The Role of World's Agricultural Lands for Future Food Security

Srinivas C. Rao, Jean L. Steiner, and Herman S. Mayeux
USDA-ARS, Grazinglands Research Laboratory
El Reno, Oklahoma

ABSTRACT

Food security has been described by the Food and Agricultural Organization as existing when "all people, at all times, have physical and economic access to sufficient, safe, and nutritious food to meet their dietary needs and food preferences for an active and healthy life." The Johannesburg Declaration on Sustainable Development framed the issue of the basic human right of universal food security in broad terms of poverty eradication, with the specific goal of at least halving the number of undernourished people in the world by the Year 2015. Lack of food security poses a particular burden on peoples and nations in the dryland regions of the world, particularly in tropical areas of Africa and Asia that are experiencing rapid population growth and/or high population density. Global food demand is expected to more than double by 2050 because of population growth and increased per capita consumption. While the challenge cannot be met through increased agricultural production alone, increased production is essential as part of the solution. However, in many cases, production capacities of dryland countries are deteriorating in the face of rapid population growth, misdirected agricultural practices, and widespread land degradation. Subsequent chapters in this publication highlight these issues in more detail and identify basic principles of soil and water management and plant growth under drought conditions. The authors also identify promising new technologies such as biotechnological approaches to develop improved drought and pest-resistant crop varieties, emerging climate forecast capabilities that might be applied to risk management for dryland farmers, and remote sensing approaches for better inventory and monitoring of environmental conditions in vast dryland areas to better target remediation. The principles and technologies presented in this publication, along with necessary international, national, and local policies to enhance capacity, have the potential to contribute to improved food security and livelihoods for the people in dryland regions of the world.

INTRODUCTION

Food security was recognized as a basic human right by the United Nations (UN) in the Universal Declaration of Human Rights of 1948. Commitment to this principle has been reaffirmed in numerous later U.N. Conventions and Resolutions, including the Millennium Declaration of 2000 in which freedom from hunger was identified as a fundamental value underlying international relations in the 21st century. The Johannesburg Declaration on Sustainable Development framed the issue of the basic human right of universal food security in broader terms of poverty eradication, with the specific goal of at least halving the number of undernourished people in the world by the Year 2015. Food security has been described by the Food and Agricultural Organization (www.fao.org/spfs/) as existing "when all people, at all times, have physical and economic access to sufficient, safe, and nutritious food to meet their dietary needs and food preferences for an active and healthy life."

Lack of food security poses a particular burden on people and nations in the dryland regions of the world, particularly in tropical areas of Africa and Asia that are experiencing rapid population growth and/or high population density. Food security must be addressed at global, national, community, and household levels. At all levels, the lack of food security is related to a complex array of issues including extreme poverty, poor health, political instability, and extreme climates. The problem cannot be solved through increased agricultural production alone. However, agricultural production is essential as part of the solution.

Per capita food supply in an agrarian society results from the amount of land under production, production levels achieved per unit area, and numbers of people to be supported per unit area of production. In the face of ongoing population increases, competition from other sectors reduces land available for agriculture. Negative threats to productivity include land degradation, lack of access to inputs and markets, and variable and changing climate. The challenge is monumental.

Population

The world's population more than doubled in the last half century, reaching 6 billion in 1999. World population is projected to grow from 6 billion to 9.3 billion by 2050. With a global population growth rate of 1.3% per year, almost all of the growth will be in less developed countries. The very poorest countries are expected to triple in population from 600 million in 1995 to 1.8 billion by 2050. About 45% of the total growth is anticipated in China, India, Pakistan, Bangladesh, and Indonesia (UNFPA, 2003). China and India, the world's two most populous countries, provide examples of how even modest growth rates produce large numbers of additional people when the population base is large. Many of the countries under pressure from high and increasing population also face serious threats from high and increasing prevalence of HIV/AIDS. This is now the leading cause of death in sub-Saharan Africa and the fourth highest cause of death globally. More than 40 million children likely will be orphaned by HIV/AIDS by 2010, further exacerbating problems of poor health, poverty, hunger, and demands on governments and international programs.

Total food demand is expected to increase by 50% by 2015 and by more than 110% by 2050 (Labunets, 2003) through population growth and increased per capita consumption. However, in many cases, food production capacities of drylands are deteriorating in the face of rapid population growth, inadequate agricultural systems, and widespread land degradation. The bottom line is food security cannot be achieved if two natural resources essential to agricultural production—land and water—continue to become increasingly degraded and polluted. Although some experts are confident that a future world of nearly 10 billion people will be able to feed itself, others share grave concerns about the future sustainability of current farming practices, especially in poor, food-deficit countries with growing population.

World's Agricultural Lands

The area of global cropland is roughly 1.5 billion hectares, of which about 17% is irrigated and the rest is rainfed (Hofwegen and Svendsen, 2000). There is widespread concern about declining availability of agricultural land. The reduced per capita land base associated with population growth is aggravated by land degradation and conversion of agricultural land to urban or other uses. Between early 1960s and the late 1990s, world cropland grew by only 11%, while world population almost doubled. As a result, cropland per person fell by 40%, from 0.43 ha to only 0.26 ha (FAO, 2002).

Irrigated Lands

Irrigated land accounts for about 17% of the world's cropland, and produces about 40% of the world's food (FAO, 1988). Asia has highest percent of irrigated land (66%). About 70% of the grain in China and 50% of the grain in India is harvested from irrigated lands (Brown, 1999). One of the major questions on the future of irrigation is whether there will be sufficient fresh water to satisfy the growing needs of agricultural and nonagricultural users. On a worldwide basis, agriculture accounts for about 70% of fresh water withdrawals, followed by 22% for industry and 8% for human consumption (Engleman and Leroy, 1993). Among the countries likely to experience major water shortages in the next 25 yr are Ethiopia, India, Kenya, Peru, and China (Gardner-Outlaw and Engleman, 1997). The rate of expansion of agricultural lands under irrigation is slowing. Already, most of Africa and the Middle East, much of western USA and northern Mexico, parts of Chile and Argentina, and nearly all of Australia suffer water shortages (Hinrichsen, 1998). Significant areas of irrigated agricultural lands (10–15%) are degraded to some extent by waterlogging and salinization (Alexandratos, 1995). The FAO estimates that salt build-up has severely damaged 30 million ha of the world's 240 million ha of irrigated land. Salinity has cut yields on nearly one-quarter of China's irrigated lands and 21% of Pakistan's irrigated lands (Hinrichsen, 1998). Another major pressure that leads to declining irrigation is high energy costs associated with delivering water to crops.

Renewable fresh water resources are subject to severe competition between agriculture, industrial and residential uses, while demand is increasing for all these uses. Falkenmark et al. (1998) reported that a unit of water invested in industry could give a financial return 100 times higher than water used for irrigation. In China, water invested in industry returns 70 times more economic benefit than the same water unit invested in agriculture (Brown and Halweil, 1998). This shift of water is not happening in China alone, but rather worldwide and in developing countries particularly. Water invested in industry also generates larger employment than water invested in agriculture. The shift of water from agriculture to the growing cities and industry may hinder the future global food production. This means that dryland agriculture will be increasingly important in meeting food requirement in the future.

Drylands

Almost 40% of the earth's total land surface (6.1 billion ha) is dry. Out of this, about 0.9 billion hectares are hyper arid deserts. The remaining hectares are arid, semi-arid, and dry subhumid lands that are collectively referred to as drylands. It is estimated that 70% of potentially productive drylands are threatened by various forms of degradation, impacting the well-being and future of one-sixth of the world's population (Harahsheh, 2002). The environmental conditions of the world's drylands and unpredictability of rainfall make these areas marginal for intensive agriculture. When cropping is attempted, the risk of crop failure is high due to moisture deficit. Land degradation in drylands—via wind and water erosion, loss of soil fertility, salinization, groundwater depletion and loss of vegetation—results in decline of both economic and environmental potential in these regions. Dregne (1977) estimated that only 18% of the world's drylands had experienced slight degradation, whereas moderate and severe to very severe degradation had taken place on 54 and 28% of the land, respectively. Eswaran et al. (2001) combined analysis of vulnerability of land to desertification with maps of population density and determined that 44 million km^2 of land (34% of global land area) with more than 2.6 billion inhabitants (44% of the global population) were at risk to the pressures of desertification. Of these lands, 7.9 million km^2 of land with about 1.4 billion inhabitants were at very high risk of degradation.

The earth's drylands are found in more than 110 nations and are a vital part of the earth's human and physical environment. Agriculture remains the dominant activity in the dryland regions. Within the developing regions, 73% of Africa's agricultural lands are moderately to severely degraded. Asia has the largest area of land-affected desertification, 1.4 million ha, of which 71% is moderately to severely degraded. Nearly 75% of Latin America's drylands are in this state (Sultan, 1998). In addition to productivity losses and increasing poverty, drylands degradation results in significant reduction in carbon (C) storage in soils and increased dust in the atmosphere, both of which have implications for global atmospheric processes and global change. People play a considerable role in degrading their own land. The stresses of poverty and overpopulation push them to overgraze and overcultivate. With limited access to resources and in a desperate struggle to survive, they are driven to destroy forest belts, practice poor irrigation, and use inappropriate agricultural practices such as slash and burn and shortened fallow periods. Progressive degradation under the pressure of relentlessly expanding population is the most common image of the future of the world's drylands.

CHALLENGES AND PROSPECTS FOR FUTURE FOOD SECURITY

Prospects for food supply and demand in the future is a matter of serious concern. Some of the current farming practices have high environmental cost. It's becoming quickly apparent that the way people have grown and continue to grow food crops have led to widespread ecological degradation. Water shortages, and degradation of agricultural lands through soil erosion, salinization, and water logging, pose a serious threat to food production (Alexandratos, 1995). According to a report by the International Food Policy Research Institute (IFPRI) and World Resources Institutes (WRI), world food production is "at risk from farming methods that have degraded soils, parched aquifers, polluted waters, and caused the loss of animal and plant species" (Wood et al., 2001). The report emphasized that since agricultural lands dominate in populated areas, in addition to food production, we must also rely on agricultural lands for many goods and services, such as clean water and habitat for other species.

Irrigated Lands

Irrigation is vitally important in meeting the food and fiber needs for a rapidly expanding world population. One of the major questions on the future irrigation is whether there will be sufficient fresh water to satisfy the growing needs of agricultural and nonagricultural users. Many fresh water sources—underground aquifers and rivers—are stressed beyond their limits. Large areas of irrigated crops are grown using groundwater from aquifers that are being depleted (Postel, 2001). The great global challenge for the coming years will be how to produce more food with less water. The first line of attack is to increase irrigation efficiency. At present most farmers irrigate their crops by flooding their fields or channeling the water down parallel furrows, using gravity to move the water across the land. The plants absorb only a small fraction of water, while the rest drains into rivers or aquifers, or evaporates. Drip irrigation ranks high among irrigation technologies that enables farmers to deliver water directly to the plant roots thereby eliminating waste. Studies in countries like India, Israel, Jordan, Spain, and the USA have shown that drip irrigation reduces water use by 30 to 70% and increases crop yield by 20 to 90% compared with flood irrigation (Postel, 2001). However, higher cost of these technologies compared to simple flood methods has been a barrier to their spread. Improved timing and scheduling of irrigation to more precisely match the water needs of the plant can also reduce the water withdrawal for irrigated agriculture.

Most major irrigation schemes throughout the world suffer to some degree from the effects of salinity and water logging. Irrigated soils generally require some amount of leaching to move salts below the root zone. However, in areas with shallow groundwater, regional salinity problems often arise when excessive leaching causes the groundwater to move near the surface. This results in upward movement of water toward the surface, driven by evaporation, and carrying salts into the root zone. Through complex and diverse mechanisms, many formerly productive irrigated lands have become saline wastelands. The proper operation of viable, permanent irrigated lands requires careful monitoring and management of the levels and distribution of salinity within the root zone and within regional groundwater supply. Salinity adversely influences crop establishment, reduces plant growth rates, reduces yields, and, in severe cases, results in total crop failure (Rhoades and Loveday, 1990). Rhoades et al. (1988) reported that salinity affects on the emergence and establishment can be overcome for some crops by irrigating with diluted saline water during crop establishment and then with undiluted saline water later in the growing season.

The first requisite for the reclamation of salt-affected soil is adequate drainage. Salinity of soils can often be reduced to an acceptable level by leaching. In very saline soils, it may be required to set aside cropping temporarily and accelerate leaching of the salt from the root zone. However, in shallow, saline water table areas, this is not an option for remediation. Acceleration of evaporation (to lower a saline water table) through intensification of cropping systems or planting of deeper rooted and salt-tolerant plants may be required. Some of these procedures may help in reducing the degradation of irrigated agricultural lands. In some cases, it may not be economically feasible to maintain low salinity. In such instances, judicious selection of crops that can produce satisfactory yields under saline conditions and use of management practices to minimize the impacts of salinity may make the difference between success and failure. Recently, a team of scientists at Purdue University discovered the protein (AtHKT1) that is thought to act as a transporter in plant tissue, binding with the salt ion and moving it into the plant cells. This finding provides basic knowledge that may have future potential to block the activity of this protein through genetic modification (Hasegawa, 2002). Molecular genetics researchers are identifying genes that provide tolerance to salinity that may be of great benefit in maintaining food production on irrigated agricultural lands under threat from salinity.

Drylands

Continued reliance on irrigation, high yielding varieties, and fertilizers alone will not bring about the required increases in food grain production needed to meet food requirements. Improving productivity and stability of production in rainfed areas will, therefore, be crucial in meeting the needs of the increasing population. For decades, drylands have been the neglected stepchild of development because of the harsh, but compelling logic that investments in wetter agricultural areas with richer soils would yield far greater returns. Instead of the high input approach favored for rich lands, a more appropriate perspective for drylands would be to search for technologies and adaptations that require low external inputs and minimize vulnerability to crop failure.

Dryland areas are very diverse in their agro-climatic conditions, and hence in their potential for agricultural growth. In many dry regions, productivity is declining and living conditions are deteriorating. Land degradation in arid, semi-arid and dry subhumid land areas is generally attributed to a combination of climatic variations and human activities. Some of the human activities that can cause degradation of dry regions include: (i) cultivation of fragile soils, (ii) reduction in the fallow period, (iii) overgrazing that exposes soil to wind and water erosion, (iv) agricultural practices that result in the net export of soil nutrients, and (v) poor irrigation practices. Drylands issues are complex. However, solutions can be developed if the affected people are given increased access to resources to lead the processes of change.

To prevent future degradation of fragile drylands, there is a need to monitor and assess the trends of the world's drylands. Remote sensing from space could provide the necessary information to monitor the degradation. Landsat and other satellite imagery have proven uniquely effective for measuring and determining changes in the global landscape. Agencies of the U.S. government, especially the Agency for International Development and the National Aeronautic and Space Administration, have explored using the Landsat system as a primary land survey tool to monitor degradation. Adaptation of latest remote sensing technologies may play an important role in combating degradation of dryland agricultural lands through earlier detection of problems and better targeting of remediation efforts.

Sparse precipitation and highly variable climate conditions make the risk of crop failure high. In the future, improved seasonal climate forecasts may allow farmers to respond more proactively and less reactively to rainfall variability. Experience in Australia, parts of Africa, the USA, and South America has shown that the emerging ability to forecast future seasonal rainfall and temperature may provide increased potential to manage risks, enhance productivity, and improve farmers' livelihoods (Meinke et al., 2002). However, researchers and developers of such technologies must work closely with farmers and extension agents to adapt climate forecast information to specific farm management issues in different regions. For climate forecasting to be successful, there will be a need to establish national and regional 'agro meteorological teams', that acquire forecast data, make the necessary calculations, and communicate them to the farmers and their advisors in a timely and useful fashion.

While climate and soils provide great management challenges, drylands still are critical to provide food for local populations. Sustainable practices for use on drylands are constrained by low rainfall patterns. In dryland agricultural areas, the amount and distribution of rain during the rainy season governs the quality and length of the growing season. To meet crop water needs and secure an adequate length of growing period, practices must focus on: (i) conserving water in the soil profile by allowing adequate opportunity for rain water to infiltrate into the soil and (ii) shaping the land surface to minimize runoff received during the periods of high volume rainfall storms.

The greatest deterrent to a high rate of water absorption is the tendency for soil to puddle at the surface and form a seal or crust against water intake. Tillage has often been used to roughen the surface to retain water longer for increased infiltration. However, the positive impact of tillage on infiltration only persists a short while and over time decreases soil organic matter, often leading to future greater problems with soil crusting. Retention of crop residues in dryland agriculture not only improves fertility levels and soil structure, but also can improve water infiltration and moisture storage in soil.

Application improved land and water conservation methods in watershed management has been widely accepted as a practical solution for sustaining agricultural production and for increasing ecological security of rainfed dryland areas. No-till and reduced-till have become important soil-crop management systems on dryland farms. These systems are important because they protect soil from erosion, and often increase soil organic matter, improve precipitation-storage efficiency, increase biological activity, and increase the number of crop options for dryland rotations (Vigil et al., 1995). Retaining crop residues on the soil surface after grain harvest can result in storing more precipitation water in the soil by reducing storm runoff, increasing infiltration, and decreasing evaporation (Bond and Willis, 1969; Unger, 1983). More detailed discussions on the role of tillage and management of crop residues in increasing soil water storage are presented in following chapters.

Careless management of woodlands, grasslands, and arid areas has resulted in significant degradation of vegetation cover and soil quality. Associated with these activities is a net C loss from terrestrial biomes and soil to the atmosphere. Over time, human activities have altered the amount of C that flows through and is stored in various reservoirs. Drylands, as an ecosystem with extensive surface across the globe, can restore a large amount of C, most of it in the soil rather than in vegetation. A detail discussion on C sequestration in dryland agriculture is presented in a later chapter.

Crops that are suited to irrigate or high rainfall areas are generally unsuited for dryland conditions. Creswell and Martin (1998) described plant characteristics that enhance successful production in dryland agriculture, including (i) a short stem with limited leaf surface to minimize transpiration rates and extend the stored water supply longer into the season, (ii) a deep, prolific root system for better moisture utilization, and (iii) fast maturing crops that can develop prior to the onset of hot and dry conditions and mature before soil moisture is completely exhausted. The other important factor for sustainable production is plant adaptation to drought situations. Shantz (1956) discussed four types of plant adaptation: (i) drought escaping: short season, short height crops that respond to a brief water supply and complete their life cycle quickly; (ii) drought evading: short season, fast growing root system, and high ratio of grain to straw; (iii) drought resistant: crops with extensive root system that obtain soil moisture during repeated water stress, and maintenance of cell membranes and maintaining adequate cell water content when tissue moisture stress occurs; and (iv) drought enduring: crops go into temporary dormancy under water stress and can resume growth as water and favorable temperatures are again available.

Agriculture in semi-arid lands is limited mainly due to shortage in water resources on one hand, and high irradiance throughout the year on the other hand. Future gains in agricultural productivity in drylands may depend heavily on the use of biotechnology to improve the health and stabilize production of agriculturally important plants. Research on environmental interactions is emerging rapidly as a powerful means of increasing resistance to stress and disease. Basic research on genetic control of biochemical responses to water stress in drought-hardy native plants can suggest how crop plants might be modified genetically for more consistent performance in years of drought. To meet the food security for the growing population, it is necessary to capitalize on recent research breakthroughs to accelerate development of varieties that are more resistant to drought, diseases, and insects. However, as Steiner et al. (1988) discussed, most production in dryland regions is far below the genetic potential of currently available crops, so a broad approach toward adapting and supporting improved agronomic practices, with a strong emphases on soil and water conservation, is also required.

REFERENCES

Alexandratos, N. 1995. World agriculture: Towards 2010, FAO Study. John Wiley & Sons, Chichester, UK and FAO, Rome.

Bond, J.J., and W.O. Willis. 1969. Soil water evaporation: Surface residue rate and placement effects. Soil Sci Soc. Am. Proc. 33:445–448.

Brown, L.R. 1999. Feeding nine billion. p.115–132. *In* L. Starke (ed.) State of the world. W.W. Norton and Co., New York.

Brown, L.R., and B. Halweil. 1998. China's water shortages could shake world food security. http://www.worldwatch.org/mag/1998/114.

Creswell, R., and F.W. Martin. 1998. Dryland farming: Crops and techniques for arid regions. Educational Concerns for Hunger Organization, Ft. Myers, FL. Available at http://www.echonet.org// tropicalag/technotes/drylandF.pdf (verified 23 May 2003).

Dregne, H.E. 1977. Desertification of arid lands. Econ. Geogr. 53:322–331.

Engleman, R., and P. Leroy. 1993. Sustaining water: Population and the future of renewable water supplies. Population Action Int., Washington, DC. Available at http://www.cnie.org/pop/pai/water-11.html (verified 30 Apr. 2003).

Eswaran, H., P. Reich, and F. Beinroth. 2001. Global desertification tension zones. p. 24–28. *In* D.E. Stott et al. (ed.) 2001. Sustaining the global farm. Selected papers from the 10th Int. Soil Conservation Organization Meet., West Lafayette, IN. 24–29 May 1999. Available at http:// topsoil.nserl.purdue.edu/nserlweb/isco99/pdf/ISCOdisc/tableofcontents.htm (verified 23 June 2003).

Falkenmark, M., W. Klohn, J. Lundqvist, S. Postel, J. Rockstrom, D. Seckler, H. Shuval, and J. Wallace. 1998. Water scarcity as a key factor behind global food insecurity: Round table discussion. Ambio 27(2). Royal Swedish Academy of Sciences.

Food and Agricultural Organization. 1988. World agriculture toward 2000. An FAO study. Bellhaven Press, London.

Food and Agriculture Organization. 2002. World agriculture towards 2015/2030 [Online]. FAO study, Rome. Available at http://www.fao.org/docrep/004/y355e/y3557eo8.htm (verified 15 May 2003).

Gardner-Outlaw, T., and R. Engleman. 1997. Sustaining water, easing scarcity: A second update. Population Action Int., Washington, DC.

Harahsheh, H. 2002. GIS development [Online]. Available at www.gisdevelopment.net/interview/ previous/ev029.htm (verified 23 June 2003).

Hasegawa, P.M. 2002. Protein points way to salt-tolerant crops [Online]. Available at www.newswise. com/articles/2002/1/salinity.pur.html (verified 23 June 2003).

Hinrichsen, D. 1998. Feeding a future world [Online]. Available at http://www.peopleandplanet.net/doc. php?id=386 (verified 23 June 2003).

Hofwegen, P.V., and M. Svendsen. 2000. A vision for food and rural development [Online]. Available at http://www.worldwatercouncil.org/Vision/Documents/WaterforFoodVisionDraft2.PDF (verified 23 June 2003).

Labunets, H. 2003. Untitled document [Online].Available at www.worldfoodprize.org/youth/papers.html/ specialized%2 (verified 20 June 2003).

Meinke, H., J. Hansen, R. Selvaraju, S. Gadgil, A.R. Khan, and K.K. kanikicharla. 2002. Management responses to seasonal climate forecasts in cropping systems of South Asia's Semi-Arid Tropics [Online]. Available at http://iri.columbia.edu/application/sector/agriculture/SIndia (verified 23 June 2003).

Postel, S. 2001. Growing more food with less water. FAO. Available at www.sciam.com/issue.cfm?issue date Feb-01 (verified 25 June 2003.)

Rhoades, J.D., and J. Loveday. 1990. Salinity in irrigated agriculture. p.1089–1142. *In* B.A. Stewart and D.R. Nielsen (ed.) Irrigation of agricultural crops. Agron. Monogr. 30. ASA, CSSA, and SSSA, Madison, WI.

Rhodes, J.D., F.T. Bingham, J. Letey, A.R. Dedrick, M. Bean, B.J. Hoffman, W. Alves, R.V. Swain, P.G. Pecheco, and R.D. Lemert. 1988. Reuse of drainage water for irrigation: Results of Imperial Valley study. I. Hypothesis, experimental procedures and cropping results. Hilgardia 56:1–16.

Shantz, H.L. 1956. History and problems of arid lands development. p. 3–25. *In* G.F. White (ed.) The future of arid lands. Publ. 43. Am. Assoc. Adv. Sci., Washington, DC.

Steiner, J.L., J.C. Day, R.I. Papendick, R.E. Meyer, and A.R. Bertrand. 1988. Improving and sustaining productivity in dryland regions of developing countries. Adv. Soil Sci. 8:79–122.

Sultan Ali F.H. 1998. Drylands: A call for action. IFAD [Online]. Available at www.ifad.org/pub/dryland/e/drylands.pdf (verified 23 June 2003).

United Nations Fund for Population Assistance. 2003. Population numbers and trends [Online]. Available at http://www.unfpa.org/ (verified 18 July 2003).

Unger, P.W. 1983. Water conservation: Southern Great Plains. p. 35–55. *In* H.E. Dregne and W.O. Willis (ed.) Dryland agriculture. Agron. Monogr. 23. ASA, CSSA, and SSSA, Madison, WI.

Vigil, M.F., D. Nielsen, R. Anderson, and R. Bowman. 1995. Taking advantage of benefits of no-till. Conservation Tillage Fact Sheet 4-95. USDA-ARS and USDA-NRCS, Akron, CO.

Wood, S., K. Sebastian, and S.J. Scherr. 2001. Pilot analysis of global ecosystems: Agroecosystems. IFPRI and World Resources Int., Washington, DC. Available at www.ifpri.org (verified 18 July 2003).

2 Dryland Agriculture: Long Neglected but of Worldwide Importance

Bobby A. Stewart

Dryland Agriculture Institute
West Texas A&M University
Canyon, Texas

Parviz Koohafkan

Land and Water Development Division
Food and Agriculture Organization
of the United Nations, Rome

ABSTRACT

Dryland areas occur widely in all continents of the world. Thirty-eight percent of the world's land surface is classified as semi-arid or dry subhumid, and there is an additional 7% arid land used primarily for grazing animals. Dryland farming is the growing of crops in areas where water supply constitutes the major constraint and is widely practiced in semi-arid and dry-subhumid regions. Soil degradation is a widespread problem in drylands and largely results from wind and water erosion, organic matter depletion, chemical deterioration, and salinization. In the worst cases, desertification occurs. Conservation agriculture is the integration of practices that avoids mechanical soil disturbance, maintains a soil cover by a growing crop or residues of previous crops, and rotates crops. These practices reduce soil degradation and in some cases even restore many of the favorable chemical, physical, and biological properties present in the soil initially following the introduction of crop production. Conservation agriculture is gaining acceptance in many countries of the world but has been most successful in favorable rainfall regions where it is relatively easy to maintain soil cover and rotate a wide variety of crops including legumes. The sparse rainfall and high temperatures in dryland regions are major constraints along with the demand of crop residues for feed and fuel, but adoption of conservation agriculture may be the key to sustainable crop production in marginal areas. Preliminary estimates are that the average yield of cereals in dryland regions can be increased 30 to 60% annually by increasing crop water use by 25 to 35 mm achievable using conservation agriculture.

INTRODUCTION

Dryland agriculture is a commonly used term but one that does not have a common definition. Some use it synonymously with rainfed agriculture. Rainfed agriculture simply means that there is no irrigation. Dryland agriculture is that part of rainfed agriculture where water is the most limiting factor. Mathews and Cole (1938) stated that dryland farming in its broadest aspects is concerned with all phases of land use under semi-arid conditions. Not only how to farm, but how much to farm and whether to farm must be taken into consideration. Stewart and Burnett (1987) characterized dryland agriculture as rainfed systems that emphasize water conservation, sustainable crop yields, limited inputs for soil fertility maintenance, and wind and water erosion constraints. Oram (1980) also distinguished between rainfed farming and dryland agriculture. He stated that successful dryland agriculture is husbandry under conditions of moderate to severe water stress during a substantial portion of the year that requires special cultural techniques and adapted crops and systems. More recently, ACIAR (2002) stated that dryland cropping refers to those agricultural areas where the average water supply to the crop limits potential yield to <40% of full (water-unlimited) potential.

EXTENT OF DRYLANDS

The Food and Agriculture Organization (FAO) has defined drylands as those regions climatically classified as *arid, semi-arid, or dry subhumid* based on length of growing period for annual crops (FAO, 2000). The growing period begins when monthly precipitation exceeds half of the monthly potential evapotranspiration. The regions where the monthly rainfall never exceeds half of the potential evapotranspiration have zero growing days and are not included in the drylands. They are classified as hyper-arid areas with no agricultural potential.

Arid regions have 1 to 59 growing days, semi-arid have 60 to 119, and dry subhumid regions have between 120 to 179 growing days. Combined, these regions account for 45% of the worlds land area—7% arid, 20% semi-arid, and 18% dry subhumid. The distribution of these areas among the different regions of the world is presented in Table 2–1. The *hyperarid* lands are not included in Table 2–1 but

Table 2–1. Percent of world land area (134.9 million km^2) in various regions, percent of land in regions for different dryland areas, percent of world population (6.2 billion) in various regions, percent of population in regions living in dryland areas, and percent of population in regions engaged in agriculture.

Regions	World land area	Arid	Semi-arid	Dry sub-humid	Total world population	Population in drylands	Agricultural population
	%						
Asia and Pacific	21.5	6	15	17	56	44	59.5
Europe	5.4	12	28	23	12	18	13.5
North Africa and Near East	9.5	4	11	5	5	44	44.3
North America	14.8	12	28	23	5	19	3.0
North Asia and East of Urals	15.6	11	51	33	4	89	17.4
South and Central America	15.4	11	6	10	8	24	23.0
Sub-Saharan Africa	17.7	6	13	19	10	36	65.0
World (Total)	100	7	20	18	100	38	45.8

(Data constructed from values obtained from FAO, 2000.)

make up an additional 19% of the world's land area. The growing period classification system is based on agro-ecological factors and generally works well for assessing the potential of an area for growing crops, but there are notable exceptions. For example, in much of the Great Plains of the USA, one of the largest dryland cropping regions of the world, there is not a single month of the year when average precipitation exceeds half of the reference evapotranspiration. Based on the growing day classification, this area is classified as hyperarid with no agricultural potential. Dryland farming can be practiced in this region because management practices have been developed that allow the accumulation of 100 to 200 mm of plant-available water in the soil during fallow periods to supplement precipitation received during the growing season. These cropping systems have a cropping intensity of less than one, meaning that a crop is not harvested every year. Examples are wheat (*Triticum aestivum* L.)-fallow resulting in one crop every 2 yr; wheat-sorghum [*Sorghum bicolor* (L.) Moench]-fallow resulting in two crops every 3 yr; and wheat-sorghum-sorghum-fallow resulting in three crops every 4 yr. The average annual precipitation for the region where dryland farming is practiced ranges from about 400 to 600 mm, but the amount for any given year for a specific location varies from about 50% of average annual to about 200%. The variation in yields is even greater, ranging from zero to about three times the average yield. Drought conditions occur every year but the extent and severity vary greatly.

Dryland areas like the U.S. Great Plains may be better characterized by a climatic aridity index. One such index proposed by the United Nations Conference on Desertification (UNESCO, 1977) defines bioclimatic zones by dividing the annual precipitation (P) by the annual potential evapotranspiration (PET). Climatic zones were defined at hyperarid (P/PET = <0.03), arid (0.03<P/PET<0.20), semi-arid (0.20<P/PET<0.50), and subhumid (<0.50P/PET<0.75). By this classification, most of the U.S. Great Plains where dryland cropping is practiced is semi-arid.

KEY CHARACTERISTICS OF SEMI-ARID LANDS

Bowden (1979) divides semi-arid lands into two broad physical-climatic regions—the tropical semi-arid territory close to the equator and the steppe located in the mid-latitudes. Each region has peculiarities of climate, settlement, and resource development that can vary as much internally as externally. In contrast to the frost-free, semi-arid tropics, the semi-arid mid-latitude lands are characterized by definite warm and cold seasons. Bowden (1979) used annual precipitation between the limits of 250 and 550 mm to define the steppe. In these conditions, crop production is marginal and the growing season is often shortened by unseasonal frost. The summer precipitation is mainly evaporated or transpired during the warm season, but winter rainfall and snow usually contribute to stored water in the soil profile. The dominant vegetation under native conditions was grass, and plowing of the grassland was difficult without strong draft animals and the steel plow. Therefore, except beside streams where shallow groundwater was available, almost none of the world's steppe land was extensively farmed 150 yr ago (Bowden, 1979). Mechanization led to a large expansion of agricultural production in the semi-arid areas of North America and Eurasia.

Bowden (1979) lists four keys that are unique to semi-arid lands.

- Key 1. No growing season is or will be nearly the same in precipitation amount, kind, or range, or in temperature average, range, or extremes, as the previous growing season. Although this key is critical in any rainfed system, it requires absolute attention in dryland farming. Crop cultivation requires an adjustment every year, which leads to the second key.
- Key 2. Crops cannot be planned or managed in the same manner from season to season. Most of the world's agricultural practices in either humid or arid areas have some annual predictability. In semi-arid climates, however, even highly mechanized, technically advanced, commercial farms such as those in the High Plains of North America or the out back of Western Australia do not have sufficiently stable production for the individual or government to rely on a given production figure for the following season.
- Key 3. Soil and water resources are changed once agriculture is introduced into a semi-arid region. For example, the soils of most semi-arid lands developed under grass on relatively flat topography. The competition for water and nutrients to produce crops requires removal of the protective grass cover. Because the crops are annual and dependant on precipitation, little or no vegetative cover is produced during severe drought years leaving the soil highly vulnerable to wind erosion.
- Key 4. There is abundant sunshine due to many cloud-free days. This has potential benefits and is shared with most arid climates. Abundant sunshine means higher temperatures that induce rapid growth, but it also creates a situation that demands careful management of soil water. Warm seasons, high sun and cloud-free conditions stimulate growth, but also increase evaporation and transpiration. It is possible for a grain crop to mature rapidly due to several weeks of rainless conditions and desiccate just days before ripening. It is equally possible for a few millimeters of precipitation to occur at almost the last moment and produce a good grain crop.

An understanding of all four characteristics is vital, but perhaps Key 3 deserves the most attention because it concerns the resource base that is affected by human activities and often changes rapidly for the worse. Sustainability of the soil resource base in dryland regions is a major concern and this is particularly true for lands that are cultivated. Land degradation is very common in most dryland areas and can lead to desertification. Desertification is defined as land degradation in arid, semi-arid, and dry subhumid areas caused by climatic variability and human activities (FAO, 2000). Whenever an ecosystem like a grassland prairie in a semi-arid region is transformed into an agroecosystem for the purpose of food and fiber production, several soil degradation processes are set in motion, particularly if raindrops fall directly onto the soil surface without vegetation, crop residues, or mulches present. Some examples are soil organic matter decline, wind and water erosion, deterioration of soil structure, salinization, and acidification as depicted in Fig. 2–1. In an ideal situation, soil conservation practices such as minimum or zero-tillage, crop rotations including legume crops, application of inorganic and organic fertilizers, crop residue management, terracing, and others that reduce degradation processes and maintain soil productivity are used to offset the degradation processes.

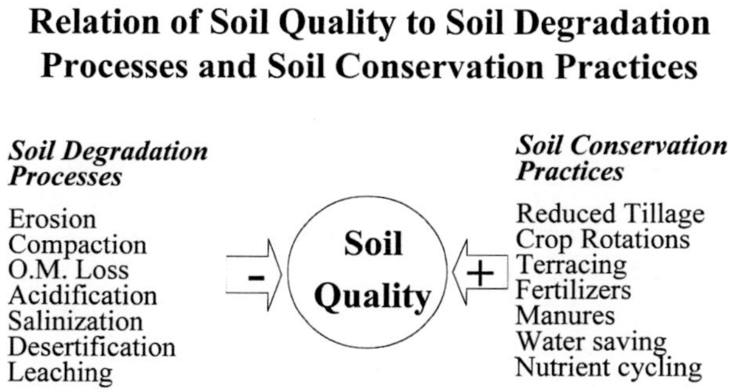

Fig. 2–1. Degradation processes and soil conservation processes that often occur simultaneously in cultivated soils.

In dryland areas, the degradation processes are often much more dominant than the soil conservation practices so the soil resource base can quickly degrade. In general, degradation processes proceed more rapidly as the climate becomes hotter and drier while the soil conservation practices become more difficult to implement under these conditions. As the soil degrades, the infiltration rate and water-holding capacity decrease making the already limited water resource even less effective resulting in a downward spiral of soil quality and crop production.

THE "DUST BOWL"—A CLASSIC EXAMPLE

Following World War I of 1914 to 1918, high wheat prices, coupled with the development of power machinery, led to the rapid expansion in cultivated land and large-scale cropping. Settlers pushed west from the subhumid eastern edge of the Great Plains because there were vast acreages of land that had not been developed. Early-day conservationists had warned of the erosion that would take place in many parts of the Great Plains if the land were cultivated. As agriculture expanded, however, the semi-arid Great Plains was opened to homesteading, land was broken from sod, and large-scale farming developed. This expansion largely occurred during the decade of 1915 to 1925 when the average annual precipitation was higher than the long time average. The more favorable precipitation, coupled with more than adequate plant nutrients coming from the decomposition of soil organic matter from the newly plowed grassland soils, resulted in high wheat yields and prosperous farmers. Unfortunately, the decade of higher than average precipitation was followed by drier years and the decade of the 1930s became known as the "Dust Bowl" era. The soil organic matter content of the soils declined rapidly after cultivation began. The decline was hastened by the fact that much of the land was fallowed for 15 or more months between wheat crops to store water in the soil profile for the subsequent crop. During the fallow period, the land was often tilled 10 to 12 times for weed control. The long and severe drought of the 1930s resulted in wind erosion that was so severe that national attention was focused on the Great Plains.

Burnett et al. (1985) reported that 43% of the area had serious erosion damage. It was estimated that about 2.6 million hectares in the Southern Plains were removed from cultivation because of erosion. In retrospect, it is clear that much of the land should never have been cultivated. Short-term gains, however, nearly always take precedent over long-term consequences. Even though the region had become devastated during the 1930s, favorable rainfall years returned during the 1940s when wheat prices were again high because of World War II and the interest in plowing grasslands resumed. Again, droughts followed and wind erosion in some areas became almost as serious in the 1950s as in the 1930s. The economy of the region, however, was much better because irrigation development increased rapidly during the 1950s. Dryland farmers also began using stubble-mulch tillage that left more crop residues on the soil surface that significantly reduced wind erosion and increased soil water storage. Therefore, the "Dust Bowl" of the 1930s did not return but the fragility of the region was reaffirmed. The history of crop production in the Great Plains offers much to be learned about the sustainability of crop production in semi-arid regions.

SUSTAINABLE AGRICULTURE

There are many different concepts of sustainable agriculture, but none is generally accepted. Sustainability, to many, conveys the idea of a balance between short-term human needs and long-term environmental concerns. The American Society of Agronomy defined sustainable agriculture as one that, over the long term, enhances environmental quality and the resources base on which agriculture depends, provides for basic human food and fiber needs, is economically viable, and enhances the quality of life for farmers and society as a whole (Schaller, 1990). Similarly, the U.S. Congress defined sustainability in the 1990 Farm Bill as an integrated system of plant and animal practices having a site-specific application that will over the long term: (i) satisfy human food and fiber needs; (ii) enhance environmental quality and the natural resource base upon which the agriculture economy depends; (iii) make the most efficient use of nonrenewable resources and on-farm resources and integrate, where appropriate, natural biological cycles and controls; (iv) sustain the economic viability of farm operations; and (v) enhance the quality of life for farmers and society as a whole. Ruttan (1989) proposed, as a guide to research, that the definition of sustainability should include (i) the development of technology and practices that maintain and/or enhance the quality of land and water resources, and (ii) the improvement in plants and animals and the advances in production practices that will facilitate the substitution of biological technology for chemical technology. The first part of the Ruttan definition is similar to that of the U.S. Congress, but the second part of the Ruttan definition emphasizes more reliance on biological technology as opposed to chemical technology. Sustainability should be considered dynamic because, ultimately, it will reflect the changing needs of an increasing global population. The common thread among all definitions of sustainability, however, is that quality of the resource base should be *enhanced.*

Enhancing the quality of the resource base at the same time that crops are being produced is a challenge for any agricultural system, but it is indeed challenging for systems located in semi-arid regions. The reasons for the difficulty were partially presented earlier in the discussion of Fig. 2–1. The soil degradation processes of soil organic matter decline and wind and water erosion are very pronounced in dryland regions and it is difficult to offset these effects with soil improvement practices. Early agriculture often depended on shifting agricultural systems such as bush-fallow. These systems were sustainable because crops were only grown for a few years and then the partially degraded soil was allowed to return to "bush" for a number of years to replenish itself. In most cases, this is no longer feasible because of land pressure and economic conditions. Clean fallow has been practiced in many semi-arid regions as a way to store water in the soil for use by a subsequent crop. Although this practice is successful for increasing crop yields and greatly reducing risk of crop failure, it often causes serious soil degradation as already discussed in relation to the Dust Bowl. Sustaining the soil resource in semi-arid regions is clearly a formidable task.

Sustainability involves technical, economic, and social conditions. All three conditions are highly important and interrelated. It may be entirely possible to have an agroecosystem that is technically sustainable that is neither economically feasible nor socially acceptable. In this chapter, sustainability will be addressed only in terms of how agricultural practices and systems affect the productivity and sustainability of the soil resource base.

CONSERVATION AGRICULTURE

The most successful sustainable agroecosystems are those that use some form of conservation agriculture. The goal of conservation agriculture is to conserve, improve, and make more efficient use of natural resources through integrated management of available soil water and biological resources combined with external inputs (FAO, 2002). Conservation agriculture contributes to environmental conservation as well as to enhanced and sustained agricultural production. It can also be referred to as resource-efficient agriculture. Conservation agriculture is based on three principles:

- Avoiding mechanical soil disturbance
- Maintaining a permanent soil cover, by crop residues and crops
- Crop rotation.

Conservation agriculture is practiced on 45 million ha, mostly in South and North America (FAO, 2002). However, this accounts for only about 3% of the 15 billion ha of arable land worldwide. Conservation agriculture has been most successful in South America, particularly Brazil, where economic and environmental pressures are great. Conservation agriculture is practiced from the humid tropics to almost the Arctic Circle and on all kinds of soils. However, FAO (2002) stated "So far the only area where the concept has not been successfully adapted is the arid areas with extreme water shortage and low production of organic matter. In these areas both humans and animals compete with the soil for crop residues." Again, the difficulty of using conservation agriculture in dryland regions agrees with the discussion above of Fig. 2–1. Maintaining a permanent soil cover and utilizing crop rotations are the principles of conservation agriculture that are most difficult to practice under dryland conditions. The fact that it is difficult to carry out all the principles of conservation agriculture in dryland regions must not deter efforts to adapt the concepts as far as feasible in these regions. The dryland regions are where the benefits of conservation agriculture are critically needed and where soil degradation can be disastrous without the application of some of the concepts of conservation agriculture.

APPLYING CONSERVATION AGRICULTURE CONCEPTS IN DRYLAND REGIONS

The greatest problem with applying conservation agriculture concepts in dryland regions is the lack of crop residues. Crop residues are lacking because the limited and highly variable precipitation limits biomass production. In many cases, the situation is made even worse because of the use of crop residues for animal feed and fuel. The removal of crop residues accelerates the already fast decline of soil organic matter common in dryland areas. This lowers the soil water-holding capacity and fertility and results in even lower yields and a downward spiral of crop productivity and soil quality.

The long-term future of many dryland regions depends on stopping, or reversing, the downward spiral of crop productivity and soil quality. This is a particular challenge in many of the developing countries where yields are low and the demand for crop residues is great. FAO (1996) reported that the 1988 to 1990 average yields of wheat, maize (*Zea mays* L.), and sorghum in developing countries in semi-arid regions were 1100, 1130, and 650 kg ha^{-1}, respectively. These low grain yields result in relatively low amounts of crop residues.

There are numerous studies, however, that show even small amounts of crop residues can be beneficial for controlling wind erosion and increasing soil water storage. Fryrear (1985) established the relationship between soil loss by wind erosion and the percent of soil cover. Covering 20% of the surface reduced soil losses by 57%, and a 50% cover reduced soil losses by 95% compared to soils with no cover. Fryrear concluded that the cover could be any nonerodible material such as large clods, gravel, cotton (*Gossypium hirsutum* L.) gin trash, or any diameter stick between 3.1 and 25.4 mm in size. For 50% groundcover, approximately 1 Mg ha^{-1} of wheat straw is required (Van Doren and Allmaras, 1978). To achieve the same rate of cover, however, an estimated 3 Mg ha^{-1} of sorghum stalks and 9 Mg ha^{-1} of cotton stalks are required (Unger and Parker, 1976). Therefore, under semi-arid conditions when grain yields are only in the range of 1 Mg ha^{-1}, wheat or other small grain crops are likely to be the only crops that will furnish sufficient crop residue to significantly impact wind erosion. Vertically oriented residues are also more effective for controlling wind erosion than flat oriented residues. Water erosion is also a problem in semi-arid regions and Bilbro et al. (1994) reported a similar relationship between soil cover and water erosion as for wind erosion.

Residue cover can increase the efficient use of limited precipitation in several manners. Plant residues on the soil surface can cause a reduction in runoff, principally by protecting soil surfaces that are prone to crusting from raindrop action. Cornish and Pratley (1991) reported that fallow efficiencies for numerous clay soils in Queensland, Australia were increased from about 21 to 29%, almost entirely be-

Table 2–2. Progress in wheat-fallow systems at Akron, CO (Greb et al., 1979).

Years	Tillage	No. tillage operations	Fallow water	Fallow efficiency	Wheat yield
			mm	%	Mg ha^{-1}
1916–1930	Maximum tillage, plow, and harrow	7–10	102	19	1.07
1931–1945	Shallow disk, rodweeder	5–7	118	24	1.16
1946–1960	Begin stubble mulch in 1957	4–6	137	27	1.73
1961–1975	Stubble mulch, herbicides in 1967	2–3	157	33	2.16
1976–1990	Minimum till, projected no-till	0–1	183	40	2.69

cause of reduced runoff when some surface residues were left on the soil surface. Fallow efficiency is the percent of precipitation occurring during the fallow period that is stored in the soil profile at the end of the fallow period. Crop residues on the soil surface also reduce evaporation of water from the soil. The residues reduce the energy available for evaporation because of its poor thermal conductivity. It also increases the resistance to vapor transfer from the soil to air. These conditions have the greatest effect when there are frequent rains so that the soil surface does not totally dry between precipitation events. This allows movement of water deeper into the soil where it is much less subject to evaporation. Observations by the authors also suggest that on soils where there is sufficient cracking that a major advantage of no-till is that the cracks are not destroyed and relatively small precipitation events can result in rainfall rapidly moving 30 cm or deeper into the soil compared to only 5 to 10 cm in tilled soil where the cracks have been destroyed. Most of the water retained near the surface will be evaporated in a few days unless additional precipitation occurs.

The amount of crop residues remaining on the surface depends on several factors. Assuming that the residue is not used for animal feed or household fuel, the most important factor is the kind and number of tillage operations. Moldboard tillage will bury approximately 95% of the residue with one operation. In comparison, disk tillage will bury about 90%, chisel plows about 50%, and under cutters (sweeps 50–75 cm wide) about 25%. Even the sweep plows that are often used during long fallow periods will bury most of the crop residues because there are usually three or more tillage operations during the fallow period when tillage is used exclusively for weed control. The greatest amounts of crop residue are left on the soil surface when herbicides are used for weed control and tillage can be greatly reduced or completely eliminated.

Greb et al. (1979) were among the earliest researchers to clearly show the benefits of reduced tillage on water conservation in semi-arid regions. In a wheat-fallow system that included a 15 to 16 mo fallow period, they found water storage in the soil profile during the fallow period increased every time the number of tillage operations decreased (Table 2–2). This is due to two factors. The first is that the soil is dried to the depth of tillage each time it is tilled. The second is that the less the soil is tilled, the greater the amount of mulch that remains on the surface to reduce the evaporation potential. More importantly, Greb et al. showed that only a few millimeters of additional soil water storage at time of seeding increased wheat grain yields dramatically. An increase of 55 mm of soil water storage at seeding doubled the grain yield of wheat from 1.07 to 2.16 Mg ha^{-1} (Table 2–2). The 55 mm increase was achieved by increasing the fallow efficiency from 19 to 33%. The reason that such a small increase of soil water increases grain yield so much is because the threshold amount of evapotranspiration required for grain production is already met and the additional water increases grain production directly. Musick and Porter (1990), Rhoads and Bennett (1991), and Krieg and Lascano (1990) reviewed the literature regarding the water-use efficiency of wheat, maize, and sorghum, respectively. While the efficiencies varied considerably depending on yield levels and climatic conditions, some reasonable guidelines can be developed from the many studies that have been conducted worldwide. We suggest as a general guide that 1.7 kg of maize grain can be produced in dryland regions for each additional cubic meter of water used for evapotranspiration, 1.5 kg of sorghum, and 1.3 kg of wheat. The threshold values, however, are also different for each of the crops with wheat usually having the lowest and corn having the highest.

Based on the water-use efficiency values above, the average yield of wheat could be increased by 325 kg ha^{-1} by increasing seasonal evapotranspiration by 25 mm. This could potentially increase the average wheat yield of developing countries by 30% because it was reported earlier that the average yield of wheat was only 1100 kg ha^{-1} in these countries. Likewise, average maize yield could be potentially increased by 425 kg ha^{-1} and sorghum by 375 kg ha^{-1} that would be increases of 38 and 58%, respectively. Increasing evapotranspiration by 25 mm in semi-arid cropping regions over the next several years appears to be a reasonable goal, particularly in areas with average annual precipitation of 350 mm or greater.

Unger (1978) also showed that relatively small amounts of wheat residue left on the soil surface could significantly increase soil water storage during fallow (Table 2–3). Although the greater the amount of wheat residue left on the surface, the greater the amount of soil water storage, there was a significant increase in both soil water storage and subsequent grain sorghum yield from only 1 Mg ha^{-1}. This amount of wheat residue is commonly available even when grain yields are as low as 500 kg ha^{-1}. Therefore, even though there are not nearly as many crop residues available in semi-arid regions as desired, some benefits can be achieved by properly managing them. An upward spiral of crop productivity and soil quality can result but significant progress will take years, or even decades. In contrast to irrigation where dramatic increases can be achieved immediately, increasing yields in semi-arid regions by increasing the efficient use of limited precipitation is a long-term process. Policy makers must recognize this and more importantly, they must be willing to commit resources and develop strategies to promote soil and water management practices in semi-arid regions.

An example of the benefits that can accrue over a long time from improved water management practices is the data shown in Fig. 2–2. These are farmer yields of wheat grain for Deaf Smith County, Texas. The average annual precipitation for the county is about 450 mm and the annual potential evapotranspiration is about 1800 mm. Therefore, drought is a very common occurrence and severe water stress occurs every year. The yearly precipitation amounts show a range from <200 mm to >800 mm. A 10-yr moving average (each yearly point is the average precipita-

Table 2–3. Straw mulch effects on soil water storage during an 11-mo fallow, water storage efficiency, and dryland grain sorghum yield at Bushland, TX. (Adapted from Unger, 1978.)

Mulch rate	Water storage†	Storage efficiency‡	Grain yield	Total crop water use§	WUE¶
Mg ha^{-1}	mm	%	Mg ha^{-1}	mm	(kg m^{-3})
0	72 c#	22.6 c	1.78 c	320	0.56
1	99 b	31.1 b	2.41 b	330	0.73
2	100 b	31.4 b	2.60 b	353	0.74
4	116 b	36.5 b	2.98 b	357	0.84
8	139 a	43.7 a	3.68 a	365	1.01
12	147 a	46.2 a	3.99 a	347	1.15

† Water use determined to 1.8-m depth; precipitation averaged 318 mm during the fallow period.
‡ Storage efficiency is percent of precipitation occurring during the 11-mo fallow period that was stored in the soil at the end of the fallow period.
§ Growing season precipitation plus change in soil water during the growing season.
¶ Water-use efficiency (WUE) based on grain produced, growing season precipitation, and soil water change during the growing season.
# Column values followed by the same letter are not significantly different at the 5% level (Duncan's multiple range test).

tion amount for the year shown plus the nine previous years) line of annual precipitation is also shown and although there is some variation, the average annual amount has remained relatively stable. The county average wheat yields for each year are also shown in Fig. 2–2 and there is also a line showing the 10-yr moving averages. It is noteworthy that the moving yield average and moving precipitation average closely paralleled each other until the early 1970s. Since that time, the moving average grain yield increased essentially every year and the average yield has

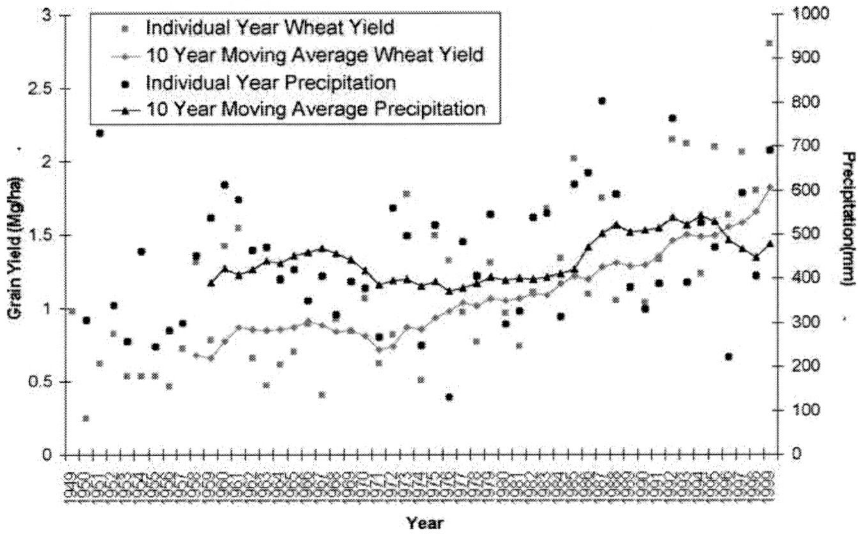

Fig. 2–2. Relationship between annual precipitation and average wheat yields for Deaf Smith County, Texas, showing gradual increase in yield and improved precipitation use efficiency (Unpublished data, Zhen Wu, W.A. Colette, and B.A. Stewart, West Texas A&M University).

more than doubled. No single factor is responsible but it clearly shows that the use efficiency of the precipitation has dramatically increased. Water management is the first factor that must be addressed in dryland regions because other improved technologies such as improved cultivars and fertilizers are usually not beneficial without improved water management. The early 1970s was when the cost of oil and other energy sources increased rapidly and there was a concerted effort by researchers, extension personnel, and industry representatives to promote less tillage and more herbicide usage. This increased the amounts of crop residues remaining on the soil surface resulting in more soil water storage during fallow periods and higher crop yields. There is no doubt that reduction in tillage has been a significant factor in the increasing yield of wheat illustrated in Fig. 2–2. The trend toward less tillage continues so it is anticipated that crop yields will also continue to increase, but at a slow rate. The quality of the soil resource base will also be enhanced if this trend continues because more crop residues will be produced and with reduced tillage, a higher percentage of the carbon in the crop residues will be retained in the soil as organic matter. The increased level of soil organic matter will improve soil physical properties and continue the upward spiral of crop production and soil quality.

CONCLUSION

Conservation agriculture is based on the principles of avoiding mechanical soil disturbance, maintaining a permanent soil cover by crop residues or growing crops, and crop rotation. While conservation agriculture has been highly successful in humid areas, its suitability for dryland regions has been questioned. However, there is ample evidence that shows some of the principles of conservation agriculture can be applied to dryland regions and crop productivity and soil quality can be enhanced as a result. Success will not be easy, and certainly not quick. Because water is so limiting in these areas, the amount of crop residues is insufficient to maximize water-use efficiency and quickly change soil organic matter content. Reduction in tillage often requires the use of herbicides for weed control. This requires higher management skills and increased input costs, both of which are often lacking in dryland regions. The residues, particularly in developing countries, are often removed from the soil for use as animal feed or household fuels and this further complicates the situation. The long-term sustainability of dryland soils, however, may be significantly enhanced by reduced tillage that leaves more crop residues on the soil surface. Researchers, change agents, and policy makers must promote these principles to the fullest extent feasible and develop strategies and policies that can be implemented successfully by the farmers. Although progress will be slow in dryland regions, there is ample evidence indicating the long-term results will be positive. Increasing crop evapotranspiration by 25 mm in semi-arid regions is a reasonable goal, and this will potentially increase wheat yields by 375 kg ha^{-1}, which would raise the average wheat yields in developing countries in semi-arid regions by 30%.

REFERENCES

Australian Centre for International Agricultural Research. 2002. Improving water-use efficiency in dryland cropping. http://www.aciar.gov.au/web.nsf/att/ACIA-5K36CC/$file/2-3.PDF (verified 16 Oct. 2003.

Bilbro, J.D., B.L. Harris, and O.R. Jones. 1994. Erosion control with sparse residue. p. 30–32. *In* B.A. Stewart and W.C. Moldenhauer (ed.) Crop residue management to reduce erosion and improve soil quality. Conserv. Res. Rep. 37. Agric. Res. Serv., U.S. Dep. Agric., Washington, DC.

Bowden, L. 1979. Development of present dryland farming systems. p. 45–72. *In* A.E. Hall et al. (ed.) Agriculture in semi-arid environments. Springer-Verlag, Berlin.

Burnett, E., B.A. Stewart, and A.L. Black. 1985. Regional effects of soil erosion on crop productivity? Great Plains. p. 285–304. *In* R.F. Follett and B.A. Stewart (ed.) Soil erosion and crop productivity. ASA, CSSA, and SSSA, Madison, WI.

Cornish, P.S., and J.E. Pratley. 1991. Tillage practices in sustainable farming systems. p. 76–101. *In* V. Squires and P.G. Tow (ed.) Dry farming—Australia. Sydney Univ. Press, Sydney, Australia.

Food and Agriculture Organization. 1996. Prospects to 2010: Agricultural resources and yields in developing countries. p. 26–36. *In* Vol. 1 Technical Background Documents 1–5. World Food Summit. FAO of the United Nations, Rome.

Food and Agriculture Organization. 2000. Land resource potential and constraints at regional and country levels. World Soil Resources Rep. 90. FAO of the United Nations, Rome.

Food and Agriculture Organization. 2002. Intensifying crop production with conservation agriculture.FAO of the United Nations, Rome. Available at http://www.fao.org/ag/ags/AGSE/main.htm (verified 16 Oct. 2003).

Fryrear, D.W. 1985. Soil cover and wind erosion. Trans. Am. Soc. Agric. Eng. 28:781–784.

Greb, B.W., D.E. Smika, and J.R. Welsh. 1979. Technology and wheat yields in the Central Great Plains: Experiment station advances. J. Soil Water Conserv. 34:264–268.

Krieg, D.R., and R.J. Lascano. 1990. Sorghum. p. 719–739. *In* B.A. Stewart and D.R. Nielsen (ed.) Irrigation of agricultural crops. Agron. Monogr. 30. ASA, CSSA, and SSSA, Madison, WI.

Mathews, O.R., and J.S. Cole. 1938. Special dry-farming problems. p. 679–692. *In* Soils and men. Yearbook of Agriculture. U.S. Dep. of Agric., Washington, DC.

Musick, J.T., and K.B. Porter. 1990. Wheat. p. 597–638. *In* B.A. Stewart and D.R. Nielsen (ed.) Irrigation of agricultural crops. Agron. Monogr. 30. ASA, CSSA, and SSSA, Madison, WI.

Oram, P. 1980. What are the world resources and constraints for dryland agriculture? p. 17–78. *In* Proc. on the International Congress for Dryland Farming, Adelaide, South Australia. Dep. of Agric., Adelaide, Australia.

Rhoads, F.M., and J.M. Bennett. 1990. Corn. p. 569–596. *In* B.A. Stewart and D.R. Nielsen (ed.) Irrigation of agricultural crops. Agron. Monogr. 30. ASA, CSSA, and SSSA, Madison, WI.

Ruttan, V.W. 1989. Sustainability is not enough. Better Crops Plant Food (Spring):6–9.

Schaller, N. 1990. Mainstreaming low-input agriculture. J. Soil Water Conserv. 45:9–12.

Stewart, B.A., and E. Burnett. 1987. Water conservation technology in rainfed and dryland agriculture. p. 355–359. *In* W.R. Jordan (ed.) Water and water policy in world food supplies. Proc. of the Conf., College Station, TX. 26–30 May 1985. Texas A&M Univ., College Station, TX.

Unger, P.W. 1978. Straw-mulch effect on soil water storage and sorghum yield. Soil Sci. Soc. Am. J. 42:486–491.

Unger, P.W., and J.J. Parker, Jr. 1976. Evaporation reduction from soil with wheat, sorghum, and cotton residues. Soil Sci. Soc. Am. J. 40:954–957.

United Nations Educational, Scientific and Cultural Organization. 1977. World map of desertification. A/Conf. 74/2. FAO of the United Nations, Rome.

Van Doren, D.M., Jr., and R.R. Allmaras. 1978. Effects of residue management practices on the soil physical environment, microclimate, and plant growth. p. 49–83. *In* W.R. Oschwald (ed.) Crop residue management systems. ASA Spec. Publ. 31. ASA, Madison, WI.

3 USDA-ARS Research and Development for Sustainable Dryland Agriculture

Srinivas C. Rao and Herman S. Mayeux

USDA, ARS, Grazinglands Research Laboratory
El Reno, Oklahoma

Allen R. Dedrick

USDA, ARS, National Program Staff
Beltsville, Maryland

ABSTRACT

Continued or enhanced competition for water and rising costs of irrigation underscore predictions that additional food required by a growing population must be provided in large part by more efficient and productive drylands. Much progress was made in the last century by USDA-ARS and its partners in improving water conservation, water-use efficiency, and the productivity of dryland crops. Conservation tillage with surface residue cover enhances water infiltration, reduces evaporation, and protects soils from erosion. New crop cultivars with short statures, deep and extensive root systems, and reduced growing seasons will play an important role in avoiding drought. New technologies such as remote sensing of soil water content and long-term weather outlooks present numerous opportunities for reducing risks in dryland agriculture. Advances in biotechnology have only begun to provide novel approaches to the development of drought-tolerant agricultural crops. These and other technologies continue to provide opportunities to achieve efficiency and profitability in dryland agriculture, and will continue to be the focus of the USDA-ARS research program.

INTRODUCTION

Food security, economic prosperity, and environmental quality in the USA are dependent upon developing and maintaining sustainable agricultural production systems. Globally, these issues are also linked, as the USA has historically been a major supplier of commercial food products for world trade and humanitarian food aid for countries in crisis. Currently, the USA produces one third of the world's coarse grains and accounts for two thirds of the world trade in those commodities. As a result, U.S. agriculture is one of the few industrial sectors that consistently run a trade surplus. In recent years, the US has also provided about half of the world food aid in cereals (FAO, 1999). World demand for these commodities is expected to continue increasing, with net cereal imports by less developing countries nearly doubling by 2020 to 192 Tg and 60% of those imports being supplied by the USA (Pinstrup-Anderson et al., 1999). Food, feed, and fiber demands in the USA are also expected to increase. By the mid-21st century the U.S. population is expected to increase by 50% while availability of high-quality farm land for agricultural production is projected to decline by 13% due primarily to urban encroachment on nonirrigated farm land (Lawrence, 2000).

Since the Dust Bowl years of the 1930s, increasing food, feed, and fiber demands were often met by increasing the amount of irrigated area. As a result, the number of irrigated acres in the U.S. High Plains alone increased from <1 M ha in 1949 to >6 M ha in 1980. Since 1990, regional losses and gains in irrigated areas were recorded, but the total area of irrigated cropland in the USA has not increased dramatically since 1969. As a result, <25 M ha of about 177 M ha of cropland in the USA is irrigated. There have been small increases (~2–3 M ha) in both the extent and percent of all farmland that is irrigated (Howell, 2001). Most of the net increase has occurred in the Mississippi Delta from southern Missouri to northeast Louisiana, in the western High Plains, and the Central Valley of California. Declines in irrigated cropland have occurred in southern Florida, the Rice Belt in Texas and Louisiana, and in portions of the Central Plains, primarily the Oklahoma Panhandle and southwest Kansas.

The lack of expanded irrigation area undoubtedly reflects one of the most important questions regarding the future of irrigated agriculture—whether there will be sufficient water to satisfy the growing needs of both agriculture and nonagriculture users. Currently, some 80% of all water used each year goes to irrigate agriculture, which in turn produces 30 to 40% of the world food crops on just 17% of all arable land (Hoffman and Hussain, 1998). This level of productivity and return on investment is certainly impressive. However, in many regions such as the U.S. High Plains, the major source of water for irrigated agriculture is groundwater (McGrath and Dugan, 1993) and historically withdrawal for irrigation has exceeded recharge. As a result, the water level in High Plains aquifers in parts of Texas, Oklahoma, and southwestern Kansas has declined more than 30 m (Luckey et al., 1981) and irrigated cropland in this region declined by 130 000 ha between 1992 and 1997.

Rosegrat et al. (1999) reported that rapid growth in water demand, the high cost of developing water sources, and continued depletion of water resources threaten future food production. The extent of irrigated croplands is declining because water resources are being depleted in many regions, pumping costs are increasing as energy costs rise, and the value of agricultural products has not risen. Consequently, producers are increasingly switching back to dryland production in those areas where rainfall amounts and patterns permit.

Approximately 152 M ha of U.S. cropland is rain-fed, with about 40% receiving <500 mm of precipitation each year. These areas, classified as subhumid, semi-arid, or arid generally lie west of the 100th Meridian (Fig. 3–1). Globally, 96% of all cropland is classified as dryland. Agriculture in these areas provides about 67% of all crop production and about 50% of the economic value of all crops. The continued demand for food, feed, and fiber, coupled with small increases in addi-

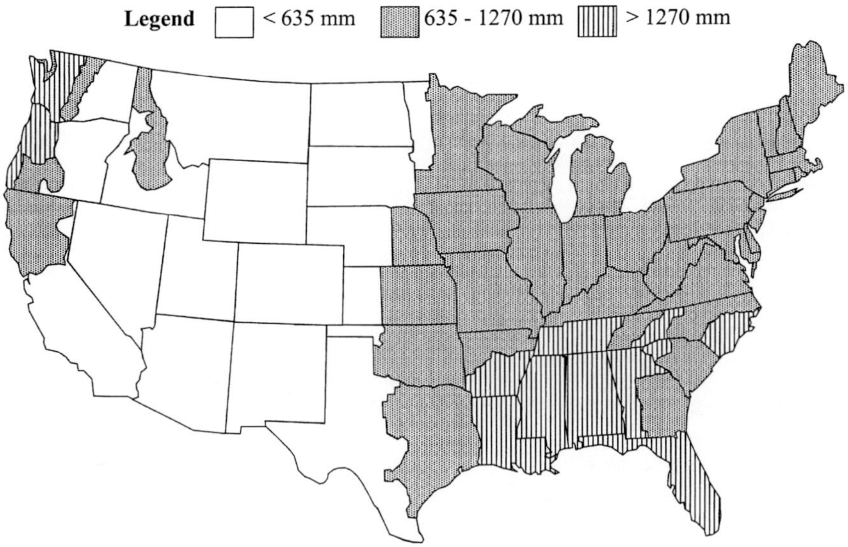

Fig. 3–1 Average annual precipitation in the USA from 1931 to 2000.

tional irrigated lands, means that dryland agriculture will be increasingly important in meeting humankind's future requirements. However, the relatively harsh dryland environmental conditions and unpredictability of rainfall make many of these areas marginal for agricultural production. For example, widespread drought in the south central USA in 1966, 1998, 1999, and again in 2000 raised serious concerns about our nation's vulnerability to extended periods of water shortage. Furthermore, annual economic losses to drought in the USA alone are often measured in billions of dollars.

Reducing the risks associated with dryland crop production is one of the critical missions of the USDA-ARS research programs. Portions of several national programs focus on four critical areas to accomplish this mission. They include:

1. Tillage and crop residue management practices to improve soil water storage.
2. Soil water monitoring and development of long-term weather forecasting.
3. Crop selection and rotations.
4. Physiology and genetics of stress tolerance in crops.

The remainder of this chapter will review some of the highlights and accomplishments associated with ARS research that addresses these issues.

TILLAGE AND CROP RESIDUE MANAGEMENT

Dryland agriculture is practiced in subhumid and semi-arid regions where water deficiencies result from relatively low and erratic precipitation, high evaporation rates, and sometimes inefficient capture and storage of precipitation as soil

Table 3–1. Soil water storage during fallow as influenced by straw mulch rates at four Great Plains locations (Greb et al., 1979).

		Soil water gain (t ha^{-1} of mulch)			
Location	Years	0	2.2	4.4	6.6
		cm			
Bushland, TX	3	7.1	9.9	9.9	10.7
Akron, CO	6	13.4	15.4	16.5	18.5
North Platte, NE	7	16.5	19.3	21.6	23.4
Sidney, MT	4	5.3	6.9	9.4	10.4
Average		10.7	12.7	14.5	15.7
Gain by mulching			2.0	3.8	5.0

water, especially during periods when crops are not growing. It is a dynamic and highly complex production system for which the major limitation on food, feed, and fiber production is a deficiency of water (Willis and Dregne, 1983). Research addressing on-site conservation of soil water under dryland conditions was initiated at many U.S. locations after the major drought and associated dust storms that plagued the Great Plains and surrounding areas during the 1930s. Improved water conservation practices received much attention as an approach to stabilizing agricultural production in this region and as a result, most crops in the Great Plains are grown without irrigation.

In recent years, ARS research has emphasized the role of conservation tillage on soil and water conservation, tillage effects on soil quality, and crop yield responses. Defined as any tillage or planting system that maintains at least 30% of the soil surface covered by the previous crop's residue after planting, conservation tillage is often justified as an approach for reducing soil erosion. For instance, van Doren and Allmaras (1978) reported that <1 Mg ha^{-1} of wheat (*Triticum aestivum* L.) straw provided 30% cover of the soil surface and reduced soil losses by 70%.

ARS research at numerous locations in the Great Plains has shown that retaining crop residues on the soil surface by use of reduced- or no-tillage practices also improved soil water and consequently, crop yields (Tanaka, 1989; Unger, 1984; Wilhelm, 1989). Greb (1983) reported that the net soil water gain during fallow is influenced by the quantity of residue available at the beginning of fallow. Results from four Great Plains locations documented increases of 2 cm of net soil water storage with 2.2 tons ha^{-1} of mulch and 5 cm with 6.6 tons ha^{-1} of mulch (Table 3–1). Crop residues reduce evaporation by insulating and cooling the soil surface, reflecting solar energy, suppressing wind speed near the soil surface, and providing a barrier against water vapor movement. No-tillage also maintains soil pores that not only provide more rapid infiltration, but also allow water to be stored deeper in the soil profile where it is less subject to evaporation. Collectively, the tillage and residue management practices developed through research such as this are essential for future improvements in dryland agriculture.

SOIL WATER AND WEATHER FORECASTING

Dryland agriculture is extremely vulnerable to climate variability and as a result, inadequate and untimely precipitation are perhaps the most frequent weather anomalies encountered by producers using these practices. ARS soil water research has been designed to address several management decisions including: reducing plant population to match available water supply; increasing row spacing to encourage root extension into untapped stored soil water; incorporating short-duration crops into more diverse rotations to avoid drought; and improving weed management to prevent competition for available soil water.

Climatic extremes often cause significant and widespread losses that impact consumers as well as farmers, ranchers, and all other sectors associated with or affected by agriculture. If soil water content could be more accurately estimated, the risk associated with variability in precipitation could be reduced. ARS research at Beltsville, MD and El Reno, OK have demonstrated that root-zone water content can be accurately estimated by incorporating remotely sensed estimates of surface soil water content into simulation models that include relevant soil and vegetation information (Starks and Jackson, 2001). When operational, this type of remote sensing/modeling approach could provide almost real-time monitoring of soil water throughout the growing season, thus allowing producers to make informed decisions on whether sufficient soil water is available for more water-sensitive crops or options such as double cropping. Furthermore, when integrated with the long-term weather forecasts or climate outlooks currently distributed by the National Oceanic and Atmospheric Administration's (NOAA), Climate Prediction Center (Schneider and Garbrecht, 2002), future soil water supplies could be predicted up to several months in advance.

ARS dryland agricultural research addressing soil water and weather relationships is providing assessments and predictions that are potentially useful to not only producers making a wide variety of soil and crop management decisions such as crop selection and fertilizer rates and dates of application, but also for water resource specialists who use the information to improve management of surface and groundwater resources. Additional research is needed to improve interpretation and encourage adoption of the large-scale forecasts for agricultural applications at the farm scale, but with continued progress there should be an increased stability and sustainability of agricultural production systems through proactive adaptation and risk-based planning. Steiner (2004, this publication) presents a detailed discussion of this subject.

CROP SELECTION AND ROTATIONS

Crop selection and choice of cultivars are among the most fundamental but important decisions made by producers for dryland agriculture. Several ARS studies have shown that cultivars that are productive and well adapted for irrigated conditions or humid, high-rainfall areas are not necessarily as productive under dryland conditions. Crop selection, both species and cultivar, depends on many factors, but for dryland farming, certain agronomic characteristics have been shown to be especially relevant. Generally, crops developed for sustainable production under dryland conditions should have a short stature with limited leaf area to minimize transpiration; be endowed with deep, dense root systems to enhance soil water extraction and use; and be early maturing with the capability to complete development prior to the onset of hot and dry conditions and to mature before plant available soil moisture is exhausted.

The ARS efforts to develop cultivars that are more genetically suitable for dryland agriculture played an important role in the dramatic increases in crop yields observed over recent decades. Data from USDA's National Agricultural Statistics Service (NASS) indicate that through these efforts, national average wheat yields have doubled since 1950. Sorghum (*Sorghum bicolor* L.) yields at Bushland, TX actually increased 139% from 1956 to 1997, with 46% of that increase attributed to the use of improved hybrids (Unger and Baumhardt, 1998). These yield increases were not related to changes in precipitation amounts or patterns, but some portion of the remaining 54% can be attributed to improved conservation practices and an associated increase in plant-available root zone soil water. The relatively recent development of corn (*Zea mays* L.) hybrids that perform well under dryland conditions resulted in an increase in land devoted to dryland corn production from 10 000 ha to more than 200 000 ha during the last 7 to 10 yr in the four-state region that includes Colorado, Kansas, Wyoming, and Nebraska. Similarly, the area planted to dryland hybrid sunflower (*Helianthus annuus* L.) increased by 260 000 ha.

Currently, winter wheat is still the dominant crop in the southern and central Great Plains, and the conventional wheat-fallow production system, developed in the 1930s to minimize crop failures due to erratic precipitation (Hinze and Smika, 1983), is the predominant management strategy. However, reduced or no-till practices coupled with herbicidal control of weeds have led to greater precipitation storage efficiencies, and therefore better success with more frequent cropping systems (Peterson et al., 1993). Research at several ARS locations continues to focus on overcoming constraints to the production of forage or short-duration crops during the summer fallow period, even for conventional, continuous wheat production systems.

Providing timely, relevant information on crop production options has been a primary objective of agricultural research in the USA for more than a century. In the early part of the 20th century, research locations were established throughout the Great Plains to develop crop rotations and better tillage practices for dryland conditions (Chilcott, 1927). In those portions of the Central and Northern Great Plains where agriculture remains dominated by winter and spring wheat, improved water-use efficiency gained through reduced tillage, improved genetics and better weed control, provide opportunities for more diverse crop rotations including oil seeds, forages, legumes, and even specialty crops.

Crop diversification remains an important step toward the goal of increasing profitability and sustainability of dryland agriculture. In recent years, several alternative crops including field peas [*Pisum sativum* var. *arvense* (L.) Poir.], sunflower, proso millet (*Pennisetum squamulatum* L.), and canola (*Brassica napus* L.) have been evaluated by the USDA-ARS at Akron, CO for their agronomic and economic potential under dryland systems. Some of these crops appear to offer considerable potential as alternatives to winter wheat (Table 3–2).

GENETICS AND PHYSIOLOGY OF DROUGHT TOLERANCE

Plant response to water deficits at the cellular and genetic level have both economic and evolutionary importance, directly affecting plant productivity in agri-

Table 3–2. Dryland yield, typical market prices, cost of production, and estimated returns of alternative crops grown at the Central Great Plains Research Station, Akron, CO (Vigil et al., 1997).

Crop	Yield under dryland conditions	Market price	Cost to† produce	Estimated range of net returns	Estimated average net return
	kg ha^{-1}	\$ kg^{-1}		\$ ha^{-1}	
Field pea	784–2354	0.13–0.48	180–190	−86–960	?‡
Garbanzo bean	336–1792	0.33–0.55	212–234	−123–775	?
Field bean§	336–1232	0.30–0.61	224–234	−130–535	?
Canola crambe	560–2352	0.19–0.28	229–239	−128–445	57
Triticale (forage)	2240–6720	0.07–0.09	167–279	5–387	66
Proso millet	1120–2800	0.13–0.26	182–202	−59–558	42
Sunflower	840–1792	0.20–0.29	197–247	−81–316	62
Winter wheat	1680–4032	0.13–0.17	192–254	−32–518	91
Corn	1881–4480	0.08–0.15	212–227	−22–568	123

† Includes cost of seed, fertilizer, chemical weed control, land preparation based on custom rates, planting, and harvest.
‡ Not enough data to make reliable estimates.
§ Field beans include pinto, great northern, kidney, black turtle, and navy bean.

culture and survival under natural environments. Ramanathan (1988) argued, based on predictions of global environmental change, that developing crops that are more tolerant to water deficits while maintaining productivity will become a critical requirement early in the 21st century. Understanding how plants tolerate water loss is a vital prerequisite for developing strategies for improving biomass and seed production under drought conditions. Through the efforts of ARS and other scientists, much has been accomplished in this area even though the specific emphasis has changed over time. Currently, the focus is on discovery of those genes that confer tolerance that is expressed in vegetative structures during the response of various plants to water deficits (Ingrams and Bartels, 1996; Bray, 1997; Shinozaki and Yamaguchi-Shinozaki, 1999; Xiong and Zhu, 2002; Zhu, 2002). From this work, our knowledge of stress tolerance has improved immensely, but this approach is not without limits, especially if we work with a narrow suite of crop species. The limited tolerance of most crops implies that the genetic information necessary for expanding their tolerance may not be exploitable or perhaps it's not even present. Furthermore, the response to drought by sensitive species may not be a consequence of low tolerance but rather a consequence of the damage incurred and the plant's ability to repair that damage.

Desiccation tolerance may also prove to be a key phenomenon in efforts to enhance or provide drought tolerance to crop species. Desiccation tolerance is proving to be as unique as it is important to survival of certain plants native to desert environments. It is physiologically and genetically distinct form drought tolerance, although similarities do exist. Desiccation represents an extreme state or environment that requires a more complex cellular response for recovery than that of a drought-sensitive plant exposed to a relatively small water deficit (Phillips et al., 2002). Nevertheless, an understanding of desiccation tolerance can have a major impact on our ability to modify crops for increased drought tolerance. A good analogy is the advances in our understanding of salt tolerance and the modification of crops for increase salt tolerance that arose directly from studies of halophytic plants (Nelson et al., 1998). The isolation, manipulation, and engineering of a single gene (a vacuolar Na+/H+ antiport) had a significant impact (Apse et al., 1999). But, the uniqueness of desiccation tolerance with regard to other plant stresses can also be seen by the fact that halophytes and cold-tolerant species, even when hardened, are not tolerant of desiccation.

Current ARS studies focus on two extremes within the plant complexity spectrum, a desiccation-tolerant bryophyte, tortula moss [*Tortula ruralis* (Hedw.) Gaertn et al.], and a desiccation-tolerant grass, (*Sporobolus stapfianus* Gandoger). These plants represent the two evolutionary ends of the spectrum with regard to the mechanisms by which they achieve desiccation tolerance. The prevailing evidence suggests that two basic strategies for vegetative desiccation tolerance exist (Oliver and Bewley, 1997). The desiccation-tolerant bryophyte appears to rely on a limited extent on cellular protection but more so on a repair and reconstitution mechanism, induced upon rehydration. This is perhaps the more primitive mechanism and one that was operative as plants invaded the land (Oliver et al., 2000). The desiccation-tolerant angiosperms appear to rely almost totally on an inducible cellular protection mechanism that requires a certain amount of time to become expressed in the leaf tissues. An understanding of both of these mechanisms might provide two strategies, employing genes that protect cells from the damage of dehydration and genes that repair the damage. Scientists at Lubbock, TX have identified 18 genes from *T. ruralis*, designated rehydrins, that appear to be involved in the repair of cellular damage (Scott and Oliver, 1994; Oliver et al., 1997; Wood et al., 1999). Interestingly, several of the 152 *T. ruralis* express sequence tags (ESTs) generated there were significantly similar to unidentified desiccation-tolerance genes isolated from a desiccation-tolerant dicot (*Craterostigma plantaginium).*

Studies of the desiccation-responsive genes of the tolerant grass, *S. stapfianus*, focus on a single gene that exhibits properties that might be useful. SsRab2 is a gene that is not induced by abscisic acid (ABA) but is induced during desiccation and again upon rehydration of the grass. SsRab2 encodes a small binding protein that is, in other systems, associated with vesicular trafficking. This process is involved in maintaining membrane integrity and hence may be of importance to both cellular protection and repair (O'Mahony and Oliver, 1999). Genomic level investigations have been initiated for both species, in the hopes of identifying not only individual genes that are important in dehydration tolerance but also gene networks and unique signaling pathways. Preliminary studies utilizing cDNA microarray technology suggest that this could be a successful strategy and that novel approaches to crop improvement for drought tolerance will be forthcoming.

SUMMARY AND CONCLUSIONS

Much progress was made in the last century in improving water conservation, water-use efficiency, and the productivity of dryland crops. Technologies from a variety of scientific disciplines enhanced the productivity and sustainability of dryland agriculture in recent decades. These technologies include minimum tillage practices that maintain surface residue cover to enhance water infiltration, reduce evaporation, and protect the soil from erosion. Producers now use short-stemmed cultivars with short life cycles, deep or extensive root systems, and limited leaf surface area. We continue to address dryland production problems with research on tillage, alternative crops, and the breeding of more water-efficient cultivars, but we also bring new technologies to bear.

Remote sensing of soil water content and the seasonal weather outlooks present numerous opportunities for dryland agriculture. Both reduce risk associated with limited rainfall. When implemented, these technologies will increase stability and sustainability of agricultural production system through pro-active adaptation and risk-based planning. Advances in biotechnology have only begun to provide novel approaches to development of drought-tolerant agricultural and horticultural crops.

Continued or enhanced competition for water and rising costs of irrigation underscore predictions that the additional food required by a growing population must be provided in large part by a more efficient and productive dryland agriculture. Advances in plant genetics and agronomic conservation technologies, when considered in concert, continue to provide the greatest opportunities to achieve sustainability and profitability in dryland agriculture and will continue to be the focus of the ARS research program.

REFERENCES

Apse, M.P., G.S. Aharon, W.S. Snedden, and E. Blumwald. 1999. Salt tolerance conferred by overexpression of a vacuolar Na+/H+ antiport in Arabidopsis. Science (Washington, DC) 285:1256–1258.

Bray, E.A. 1997. Plant responses to water deficit. Trends Plant Sci. 2:48–54.

Chilcott, E.C. 1927. The relationships between crop yields and precipitation in the Great Plains area. USDA Misc. Circ. 81. U.S. Gov. Print. Office, Washington, DC.

Food and Agriculture Organization. 1999. Food outlook, No. 4. p.13. Global Information and Early Warning System. FAO of the United Nations, Rome.

Greb, B.W. 1983. Water conservation: Central Great Plains. p. 57–70. *In* H. Dregne and W. Willis (ed.) Dryland agriculture. Agron. Monogr. 23. ASA, CSSA, and SSSA, Madison, WI.

Greb, B.W., D.E. Simka, and J.R. Welsh. 1979. Technology and wheat yields in the Central Great Plains: Experiment station advances. J. Soil Water Conserv. 34:264–268.

Hinze, G.O., and D.E. Smika. 1983. Cropping practices: Central Great Plains. p. 387–395. *In* H.E. Dregne and W.O. Willis (ed.) Dryland agriculture. Agron. Monogr. 23. ASA, CSSA, and SSSA, Madison, WI.

Hoffman, M., and S. Hussian. 1998. The world water and climate atlas [Online]. Available at http://www.worldbank.org/html/cgiar/press/watatlas.html (verified 25 May 2003).

Howell, T.A. 2001. Enhancing water use efficiency in irrigated agriculture. Agron. J. 93:281–289.

Lawrence, A. 2000. America's disappearing farmlands—A citizen's crusade to stop America's disappearing land [Online]. Available at http://www.ardicountry.com/farmland.html (verified 26 May 2003).

Luckey, R.R., E.D. Gutentag, and J.B. Weeks. 1981. Water level and saturated-thickness changes, predevelopment to 1980, in the High Plains aquifer in parts of Colorado, Kansas, Nebraska, New Mexico, Oklahoma, South Dakota, Texas, and Wyoming. U.S. Geological Survey Hydrologic Investigations Atlas HA-652. U.S. Gov. Print. Office, Washington, DC.

McGrath, T., and J. T. Dugan. 1993. Water-level changes in the High Plains Aquifer—Predevelopment to 1991. U.S. Geol. Survey Water-Resour. Investigation Rep. 93-4088. U.S. Gov. Print. Office, Washington, DC

Nelson, D.E., B. Shen, and H.J. Bohnert. 1998. Salinity tolerance—Mechanisms, models and the metabolic engineering of complex traits. Genetic Eng. (New York) 20:153–176.

Oliver, M.J., and J.D. Bewley. 1997. Desiccation tolerance of plant tissues: A mechanistic overview. Hortic. Rev. 18:171–214.

Oliver, M.J., Z. Tuba, and B.D. Mishler. 2000. Evolution of desiccation tolerance in plants. Plant Ecol. 151:85–100.

Oliver, M.J., A.J. Wood, and P. O'Mahony. 1997. How some plants recover from vegetative desiccation: A repair based strategy. Acta Physiol. Plant. 19:419–425.

O'Mahony, P., and M.J. Oliver. 1999. Characterization of a desiccation responsive small GTP binding protein (Rab2) from the desiccation-tolerant grass *Sporobolus stapfianus*. Plant Mol. Biol. 39:809–821.

Peterson, G.A., D.G. Westfall, and C.V. Cole. 1993. Agroecosystem approach to soil and crop management research. Soil Sci. Soc. Am. J. 57:1354–1360.

Phillips, J.R., M.J. Oliver, and D. Bartels. 2002. Molecular genetics of desiccation and tolerant systems. p. 319–341. *In* M. Black and H.W. Pritchard (ed.) Desiccation and plant survival. CABI Publ., Wallingford, UK.

Pinstrup-Anderson, P., R. Pandya-Lorch, and M.W. Rosegrant. 1999. World food prospects: Critical issues for early twenty-first century. Food Policy Rep. IFPRI, Washington, DC.

Ramanathan, V. 1988. The greenhouse theory of climate change: A test by an inadvertent global experiment. Science (Washington, DC) 240:293–299.

Rosegrat, M.W., W.C. Ringler, and R.V. Gerpacio. 1999. Water and land resources and global food supply. p. 167–186. *In* G.H. Peters and J.V. Braun (ed.) Food security, diversification, and resource management: Refocusing the role of agriculture? Proc. 23rd Int. Conf. of Agric. Economists, Sacramento, CA. 10–16 Aug. 1997. Aldershot, Hants, England and Asgate, Brookfield, VT.

Schneider, J.M., and J.D. Garbrecht. 2002. A blueprint for the use of NOAA/CPC precipitation climate forecasts in agricultural applications. p. J71–J77. *In* Proc., Am. Meteorol. Soc. 3rd Symp. on Environmental Applications, Orlando, FL. 13–17 Jan. 2002. Am. Meteorol. Soc., Boston, MA.

Scott, H.B. II, and M.J. Oliver. 1994. Accumulation and polysomal recruitment of transcripts in response to desiccation and rehydration of the moss *Tortula ruralis*. J. Exp. Bot. 45:577–583.

Shinozaki, K., and K. Yamaguchi-Shinozaki. 1999. Gene expression and signal transduction in water-stress response. Plant Physiol. 115:327–334.

Starks, P.J., and T.J. Jackson. 2001. Estimating soil water content in tallgrass prairie using remote sensing. J. Range Manage. 55:474–481.

Steiner, J.L., J.M. Schneider, J.D. Garbrecht, and X.J. Zhang. 2004. Climate forecasts: Emerging potential to reduce dryland farmers' risks. p. 47–66. *In* S.C. Rao and J. Ryan (ed.) Challenges and strategies for dryland agriculture. CSSA Spec. Publ. 32. CSSA and ASA, Madison, WI.

Tanaka, D.L. 1989. Spring wheat plant parameters as affected by fallow methods in the northern Great Plains. Soil Sci. Soc. Am. J. 53:1506–1511.

Unger, P.W. 1984. Tillage and residue effects on wheat, sorghum, and sunflower grown in rotation. Soil Sci. Soc. Am. J. 48:885–891.

Unger, P.W., and R.L. Baumhardt. 1998 Factors related to dryland grain sorghum yield increases:1939 through 1997. Agron. J. 91:870–875.

Vigil, M.F., D. Nielsen, and R. Anderson. 1997. Cost of production and yields of alternative dryland crops. Conserv. Tillage Fact Sheet 2-97. USDA-ARS and USDA-NRCS, Washington, DC.

Willis, W.O., and H.E. Dregne. 1983. Preface. p. xiv. *In* W.O. Willis and H.E. Dregne (ed.) Dryland agriculture. Agron. Monogr. 23. ASA, CSSA, and SSSA, Madison, WI.

Wood, A.J., R.J. Duff, and M.J. Oliver. 1999. Expressed sequence tags (ESTs) from desiccated *Tortula ruralis* identify a large number of novel plant genes. Plant Cell Physiol. 40:361–368.

Xiong, L., and J.-K. Zhu. 2002. Molecular and genetic aspects of plant responses to osmotic stress. Plant Cell Environ. 25:131–139.

Zhu, J.K. 2002. Salt and drought stress signal transduction in plants. Annu. Rev. Plant Biol. 53:247–273.

4 A Grey to Green Revolution in the Semi-Arid Tropics of Asia and Africa

William D. Dar and Eric M. McGaw

International Crops Research Institute for the Semi-Arid Tropics (ICRISAT), Patancheru, India

D.V.R. Reddy

Donald Danforth Plant Science Center St. Louis, Missouri

ABSTRACT

The mission of the International Crops Research Institute for the Semi-Arid Tropics (ICRISAT) is to conduct agricultural research to help the poorest of the poor in the semi-arid tropics, one of the harshest agro-ecoregions in the world. Nearly one sixth of the world's population lives in the semi-arid regions of 48 developing countries in Africa, Asia, and Latin America. Risks are pervasive and greater here than in any other important food production system. Although the Green Revolution was a great achievement, it was unable to solve the problems of food insecurity for hundreds of millions of poor people. ICRISAT's main strategy is to adapt productive systems to the natural variability of the environment, rather than the other way around. We seek to empower farmers to use available farm resources to develop their own capacities and become self-reliant. ICRISAT has begun a program of global consultations with an array of partners, including the private sector, on the new priorities for agricultural research. A result of this consultative process is that ICRISAT has revised its regional research strategies. In an important departure from the past, the primary development strategy is not crop- or technology-based, but starts with the identification of real opportunities for smallholder farmers. It then seeks to develop those opportunities into tangible benefits at the farm level by pursuing science for development, the creation of new partnerships, and by linking farmers to markets.

INTRODUCTION: THE CURRENT SITUATION

The world today faces greater challenges than ever before in balancing food and environmental security. The International Crops Research Institute for the Semi-Arid Tropics (ICRISAT) focuses on agricultural research to help the poorest of the poor in the semi-arid tropics (SAT), one of the harshest agro-ecoregions in the world. The SAT is characterized by extreme poverty, persistent drought, infertile soils, erratic rainfall, growing desertification, and soil degradation. Another serious constraint is that agricultural production cannot keep pace with the rapid population growth that typifies the SAT.

Nearly one sixth of the world's population lives in the semi-arid regions of 48 developing countries in Africa, Asia, and Latin America. They are poor and malnourished, deprived of basic health and nutrition, and live on less than a dollar a day (Ryan and Spencer, 2001). Most are subsistence farmers who need to adopt agricultural practices that give maximum returns. Risks are pervasive and greater here than in any other important food production system. Food security remains a dream to these farmers.

The Green Revolution was unquestionably one of the great achievements of the last century. Without it, at least a billion people would be hungry today. It showed the importance and power of modern technology in enhancing agricultural productivity, thereby addressing the issue of food security. Nonetheless, it was unable to solve the problems of food insecurity for 48% of the population of South Asia, 35% of sub-Saharan Africa and 17% of Latin America. Food prices, though currently at an all-time low, are not helping the poor to overcome hunger. Even subsistence farmers usually need to purchase significant portions of their daily food. Unless agriculture generates adequate income, improvements in the livelihoods of SAT farmers are unlikely.

ICRISAT's research is directed at developing appropriate technologies for sustainable agriculture in SAT regions, mainly targeted to benefit millions of marginal farmers. The Institute's main strategy is to adapt productive systems to the natural variability of the environment, rather than the other way around. We seek to empower farmers to use available farm resources to develop their own capacities and become self-reliant. This is aided through improved integrated soil and water management strategies and by adopting tools and techniques for the sustainable utilization of natural resources. ICRISAT takes the following approach to achieve this.

CULTURAL PRACTICES

Harvesting Water

Integrated watershed management is one of the most important requirements for maximizing yields from dryland crops. Watersheds, or small valleys, are especially strategic intervention points because they are the concentration zones for water and nutrient flows. If properly managed, these precious areas can be leveraged to increase crop productivity. However, the risk is equally high if they are mismanaged, because erosion and nutrient depletion proceeds rapidly along the slopes.

Improving the management of watersheds in marginal areas requires working closely with local communities, because many farmers usually share these lands. We have found that governmental and nongovernmental organizations (NGOs) are enthusiastic about teaming up with research organizations like ICRISAT and its national research partners to provide a community interface. Together, we make a holistic evaluation of the problems and the opportunities particular to each watershed.

Certain principles can be quickly tailored to generate watersheds that benefit many smallholders. The use of inexpensive, small earthen dams to arrest gully erosion and transform these gullies into seasonal reservoirs instantly converts a liability into an asset. Not only does this recharge groundwater, preventing wells from going dry, it also extends the potential growing season in these areas by making small-scale irrigation possible even during the dry season.

By enabling double-cropping and dry-season cropping on limited areas, these techniques allow farmers to diversify their crops by adding legumes, high-value crops and livestock to their farming systems. Legumes break cereal pest and disease cycles and add nitrogen (N) to the soil. Animal manure likewise benefits the soil. Diversified and high-value crops reduce risk, increase farmers' incomes, and give them an incentive to invest in land improvement. This approach is being applied by ICRISAT in partnership with several benchmark communities in southern India (Wani et al., 2001), northern Thailand, and northern Vietnam.

Other important methods appropriate for smallholders include the use of bunds and tied ridges to increase moisture percolation into the root zone, and *zai* holes in which farmers insert organic matter to prevent erosive losses. By simply leaving some crop straw on the soil surface, farmers can significantly reduce the hazard of wind and water erosion, increase moisture percolation, and reduce evaporative losses, while providing better growth conditions for soil fertility-enhancing biota. Our sister organization, the World Agroforestry Center, is investigating the role of trees in improving dry tropical agroecosystems.

A model of successful application of integrated watershed management was developed taking into account social, political, and institutional factors. The main components were:

- Low-cost soil and water conservation structures
- Environment-friendly integrated nutrient management
- Environment-friendly and cost-effective plant protection measures
- Crop diversification with legumes to avoid air and water pollution

Application of this model in a village called Kothapally in Andhra Pradesh, India has resulted in:

- Improved groundwater levels (5–6 m)
- Improved vegetative cover
- Increased productivity from 1500 kg ha^{-1} to 3300 kg ha^{-1} of maize and from 1070 to 2600 kg ha^{-1} of sorghum
- More than fivefold increase in income over traditional practices

One of the distinct advantages of this approach is that the model allows flexibility. We are implementing this approach in partnership with several benchmark communities in southern India, northern Thailand, and northern Vietnam (Ramakrishna et al., 2002). The knowledge we gain from these benchmark sites will be used to scale up and disseminate the technology ever more widely, notably to the dry areas of Africa.

Soil Fertility

Many areas in the dry tropics are severely constrained by low soil fertility, especially in Africa. The low biomass productivity of water-starved areas, coupled with high soil temperatures that promote oxidation of organic matter, render these areas innately deficient in soil organic matter. Adding to this problem, these soils are often overcultivated and crop residues are fed to cattle or used for construction material rather than being returned to the soil.

Although conventional wisdom indicates that fertilizer application is too risky in low-rainfall environments, our joint research indicates otherwise. We are finding that soil nutrients, in the proper amounts, balance and timing actually reduce risk by increasing crop water-use efficiency through several means:

- Stimulate root growth so that plants find and use more of the rainwater that falls on the land
- Hastening crop maturity to avoid drought
- Increasing harvest index (grain filling), thus enhancing yield

Soft drink bottle caps were used to apply fertilizer into planting holes along with the seed. The rates were affordable and contributed to increasing the yields of pearl millet [*Pennisetum glaucum* (L.) R.Br.] and sorghum [*Sorghum bicolor* (L.) Moench] in Niger. Currently, we are assisting FAO to disseminate this technology to farmers across Niger. At least 5000 farmers have used it so far, and it is having a major impact on food security. We are having similar success with a distribution and application system of convenient, 'farmer-size' small bags of fertilizer in the dry areas of Zimbabwe and Malawi.

Pearl millet is drought tolerant. The yields of this crop have been declining in Africa at a rate of nearly 3% per annum since the mid-1980s. An innovative microfertilizer technique has yielded encouraging results.

DEVELOPMENT OF SUITABLE CULTIVARS

Early Maturity

No trait of improved varieties has been more popular with farmers than early maturity, even though it constrains yields (since the crop has less time to produce biomass). Early-maturing varieties stabilize yields by avoiding drought, and enhance income by capturing peak prices before the main crop hits the market, allowing farmers to quickly pay back high-interest loans they often take to cover the costs of labor and inputs. ICRISAT's breeders, working jointly with partners, have been able to greatly compress the life cycle of all five of the institute's mandate crops: sorghum, millet, groundnut (*Arachis hypogaea* L.), pigeonpea [*Cajanus cajan* (L.) Millsp.] and chickpea (*Cicer arietinum* L.).

Early-maturing sorghum and millet varieties have been enthusiastically adopted across the dry tropics for these reasons. For example, production of the millet variety SOSAT C88 is currently spreading rapidly across northern Nigeria. Farmers value its combination of high yield, vigorous growth, resistance to downy mildew and excellent taste. Farmers themselves had a hand in selecting early-maturing pearl millet Okashana 1 in Namibia through a participatory process where researchers invited farm cooperatives to choose the most promising lines within breeding nurseries. It now covers half the pearl millet area of the country.

Early-maturing sorghum ICSV 111 IN, or *Kapaala*, has many advantages. It is good to eat and suitable for making beer. It is also adaptable to an array of cropping systems and escapes drought because it matures early. As a result, it has become one of the most popular sorghum varieties in Ghana.

Short-duration chickpea genotypes developed by ICRISAT and partners escape drought and resist fusarium wilt. Some of the short-duration cultivars are *Sweta* (ICCV-2), *Kranti* (ICCC 37), and *Bharati* (ICCV 10). ICCV 96029, a super-early chickpea that matures in 75 to 80 d in peninsular India, provides the opportunity to cultivate chickpea in dry areas as well as in fallow lands. Development of such varieties will lead to increased productivity and income (Jagdish Kumar and Abbo, 2001).

Another legume transformed by early maturity is pigeonpea, which has been extended into new areas and inserted into new rotations, generating ever-higher value for farmers. New varieties can be sown after sorghum in central India, in rotation with wheat in the Indo-Gangetic Plains (Dahiya et al., 2001), and after wheat in the Gezira irrigation zone of central Sudan.

Our socioeconomists documented how early-maturing pigeonpea in central India substantially raised farmers' net incomes while diversifying their operations and making them more sustainable. Likewise, extra-short duration, relatively photoperiod-insensitive bush-type pigeonpeas bred by ICRISAT and its partners have shown high adaptation to wide latitudes ranging from 7 to 46°N (Saxena, 2000), and produce multiple flushes of high-value green peas for vegetable markets or for export (Faris et al., 1987).

Hybrids

Hybrids such as corn or maize are proven tools that farmers can use to grow their way out of poverty. However, conventional wisdom holds that hybrids only express their potential in favorable environments, and are therefore unprofitable and unattractive to the seed industry in marginal areas.

Our partnership-based work has shattered this belief. We have found that hybrid vigor actually increases stress resistance; it makes both shoots and roots grow more vigorously, so plants express greater resilience against a wide range of challenges, including drought, pest and disease attacks, and weed competition. Hybrid vigor increases yields by 20 to 30%. This is profitable even when mean yields are low because these crops have relatively high seed multiplication ratios (1:400 for millet, 1:200 for sorghum, 1:30 for pigeonpea). Consequently, the cost of seed can be kept correspondingly low for farmers.

Hybrids are also more responsive to improved management, providing an inbuilt incentive to farmers to invest in soil fertility, weed control, and related agronomic practices. By providing farmers with yet another option to help them grow their way out of poverty, we are creating situations that naturally foster greater investment in the land and a more sustainable path to the future.

Hybrids of our two cereal crops, sorghum and millet, are now sown across substantial areas of the dry tropics of India, where highly developed technical skills and infrastructure can support this work. Research partnerships with Haryana Agricultural University and the USAID Title XII International Sorghum/Millet Collaborative Research Support Program (INTSORMIL) resulted in the development of the early-maturing public sector hybrid HHB 67, which allows farmers to grow a crop of mustard during the postrainy season. We also continue to explore systems requiring less technical expertise, such as topcross hybrids for Africa.

Converting Adversity into Opportunity

In Sudan, prices for their staple food, faba bean (*Vicia faba* L.), were climbing beyond the reach of the poor. Faba bean requires cool temperatures to grow and is unsuited to the warm climate that typifies Sudan. Joint research between Sudan and ICRISAT identified that early-maturing pigeonpea ICPL 90028 can be grown on 50 000 ha of drylands in the massive Gezira irrigation scheme south of Khartoum (S.N. Silim, personal communication, 2001).

In contrast with heat-loving pigeonpea, chickpea is by nature a cool-season crop. Short-duration, heat-tolerant varieties are grown on an estimated 14 million hectares of rainfed rice (*Oryza sativa* L.) fallows across South Asia, mostly lying gray and barren after the rice harvest. These areas can now be greened with a lush carpet of chickpea. Currently chickpea is grown on nearly 10 000 ha in the Barind region of Bangladesh, saving the country nearly $3 million in imports annually (Musa et al., 1998).

ADAPTATION OF ECO-FRIENDLY AND COST-EFFECTIVE PEST MANAGEMENT TECHNOLOGIES

Pest outbreaks during the SAT's brief rainy season can be just as punishing to crops as in the wetter agro-ecosystems. These scourges can often be prevented or reduced substantially using local or shared knowledge, combined with cutting edge research. ICRISAT is helping farmers find and share the knowledge they need to minimize losses due to pest attack.

Insects damage legumes much more heavily than cereals. Pod borer (*Helicoverpa armigera)* causes more than $1 billion in losses to pigeonpea and chickpea crops each year. Integrated pest and disease management practices have made it possible to reduce and sometimes even eliminate pesticide sprays, substituting eco-friendly control practices and biopesticides such as the nuclear polyhydrosis virus and preparations from neem leaves. By monitoring pest populations and spraying only on a need basis, farmers save money and reduce health risks. By encouraging natural enemies such as wasps and birds, the ecosystem benefits and crops are protected. Some pigeonpea farmers even shake the plants to knock the pest larvae onto plastic sheets dragged between the rows, collecting the larvae to feed to their chickens. These measures are widely being used by farmers to control pod borer.

The pod borer problem in cotton cannot be disentangled from the problem in pigeonpea, because the two crops are grown in the same areas. Therefore, strategies need to be applied for both crops simultaneously.

A knowledge-based approach can be equally effective against plant diseases. Smallholder farmers in Nepal had given up cultivation of chickpea because of repeated devastation from Botrytis grey mold. Integrated control measures, which include resistant cultivars, have induced the farmers to take up cultivation of chickpea in Nepal once more (Pande et al., 2003).

EXPLOITING THE POTENTIAL OFFERED BY RECENT ADVANCES IN BIOTECHNOLOGY

Marker-assisted breeding is being used to develop sorghum cultivars with resistance to *Striga*, and pearl millet cultivars with resistance to downy mildew.

Biotechnology also holds much promise for integrated pest management. If the *Bt* gene can be incorporated into the plant genome of pigeonpea and chickpea, this will finally give us a powerful but environmentally benign way to prevent damage from the pod borer. This gene is harmless to humans, birds, and other animals. We are also using biotechnology to create resistance against viruses like peanut clump virus, for which there is no known resistance within the groundnut gene pool.

FOSTERING NEW PARTNERSHIPS

Seed Supply

ICRISAT is partnering private and public sector organizations to evolve strategies that ensure adequate seed supply at the right time. An understanding of institutional policies, strengths and weaknesses of farmers' seed systems, and private and public sector agencies is vital for improving the seed supply.

Public-Private Partnerships

Alarmed by continuing declines in funding, both public and private sectors are exploring closer ties. For example, 14 companies have collectively pledged more than $100 000 annually for the next 5 yr to help support applied plant breeding research at ICRISAT.

ENCOURAGING PARTICIPATION BY WOMEN

By focusing on women's crops and activities, researchers and development agencies can ensure that their efforts benefit those who need them most—the women and children who depend on them. Women produce 80% of sub-Saharan Africa's food, and 60% of Asia's. Responsibilities for certain crops, including management practices and post-harvest activities, are primarily borne by women.

Groundnut is mainly a women's crop, especially in Africa. The kinds of research-for-development interventions that can make a difference are illustrated by our partnership work in Malawi, where women are the main cultivators of this crop.

About one third of Malawian children under five are malnourished and underdeveloped. Child blindness is common, partially due to vitamin A deficiency, which is strongly associated with insufficient fat in the diet. The human body requires fat to absorb vitamin A, such as the oil provided by groundnuts.

A project to spread new varieties together with improved crop management practices was launched in 1998 jointly with Plan International, an NGO. To date, the project has reached 9200 households in 291 villages around Lilongwe and Kasungu. Each family received 5 kg of seed free under an agreement that, after harvesting the crop, they return 10 kg for distribution to farmers. In this way, farmers and communities are learning how to work together to multiply and share these valuable new lines, a practice that will bear fruit again and again as new varieties are released.

In India, studies showed that adoption of improved technology packages helped provide new income channels through ‘task specialization’. For example, the introduction of chickpea into the rice fallows of the Barind zone of northwestern Bangladesh provided a new income stream for women who harvest the top twigs for consumption as a fresh vegetable. Similarly, early-maturing pigeonpea varieties in Kenya are enabling women’s cooperative members to harvest high-value green peas for export to the United Kingdom, a highly remunerative activity generating much needed hard currency.

Women farmers are also becoming pioneers as millet seed producers and marketers, helping to spread new varieties and overcome the sorghum and millet seed supply bottleneck in Zimbabwe.

CONCLUSION: THE NINE CRITICAL DRIVING FORCES FOR FOOD SECURITY

Notwithstanding the foregoing experiences, we will conclude this chapter by citing what our sister Center, the International Food Policy Research Institute, considers the nine critical driving forces for sustainable food security. The success of the Grey to Green Revolution must be viewed in the context of these nine forces.

1. Accelerating Globalization and Further Trade Liberalization

Without the right policies and institutions at the national and international levels, globalization may either bypass or harm many poor people in developing and developed countries alike. Policymakers will need to guide the globalization process so that it benefits poor people, improving their food and nutrition situation while preserving their natural resources.

2. Sweeping Technological Changes

Technological advances in molecular biology, energy, information, and communications have the potential to help poor people achieve food security and make natural resource management more sustainable. If, however, researchers and policymakers continue to focus on the needs of people in rich countries, the result will be scientific and technological apartheid.

3. Degradation of Natural Resources and Increasing Water Scarcity

Degradation of natural resources is rampant in areas with poor soils, irregular rainfall, dense population and stagnant productivity. Environmental degradation contributes to poverty, but also often results from it. Food security solutions that fail to address natural resource issues effectively cannot be sustainable.

4. Health and Nutrition Crises

Malaria, tuberculosis, micronutrient deficiencies, HIV/AIDS, and other chronic diseases are all compromising food and nutrition security in many developing countries. These global health crises not only destroy human lives, but also slam the door on opportunities. They impoverish millions of people, raising health costs and causing severe shortages of productive workers.

5. Rapid Urbanization

By 2020 many more people in the developing world will live in urban areas where they will make heavy demands on the capacities of these cities to provide jobs, education, health care, and food. Although current policies must continue to focus on the countryside, where the majority of poor and food-insecure people still live, future policy action must pay increasing attention to growing poverty, food insecurity, and malnutrition in urban areas.

6. The Changing Face of Farming

The nature of farming is changing rapidly in many developing countries because of the aging of the farm population, the feminization of agriculture, labor shortages, depleted assets resulting from the HIV/AIDS crisis, and the decreasing cost of capital relative to labor. Small-scale family farms, traditionally the backbone of much of developing-country agriculture, are threatened.

7. Continued Conflict

Violent conflicts continue to cause severe human misery in many developing countries, especially in sub-Saharan Africa. Although humanitarian assistance can provide food and shelter for the many millions of refugees and displaced persons, policymakers must deal with the underlying causes and the effects on the people in wartorn areas. Achieving sustainable food security is not possible amidst conflict.

8. Climate Change

Many scientists and policymakers believe that climate change is leading to more frequent and more severe natural disasters. More research is needed on this hypothesis, as it has profound implications for food security. Future agricultural policies must focus on finding ways to keep agriculture productive as climate change continues.

9. Changing Roles and Responsibilities of Key Actors

National governments in many developing countries have found themselves playing a new and diminished role in the last couple of decades. Local governments, business and industry, NGOs, and other segments of civil society are now undertaking many activities previously performed by national governments. At the global level, transnational corporations and broad NGO coalitions are becoming increasingly prominent in policy debates. National governments must not risk losing their capacity to perform the functions that only they can perform, such as ensuring the rule of law and developing nationwide infrastructure.

It is true that the Green Revolution resulted in quantum increases in the production of cereal crops. Some believe that this benefited only rich farmers who could afford the inputs. Nonetheless, ample evidence exists that millions of marginal farmers have benefited from the technologies developed to bring about the Green Revolution.

It must be emphasized, however, that the Green Revolution has not solved the food problem. The new technologies did not benefit the marginal rural farmlands of the poor people living in the SAT. In this chapter, we have proposed options to overcome this inequity. There is a dire need for a Grey to Green Revolution, helping to green grey areas characterized by marginal environments, yearly climatic variations, high risks, and scarce capital for the poor. The Grey to Green Revolution not only involves increasing crop productivity but also leads to empowering the poor to build their own capacities, self-confidence, and self-reliance. Our investments in achieving this goal are paying off. To cite one example, a study by the Asian Development Bank in 2000 concluded that "Investments in infrastructure, agricultural technology and human capital are now as productive in many rainfed areas as irrigated areas and have a much greater impact on poverty alleviation."

CONCLUSION

ICRISAT has begun a program of global consultations with partners and colleagues from national agricultural research and extension systems, subregional organizations, donor organizations, international agricultural research centers, advanced research institutes, NGOs, and the private sector on the new priorities for agricultural research in the SAT (Ryan and Spencer, 2001). A result of this consultative process is that the ICRISAT global team has begun to revise its regional research strategies to target real development impacts on human and environmental well-being, in both the short and medium term. In an important departure from the past, the primary development strategy is not crop- or technology-based, but starts with the identification of real opportunities for smallholder farmers. It then seeks to develop those opportunities into tangible benefits at the farm level by pursuing science for development, the creation of new partnerships, and by linking farmer-producers to the market chain.

Many SAT farmers face food production deficits virtually every year due to the poor agricultural potential of their natural resource base, undeveloped markets, and poor infrastructure. Commercial agriculture may be too ambitious for this group—we need technologies that specifically target food security in the poorest households and provide diversified options for generating income through access to alternative crops, new varieties with marketing potential, and through organizational and institutional development. This can be linked with the development of rural markets to move food from surplus to deficit areas, and with the commercialization of production for sale to urban or export markets. In areas with better farming conditions and market access, research programs should focus on market-orientated production and value addition. But technologies must offer competitive returns to labor and capital compared with alternative income-earning opportunities.

The lessons learned to date are that technologies need to be matched not only with the crop or livestock enterprise and the biophysical environment, but also with the market and investment environment. In eastern, western, and southern Africa, ICRISAT, in partnership with the subregional organizations, has been promoting an increased awareness of rural seed distribution systems to ensure that farmers have access to crop varieties that will improve household subsistence, and that processors have access to varieties that meet market needs. The lessons learned are that plant breeders and natural resource management scientists must integrate their work with the change agents (both public and private) so that flexible cropping systems are developed, with target groups that can respond to rapid changes in market opportunities. In the longer term, carefully prioritized biotechnology work that acknowledges consumer concerns will underpin these activities.

With this integrated approach in mind, ICRISAT has expanded the integrated natural resource management paradigm to acknowledge the role crops and genetic improvement will have in achieving the Grey to Green Revolution. The expanded version of the term includes both genetic and nongenetic solutions—integrated genetic and natural resource management.

REFERENCES

Dahiya, S.S., Y.S. Chauhan, S.K. Srivastava, H.S. Sekhon, R.S. Waldia, C.L.L. Gowda, and C. Johansen. 2001. Growing extra-short duration pigeonpea in rotation with wheat in the Indo-Gangetic Plain. Natural Resources Management Program Rep. No. 1. ICRISAT, Patancheru, Andhra Pradesh, India.

Faris, D.G., K.B. Saxena, S. Mazumdar, and U. Singh. 1987. Vegetable pigeonpea. A promising crop for India. ICRISAT, Patancheru, Andhra Pradesh, India.

Jagdish Kumar, and S. Abbo. 2001. Genetics of flowering time in chickpea and its bearing on productivity in semiarid environments. Adv. Agron. 72:107–137.

Musa, A.M., M. Shahjahan, J. Kumar, and C. Johansen. 1998. Farming systems in the Barind tract of Bangladesh. Paper presented at the Farming Systems Symposium at the International Congress on Agronomy, Environment and Food Security for the 21st Century, New Delhi. 23–25 Nov. 1998. Indian Soc. of Agron., New Delhi.

Pande, S., J.N. Rao, C. Johansen, R.K. Neupane, and P.C. Stevenson. 2003. Rehabilitation of chickpea through integrated pest management of Botrytis Grey Mold in Nepal. *In* Proc. of the 8th International Congress of Plant Pathology, Christchurch, New Zealand. 2–7 Feb. 2003.

Ramakrishna, A., T.D. Long, H.M. Tam, and S.P. Wani. 2002. Sustaining success and learning from the experience: A case study of Thanh Ha watershed, Vietnam. Presented at the Int. Symp. on Sustaining Food Security and Managing Natural Resources in Southeast Asia—Challenges for the 21st Century, Chiang Mai, Thailand. 8–11 Jan 2002. Univ. of Hohenheim, Germany.

Ryan, J.G., and D.C. Spencer. 2001.Future challenges and opportunities for agricultural R&D in the Semi-Arid Tropics. ICRISAT, Patancheru, Andhra Pradesh, India.

Saxena, K.B. 2000. Pigeonpea. p. 82–112. *In* S.K. Gupta (ed.) Plant breeding: Theory and techniques. Agrobios, Jodhpur, India.

Wani, S.P., P. Pathak, H.M. Tam, A. Ramakrishna, P. Singh, and T.K. Sreedevi. 2001. Integrated watershed management for minimizing land degradation and sustaining productivity in Asia. *In* Z. Adeel (ed.) Integrated Land Management in Dry Areas: Proc. of a Joint UNU-CAS Int. Workshop, Beijing, China. 8–13 Sept. 2001. United Nations Univ., Tokyo, Japan.

5 Climate Forecasts: Emerging Potential to Reduce Dryland Farmers' Risks

Jean L. Steiner, Jeanne M. Schneider, Jurgen D. Garbrecht, and Xunchang J. Zhang

USDA, ARS
El Reno, Oklahoma

ABSTRACT

Agricultural management strategies are needed to improve lives of people in harsh, dryland regions. If the upcoming season's climate was predictable, farmers could tailor practices to match anticipated climate, reducing risks during adverse seasons, while investing more to benefit from favorable seasons. Such a possibility has long been a dream, but there is reason for optimism that our ability to predict climate is improving. In this chapter we describe climate forecasts and discuss potential applications at the farm level. Climate forecasts include local early indicators of future climate, correlation of local climate to global processes, and dynamic modeling of climate processes. Operational forecasts offer potential to guide production decisions, such as crop species or cultivar selection, fertility management, area to be planted, pest management, intensity and timing of grazing and purchase, sale, or movement of animals. Management decisions related to marketing, labor, and diversification, and regional decisions relating to input supply, markets, transportation, storage, or community health services could also be guided by climate forecasts. Forecasts have sufficient utility to guide decision-making in some regions for some seasons. To move forward, continued improvement and evaluation of forecasts skill are needed. Improvements in forecasting tools for regions that gain little from current forecasts and forecasts of extreme events should be a focus for further work. Uncertainty analysis for scenario simulation, tools to assess tradeoffs within a whole farm context, and better methods to communicate probabilistic outcomes are needed. Perhaps most critical is engaging farmers as partners in development of new tools to support decision-making on-farm and using seasonal climate forecasts within the context of overall risk analysis and management of an agricultural system.

INTRODUCTION

The arid and semi-arid dryland regions of the world provide some of the harshest environments for human sustenance and multiple investments are needed to improve lives of people in these regions (Steiner et al., 1988). One investment that may have potentially large payoff is in the area of development, adaptation, and implementation of climate forecasting systems for agricultural management in dryland regions. Because these regions have marginal and unreliable precipitation, prevailing agricultural systems are highly conservative. Farmers in these regions must minimize risk of crop loss or loss of the costs of agricultural inputs. Such losses can jeopardize economic and food security of the household. The conservative systems, however, do not allow the farmers and rural communities to maximize benefits during more favorable years. If the upcoming season's climate was more predictable, farmers could tailor practices to match anticipated climate, reduce economic and crop failure risks during adverse seasons, while investing in higher production inputs to benefit from more favorable seasons.

While knowing the next season's climate has long been a dream of people in harsh dryland regions of the world, today there is reason for optimism that our ability to predict future seasonal climates is improving. A comprehensive summary of research and applications related to seasonal climate forecasts with application to agriculture and natural resource management were reported in Hammer et al. (2000); readers are referred to this resource for more detailed information. In this chapter, our objective is to present an overview of operational climate forecasts and discuss evaluation and interpretation of these forecasts for relevance at the farm scale. We then illustrate potential forecast applications for farm management decisions within the context of cropping/grazing systems in the Southern Great Plains of the USA.

OVERVIEW OF CLIMATE FORECASTING

Climate forecasting is an age-old concept, and initiatives to improve and use such forecasts are underway in many regions of the world. For example, brightness of stars in the Pleiades constellation near winter solstice was used traditionally as an indicator of variation in summer rainfall and autumn harvest in the Andes. Orlove et al. (2000) recently reported that poor visibility of the Pleiades in June was caused by increased subvisual high cirrus clouds. This phenomenon was observed during El Niño years, and was often associated with low rainfall during the subsequent growing season. This finding provided a different perspective of traditional knowledge about the climate system and presents an opportunity to extend traditional knowledge into modern technologies. Climate forecasts will be discussed in terms of local early indicators of future climate, correlation of regional climate to global processes, dynamic modeling of climate processes, and available operational climate forecast systems.

Early Indicators in Local Climate Records

Pioneering research conducted in the 1970s by J. I. Stewart and others (Stewart and Hash, 1982; Stewart and Kashasha, 1984; Stewart and Faught, 1984; Stewart, 1988) developed the concept of response farming. The basis of response farming was the identification of correlations between date of onset of the rainy season with both the length of the growing season and total seasonal precipitation. Such relationships gave an early indication of the type of season to be expected. Man-

Table 5–1. Probability of seasonal rainfall based on April precipitation in Mildura, Victoria, Australia. (V. Sadras, personal communication, 2002).

Seasonal rainfall	"DRY" April†	"WET" April
mm	%‡	%
50	4	0
100	16	4
150§	**40§**	**8§**
200	60	36
250	96	68
300	100	88
350		96
400		96
450		100

† "Dry" indicates April rainfall <13 mm, the median rainfall for this location, and "Wet" indicates April rainfall >13 mm.
‡ Based on a 25-yr precipitation record, each year represents 4% probability.
§ Assuming that 150 mm of seasonal rainfall was determined to be a critical threshold for the production system, then the odds of having that value or less would be 40% following a "Dry" April but only 8% following a "Wet" April.

agement responses were then developed to match the most probable type of growing season. With early onset, and in anticipation of a good rainy season, longer growing season crops could be planted and higher level of inputs could be purchased. With late onset indicating higher probability of low rainfall, a conservative management system could be followed to ensure food security and minimize economic risks. The approach was originally developed in Kenya and was later extended to Sub-Saharan West Africa, the Mediterranean region, and parts of Asia. Response farming is generally most applicable to Mediterranean and monsoonal climates, where virtually all of the annual precipitation comes in the rainy season; it is less applicable to continental climates where precipitation may come in any season, though usually there are months or seasons with higher and more reliable precipitation. Mediterranean, monsoonal, and continental climate patterns are illustrated in Steiner et al. (1988). McCown et al. (1991) evaluated response farming in Kenya and found that while some economic benefit and risk reduction could be realized through variable management under response farming, the benefit was smaller than that realized by adoption of a few simple fixed management changes, relative to prevailing practices.

Stewart's work provided the basis for later research by Sadras et al. (2003) who developed systems for the southeast Australia Mallee region to adjust seasonal management based on April precipitation. The participatory research identified April precipitation as relevant to farmers' decision-making. Sadras (2002) developed correlation relationships between April precipitation and growing season precipitation (Table 5–1) for subregions, and simulated profitability for "conservative" and "risky" strategies across a 40-yr climate record (Fig. 5–1). The figure presents profit differences of "conservative" vs. "risky" strategies in such a way that a positive difference indicates benefit to the conservative strategy while a negative difference indicates benefit to the riskier strategy. In many years, the profit difference between the two systems was relatively modest, but a manager following a set, rather than responsive strategy, would miss the opportunity for much larger profit to the intensive

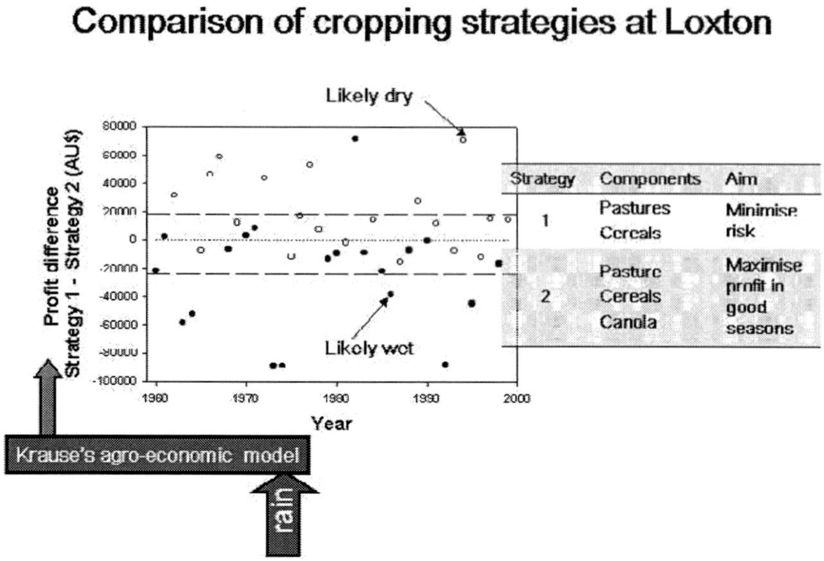

Fig. 5–1. Estimated impact of cropping strategy on profitability in wet and dry forecast seasons. Source: Sadras et al., 2003.

system during several of the wet forecast years or would miss the far higher profit for the conservative system in several of the dry forecast years. The "conservative" and "risky" systems varied by subregion and by soil type, allowing farmers to focus on the systems most relevant to a particular location. Particularly for drier sites, adoption of a dynamic cropping strategy by selecting a conservative regime when April precipitation was below the median and a riskier (more intensive) regime when April precipitation exceeded median indicated overall improved economic return (simulated), particularly during the most extreme years.

Correlation of Regional Climate to Global Processes

Another body of research has explored correlation of historic climate at a particular location or region to ocean-atmosphere patterns observed elsewhere on the globe. Tropical Pacific Ocean patterns have exhibited numerous and diverse links with weather and climate in many parts of the globe. Such linkages across large distances in the earth: atmosphere system are often called "teleconnections".

One such teleconnection described early in the 20th century was the Southern Oscillation, based on the difference in atmospheric pressure at Darwin, Australia, and Tahiti. This pattern was shown to be correlated to Indian monsoonal rainfall and later to precipitation in many parts of Australia and other parts of the globe. For example, Hutchinson (1992) described how the Southern Oscillation Index (SOI) was correlated with year-end rainy season precipitation in Somalia, with an absence of high precipitation seasons in years with a high SOI, but great variability of precipitation in low SOI years. Stone et al. (1996a) described five phases of the SOI and developed correlations between SOI and precipitation patterns in various regions of Australia. A more recently described atmospheric pressure pattern is the North Atlantic Oscillation that shows correlations with temperature and precipitation patterns in eastern North America, northern Europe, and northern Asia.

The best-known teleconnection is linkage of sea surface temperature (SST) in the equatorial Pacific with global precipitation and temperature patterns. This phenomenon is called El Niño during times when the SST is high relative to the long-term average, and La Niña when SST is lower than average. These SST patterns are correlated to temperature and precipitation patterns in many parts of the world (e.g., Phillips and McIntyre, 2000). Because the SST and atmospheric pressure patterns are linked phenomena, the term El Niño-Southern Oscillation (ENSO) is commonly used.

More recently, the Indian Ocean Dipole (IOD) was described which might provide better insight into the monsoonal precipitation on the Indian subcontinent or Western Australia. The correlation of the IOD and the ENSO to monsoonal precipitation in India is of opposite sign, so the impacts of both must be considered in forecasting precipitation for that region.

Although this chapter focuses most strongly on precipitation, forecasting temperature is also important in reducing risk in many agricultural systems. Stone et al. (1996b) described SOI correlation with frost dates in the Australian spring wheat [*Triticum aestivum* (L.)] belt, with the potential that later planting dates or varieties with greater frost tolerance could be selected when the risk of a late frost was above average. Lobell and Asner (2003) reported that U.S. corn (*Zea mays* L.) and soybean [*Glycine max* (L.) Merr.] yields across major production regions were correlated with maximum temperature since the 1980s. Generally negative correlations of yield to temperature were found in the Southeastern USA and much of the Midwestern corn-soybean belt while positive correlations were seen further west and north in the Great Plains. Such correlations have management and marketing implications for farmers who could select different options based on forecasts of higher or lower than normal temperature.

Dynamic Climate Models

Increasing knowledge of atmospheric and global processes is leading to rapid improvements in dynamic global circulation models (GCM) that simulate the global system and can reproduce many of the patterns observed through the years. Climate forecasting is a rapidly advancing field of research and operation, based on the rapid advances in basic knowledge of oceanic and atmospheric processes, improved remote sensing and environmental monitoring technologies, and increased computing power. Forecasts can use GCMs in two ways, either for direct forecast of precipitation and climate patterns in particular regions, or in a hybrid format. The hybrid format forecasts sea surface temperature and atmospheric pressure patterns (as a forecast of the El Niño/Southern Oscillation [ENSO] phase) and then uses statistical correlations of ENSO patterns to forecast precipitation and temperature patterns for various parts of the world. Whether in the GCM or hybrid mode, incorporation of other ocean temperature and atmospheric signals, as well as land surface moisture and snow cover impacts, may improve forecasts in the future for regions of the globe that have weak or undetectable correlation to the ENSO signal.

Operational Climate Forecasts

Operational forecasts are being made by various groups around the world. A widely used forecast of global and regional climate is made by the International Research Institute for Climate Prediction (http://iri.columbia.edu/climate/). In Australia, the Queensland Department of Primary Industry and the Commonwealth Bureau of Meteorology release seasonal climate forecasts based on ENSO and SOI signals for Queensland and the Australia continent (http://www.bom.gov.au/climate/ahead), or globally (http://www.longpaddock.qld.gov.au/index.html). The U.S. National Oceanic and Atmospheric Administration's Climate Prediction Center (http://www.noaa.gov/climate.html) releases 3-mo seasonal climate forecasts covering the coming year for the USA.

All of these forecasts are statements of probability of future climate states, relative to the normal distribution of climate for that particular region and that particular season. The forecasts are not a prediction of one particular future climate pattern, but a statement of probability of a range of possible outcomes across a season. Regardless of whether the forecast is above or below normal, any outcome within the probability distribution may be realized. Additionally, because of high variability within seasons, wet seasons may exhibit short-term, dry periods or vice verse. Because of the inherent uncertainty, economic value of a forecast increases as the skill of the forecast increases and varies depending on what application is made with the forecast (e.g., Gadgil et al., 1995).

Hammer et al. (1996) reported that tactical management based on five phases of the SOI increased profit and reduced risk compared to fixed management in Australian wheat regions. The tactical responses included selection of cultivar maturity and N-fertilization strategy based on forecast frost dates and seasonal precipitation. In simulation studies focused on Zimbabwe, Phillips et al. (1998) emphasized the relative importance of forecasting favorable seasons and managing for enhanced productivity, compared to forecasting adverse seasons. This may be a reflection of the risk-adverse management systems developed to avoid total crop failure during drought years.

Because forecasts are a relatively new product, and the forecasts are being released to new user groups outside the traditional meteorology community, new methods for evaluation are needed. Schneider and Garbrecht (2003a, 2003b) developed indices to evaluate seasonal forecasts for agricultural applications. In their system, "usefulness" addresses the question of how often, and by how much, the forecasts predict departures from normal. The "dependability" index assesses how often the forecasts predict the direction of precipitation departures from normal, and is assessed separately for wet and dry forecasts. "Effectiveness" combines usefulness and dependability to define the frequency of forecasts offering dependable predictions of useful departures, answering the question "How often can I do better using these forecasts?". Their "effectiveness" index for the National Ocianic and Atmospheric Administratio/Climate Prediction Center (NOAA/CPC) forecasts for the continental US (Fig. 5–2) is highest in the Desert Southwest and Florida, with

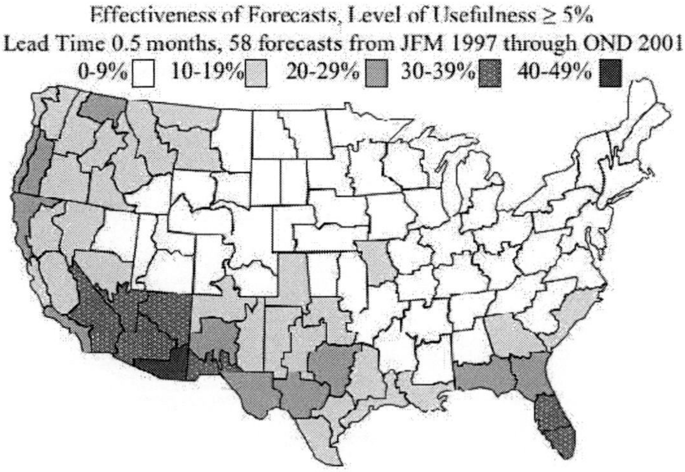

Fig. 5–2. Effectiveness index for NOAA/CPC forecasts for 3-mo total precipitation.

good results in the Pacific Northwest, northern Rocky Mountains, and along the Gulf Coast from Texas to the coastal Carolinas. The forecasts are available for 3 to 12 mo out, but the skill level declines rapidly after 6-mo out.

For the regions with high effectiveness, these forecasts may have considerable water resource and agricultural implications. Given increasing forecast skills for some regions and seasons, the question remains of how to downscale and interpret the impact of forecasts for applications at a local level. As illustrated in Fig. 5–3, the probability distribution of forecast precipitation, relative to normal, will usually not exhibit the same deviation from normal as the probability distribution of an outcome simulated for forecast and normal climate scenarios.

APPLICATIONS TO AGRICULTURE

Climate forecasts are based on average precipitation across relatively large regions (e.g., in the contiguous USA, the forecasts are made for 102 climate divisions). For application to decision-making at a farm level, it is important to know how climate at that particular location relates to the climate in the forecast division (Fig. 5–4). If the local climate distribution differs significantly from the forecast division, then the seasonal forecast may need to be interpreted relative to the local normal, rather than the regional normal. A critical early step in this process is engaging the user community to determine their understanding of climate and weather, and find out how they might want to apply climate forecasts to their system (e.g, Letson et al., 2001).

Because a climate forecast is a probabilistic statement, interpretation must deal with the uncertainty associated with the forecast. A climate forecast for precipitation is a surrogate for information about the likely amount of crop water use (Fig. 5–5), which is associated with large impacts on yield (Stewart and Steiner, 1990).

There is considerable uncertainty in the relationship between crop yield and soil water content due to amount and distribution of growing season precipitation (Fig. 5–5b). Uncertainty also exists in the relationship between crop yield and growing

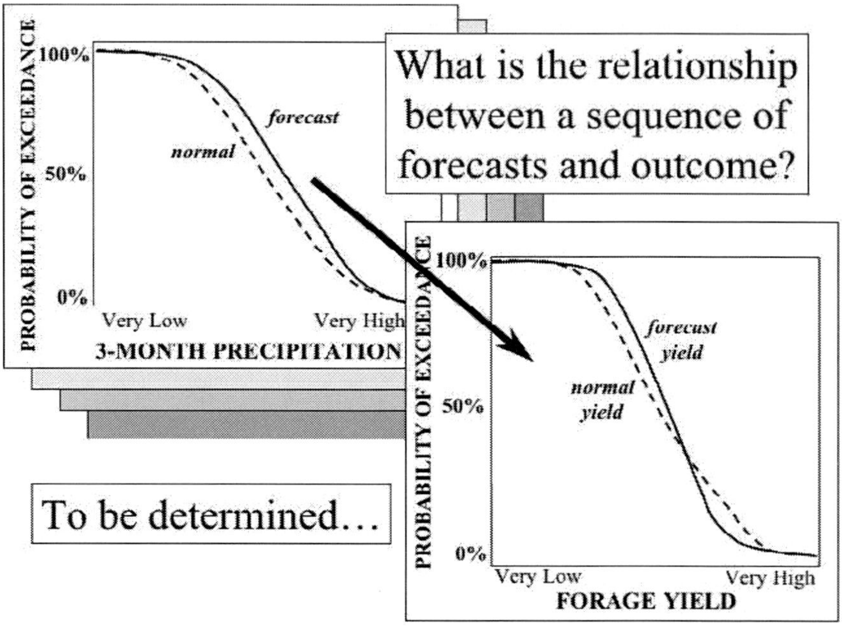

Fig. 5–3. Interpreting the impact of seasonal climate forecasts for farm-level management decision remains an area of uncertainty. Source: Schneider, 2002.

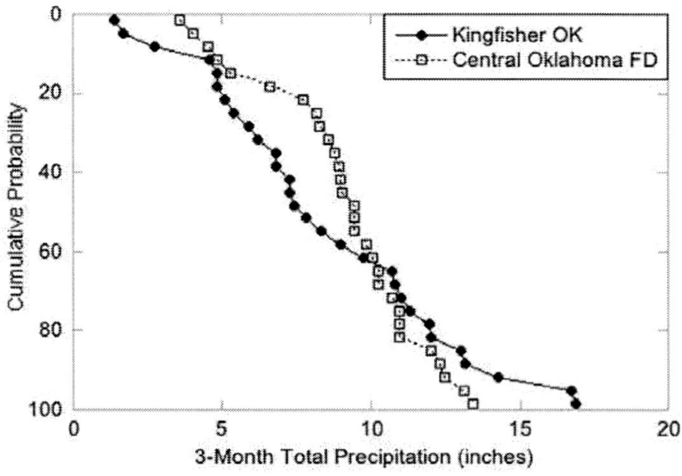

Fig. 5–4. Differences between 30-yr July-August-September precipitation distribution averaged across the Central Oklahoma Forecast Division and for a single station within that division, Kingfisher, OK.

season precipitation due to differences in precipitation effectiveness and soil water supply (Fig. 5–5c).

Since soil water depletion is a major component of crop water use and is highly variable at planting time in many regions, opportunities to integrate measurement of soil water content at planting with use of climate forecasts should be investigated. Robinson and Butler (2002) found that pre-plant soil water content provided the best forecast of dryland crop yields in the northern Australian grainbelt, but relatively few farmers accurately measured soil water content prior to planting. Prior to turning to seasonal climate forecast to reduce risks, there usually will be greater return to first analyzing risks associated with the current management system, adopting good agronomic practices, and implementing relatively straightforward monitoring (such as soil water or soil nutrient contents) into decision-making processes. Good farm managers who have done this may realize additional risk reduction through use of seasonal climate forecasts.

Crop growth models are often used to simulate probable production or profitability outcomes associated with a climate forecast. Most crop models require daily weather data. For analysis of performance of a management scenario in a variable climate, crop models are run over a number of years to determine a range of outcomes associated with a range of climate conditions. Researchers generally use long-

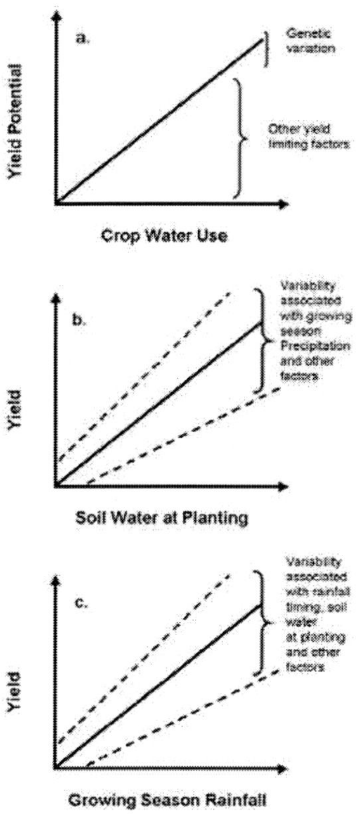

Fig. 5–5. Conceptional relationship of crop yield to (a) crop water use, (b) soil water at planting, and (c) growing season precipitation.

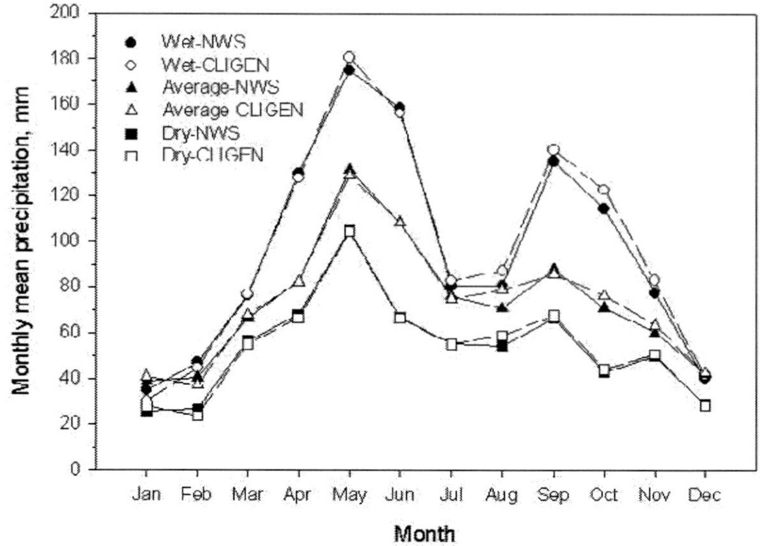

Fig. 5–6. Tercile precipitation for a station in central Oklahoma illustrating the ability of CLIGEN to replicate National Weather Service (NWS) monthly mean observed precipitation data. Source: Zhang, 2003.

term historical weather data or weather generators that produce daily values based on the mean and standard deviation of historical climate records for a location.

To describe the system response to a management scenario in a variable climate, some researchers have used tercile analysis. To do this, alternative scenarios are contrasted for the driest, average, and wettest one-third of the climate years on record. If a climate forecast was available, a farmer might select the scenario that performed best for the driest tercile when the forecast was for higher probability of below-normal conditions and the scenario that provided the best outcome for the wettest tercile when the forecast was for a higher probability of above-normal precipitation. Zhang (2003) showed that the CLIGEN weather generator (Nicks et al., 1995) produced the same distribution of monthly mean precipitation as historical climate data in Oklahoma (Fig. 5–6). Subsequent analyses using the generated climate years and the Water Erosion Prediction Project (WEPP) model showed that yield distribution of winter wheat was responsive to dry, average, and wet precipitation regimes, but was only responsive to initial stored soil water in central Oklahoma in the driest years (Fig. 5–7). While this simulation approach demonstrates crop yield sensitivity to annual wet or dry climate conditions, the potential of the tercile approach is more limited for risk-based decision-making based on seasonally issued forecasts and seasonally sensitive agronomic productivity. Also, the selection of climate conditions associated with one tercile limits the flexibility to reflect the risk of the full range of possible outcomes provided by the forecasts.

Another approach uses analog climate years, based on a climate indicator. For instance, the Queensland Center for Climate Applications contrasted scenarios for the five phases of the SOI index (Stone et al., 1996a) by selecting all years in the

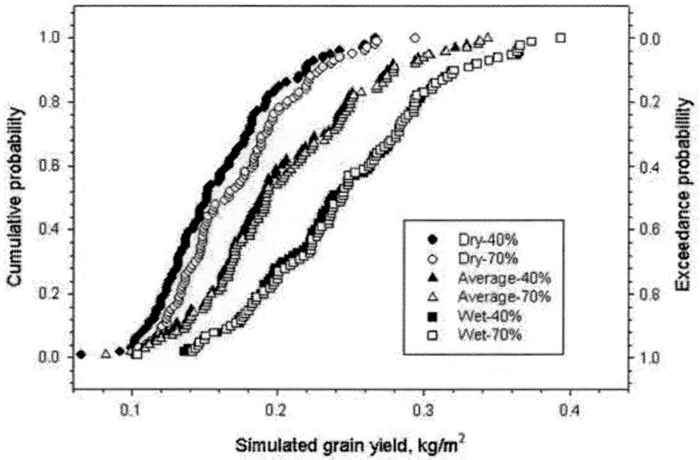

Fig. 5–7. Tercile analysis of probability distribution of WEPP-simulated yield in dry, average, or wet years with 40 or 70% stored soil water at planting. Source: Zhang, 2003.

historical record that match the current phase of the SOI as analogs for the probable climate for the upcoming season. A third approach that is currently under consideration by the authors is modification of weather generators to produce a full range of possible climate sequences that reflect the frequency distribution of the seasonal climate forecast. These generated alternative climate sequences are then fed into an agronomic model to estimate the range of agronomic responses that correspond to the seasonal climate forecasts. The frequency distribution of the agronomic responses then provide the necessary information to establish the production risk associated with that forecast, which can be used in crop enterprise budgets to compare alternative crops or assess the profitability of a certain scenario. Carberry et al. (2002) have worked with Australian farmers who have had some successes in using seasonal climate forecasts in farm level decision-making. Their system, FARMSCAPE, combined soil monitoring and simulation with the climate forecasts, and involved farmers, advisors, and researchers working together closely. Their experience indicated that seasonal climate forecasts without the other tools provided little benefit.

Management Decisions Impacted by Climate Variability

There are numerous levels of decision-making that could be guided by climate forecasts. These include agronomic, crop/livestock, household economic or business decisions, as well as regional-level decisions. Agronomic decisions may include things such as crop selection, for example, maize (*Zea mays* L.) vs. sorghum [*Sorghum bicolor* (L.) Moench] vs. millet [*Pennisetum glaucum* (L.)] as a summer crop depending on the probability distribution of growing season precipitation. For a given species, selection of a long vs. short season cultivar could be guided by precipitation or temperature forecasts. Greater planting density and narrow rows have the potential to capture more radiation and potentially produce higher yields in good seasons, but may be more drought prone due to more rapid depletion of stored soil water. Fertility levels can be adjusted based on anticipated precipitation to reduce risks associated with yield reduction and economic loss. In some regions, the amount of area to be planted may be adjusted based on seasonal forecasts, or crops could be planted in heavier soils if the forecast is for dry conditions or on more freely draining soils if a wet season is anticipated. There is also potential to anticipate the pressure associated with some crop pests based on forecasts (e.g., Maelzer and Zalucki [2000] reported correlation of *Helicoverpa* species infestation with SOI, up to 6–15 mo in advance).

In crop/livestock systems, decisions may relate to planning for future stocking rates; management of a particular forage crop for grazing, haying, or in some instances grain harvest; intensity and timing of grazing on different areas; the need for supplemental feed; and to guide purchase, sale, or movement of animals based on anticipated forage/feed availability.

At the household level, business decisions could include marketing or hedging based on climate forecasts in the local area as well as in major global production areas for a particular crop. Forecast of unfavorable seasons might lead to decisions to diversify farm enterprises. In some cases, climate forecasts might influence decisions about the need for off-farm income relative to the need for on-farm labor and food security.

At the regional level, climate forecasts could guide decisions such as anticipated need for inputs (fertilizers, seeds of different crop species and varieties), market capacity, storage, and transportation needs; community health service requirements associated with climate variability (e.g., Bi et al., 1998); or drought preparedness planning and implementation (Dilley, 2000; Finan and Nelson, 2001).

Decision Points in a Cropping/Grazing System in the Southern Great Plains

To illustrate potential applications of climate forecasting to agricultural decision-making, we have selected a major cropping/grazing system common to the Southern Great Plains of the USA, with winter wheat, cold- and warm-season perennial grasses, summer annual crops, and beef cattle (*Bos taurus*) as major components. The beef cattle system we will discuss is the "stocker" phase of beef production in the USA. The beef system in the USA dominantly consists of three phases: cow/calf production, stocker growth, and feedlot finishing. These phases often occur with different owners for each phase and often take place in geographically separate regions (Fig. 5–8). Cow/calf production is predominantly located in the southeastern USA, but is also important in rangelands of the semi-arid and arid west. Stocker animals are weaned at about 8 mo of age and frequently are transported to other regions for additional weight gain, utilizing perennial grasses and other forages. The Southern Great Plains is the destination for large numbers of these animals, generally being shipped into the area in the fall. The stocker cattle are grazed on native and introduced perennial grasses as well as annual forages. An important forage in the Southern Great Plains is winter wheat, often grown as a dual-purpose crop that provides fall and winter forage for grazing as well as a sub-

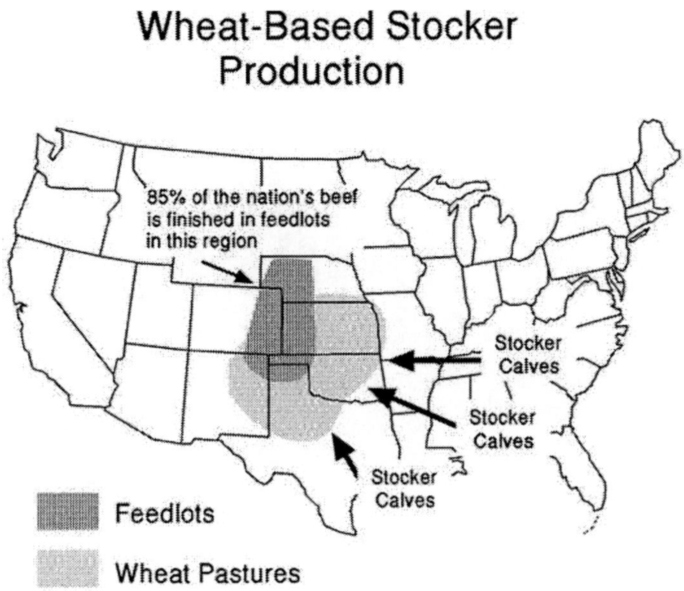

Fig. 5–8. Beef production systems in the USA, including cow/calf production, stocker grazing, and confined finishing. Source: J.A. Daniel, USDA-ARS, El Reno, OK, unpublished data, 1999.

sequent grain crop. The economic return to wheat farmers from stocker grazing can equal the economic returns of the grain crop.

The Great Plains is a subhumid to semi-arid region that extends from central Canada to central Texas. There is a strong east-west annual average precipitation gradient of approximately 100 mm decrease with each 160 km from roughly the 100th meridian toward the Rocky Mountains. The climate at El Reno, OK, in the subhumid region illustrates year-round distribution of precipitation with the peak in May and June, and relatively low precipitation in the hottest months of July and August, when potential evapotranspiration greatly exceeds precipitation (Fig. 5–9). The dominant native prairie species are warm-season grasses, but considerable opportunity exists to grow cool-season perennials and annual crops. Winter temperatures can present favorable or unfavorable conditions for plant growth, with extreme variability within and between years.

This system is summarized in Table 5–2 which identifies numerous decisions that are required throughout the year, often with multiple, complex factors involved and tradeoffs across five major enterprises that comprise the system. Some decisions could be strongly impacted by a seasonal climate forecast (see underlined decisions in Table 5–2) but even those would also be influenced by additional factors. Some of the climate-sensitive decisions might be guided using existing crop models (e.g., decision whether or not to plant a summer annual following wheat harvest), while others would require whole farm models that incorporate crop, livestock, and marketing issues. Additional factors that have a large impact on economic viability or quality of life for a farm family may not be largely influenced by climate.

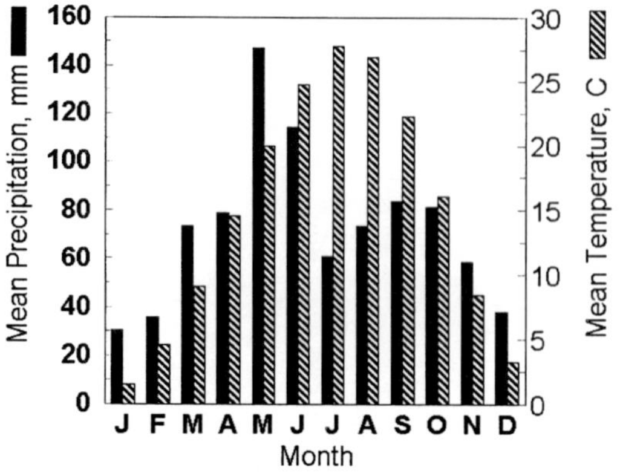

Fig. 5–9. Monthly mean precipitation and temperature for Canadian County, Oklahoma, 1971 to 2000.

Continuous monoculture of winter wheat is fairly common in much of the Southern Great Plains. These lands, often intensively tilled, have low organic carbon (C) level soils and the surface is often left bare in the summer when intensive convective storms present a great risk of erosion. The soils are also subject to erosion by wind, further degrading the soil and presenting air quality and visibility problems in the region. The farmers choose continuous wheat cropping because of their reliance on the dual purposes of the wheat to maintain economic returns. Planting a short-season summer annual following harvest of wheat grain in June would provide cover to protect the soil from erosion as well as providing C and potentially nitrogen (N) to the soil. Additionally, it would provide a high quality forage in August and September to supplement warm-season perennial pastures that have low forage quality at that time. The feasibility of double cropping depends on availability of soil water for germination and establishment. Additionally, July and August are the least reliable months for rainfall in this region. If summer cropping is implemented, there is a need for precipitation in September or October to recharge the soil water for planting and establishment of the next wheat crop. The likelihood of success would be enhanced by recharge of soil water in late May and early June when the wheat crop was maturing and using little water. At wheat harvest in June, soil water content could be measured and seasonal climate forecasts for summer and fall would be available. The summer forecasts for this region have very low utility at this time (Schneider and Garbrecht, 2003b) so crop models using normal precipitation distributions might be most suited for evaluating alternative scenarios (plant a summer annual or don't plant a summer annual). However, the fall climate forecasts have better skill, particularly in El Niño years, so a forecast of higher-than-normal odds of high fall precipitation might increase confidence in the decision to plant a summer annual. A seasonal forecast for fall precipitation that is above or below normal odds might also influence decisions about stocking rates and delivery dates of stockers for fall/winter grazing.

Based on the acceptable dependability of seasonal forecasts in some regions and some types of seasons (Schneider and Garbrecht, 2003b), as well as the rapidly advancing state of knowledge in the ocean:atmosphere:climate arena, we believe these forecasts are good enough to help guide management decision-making to reduce risks associated with climate variability. In our research, we are focusing on how to apply these climate forecasts to tactical decision-making at the farm level. This will require a broad approach to evaluating and managing risks, such as described by Carberry et al. (2002). The wheat-stocker-grass system described above is one of the initial systems we will examine. Additional applications are being explored in the area of soil and water conservation and water resource management.

NEXT STEPS TOWARD APPLYING CLIMATE FORECAST TO DRYLAND FARMING

The state of knowledge in the ocean:atmosphere:climate arena is rapidly growing and evolving so people focusing on applications of seasonal forecasts for decision-making will need to stay apprised of developments in this field of science, and how the new knowledge is feeding into operational forecasts. The forecasting skill of current technologies varies greatly by region and it will require ongoing research to develop effective forecasting tools for regions that currently gain little from forecasts. The strength of teleconnectic signals, as well as directions of the phases, can have large influences on the magnitude and regional distribution of the climate impacts (Izaurralde et al., 1999) The potential economic gain from improved climate forecasting indicates that investment in such research is justified (Petersen and Fraser, 2001; Jones et al., 2000). As forecasts rely more and more on dynamic models, the variability of both the initial values of driving variables as well as uncertainty within the model formulation must be considered (Palmer, 2000).

Another area that needs additional research attention is forecasting climatic extremes. Forecast of probability distributions of future climates contains the most reliable information in the middle 80% of the distribution. The distributions have little reliability in the upper or lower 10% probability. However, impacts of extreme cases are often the most critical, particularly in the areas of food security risks during drought and natural resource/environmental risks associated with either drought or floods.

Integration of climate forecasts with historical data bases and crop simulation models will allow risk-based analyses of alternative management scenarios. Two immediate tasks include (i) development of techniques to temporally and spatially downscale the climate forecasts to daily time series for particular locations needed to drive crop models, and (ii) quantify the uncertainty of climate forecasts and uncertainties of crop models including those associated with input variables, model parameters, and the models themselves. Without such uncertainty analysis reliable crop forecasts are impossible. In addition, since many management decisions involve a multitude of issues, single crop models will have limited application to many of the assessments. Advances in tools to evaluate alternatives and tradeoffs in terms of the whole farm system are needed.

Perhaps the most critical need is engagement of farmers as partners in development of new tools to support tactical decision-making on-farm and using seasonal climate forecasts in the context of overall risk analysis and management. This will require development of better methods to communicate probabilistic out-

Table 5–2. Decision points in an agricultural management calendar for Southern Great Plains (USA) cropping/grazing system (Decisions that are underlined may be influenced by seasonal climate forecast).

	Enterprises within the system †					
Month	Winter wheat	Summer perennial	Winter perennial	Summer annual	Stocker cattle	Tactical decisions to be made *Issues that may influence decision*
January	Graze					Is wheat growth adequate to support feed requirements of the cattle on the pasture? Are alternative forages available? Is supplemental feeding needed to maintain animal body weight?
February	Graze				Sell first set of stockers	Grow wheat to grain? If so, remove cattle prior to growing point emergence aboveground. *Decision impacted by cattle and grain futures markets, forage availability, and precipitation forecast.* Spring fertilizer? *What is status of soil fertility, stored soil water and prospects for seasonal precipitation?*
March	Graze out		Graze			Bale wheat in May? If so, remove cattle prior to jointing/heading.
April	Graze out		Graze			
May	End graze out or cut hay	Fertilize? Weed control? Burn?	Graze			As temperatures warm, monitor cool-season perennial growth and forage quality to determine end of grazing season.
June	Grain harvest	Graze	End grazing	Sow?		Summer double crop following wheat harvest? *What is the soil water storage and seasonal precipitation forecast? Is there adequate forage available elsewhere on the farm for anticipated needs?* Do you need to fix N with a summer crop or is there a need to build soil organic matter with a cover crop?
July		Graze? Hay? Forage quality dip		Graze? based on need for greater forage quantity or quality	Contract for cattle?	What is the anticipated carrying capacity of stocker based on cattle in the upcoming season, *based on current forage conditions, cropping plans for the autumn/winter season, and climate forecasts*? What is the purchase for cattle and what are future prices for cattle when I want to sell? Will there be adequate return to justify supplemental feeding or would a lower stocking rate with minimal supplemental feed requirement be better? Would early or late delivery be better, *based on anticipated sowing date of wheat, and anticipated condition of fall perennial forages, given the climate outlook for fall and winter?*

August		Poor forage qualtiy		Graze? Hay?	Sell or deliver to feedlot?	Is forage quality adequate to sustain gain? Is supplemental protein needed or grazing of summer annual? If forage is greater than anticipated, is there benefit in holding these cattle longer than planned, or is there more benefit in selling and stockpiling available forage for the next animals?
September	Sow for grazing		Graze	Harvest	Start to buy cattle	Determine area to plant wheat, which fields first, variety, seeding rate, fertilizer amount, *as influenced by plans for cattle enterprise and climate outlook.* As temperatures cool, monitor cool-season perennial growth to determine when grazing can start.
October	Sow for grain		Graze			
November			Graze			Is fall growth of the wheat adequate to begin grazing; are there other fall forages that should be used?
December	Graze					Is wheat growth adequate to support feed requirements of the cattle on the pasture? Are alternative forages available? Is supplemental feeding needed to maintain animal body weight?

† Approximate seasons for:

Winter wheat—October to early June, for grain production

Summer perennials

Native—June to August, forage quality dip in late July to August

Introduced—Late May to early September, forage quality dip in late July to August

Winter perennials—March to June, September to November or later with low stocking density or delayed grazing start

Summer annual—If double cropped with wheat, mid-June to early July planting.

Grazing—July-September

Hay—August or September

Cover—Terminate in August to allow recharge of September rains for wheat

Grain—August to late September, depending on species and cultivar

Stocker cattle—In general, delivered, following weaning, in mid to late fall and grazed until ~ August. However, management is highly variable. Land area per animal for spring/summer grazing is ~25% of the area required for fall/winter grazing. The area not needed for summer grazing can be harvested for hay or grain. Additional cattle can be purchased in late winter and/or mid-spring as forage availability increases.

comes for farm decision-making (Perry, 1994). It will also require assessment of climatic risks as only one of many factors that might impact decision-making. Whether at the farm level, or rural community level, uncertainty in the socioeconomic and policy arenas can inhibit adoption of climate forecasts (e.g., Eakin, 1999) or other new technologies.

REFERENCES

Bi, P., Wu, X.K., Parton, K.A., and Tong, S.L.1998. Seasonal rainfall variability, the incidence of hemorrhagic fever with renal syndrome, and prediction of the disease in low-lying areas of China. Am. J. Epidem. 148:276–281.

Carberry, P.S., A. Hochman, R.L. McCown, N.P. Dalgliesh, M.A. Foale, P.L. Poulton, J.N.G. Hargreaves, D.M.G. Hargreaves, S. Cawthray, N. Hillcoat, and M.L. Robertson. 2002. The FARMSCAPE approach to decision support: Farmers,' advisers,' researchers' monitoring, simulation, communication, and performance evaluation. Agric. Syst. 74:179–220.

Dilley, J. 2000. Reducing vulnerability to climate variability in Southern Africa: The growing role of climate information. Clim. Change 45:63–73.

Eakin, H. 1999. Seasonal climate forecasting and the relevance of local knowledge. Phys. Geogr. 20:447–460.

Finan, T.J., and D.R. Nelson. 2001. Making rain, making roads, making do: Public and private adaptations to drought in Ceara, Northeast Brazil. Climate Res. 19:97–108.

Gadgil, S., P.R.S. Rao, N.V. Joshi, and S. Sridhar. 1995. Forecasting rain for groundnut farmers—How good is good enough? Current Sci. 68:301–309.

Hammer, G.L., D.P. Holzworth, and R. Stone. 1996. The value of skill in seasonal climate forecasting to wheat crop management in a region with high climatic variability. Aust. J. Agric. Res. 47:717–737.

Hammer, G.L., N. Nicholls, and C. Mitchell. 2000. Applications of seasonal climate forecasting in agricultural and natural ecosystems. Kluwer Academic Publ., Dordrecht, The Netherlands.

Hutchinson, P. 1992. The Southern Oscillation and prediction of Der season rainfall in Somalia. J. Clim. 5:525–531.

Izaurralde, R.C., N.J. Rosenberg, R.A. Brown, D.M. Legler, M.T. Lopez, and R. Srinivasan. 1999. Different geographic distribution of winner and loser regions in strong and normal El Nino years, relative to neutral SST signals. Agric. Forest Meteor. 94:259–268.

Jones, J.W., J.W. Hansen, F.S. Royce, and C.D. Messina. 2000. Potential benefits of climate forecasting to agriculture. Agric. Ecosyst. Environ. 82:169–184.

Letson, D., I. Llovet, G. Podesta, F. Royce, V. Brescia, D. Lema, and G. Parellada. 2001. User perspectives of climate forecasts: Crop producers in Pergamino, Argentina. Clim. Res. 19:57–67.

Lobell, D.B., and G.P. Asner. 2003. Climate and management contributions to recent trends in U.S. agricultural yields. Science (Washington DC) 299(5609):1032.

Maelzer, D.A., and M.P. Zalucki. 2000. Long range forecasts of the numbers of *Helicoverpa punctigera* and H-armigera (Lepidoptera:Noctuidae) in Australia using the Southern Oscillation Index and the Sea Surface Temperature. Bull. Entomol. Res. 90:133–146.

McCown, R.L., B.M. Wifely, R. Mohammed, J.G. Ryan, and J.N.G. Hargreaves. 1991. Assessing the value of a seasonal rainfall predictor to agronomic decisions: The case of response farming in Kenya. p. 383–409 *In* R.C. Muchow and J.A. Bellamy (ed.) Climatic risk in crop production: Models and management in the semi-semi-arid tropics and subtropics. CAB Int., Wallingford, UK.

Nicks, A.D., L.J. lane, and G.A. Gander. 1995. Weather generator. p. 2.1–2.22. *In* D.C. Flanagan and M.A. Nearing (ed.) USDA-Water Erosion Prediction Project: Hillslope Profile and Watershed Model documentation. NSERL Rep. 10. USDA-ARS-NSERL, West Lafayette, IN. Available at http://topsoil.nserl.purdue.edu/nserlweb/weppmain/docs/chap2.pdf (verified 25 Nov. 2003).

Orlove, B.S., J.C.H. Ciang, and M.A. Cane. 2000. Forecasting Andean rainfall and crop yield from the influence of El Nino on Pleiades visibility. Nature (London) 403:69–71.

Palmer, T.N. 2000. Predicting uncertainty in forecasts of weather and climate. Rep. Progr. Phys. 63:71–116.

Perry, K.B. 1994. Current and future agricultural meteorology and climatology education needs of the United States Extension Service. Agric. For. Meteorol. 69:33–38.

Petersen, E.H., and R.W. Fraser. 2001. An assessment of the value of seasonal forecasting technology for Western Australian farmers. Agric. Syst. 70:259–274.

Phillips, J., and B. McIntyre. 2000. ENSO and interannual rainfall variability in Uganda: Implications for agricultural management. Int. J. Climatol. 20:171–182.

Phillips, J.G., M. A. Cane, and C. Rosenweig. 1998. ENSO, seasonal rainfall patterns, and simulated maize yield variability in Zimbabwe. Agric. For. Meteorol. 90:39–50.

Robinson, J.B, and D.G. Butler. 2002. An alternative method for assessing the value of the Southern Oscillation Index (SOI), including case studies of its value for crop management in the northern grainbelt of Australia. Austr. J. Agric. Res. 53:423–428

Sadras, V. 2002. Rainfall forecasting tool: Maximizing returns for Mallee farmers. Research Project Information. CSIRO Land and Water. Sheet No. 24. March 2002.

Sadras, V., D. Roget, and M. Krause. 2003. Dynamic cropping strategies for risk management in dryland farming systems. Agric. Syst. 76:929–948.

Schneider, J.M., and J.D. Garbrecht. 2003a. A measure of the usefulness of seasonal precipitation forecasts for agricultural applications. Trans. Am. Soc. Agric. Eng. 46:257–267.

Schneider, J.M., and J.D. Garbrecht. 2003b. Regional utility of NOAA/CPC seasonal climate precipitation forecasts. *In* Proc., Symp. on Watershed Management and Restoration, World Water and Environmental Resources Congress, June 2003. [CD-Rom computer file] Environ. and Water Resourc. Inst. and Am. Soc. Civ. Eng., Reston, VA. Available at http://www.asce.org (verified 24 Nov. 2003).

Steiner, J.L., J.C. Day, R.I. Papendick, R.E. Meyer, and A.R. Bertrand. 1988. Improving and sustaining productivity in dryland regions of developing countries. Adv. Soil Sci. 8:79–122.

Stewart, B.A., and J.L. Steiner. 1990. Water use efficiency. Adv. Soil Sci. 13:151–173.

Stewart, J.I. 1988. Response farming in rainfed agriculture. The Wharf Foundation Press, Davis, CA.

Stewart, J.I., and W.A. Faught. 1984. Response farming of maize and beans at Katumani, Machakos District, Kenya: Recommendations, yield expectations, and economic benefits. E. Afr. Agric. For. J. 44:29–51.

Stewart, J.I., and C.T. Hash. 1982. Impact of weather analysis on agricultural production and planning decisions for the semiarid areas of Kenya. J. Appl. Meteorol. 21:477–494.

Stewart, J.I., and D.A.R. Kashasha. 1984. Rainfall criteria to enable response farming through crop-based climate analysis. E. Afr. Agric. For. J. 44:58–79.

Stone, R.C., G.L. Hammer, and T. Marcussen. 1996a. Prediction of global rainfall probabilities using phases of the Southern Oscillation Index. Nature (London) 384:252–255.

Stone, R., N. Nicholls, and G. Hammer. 1996b. Frost in northeast Australia: Trends and influence of phases of the southern oscillation. J. Clim. 9:1896–1909.

Zhang, X.C. 2003. Assessing seasonal climatic impact on water resources and crop production using CLIGEN and WEPP models. Trans. Am. Soc. Agric. Eng. 46:685–693.

6 Dryland Agriculture in India

Harish P. Singh, Kapil D. Sharma, Gangireddy Subba Reddy, and Kishori L. Sharma

Central Research Institute for Dryland Agriculture
Hyderabad, India

ABSTRACT

Dryland agriculture occupies 68% of India's cultivated area and supports 40% of the human and 60% of the livestock population. It produces 44% of food requirements, thus has and will continue to play a critical role in India's food security. However, aberrant behavior of monsoon rainfall results in frequent droughts that impact resource poor farmers. Eroded and degraded soils with low water-holding capacity and multiple nutrient deficiencies, declining groundwater table, etc. contribute to low crop yields that lead to further land degradation. Managing land resources through a multidisciplinary approach in devising the most remunerative and environmentally appropriate land use characterizes Central Research Institute for Dryland Agriculture's (CRIDA) approach for maximizing crop productivity, profitability, and sustainability of dryland agriculture. Characterizing bio-physical and socio-economic resources, integrated watershed development, improvement of rainwater use efficiency, contingency crop planning, diversification of agriculture through livestock farming, alternate land uses, integrated soil-nutrient-water-crop management, and efficient farm implements can ensure long-term sustainability of dryland agriculture in India. Apart from these, evolving an institutional framework, improving credit availability and input supply systems, extension of crop insurance and launching of on-farm research cum pilot projects in farmers' participatory mode also need to be focused.

INTRODUCTION

Of an estimated 143 million ha net cultivated land in India, about 97 million ha (68%) is dryland/rainfed that produces 44% of the country's food requirements while supporting 40% of human and 60% of livestock populations (NBSSLUP, 2001). Even if the countries full irrigation potential of 139.5 million ha is realized, agricultural production in 75 million ha will continue to be solely dependent on rainfall. Physiographically, dryland agriculture encompasses the desert terrain of Rajasthan in the northwest; the plateau region of Central India; the alluvial plains of the Ganga-Yamuna River Basin; the Central Highlands of Gujarat, Maharashtra and Madhya Pradesh; the rain-shadow region of Deccan in Maharashtra; the Deccan Plateau in Andhra Pradesh and the Tamilnadu Highlands (Fig. 6–1). About 15 mil-

 Challenges and Strategies for Dryland Agriculture. CSSA Special Publication no. 32.

Fig. 6–1. Dryland agricultural region in India.

lion ha of dryland lies in the arid region which receives <500 mm rainfall; another 15 million ha is in 500 to 750 mm rainfall zone, 42 million ha is in 750 to 1150 mm rainfall zone, with the remaining 25 million ha receiving >1150 m rainfall per annum (Kanwar, 1999). About 74% of annual rainfall occurred during June to September, the southwest monsoon. This monsoon is characterized as having a high coefficient of variation (0.3–0.6) with July and August as the rainiest months. Droughts occur once in 3 to 5 yr either due to a deficit in seasonal rainfall during the main cropping season or from inadequate soil moisture availability during prolonged dry spells between successive rainfall events (Ramakrishna et al., 1999). Low yields and crop failures often lead to food and fodder scarcity resulting in a near-famine situation that further accelerate the process of land degradation. Alfisols, Entisols, Vertisols, and associated soils dominate the SAT areas (Virmani et al., 1991). These soils are generally coarse textured, highly degraded with low water retentive capacity, and have multiple nutrient deficiencies. Crops and cropping systems in the SAT are diverse depending on soil type and the length of growing season. In addition to major livestock production systems, about 93% of cultivated area under sorghum [*Sorghum bicolor* (L.) Moench], 94% under pearl millet [*Pennisetum americanum* (L.) Leeke], 79% under corn (*Zea mays* L.), 87% under pulses, 76% under oilseeds, 64% under cotton (*Gossypium hirsutum* L.), and 59% under tobacco (*Nicotiana tabacum* L.) in India predominates drylands. (FAI, 2001). Of the 97 million farm holders in these regions, 76% are small (<2 ha) and marginal, cultivating only 29% of the unconsolidated and scattered arable land. Hence, farmers' dependence on livestock complement their arable farming, as an alternative source of income, is very high. Preliminary estimates have shown that nearly two out of three heads of India's estimated cattle population of 219 million are found in the Indian drylands (FAO, 2000). Further, resource poor farmers, poor infrastructure, and low investment in technology characterize these areas.

Maintaining India's food security in the next two decades and beyond is a challenging task. Since per unit land area productivity in irrigated areas is reaching a plateau, the bulk of rising food demand has to be met by enhancing dryland productivity. Further, the lack of food and livelihood security at the farm level in resource poor regions would be a deterrent to encouraging the growing human population to remain in rural areas. The human population in India's drylands is likely to reach 600 million by 2025 from the present 410 million. Similarly, the livestock population is likely to exceed 650 million by 2025 from the present 509 million. On the other hand, the area under dryland crop production may decrease to 85 million ha by 2025 from the present 97 million ha. Thus, from such a significantly reduced cultivated area, crop production must increase from the present 0.8 to 1.0 t ha^{-1} to 2.0 t ha^{-1} by 2025. Furthermore, the quality of produce must improve to meet the global market standards. Also, the cost of production needs to be reduced in order not only to improve the farmers' net income but also remain globally competitive. It would thus appear that maintaining food security in the years ahead is a tremendous task and hence research and development efforts need to be redoubled to meet these challenges. For dryland agriculture to be viable through the century, research and development approach has to be multidisciplinary and holistic if the researchers and developers hope to maximize crop productivity and profitability while not endangering the sustainability of rainfed farming systems in the region (Singh et al., 2002). The question currently being asked is "can we meet the new challenges on sustainable dryland agriculture which raises the quality of life and is pro-nature". If the recent past is any indication, it is not an insurmountable task.

Key strategies to achieve these goals are: characterizing bio-physical and socio-economic resources using geographical information system (GIS) and remote sensing; integrated watershed development; developing strategies for improving rainwater-use efficiency through appropriate mechanisms in terms of rainwater storage, delivery and application, and contingency crop planning to minimize loss of production during drought/flood years. Diversification of agriculture by growing high value crops such as dye-yielding, aromatic and medicinal plants, spices, sericulture and livestock farming to minimize climatic risks; alternate land uses such as agroforestry, agrihorticulture, silvipasture, hortipasture, etc. to maximize returns and minimize risk could be other viable strategies for the development of these areas.

Integrated nutrient-water-crop management; building soil organic matter and soil quality enhancement; and designing appropriate farm implements for timeliness of agricultural operations also need focus. Further, developing an institutional framework for the involvement of Community Based Organizations that are capable of improving credit availability, input supply systems, launching crop insurance schemes, and on-farm research cum pilot projects in farmers' participatory mode will ensure long-term sustainability of rainfed agriculture in India. Rural agro-industry development in rainfed eco-region is necessary not only to support farm mechanization and value added products, etc. but for employment generation to ensure decent livelihoods for small/marginal farm families and landless herders.

TRENDS IN DRYLAND AGRICULTURE

Dryland agriculture in India has been practiced since time immemorial. Unlike Australia, Canada, and USA where dryland agriculture is mechanized, farmers in India have developed innovative methods to grow crops under dryland conditions. The area under dryland agriculture was 97.9 million ha between 1950 and 1951 and has decreased to 87.7 million ha in the period 1997. The net sown area, which had been stagnant for more than a decade, increased from 118.8 to 142.8 million ha during the same period (Fig. 6–2). The irrigated area nearly tripled since the 1950s. The dryland area under oilseeds and cotton registered nearly 130 and 50% increase, respectively, while it remained almost unchanged under pulses, and declined by about 25% under coarse cereals during the same period (FAI, 2001). These changes are mainly due to expansion of irrigation, improved technology, and socio-economic factors (Kanwar, 1999). The area under dryland agriculture is declining and it is expected that by 2050 it may stabilize at 75 million ha.

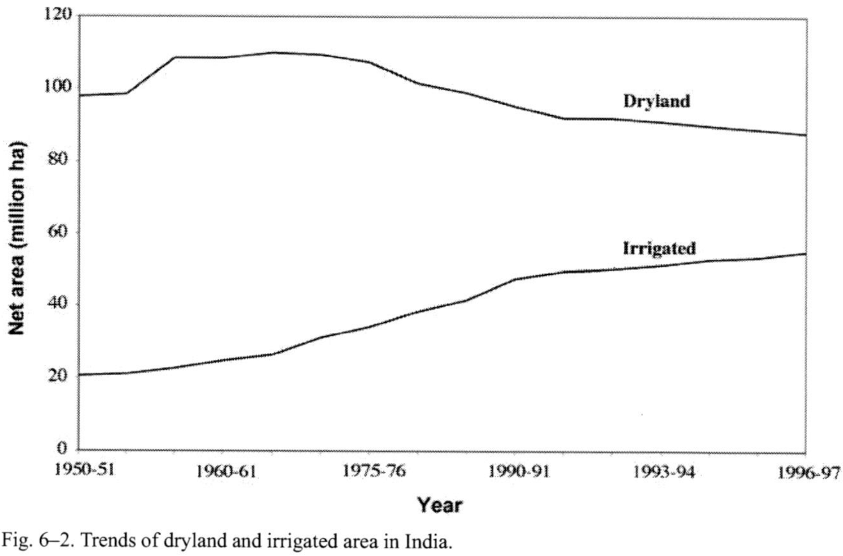

Fig. 6–2. Trends of dryland and irrigated area in India.

Since the 1950s, significant changes in the area and yield of important dryland crops have occurred. The area under coarse cereals decreased by about 11 million ha with most of this area under sorghum and pearl millet (Fig. 6–3). This indicates replacement of sorghum and pearl millet by more remunerative crops and cropping systems as well as changing food habits. The area under oilseeds increased by about 14 million ha due to increased cultivation of groundnut (*Arachis hypogaea* L.) and soybean [*Glycine max* (L.) Merr.], and growing irrigated rapeseed/mustard (*Brassica* spp.). Total area under pulses remained unchanged yet there were shifts in pulse growing areas from one agro-ecological region to another within the country. For example, chickpea (*Cicer arietinum* L.) area decreased in the northern region due to an increase in area under rice-wheat (*Oryza sativa* L.-*Triticum aestivum* L.) cropping system but increased at the same time in the central region of India. There has been a 50 to 100% increase in cotton and corn areas as more irrigation is introduced for these crops.

There has been a steady increase in the productivity of dryland crops between 1951 to 1952 and 1999 to 2000 mainly due to adoption of improved crop husbandry and dryland farming technologies (Subba Rao, 2002). Productivity of coarse cereals increased by 65%, despite a decreasing area under their cultivation. In case of sorghum and pearl millet, the major improvement in unit area productivity (120–140%) occurred in the rainy (monsoon) season crop (Fig. 6–4). However, the productivity of pearl millet fluctuates widely, perhaps due to stress caused by downy mildew and it is grown in the drier areas that have greater fluctuations in rainfall. Productivity of corn recorded threefold increase since the 1950s with the introduction of hybrids and increased use of irrigation. For oilseeds, the major increase in productivity (80%) was in rapeseed/mustard and soybean. Soybean is widely becoming the most important oilseed dryland crop in the black soil region followed by sunflower (*Helianthus annuus* L.) and safflower (*Carthamus tinctorius* L.). Trends in productivity of pulses are disappointing (Fig. 6–4). Stagnating

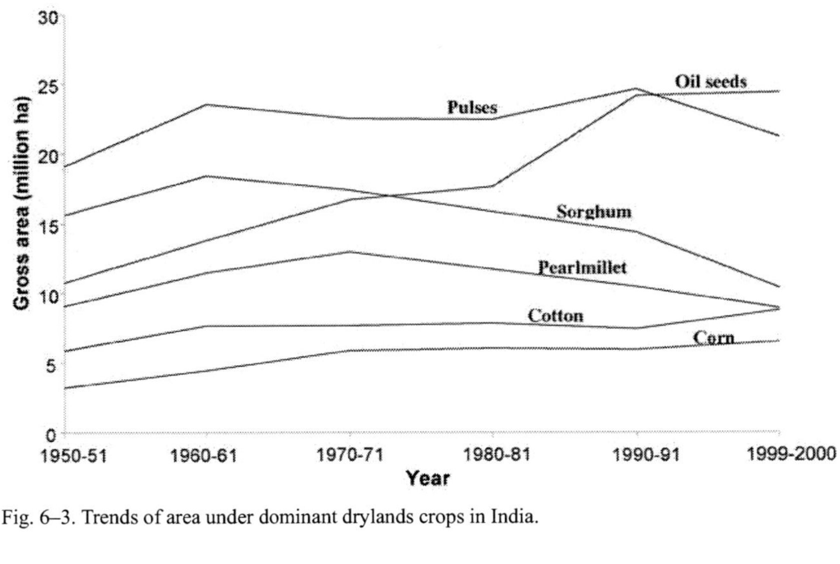

Fig. 6–3. Trends of area under dominant drylands crops in India.

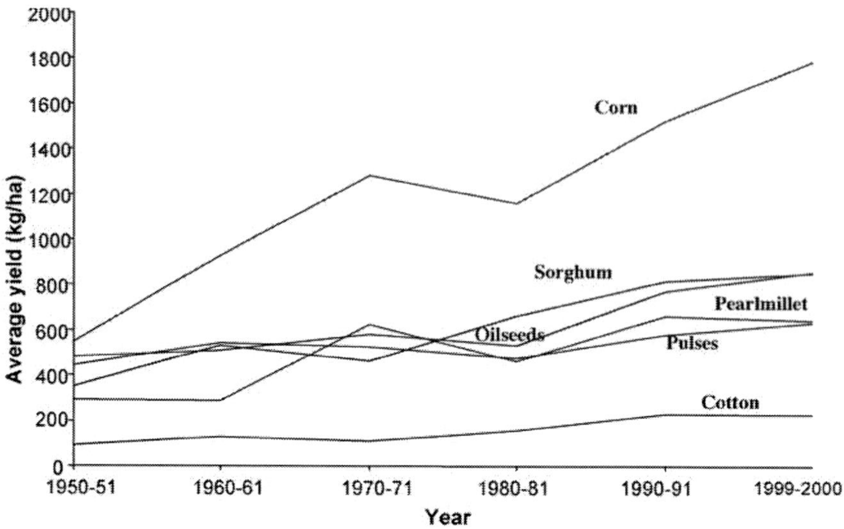

Fig. 6–4. Trends of production of dominant drylands crops in India.

pulse yields is attributed to biotic and abiotic stresses, relegating the pulse cultivation to marginal areas, and inadequate management. Cotton productivity increased by 150% since the 1950s mainly through extending irrigation to cotton growing areas in the drylands.

RESEARCH AND DEVELOPMENT IN DRYLAND AGRICULTURE

Numerous technologies have been generated over the years through a network of research institutions in the country dedicated to addressing the problems of dryland agriculture. These technologies have been refined in farmers' fields through the Operational Research Project sites, Institute Village Linkage Program, and Farm Science Centers.

Choice of Crops, Varieties, and Cropping Systems

Many traditional crops grown in drylands are of long-duration and do not match periods of water availability. In order to match crop water needs with rainfall and water available periods, extensive trials were conducted over the past three decades. A large number of improved varieties of millets, pulses, and oil seeds were evaluated for their yield patterns vis-a-vis local varieties used by the farmers (AICRPDA, 2000). Yield improvement of between 15 to 50% were recorded when traditional varieties were replaced by high yielding varieties (Table 6–1). Also a benefit of 15 to 25% in yield was demonstrated by the crop substitution strategy, which means by replacing one crop with other appropriate crop. In rainy season, the quantum and distribution of rainfall dictates the effective growing season and cropping systems for a given region. In regions, receiving 350 to 600 mm of rain-

Table 6–1. Relative potential of traditional vs. efficient crops.

Location	Traditional crop	Yield	Efficient crops	Yield
		kg ha^{-1}		kg ha^{-1}
Bellary	Cotton	200	Sorghum	2670
Varanasi	Wheat	860	Chickpea	2850
Ranchi	Upland rice	2880	Corn	3360
Indore	Greengram	1180	Soybean (yellow)	3330
	Wheat	1120	Safflower	2420
Agra	Wheat	1030	Rapeseed-mustard	2040
Hissar	Wheat	320	Sativa	1610
Udaipur	Corn	1800	Sorghum	2900
Rewa	Soybean (black)	400	Soybean (yellow)	1200

fall and 20 wk effective growing season, only single cropping is possible with all soil types except deep Vertisols. In deep Vertisols, a single post rainy season crop is possible in areas receiving 350 to 600 mm rainfall and having a 20-wk effective growing season. Intercropping (150% cropping intensity) is possible in regions having a 20 to 30 wk effective growing season and 650750 mm rainfall. In areas receiving more than 750 mm of rainfall and having an effective growing season of more than 30 wk, double cropping (200% cropping intensity) is assured (Singh and Subba Reddy, 1986).

Intercropping

Intercropping is recommended in areas receiving 600 to 800 mm annual rainfall. In such areas, at least one of the component crops succeeds in producing economic yields; even during droughts. In high rainfall areas there are greater chances of success of both the component crops and the returns are higher than for a sole crop. The ratios of rows of principal crop to component crop (row ratios) in intercropping systems were optimized to minimize competition and realizing optimum biological productivity. Most intercropping systems are additive series, where the population of a base crop is maintained nearly equal to a sole crop. Promising intercrops for different locations are shown in Table 6–2.

Double Cropping

In areas receiving more than 800 mm annual rainfall and soil moisture storage of 200 mm m^{-1}, double cropping is feasible. Such options are recommended for rainfed areas in Orissa, Bihar, Madhya Pradesh and eastern Uttar Pradesh. With some adjustment of sowing dates (early planting and harvesting of rainy season crops) double cropping is also possible in Vidharbha and Malwa plateau as well. The most efficient double cropping combinations for different environments are depicted in Table 6–2.

Contingency Crop Planning for Weather Aberrations

In dryland agriculture, drought is a common phenomenon due to either a late onset or early withdrawal of monsoon or dry spells within the cropping season. Several technologies have been developed and tested that enhance the crop production during weather aberrations. Generally, short duration crops and varieties have replaced long duration varieties (Table 3). In seasons when rainfall is inadequate after planting, mid-season corrections such as interculture, thinning, and reducing plant

Table 6–2. Dominant intercropping and double cropping systems.

Location	Annual rainfall	Soil type	Dominant cropping systems: Intercropping	Dominant cropping systems: Double cropping
	mm			
I. Rice-based production system				
Phulbani	1440	Oxisols	Pigeonpea + Rice/corn/raddish (1:1) Castor + tuber crops (1:1)	Rice/corn-horsegram
Ranchi	1370	Oxisols	Pigeonpea + corn/rice (1:1) Sesame + greengram (1:1)	Rice-linseed/chickpea
Varanasi	1080	Entisols	Pigeonpea + okra/blackgram/sesame (1:1) Rice + pigeonpea (3:1) Barley + rapeseed-mustard (6:1)	Rice/sesame-linseed/ chickpea/wheat Greengram-barley/ rapeseed-mustard
II. Corn-based production system				
Hoshiarpur	1032	Entisols/ Inceptisols	Wheat + rapeseed-mustard (1:1)	Sunhemp/pearlmillet/ corn-wheat/chickpea
Rakh Dhiansar	1100	Inceptisols	Corn + okra (1:1) Cauliflower + oat (1:1) Wheat + rapeseed-mustard (4:1) Barley + chickpea (2:2)	Corn/blackgram-wheat/ rapeseed-mustard
Arjia	862	Vertisols	Corn + blackgram (1/2:1) Groundnut + sesame (6:2)	Greengram-safflower
III. Oilseed-based production system				
Indore	964	Vertisols	Soybean + pigeonpea/corn (2:2)	Soybean-wheat/ safflower/chickpea/ rapeseed-mustard
Rewa	1048	Vertisols	Sorghum + pigeonpea (2:1)	Rice-chickpea/ lentil/wheat
Rajkot	592	Vertisols	Groundnut + castor/pigeonpea (3/4:1) Pearlmillet + castor (2/4:1)	Not feasible
IV. Cotton-based production system				
Akola	825	Vertisols	Sorghum + pigeonpea (2:1) Cotton + greengram/cowpea/ pigeonpea (2:1)	Greengram-safflower Sorghum-safflower
Kovilpatti	780	Vertisols	Sorghum + blackgram/cowpea (2:1) Cotton + blackgram (2:2)	Not feasible
V. Coarse cereal-based production system				
Solapur	561	Vertisols	Sunflower + pigeonpea (2:1/2) Chickpea + safflower (3:1) Pearlmillet + pigeonpea/mothbean (2:1)	Sorghum-chickpea
Dantiwada	621	Aridisols	Pearlmillet + greengram/clusterbean (3:1) Pearlmillet + castor (1:1)	Not feasible
Bangalore	891	Alfisols	Pigeonpea + fingermillet/groundnut (2:8)	Cowpea-fingermillet

population; plowing in plant parts for recycling; and additional application of the nitrogen (N) after alleviation of the drought have been shown to impart production stability. For late onset of monsoon followed by normal rainfall, selection of short duration varieties of pulses, oil seeds, and fodder crops are important strategies to overcome the drought situation. Under late and low rainfall conditions, leguminous crops and oilseeds showed promising results. To avoid risks from weather aberrations, intercropping is advisable so that the rainfall pattern is suitable to at least one crop and hence the risk of complete crop failure is averted.

Alternate High Value Crops

Cultivation of crops for dyes, medicines, and aromatics is economical in drylands. They also play a vital role in sustained use of degraded lands. These crops are both perennial and annuals with plants mostly of bushy stature. The advantage of bushes over larger perennials is that the former offers less competition to associated crops. The promising plants for cultivation in drylands are dyes like Indigo (*Indigofera tinctoria* L.), Henna (*Lawsonia innermis* L.), and Bixa (*Bixa orellana*

Table 6–3. Promising treatments for mitigating early and late season droughts.

		Treatment	
Location	Crop	Early season drought	Late season drought
Bangalore	Fingermillet	Kaolin spray (5%)	Increased seed rate + fertilizer doses
		$CaCl_2$ spray (2%)	Transplanting
		Defoliation	--
Agra	Pearl millet	Straw mulch at 5 t ha^{-1}	Increased seed rate and fertilizer doses
		Interculture	Transplanting
		Kaolin spray (5%)	--
Bijapur	Pearlmillet	Thinning	--
		Straw mulch at 5 t ha^{-1}	
Rajkot	Groundnut	Straw mulch at 5 t ha^{-1}	--
		Earthing up	
Ananthapur	Groundnut	Application of sand mulch at 40 t ha^{-1} during moisture stress	--
	Sunflower	Mulching with groundnut shells at 5 t ha^{-1} with one supplemental irrigation	--
Solapur	Pearlmillet	Formation of ridges and furrows	--
Phulbani	Upland rice	Deep seeding (5–7 cm) with application of FYM and fertilizer in rows	--
Ranchi	Upland rice	Straw mulch at 5 t ha^{-1}	Increased seed rate
Indore	Soybean-chickpea	--	Application of safflower stover at 2 t ha^{-1}
Bhawanipatna	Upland rice	Mulching with local weeds/farm waste since germination	--
Dantiwada	Pearl millet	$CaCl_2$ spray (5%)	--

L.). Medicinal plants like Ashwagandha [*Withania somnifera* (L.) Dunal], Senna (*Cassia angustifolia* Vahl.), Muccina [*Muccina pruriens* (L.) DC], and aromatics like curry leaf [*Murrya koeingii* (L.) Spreng], lemon grass [*Cymbophogan winki* (DC) Stapf], palmrosa [*Cymbopogan martini* (Roxb.) Wats], and sweet basil (*Ocimum basilium* L.) hold potential.

Cultural Practices and Crop Management

Timeliness

Sowing dryland crops with the onset of monsoon rainfall can significantly improve crop yield (Singh and Das, 1984). Timely sowing helps in achieving optimum utilization of seasonal rainfall, reduces the incidence of pests/diseases, and is an escape mechanism from terminal drought. Yield losses from delaying sowing by 9 to 14 d in sorghum and upland rice are as high as 43 to 137 and 36 kg ha^{-1} d^{-1}, respectively (Singh and Das, 1984). A 15-d delay in sowing of sorghum led to reduction in grain yield of 850 kg ha^{-1}. Similarly, sowing castor (*Ricinus communis* L.) during the second fortnight of July reduced bean yield of 850 to 250 kg ha^{-1}. Lack of efficient implements and adequate draft power are the major constraints to timeliness in farming operations.

Tillage

Tillage has a marked influence on the conservation of soil and rainwater. Tillage makes the soil surface more permeable and thus, supports water intake. Deep tillage (25–30 cm) helps in soil pulverization, increased rainwater infiltration, and better root growth thereby increasing crop yield (Thyagaraj et al., 1999; Vittal et al., 1983; AICRPDA, 2000). Off-season or pre-monsoon tillage has a significant impact on weed control and rainwater infiltration. Grain yields of sorghum and barley (*Hordeum vulgare* L.) were 2600 and 1570 kg ha^{-1} with off-season tillage compared to 1870 and 1370 kg ha^{-1} without off-season tillage (AICRPDA, 1986).

Studies on reduced till farming indicated that conventional tillage using recommended fertilizer and weeding, with and without off-season tillage, resulted in higher grain yields of barley, rice, lentil (*Lens culinaris* Medik.), wheat, soybean, groundnut, fingermillet [*Eleusine coracona* (L.) Gaertn] and pearl millet (AICRPDA, 1999). However, excessive tillage reduces organic carbon (C) and accentuates soil erosion.

Mulching

Mulching and crop residue incorporation contribute to the conservation of soil and water. Mulching reduces runoff from cropped fields, prevents evaporation from soil surface, and controls weeds. Incorporating sorghum stubbles at 5 t ha^{-1} to cover 69% soil surface resulted in a 0.24 t ha^{-1} soil loss and 25 mm runoff compared to a 1.58 t ha^{-1} soil loss and 83 mm runoff when this treatment was not applied (AICRPDA, 2000). Mulching also reduced soil temperature and resulted in 25% greater moisture storage in the 0 to 30 cm soil profile. Frequent cultivation between crop rows creates dust mulch and breaks the soil crust which results in reduced capillary water movement and reduced evaporative losses. Cultivation during the vegetative stage enhanced the productivity of castor, sunflower, and pigeon pea [*Cajanus cajan* (L.) Millsp.] by 15 to 20% compared to no cultivation (Subba Reddy et al., 1996). Organic wastes and crop residues such as sorghum and maize stubbles, dry grass, wheat straw, and pigeon pea stalk can also be used as surface mulch. In Vertisols, spreading of crop residues at 5 t ha^{-1} enhanced the productivity of post-rainy sorghum and sunflower by about 25% probably through efficient utilization of stored soil moisture (ICAR–ACIAR, 2001). In Alfisols, incorporation of corn residue at 4 t ha^{-1} increased crop yield in a succeeding crop by about 80% (Gajanan et al., 1999). Vertical mulching by embedding sorghum or maize stalks in rows at regular intervals across the slope is a useful practice in reducing erosion in Vertisols. Sorghum yield under vertical mulching at 5 m intervals was about 25% higher than not mulching (Itnal, 1981). Of late, work has been initiated on a mulch-cum-green manure technique at different locations around the country. Using *Gliricidia* spp. branches/lopping at 5 t ha^{-1} in sorghum + pigeon pea–castor intercropping rotation reduced runoff by 56% and soil loss by 72% and increased castor bean yield from 328 to 984 kg ha^{-1} (ICAR-ACIAR, 2001).

Integrated Nutrient Management

In dryland agriculture, integrated nutrient management (INM) is the key to sustaining soil productivity. Studies at research stations and on-farm sites have demonstrated the importance of farmyard manure (FYM), composted organic wastes, and bio-fertilizers in providing the nutrient requirements of crops and yield stability in dryland areas (Singh et al., 1999). Fifty percent of fertilizer N can be replaced using FYM or compost sources. Application of FYM at 10 t ha^{-1} along with recommended fertilizer doses stabilized the productivity of finger millet at about 3400 kg ha^{-1}. The same treatment resulted in a crop yield index of 0.66 compared to 0.36 when only chemical fertilizer was used. Continuous application of chemical fertilizers resulted in a decline in finger millet grain yield from an average of 2880 kg ha^{-1} during initial 5 yr of the study to 1490 kg ha^{-1} by the 19th year (Gajanan et al., 1999). In Vertisols, providing 50% of recommended fertilizer dose through crop residues and the remaining 50% through *Leucaena leucocephala* lopping enhanced the sorghum yield by 87, 31, and 45%, respectively compared to application of 25 kg N ha^{-1} and 50 kg N ha^{-1} of chemical fertilizers alone (AICRPDA, 1999). In Alfisols and Vertisols, about 20 kg N ha^{-1} could be supplemented through the addition of green leaves of *leucaena* or *gliricidia* (Subba Reddy, 2002). Application of FYM in set rows resulted in an additional yield increase of about 20, 30, 90, and 20% for sorghum, sunflower, castor, and pigeon pea, respectively. At the same time, water-holding capacity and organic C increased in the set rows by 8.5 and 5.7%, respectively (CRIDA, 2002).

For maintaining soil fertility, green manuring is feasible in better rainfall areas when short duration legumes are grown after the main rainy season crop (Subba Reddy, 2002). A post rainy season cover crop {horsegram [*Macrotyloma uniflorum* (Lam.) Verdc] or cowpea [*Vigna unguiculata* (L.) Walp.]} can be raised using the post-season rainfall and plowed back into the soil before flowering (Katyal et al., 1994) as a key to maintain soil fertility. A summary of INM recommendations for principal dryland crops is given in Table 6–4.

Table 6–4. Integrated nutrient management (INM) in dryland agriculture.

Location	Crop	Fertilizer N	P_2O_5	K_2O	Others
		—— kg ha^{-1} ——			
Jhansi	Cluster bean	15	60	0	Inoculation with rhizobium
	Sorghum + Dolichos	60	20	0	
Rajkot	Sorghum	90	30	0	FYM at 6 t ha^{-1}
	Pearl millet	80	40	0	FYM at 6 t ha^{-1}
	Groundnut	12	25	0	FYM at 6 t ha^{-1}
	Cotton	40	0	0	FYM at 6 t ha^{-1}
Solapur	Sorghum	50	0	0	9–10 t ha^{-1} subabul loppings can substitute 25 kg N ha^{-1}
Indore	General	(N plus P)			4–6 t ha^{-1} FYM in alternate years
	Soybean	20	13	0	FYM at 6 t ha^{-1}
Bijapur		(NP or NPK)			Mulching with tree lopping at 5 t ha^{-1}
Arjia	Corn–pigeonpea	50	30	0	50% N through organics.
	Safflower and rapeseed-mustard	30	15	0	Reduction in N by half if these crops follow legumessuch as greengram/ chickpea
Agra	Barley	60	30	0	Use of FYM plus Azotobacter
Ranchi	Soybean	20	80	40	Inoculation with rhizobium
	Groundnut	25	50	20	Inoculation with rhizobium
	Pulses	20	40	0	Inoculation with rhizobium
Dantiwada	Greengram	0	20	0	Inoculation with rhizobium
Jodhpur	Pearlmillet	10	0	0	Addition of 10 t ha^{-1} FYM
Hoshiarpur	Corn	80	40	20	Addition of FYM
	Wheat	80	40	0	
	Chickpea	15	40	0	
Akola	Cotton + greengram	25	25	0	Along with FYM to meet 25 kg N ha^{-1}

Integrated Pest Management

Crop pests are a major constraint to raising dryland productivity. Crop losses due to pest attack range from 10 to 30% to complete crop failure depending on the crop and environment. However, indiscriminate application of pesticides has led to toxicity problem for the applicator or consumer, enhanced the appearance of resistant strains of pests, resurgence of pest species, destruction of nontarget organisms such as parasites and predators, and the accumulation of residues in food products. Integrated Pest Management (IPM) encourages the most ecologically sound combinations of available pest suppression technologies to maintain a pest population below the economic threshold. Easily adaptable and economically viable IPM strategies have been developed for the control of major pests in main dryland crops (Table 6–5).

Improved Farm Implements

Timeliness and precise seed and fertilizer placement in the moist soil zone is crucial for successful crop establishment in drylands. Since the sowing of seed is to be completed in a short period of time, appropriate farm implements are necessary to cover a long land area before the seed zone dries. During the past decades, improved seed cum fertilizer drills and planters have been developed to match

Table 6–5. Integrated Pest Management in dryland agriculture.

Location	Test crop and pest	Treatment	Yield
			kg ha^{-1}
Hyderabad	Castor (Semilooper)	Farmers' practice	314
		Hand picking	350
		Dusting turmeric powder	576
		Chemical	691
	Castor (Red hairy caterpillar)	Farmers' practice	640
		Hand picking	732
		Chemical	750
	Pigeonpea (Pod borer)	Farmers' practice	203
		Bio-insecticide (extract from custard apple seed) at 5 mL L^{-1}	365
		Chemical	363
Warangal	Groundnut (Red hairy caterpillar)	Trap crop (*Calotropis gigantean/ Jatropha curcas)*	1650
		Chemical	1420
Chittoor	Groundnut (Leaf webber)	Farmers' practice	650
		Trap crop (cowpea/castor)	800

farmer's specific needs. Similarly, appropriate interculture and weeding equipments have been developed to increase production efficiency (Gyanendra Singh and Mayande, 1999).

Alternate Land-Use Systems

Any farm enterprise other than crop production is usually referred to as an alternate land-use system. Examples are growing trees and arable crops in strips, tree farming, wood lots, pastures, grasslands, ley farming, and agro-forestry systems. Among these, growing arable crops with compatible tree species has been a traditional practice by dryland farmers who view it as a way to meet the demands of food, fodder, fuel, etc. Such a system offers sustainable productivity, covering land with vegetation, and conservation of natural resources. General recommendations for alternate land-use systems based on annual rainfall and soil type are outlined in Table 6–6. The major features of land capability classes marked in the following text in roman letters have been presented in Table 6–7.

Agri-silviculture

This system is recommended for land capability class IV (Table 6–7) with annual rainfall of 750 mm. A large number of tree-crop combinations, particularly of N_2 fixing trees with sorghum, groundnut, castor, and pulses were evaluated in Alfisols and Vertisols. Short duration dryland crops such as pearlmillet, blackgram [*Vigna mungo* (L.) Hepper] and greengram [*V. radiata* (L.) Wilczek], combined with widely spaced tree rows of *Faiderbia albida* and *Hardwickia binnata*, have been found compatible in semi-arid tropical areas (Korwar, 1992).

Silvipasture

This system is recommended for land capability class V and above. Silvipasture system of alternate land use involves integrating a tree component with

Table 6–6. Alternate land-use systems in dryland agriculture.

Annual rainfall (mm)	Soil depth	Alternate land-use system	Suitable tree/grass/legume species
<500	Shallow (0–0.30 m)	Tree farming	*Prosopis cineraria, Acacia aneura, Acacia nilotica, Acacia tortilis, Pithecelliobium dulce*
	Medium (0–0.45 m)	Pasture management	*Lasiurus sindicus, Cenchrus setigerus, Sehima nervosum, Stylosanthes scabra, Clitoria ternatea*
500–750	Shallow (0–0.30 m)	Silvipastoral	*Acacia nilotica, colophosphermum mopane, Dalbergia sisso, Hardwickia binata, Cassia sturti, Albizia amara, Leucaena leucocephala, Cenchrus ciliaris, Cenchrus setigerus, Dicanthium annulatum, Panicum antidotale, Stylosanthes hamata, Macroptillium atropurpureum*
	Medium (0–0.45 m)	Hortipastoral system	*Annona squamosa, Zizyphus mauritiana, Syzigium cuminii, Emblica officinalis, Tamarindus indica, Ferinia limonia, Aegle marmelos, Cenchrus ciliaris, Panicum antodotale, Urchloa mosambicensis, Stylosanthes hamata, Macroptilium atropurpureum, Clitoria ternatea*
>750	Shallow (0–0.30 m)	Silvipastoral	3 yr *Stylosanthes hamata* and 4th yr arable crop (sorghum on heavier soils, pearl millet on lighter soils)
	Medium (0–0.45 m)	Hortipastoral	*Mangifera indica, Achras zapota*

a perennial legume or grass as a pasture to improve the productivity as well as increase fodder availability. Native pasture with babul [*Acacia nilotica* (L.) Del] and subabul (*Leucaena leucocepahala* (Lam.) De Wit] are common in dryland areas. Silvipasture systems involving palatable grasses (*buffel or anjan grass* [*Cenchrus ciliaris* L.]) and legumes {stylo [*Stylosanthes hamata* (*L.*) Taubert]} with trees such as subabul, *siris* [*Albezzia lebbeck* (L.) *Benth*], anjan *(Hardwickia binata* Roxb.), and sisso (*Dalbergia sisso* Roxb.) were found to be more productive and profitable in the drylands (Singh, 2002).

Agri-horticulture

In land capability classes II and III receiving >750 mm annual rainfall, agri-horticulture consisting fruit trees grown with arable crops are recommended. Promising fruit trees, which can be grown successfully in drylands, are jujube (*Zizipus mauritiana* Mill.), gooseberry (*Emblica officinalis* Gaertn.), custard apple (*Annona squamosa* L.), guava (*Psidium gujava* L.), tamarind (*Tamarindus indica* L.) and mango (*Mangifera indica* L.) in combination with arable crops such as cluster bean [*Cyamopsis tetragonolobus* (L.) Taub], cowpea, groundnut, and horsegram.

Alley Cropping

Alley cropping or hedgerow farming is an agro-forestry practice in which arable crops are grown between perennial hedgerows spaced at regular intervals.

Table 6–7. Salient features of land capability classes.

	Land suitable for cultivation as per USDA Classification
Class-I	Very good cultivable, deep, nearly level, productive land with almost no limitation (or very slight hazard). Soils in this case are suited for a variety of crops, including wheat, barley, cotton, maize, tomato, and bean. Need no special management practices for cultivation.
Class-II	Good cultivable land on almost level plain or on gentle slopes that have slight limitations of soil depth, salinity, texture, drainage, or erosion that reduce the choice of plants. In general these soils are suitable for wheat, barley, cotton, moderately suitable for maize, alfalfa, tomato, and slightly-suitable for bean. Recommendation is to cultivate with precaution; need simple management practices.
Class-III	Moderately-good cultivable land on almost level plain or moderate slope. These soils have limitation(s) of moderate erosion, soil depth, soil salinity, and soil texture. They have vertic characteristics or drainage problems that reduces the choice of crop. In general, these areas have varying suitability for different crops. They are unsuitable for growing vegetable crops. Recommendation is to cultivate with careful management practices; need intensive care.
Class-IV	Fairly-good land on almost level plains or moderately-steep slopes. Suitable for occasional or limited cultivation; generally unsuitable for growing a variety of crops because of strong or very strong soil salinity (S3/S4), shallow depth, erosion, fine texture, or poor excessive drainage. Suitable for selected crops and pasture. Such soils may not be economical to cultivate as they need intensive soil conservation and management practices.
	LAND UNSUITABLE FOR CULTIVATION BUT SUITABLE FOR PERMANENT VEGETATION (GRAZING)
Class-V	Land not suitable for arable farming, but very suitable for grazing; have limitations for use of implements for use of implements due to stony or rocky and marshyness.
Class-VI	Nonarable land, well suited for grazing or forestry use. Have moderate limitations, such as steep slope, severe erosion, limited soil depth, strongly gypsiferous, stony or sand-dune areas. For instance, the dense forest lands of the Himalayas (India) or gypsiferous and dunal areas of SW Iraq.
Class-VII	Fairly-well suited for grazing or forestry; not cultivable. Have severe limitation, such as very steep land subjected to erosion or very shallow, stony soils having not enough available moisture for cultivation. Need careful management for grazing and forestry.
Class-VIII	Nonarable, extremely rough, rocky, arid, wet or extremely saline land, suited only for wild life or recreation. Have very severe limitation, for instance highly eroded land, barren mountain tops (as in the Himalayas) or rocky undulating surfaces (as in the desert of south-western Iraq.

This system is recommended for land capability classes II to IV receiving 500 to 750 mm annual rainfall. Hedgerows are pruned during the cropping season to reduce their competition with crops. Short duration rainy season crops such as pearlmillet and sorghum were found to be compatible alley crops while long duration species such as castor and pigeon pea are not. Wider alleys and shorter hedgerow height trees used in wetter tropical areas were found to be better in semi-arid areas (Korwar, 1992).

Trees on Field Boundary

Growing multi-purpose trees on field boundaries is a common agro-forestry practice in drylands. Promising erect tree species are *teak* (*Tectona grandis* L. f), subabul (pollarded for fodder), palmyra palm (*Borassus flabellifera* L.), *coconut* (*Cocos nucifera* L.), and babul. In this system, the area along the field boundary is gainfully used.

Table 6–8. Recommended Inter terrace land treatments for soil and water conservation in drylands.

Annual rainfall	Soil type	Land treatment	Yield advantage
>500	Aridisols Alfisols (shallow)	Inter-plot rainwater harvesting of 1:1 cropped to uncropped land Dead furrows at 3.6 m interval	50% pearl millet 10% groundnut
500–1000	Alfisols (shallow)	Sowing across slope and ridging	10% sorghum
	Vertisols (shallow)	Compartmental bunding	25% sorghum
	Vertisols	Contour farming	35% sorghum
	Alfisols	Graded border strips	42% fingermillet
>1000	Entisols	Inter-plot rainwater harvesting	17%, corn
	Vertsols	Raised bed and sunken system	21% rice 34% soybean

Inter-Terrace Land Treatments

Adoption of inter terrace land treatments is highly variable depending on the intensity of rainfall, slope, soil type, and prevailing crop/cropping systems (Singh et al., 2000). Compartmental bunds in shallow Vertisols resulted in a 25% higher grain yield in sorghum, whereas, graded border strips in deep Alfisols enhanced the yield of fingermillet by 42%. In better rainfall areas, significant benefits due to inter terrace land treatments were recorded for crops such as upland rice, maize, and soybean (Table 6–8). In view of higher capital cost of earthen bunds and the problems related with their maintenance, research efforts were focused on vegetative barriers and live bunds during the 1980s. Although, Vetiver grass [*Vetiveria zizanoides* (L.) Nash] received the most attention, a number of plant species such as buffel or anjan grass (*Cenchrus ciliaris* L.), *Cymbopogan flexiosus*, *Pennisetum hohhennackeri*, and subabul (*Leucaena leucocephala* Lam.)were found effective in conserving soil moisture and controlling soil erosion. These crops also provide fodder and green manure. Conservation furrows at a 3.6 m horizontal interval supplemented with glyricidia [*Gliricidia maculata* (Jack.) Walp.] mulch at 5 t ha^{-1} increased sorghum grain yield to 4003 kg ha^{-1} from 1821 kg ha^{-1} under control, reduced runoff by 73%, decreased soil loss by about 50%. Apart from this, gliricidia application can supply considerable amount of N as on dry weight basis its leaves contains 3 to 4% N. About 80% of the farmers in dryland areas follow simple soil and water conservation measures like sowing across the slope, formation of conservation furrows, and key line cultivation (ICAR-ACIAR, 2001).

Rainwater Harvesting and Recycling

The advantages of harvesting rainwater in dug out ponds and using it for critical irrigation of rainy season crops and pre-emergence irrigation in post-rainy season crops has been reviewed by Singh (1986) and Singh and Khan (1999). On an average, 24 x 10^6 ha m out of 400 x 10^6 ha m of precipitation received annually in dryland areas of India, is estimated to be available as harvestable rainwater through on-farm facilities. Drylands receiving 500 to 1000 mm annual rainfall has an estimated harvestable rainwater potential of 5.54 x 10^6 ha m (Katyal, 1997). Although

Table 6–9. Effect of supplemental irrigation on dryland crop yields.

Crop	Number of replication	Yield: With irrigation	Yield: Without irrigation	Water-use efficiency
		kg ha^{-1}	kg ha^{-1}	kg ha^{-1} mm^{-1}
Wheat	18	3180	1760	28.2
Barley	4	2270	1580	13.8
Sorghum	16	1870	1110	15.1
Upland rice	4	2780	1620	23.2
Pearl millet	4	1900	1660	5.8
Finger millet		2320	1620	14.0

village level rainwater harvesting tanks have been traditionally in vogue in India, research during last 25 yr led to standardizing the variables regarding the type and area of catchment, capacity of dugouts and the utilization of stored water for supplemental irrigation on individual land holdings (Singh, 1986). Results from a large number of trials on supplemental irrigation to dryland crops in terms of yield and WUE are summarized in Table 6–9.

Integration of Livestock with Arable Crops

Traditionally dryland farmers are small subsistent landholders integrating livestock with crop production. With continuing population growth, intensifying crop and livestock mixed systems continues to play a vital role in maintaining rural livelihoods. Crop-livestock integration involves the natural resources (crops, animal, land, and water) in which these subsystems and their synergetic interactions have a significant and positive effect on the sum of their individual effects. Livestock farming systems in dryland agriculture are complex and generally based on traditional socio-economic considerations. An understanding of production factors (livestock, capital, feed, land, and labor) and processors (description, diagnosis, technology design, testing, and extension) that effect the animal production is a prerequisite for livestock integration (Fig. 6–5). Potentially important technologies, which can make a significant enhancement in productivity of both the crop and livestock within the system, can be adopted through:

- Increased fodder production as an intercrop with cereals, alley and relay cropping, forage production on bunds, food–feed cropping systems, etc.
- Better utilization of available fodder resources by timely harvesting and preserving it either as a hay or silage.
- Improving the feeding value of straw/stover by chopping, soaking with water, urea treatment, strategic supplementation of concentrate, mineral blocks or tree foliage etc., for enhanced utilization.
- Developing degraded, marginal lands through agro-forestry systems.
- Managing common property areas through user groups.
- Establishing of fodder banks in areas where surplus fodder is available and transporting it to the fodder deficit areas.
- Supplying improved seeds or saplings for fodder seed multiplication and reseeding of grazing lands.

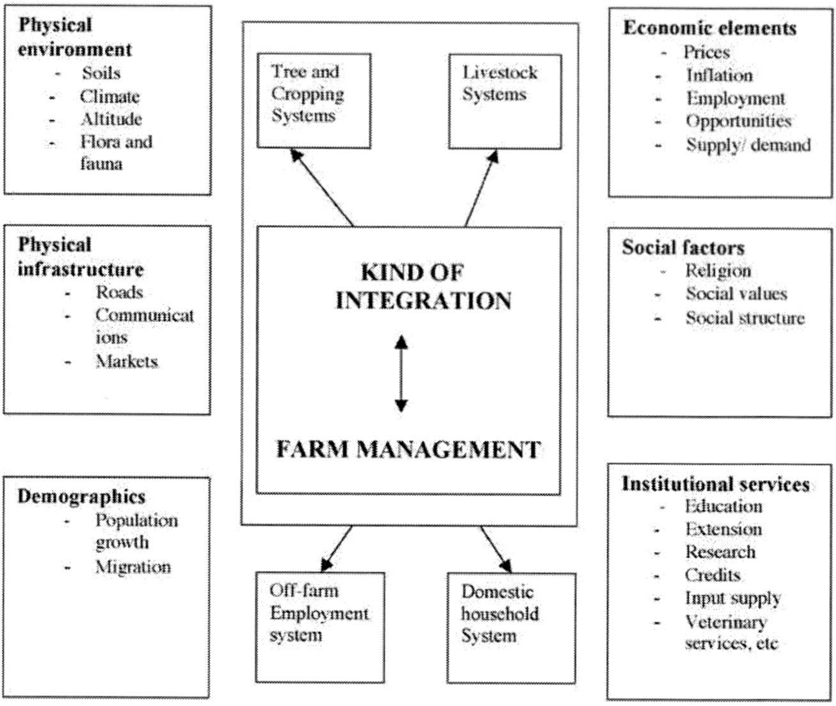

Fig. 6–5. Integration of livestock with arable crops through farm management.

- Removing low-grade animals through castration.
- Regular culling of un-economical animals.
- Protecting animals against harsh environment and adoption of preventive measures through health camps to maintain the health of livestock.

Thus, the shaping of integration process depends to a large extent on farm management. In order to achieve sustainable development in dryland areas, all components of the farming system must be taken into account as well as the context within which resource poor farmers operate (Fig. 6–5).

Integrating Dryland Technologies through Watershed Management

The concept of watershed management is important for the efficient utilization of water resources. Since rainwater conservation and utilization is the cornerstone of successful dryland farming, the watershed with distinct hydrological boundary is considered an ideal level for area development. Past experiences of watershed projects implemented in the drylands have led to improved water availability in terms of additional surface storage and enhanced recharge of groundwater. Increased water availability in wells and storage facilities have led to an increased cropping intensity of 50% over a period of 5 yr (CRIDA, 1997). Consequently these efforts resulted in adoption of improved dryland crop production technologies. Marked economic gains that have been recorded attributed to the implementation of the watershed management projects in different parts of the country.

Innovations in Technology Transfer

It is very important to integrate short-, medium-, and long-term technologies at the watershed scale when the objective is to impart stability to dryland production. This will result in ensuring peoples' participation at all stages of development.

Farmers Participation in Technology Development

Dryland farmers, by and large, have ignored technology developments in the past. Most extension efforts have been hitherto largely oriented either manipulative/passive participation or participation of material incentive/functional participation. Suitable methodologies to encourage the interactive participation of farmers in dryland agriculture are needed. Methodologies involving participatory rural appraisal, demonstration of action learning tools, focus group interactions to obtain farmers' perceptions and identify indigenous technical knowledge aimed at its utilization and refinement are necessary. Thus, progress in interactive participation will lead to voluntary participation or self-mobilization. This should be the strategy for technology transfer in the drylands.

Grass Root Extension System

Despite the growth of agricultural research and education that has led to impressive gains in productivity in several geographical areas, a vast number of villages are either yet to get exposure to improved knowledge or not adequately covered by the past and current developmental initiatives. Research and development/Nongovernmental Organizations (NGO)-farmer linkage and partnership model have proved successful in technology development and transfer in dryland India. However, this model works only if the NGO possesses the desired level of technical competence and commitment. Such NGOs are few. Another approach is that a village is or a group of villages are mobilized to act as a NGO. In India a self-help group (SHG) approach has proved successful and is now taking shape as a mass movement in many States of India. User groups and village volunteers' approaches are being implemented with some success. These approaches can be consolidated to encourage a village Local Self Governance System, a community organization. This may eventually lead to "save and nourish our own land for sustained productivity" model of sustainable development.

Using Indigenous Technical Knowledge

Ample Indigenous Technical Knowledge (ITK) exists within the farming community on subjects such as natural resources management, land management, rainwater harvesting, adopted livestock, pasture management, etc. Technologies developed using ITK and their refinement are readily accepted by the farming community.

Consortia Development

At present a multitude of government institutions, NGOs, and private sector/corporate organizations are working to improve agricultural productivity in drylands. These isolated efforts are more often handicapped for want of skilled manpower as well as a limited exposure to knowledge being acquired through research initiatives. There is a need to adopt a consortium approach that will share knowledge and experiences, with a view to address problems in the right perspective. Each agro-ecological region or subregion must establish a consortium of its own that will control land degradation and improving farm productivity on a sustainable basis. Research organizations in the region may be the focal point for this activity.

PERSPECTIVE

Future problems in dryland agriculture will be multi-faceted and unlike commodity-based efforts, a narrow strategy cannot work. A holistic approach with multidisciplinary, multi-commodity, and multi-institutional framework will be required. Land-use diversification, product processing, and value-added products, aimed at improving farmers' income will assist in curbing human migration from rural to urban areas. It is likely that the existing decline in the area of some crops such as sorghum and soybean will continue because of changing food habits. A greater shift is expected from food crops to commercial crops in the next decade, but equilibrium may be achieved with the economics of different commodities determining the choice of a cropping enterprise. Further processing of food/fruit products could emerge as an important source of income and employment. Exporting processed food will become important now that India has signed the World Trade Organization (WTO) agreement. The plants/products associated with drylands that have a comparative advantage need to be identified to realize India's share of this global trade.

Rainwater harvesting is an age-old practice in India. It is a process of collecting runoff from treated or untreated land surfaces/catchments or rooftops and storing it in an open farm pond or closed water tanks/reservoirs. Another approach is, in situ moisture storage in the soil itself. An attempt should be made to retain as much rainwater in the soil where it falls, so as to provide a favorable moisture regime to the crop. Rainwater that exceeds the infiltration and storage capacity of soil may be harvested nearby in the same field or at another convenient point in the watershed. Estimates reveal that areas receiving up to 1000 mm annual rainfall have a potential to add 6.3 million ha water equivalent through rainwater harvesting (Singh et al., 2000). On an average, 24 million ha m (Table 6–10) of the total 400 million ha m precipitation received annually in India is estimated to be available through on-farm rainwater harvesting (Katyal, 1997). Although village level rainwater harvesting tanks have been traditionally in vogue in India, research since 1980 has led to standardizing variables such as runoff characteristics, optimizing the pond size, lining materials, water lifting devices, and the most suitable crops for using the harvested water efficiently (Singh, 1986).

Effective rainwater impoundment and use would be the core strategy to increase cropping intensity and productivity. Therefore, future success attaining sus-

Table 6–10. Estimated volume of storage for small-scale rainwater harvesting structures.

Rainfall zone	Area	Rainfall for effective surface storage	Harvestable rainwater
mm	million ha	mm	million ha m
<500	52.07	5	0.78
500–750	40.26	6	1.51
750–1000	65.86	7	4.03
1000–2500	137.24	6	14.61
>2500	32.57	4	3.26

tainability goals would depend on effective rainwater harvesting and management including that received during the pre-or post-monsoon seasons. However, despite the advantages of rainwater harvesting and its reuse through dugout ponds, the adoption of this technology has been low because of high initial investments, which are beyond the capacity of most small farmers. Hence, future research on controlling evaporative and seepage losses, water lifting devices, sealing materials, and the cost of dugouts need to be carried out to make the concept feasible and economically viable to the target farmer groups.

Our current knowledge of the soil and water conservation structures and feedback from farmers revealed farmers' reluctance to spare and partition their land to accommodate conservation structures. Hence, mechanical or vegetative structures that correspond to the field boundaries may be more appropriate. Conservation structures must be simple and affordable yet effective for a given level of runoff. Experience and reported success stories in recharging wells by channeling overland flow into the percolation tanks needs to be evaluated for its hydrology and economic viability. Contour vegetative barriers offer an effective buffer against soil erosion while enhancing in situ conservation of rainwater. Their acceptance is likely to be faster if the vegetation used can provide a visible economic benefit and at the same time promote rainwater and soil conservation. Adequate attention has not been paid to the vegetative barriers native to a region. In view of the issues related to the water collection and sharing vis-à-vis the property boundaries of the farmers within a watershed, it is advantageous to have more small dugouts in a given watershed than investing in a large on-farm pond. Utilization of pond water continues to be an issue. Nevertheless, rainwater harvested and stored either on the surface or below the ground will have to be used through water-efficient irrigation techniques. Benefits of supplemental irrigation have been demonstrated.

Short-term surface storage to recharge dwindling groundwater is acceptable to the farming community. Since water is a scarce commodity, planning on its judicious use is a crucial aspect of rainwater management. Essential strategies of rainwater recycling include water-saving methods of irrigation, water-efficient crops, emphasis on moisture stress management at critical crop growth stages, and preferential irrigation of crops producing higher economic yields. Operationalizing each of these aspects needs greater emphasis during the next 25 yr.

Past efforts on breeding for drought tolerance and increased water-use efficiency have not resulted in encouraging results. Breeding for higher water-use efficiency continues to be relevant for achieving higher productivity with no change in available water. Multidisciplinary approaches involving breeders, molecular biologists, physiologists, and agronomists might provide useful results during the coming decades. Many potential crops and their wild relatives still remain unexploited. Their biodiversity could offer a potential gene pool for breeding efforts both through conventional and biotechnological tools.

Because of worldwide concern on the harmful effects of synthetic dyes, drugs, pesticides, etc. native shrub and tree species offer numerous health-safe alternatives. Many shrubs produce natural dyes and can be successfully raised under dryland agriculture. Neem [*Azadirachta indica* (A.) Juss] exhibits immense potential to yield oil, fertilizer, bio-medicine, and pesticides. These shrub and tree species extend possibilities for export in the light of current dislike for synthetic chemicals. At the same time, these plants hold promise in rehabilitating degraded lands either as commercial plantations or as components of agroforestry systems. Insecticidal properties of custard apple and jatropha (*Jatropha curcas* L.) could also be exploited for raising the income of dryland farmers.

In the 21st century, stability in crop production and income are likely to occur because of land-use diversification guided by value output rather than biological productivity. Besides introducing nontraditional crops, intercropping perennial plants is likely to provide a higher income and benefit-cost ratio. The threat posed by decreasing farm sizes can be converted into opportunities by increasing income through more profit-earning innovations. With accentuated problems of soil erosion and a resulting decrease in soil depth, a great deal of variability in productivity can be expected. A matrix of possible land uses, as influenced by resource carrying capacity is recommended for the future in Fig. 6–6.

In the process of diversification, a greater focus on location specific technologies may be relevant than a universal package across all natural and social domains. In drylands, investment capacity varies from farmer to farmer and so will be the expected income. Accordingly, packages with higher investments may be offered to some farmers, while a low input approach may be required for others. It is expected that about 40% of family income will come from activities other than row crop production such as poultry, dairying, horticulture, and sericulture. This activity mix will achieve the desired diversification while providing a profitable enterprise.

The animal component will continue to be an important enterprise in dryland farming. Fodder availability will be one of the major problems that may need attention. Some grain crops such as sorghum can be used as dual purpose in the immediate future and as a high foliage fodder with potential as a value added crop in the long term. There is an urgent need to identify annual herbal species that produce higher biomass per unit water use. Strategies need to be evolved to meet 25 to 40% of fodder requirement from top feed species and shrubs. The threat of expanding marginal lands could be contained by planting fodder and fuel-wood species. Exorbitant costs involved in rehabilitating degraded lands, necessitates appropriate technologies for modifying the planting sites only. With this approach at least 25 million ha of degraded land surface will benefit.

With expected productivity levels and land-use diversification, the demand for plant nutrients will increase several-fold. Even during the next decade, nutrient

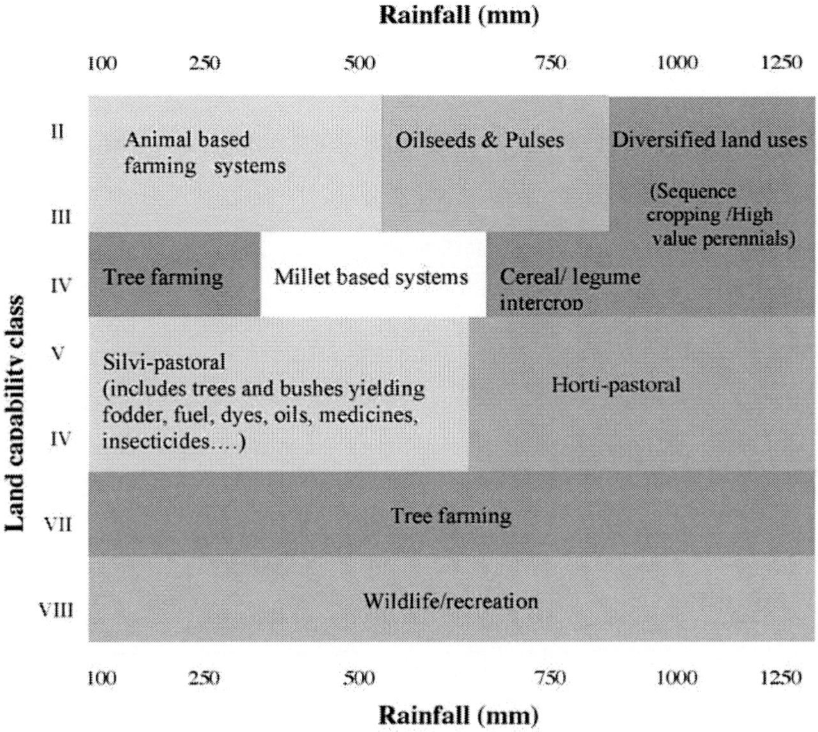

Fig. 6–6. Recommended land use for imparting sustainability to dryland agriculture.

requirements may double. More research is required on the soil moisture-nutrient interactions during the monsoon-cropping season and on enhancing the fertilizer-use efficiency by 25 to 30% from its current level. The role of rainfall frequencies in predicting fertilizer use at appropriate times may minimize the risk attached to fertilizer-use efficiency. In view of poor economic status of most dryland farmers, a considerable portion of nutrients will have to be met through organic and biological sources and used through a integrated nutrient management approach. The availability of crop residues continues to be a challenge due to other competitive uses. Alternative strategies of raising herbaceous cover crops, using the off-season rainfall, bund farming for fodder, and recycling the biomass to the soil appear promising since about 25% of the precipitation is received during the off-season and between 5 and 10% of the field area is covered by bunds.

Credit availability might improve when compared to current levels but rainfed areas would still be at a disadvantage compared to irrigated situations in attracting adequate investments. Improved marketing and price support for those products which have a wider consumption pattern. Greater participation of NGOs and building farmers associations is warranted in the present context.

Table 6–11. Yield gap for sorghum between research plots and farmers' fields.

Situation	Replication	Grain yield	Termed as
		kg ha^{-1}	
Research farm	1	3970	Potential
Farmers' fields:			
Verification trails	9	2090	Realizable
Demonstrations	39	1460	Feasible
Assured input supply	26	1060	Attainable
Rural farm	36	370	Actual

LESSONS LEARNED AND FUTURE OPTIONS

Improved crop production technologies aimed at individual farms and crops have made an impact in drylands across the country. However, many of the resource management interventions, which do not have immediate impact on the yields, need to be vigorously applied and tested in a participatory technology development mode and in a watershed perspective (Singh et al., 2000). The length of the growing season is always a limitation in dryland agriculture and hence the key issue for successful farming is timeliness of operations. Due to a wide gap in draught power available to farmers and what they need to accomplish various operations, limited farm mechanization is necessary.

The urgency of managing land resources including eco-friendly, economically advantageous and socially acceptable agricultural technologies would mitigate or alleviate many problems faced by dryland agriculture. Diffusion of improved technologies can be accelerated if introduced land-use options have clarity and economic transparency in terms of higher income from the same unit of land. This has to be the core strategy for the future of those dryland areas that have better resource endowments (Solanki and Newaj, 1999). Since land degradation continues unabated, a definite requirement for the immediate to medium-term future will be rehabilitating these lands to enable them to produce more food and biomass.

Poor adoption of technologies by the farming community continues to remain a perplexing issue even with technology packages with apparent benefits. As an illustration (CRIDA, 1997) the wide gap between the potential yield on the research farms, yield levels in verification trials, and the actual yields under real farm situations is depicted in Table 6–11.

To alleviate poor diffusion of dryland farming technologies, more on-farm research/testing is required with farmers' participation. Recent experience suggests that involving NGO's can increase the efficiencies of this process. Simultaneously a better interface between the input supply agencies and institutions that provide technological back stopping is necessary. Therefore, the future technology endeavors should include:

- Prioritization of research on eco-regional endowments with an eye to interblend local wisdom, relevance, needs, and interests of all segments of the farming community in technology development and transfer.

- Adoption of holistic approach so that soil and rainwater management technologies are integrated into 'total-factor-productivity' farming system mode and are applied/tested to/on discrete hydrological units, that is, watersheds.
- Interactive relationship between quests to maximize productivity (profitability) while not degrading the resource base against the backdrop of climatic and edaphic variables. Focus on anthropogenic interventions to extrapolate site-specific practices into area action plans.
- Focus on farmers' ambitions, preferences, and perceptions so that existing or projected technologies are likely to be accepted across a range of physical and socio-economic domains.
- Training to sharpen the skills and thinking of all the key players involved in dryland agriculture research, development, and practice.

REFERENCES

All India Coordinated Research Project for Dryland Agriculture. 1986. Annual report of AICRPDA, Central Res. Inst. for Dryland Agric., Hyderabad, India.

All India Coordinated Research Project for Dryland Agriculture. 1999. Annual report of AICRPDA, Central Res. Inst. for Dryland Agric., Hyderabad, India.

All India Coordinated Research Project for Dryland Agriculture. 2000. Annual report of the AICRPDA, Central Res. Inst. for Dryland Agric., Hyderabad, India.

Central Research Institute for Dryland Agriculture. 1997. CRIDA perspective plan—Vision 2020.CRIDA, Hyderabad, India.

Central Research Institute for Dryland Agriculture. 2002. Annual progress report. CRIDA, Hyderabad, India.

Fertilizer Association of India. 2001. Fertilizer statistics 2000-2001. FAI, New Delhi.

Food and Agriculture Organization. 2000. Year book. FAO of the United Nations, Rome.

Gajanan, G.N., B.R. Hegde, Ganapathi, Panduranga, and K. Somashekhar. 1999. Organic manure for stabilizing productivity: Experience with dryland fingermillet. All India Coordinated Res. Project on Dryland Agric., Univ. of Agric. Sci., Bangalore, India.

Gyanendra Singh, and V.M. Mayande. 1999. Role of improved farm equipment in dryland agriculture. p. 431–446. *In* H.P. Singh et al. (ed.) Fifty years of dryland agricultural research in India. Central Res. Inst. for Dryland Agric., Hyderabad, India.

Indian Council of Agricultural Research-Australian Centre for International Agricultural Research. 2001. Progress report of ICAR–ACIAR Project on tools and indicators for planning sustainable soil management on semi arid farms and watersheds in India, Central Res. Inst. for Dryland Agric., Hyderabad.

Itnal, C.J. 1981. Water conservation measures for increased production in black soils of Bijapur. All India Coordinated Res. Project on Dryland Agric., Univ. of Agric. Sci., Bangalore, India.

Kanwar, J.S. 1999. Need for a future outlook and mandate for dryland agriculture in India. p. 11–19. *In* H.P. Singh et al. (ed.) Fifty years of dryland agricultural research in India. Central Res. Inst. for Dryland Agric., Hyderabad, India.

Katyal, J.C. 1997. Research and development in rainfed agriculture: Retrospective and perspective. p. 256–270. *In* Productivity of land and water. New Age Int. Publ., New Delhi, India.

Katyal, J.C., S.K. Das, G.R. Korwar, and M. Osman. 1994. Technology for mitigating stresses: Alternate land uses. p. 291–305. *In* S.M. Virmani et al. (ed.) Stressed ecosystems and sustainable agriculture. Central Res. Inst. for Dryland Agric., Hyderabad, India.

Korwar, G.R. 1992. Alternate land use systems. p. 143–168. *In* L.L. Somani et al. (ed.) Dryland agriculture in India—State of agricultural research in India. Scientific Publ., Jodhpur, India.

National Bureau of Soil Survey and Land Use Planning. 2001. Classification of dryland regions in India. NBSSLUP, Nagpur, India.

Ramakrishna, Y.S., G.G.S.N Rao, B.V. Ramana Rao, and P. Vijay Kumar. 1999. Agro-meteorology. p. 32–60. *In* G.B. Singh and B.R. Sharma (ed.) Fifty years of natural resource management research. Indian Council of Agric. Res., New Delhi, India.

Singh, H.P. 2002. Farming systems and best practices for drought prone areas of India. *In* Farming systems and best practices for drought prone areas in Asia and the Pacific. Central Res. Inst. for Dryland Agric., Hyderabad, India.

Singh, H.P., K.D. Sharma, and G. Subba Reddy. 2002. Rainfed agriculture: Challenges of the 21st century–National perspectives. p. 3–4. *In* Proc. of the 89th Indian Sci. Congr.—Part IV. Lucknow, India.

Singh, H.P., K.L. Sharma, B. Venkateswarlu, and K. Neelaveni. 1999. Fertilizer use in rainfed areas: Problems and potentials. Fert. News 44(1):27–38.

Singh, H.P., B. Venkateswarlu, K.P.R. Vittal, and K. Ramachandran. 2000. Management of rainfed agro-ecosystem. p. 669–774. *In* Natural resource management for agricultural production in India. Indian Soc. of Soil Sci., New Delhi, India.

Singh, R.P. 1986. Farm ponds. Project Bull. 6. Central Res. Inst. for Dryland Agric., Hyderabad, India.

Singh, R.P., and S.K. Das. 1984. Timeliness and precision—Key factors in dryland agriculture. Central Res. Inst. for Dryland Agric., Hyderabad, India.

Singh, R.P., and M.A. Khan. 1999. Rainwater management: Water harvesting and its efficient utilization. p. 301–313. *In* H.P. Singh et al. (ed.) Fifty years of dryland agricultural research in India. Central Res. Inst. for Dryland Agric., Hyderabad, India.

Singh, R.P., and G. Subba Reddy. 1986. Research on drought problems in arid and semi-arid tropics. ICRISAT, Patancheru, India.

Solanki, K.R., and R. Newaj. 1999. Agroforestry: An alternate land use system for dryland agriculture. p. 463–473. *In* H.P. Singh et al. (ed.) Fifty years of dryland agricultural research in India. Central Res. Inst. for Dryland Agric., Hyderabad, India.

Subba Reddy, G. 2002 Potentials of green manuring in rainfed conditions. *In* Potentials of green manure crops in rainfed agriculture. Univ. of Agric. Sci., Bangalore, India.

Subba Reddy, G., D.Gangadhar Rao, S. Venkateswarlu, and V. Maruthi. 1996. Drought management options for rainfed castor grown in Alfisols. J. Oilseeds Res. 13(2):200–207.

Subba Rao, I.V. 2002. Land use diversification in rainfed agriculture. CRIDA Foundation Day Lecture. Central Res. Inst. for Dryland Agric., Hyderabad, India.

Thyagaraj, C.R., K.P.R. Vittal, V.M. Mayande, and K.L. Sharma. 1999. Tillage and soil management for higher productivity in drylands. p. 329–344. *In* Fifty years of dryland agricultural research in India. Central Res. Inst. for Dryland Agric., Hyderabad, India.

Virmani, S.M., P. Pathak, and R. Singh. 1991. Soil related constraints in dryland crop production in Vertisols, Alfisols and Entisols of India. p. 80–95. *In* Soil related constraints in crop production. Indian Soc. of Soil Sci., New Delhi, India.

Vittal, K.P.R., K. Vijayalaxmi, and U.M.B. Rao. 1983. Effect of deep tillage on dryland crop production in red soils of India. Soil Tillage Res. 3:377–384.

7 Drought Early Warning Systems for the Near East

Eddy De Pauw
ICARDA
Aleppo, Syria

ABSTRACT

The Near East is a region with a high degree of aridity, and is subject to frequent droughts. Agriculture is a major and sensitive sector of the region's economy, consuming most of the available water resources. Agricultural production of major grain crops is strongly affected by precipitation fluctuations. The response of the region's governments to drought has so far aimed at mitigating the worst effects of drought rather than treat it as a structural problem that can be incorporated in government policy and long-term management plans. The governments of the region are in the process of developing comprehensive dryland management strategies, in which drought monitoring systems are an important component. Important issues to be addressed are the institutional arrangements to ensure free information flow, as well as the linkages between monitoring and rapid response action at different decision-making levels. Particular drought research needs include the feasibility of drought forecasting, the characterization and spatialization of drought, and the assessment of drought vulnerability from the agroecological and livelihood perspectives.

INTRODUCTION

North Africa and West Asia, hereafter denoted 'Near East', covers a large part of the world (more than 7 200 000 km^2). The map in Fig. 7–1 indicates the level of aridity by the aridity index, the ratio of annual precipitation over annual potential evapotranspiration, calculated by the Penman method (UNESCO, 1979). Figure 7–1 demonstrates that the Near East has diverse but generally dry climates, ranging from humid, subhumid, semi-arid, arid to hyperarid. In addition, temperature regimes vary considerably, particularly as a result of differences in altitude, and to a lesser extent, oceanic/continental influences. The precipitation season for most of the region is approximately from October to April and is thus concentrated in winter.

In most of the region, evaporation exceeds precipitation. The countries of this region thus have a high degree of aridity in large parts of their territories, and are therefore highly vulnerable to drought.

With more than 90% of the land area in hyperarid, arid or semi-arid moisture regimes (areas calculated from the map in Fig. 7–1), aridity is very significant

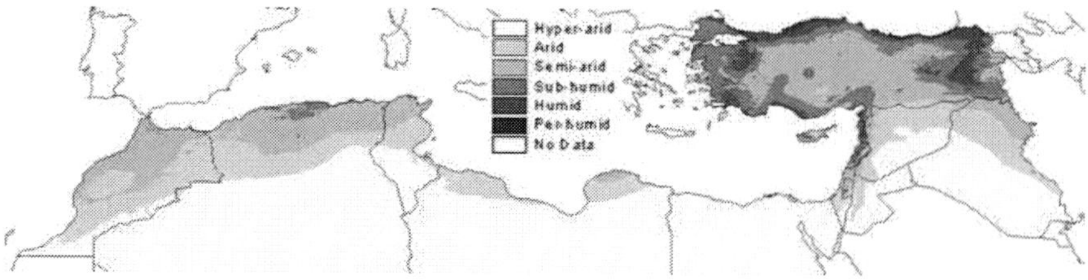

Fig. 7–1. Extent of aridity in the Near East. Source: ICARDA Geographical Systems Laboratory.

in the Near East. One exception to this general observation is Turkey, which is better endowed in surface and groundwater resources due to the orographic capture of Atlantic cyclonal precipitation, although much of the interior is semi-arid. If one excludes the hyperarid zones, which cover the driest deserts and have no potential for agricultural use, nearly 34% of the region, or about 2 460 000 km^2, is 'dryland' and has some potential for either dryland farming (semi-arid moisture regime) or for extensive rangeland (arid moisture regime).

Agriculture is an important sector of most economies in the Near East, contributing for most countries 10 to 25% of gross domestic product and providing employment to 10 to 55% of the labor force. The indirect importance of agriculture to the economies of the Near East countries is even larger, because it provides the primary goods that constitute the majority of merchandise exports.

Agriculture in the Near East is particularly vulnerable to drought. Irrigated land is comparatively scarce (Fig. 7–2a) and most of the agricultural systems depend on rainfall (Fig. 7–2b).

Although the area under irrigation is expanding, supply constraints are likely to increase. The reasons are limitations on the total size of the extractable water resources, continued population growth coupled with increasing urbanization, and competition between communities, industrial and service sectors, and agriculture

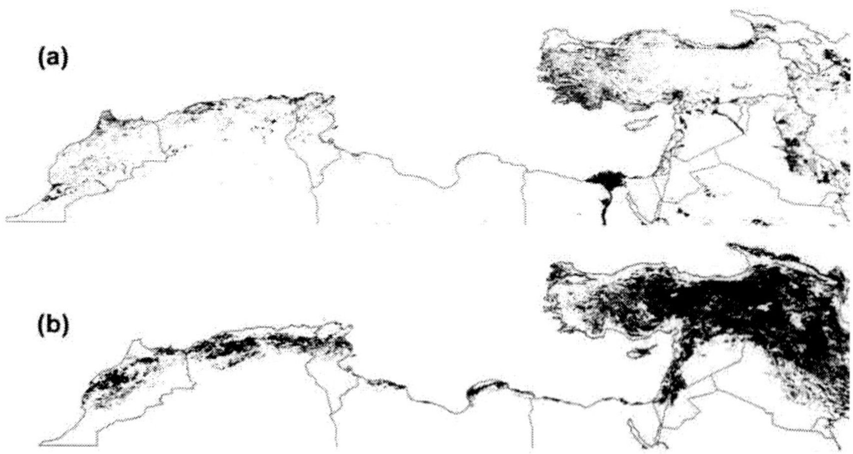

Fig. 7–2. Irrigated (a) and rainfed agriculture (b) in the Near East. Source: ICARDA Geographical Systems Laboratory.

for increasingly scarce water. Irrigated agriculture currently consumes on average 80% of all water in the Near East (Margat and Vallée, 2000), and while the key for the alleviation of drought, it is in itself highly responsive to the supply in case of a regional drought.

DROUGHT IN THE NEAR EAST

Characteristics of Drought

In its simplest definition drought is 'a deficiency of precipitation from expected or "normal" that, when extended over a season or longer period of time, is insufficient to meet demands' (Knutson et al., 1998). Drought is thus an inherent characteristic of climates with pronounced precipitation variability.

Irrespective of the degree of aridity, precipitation variability is considerable in the Near East. Figure 7–3 shows the annual rainfall variations for three stations in different moisture regimes across a considerable annual precipitation gradient (150–1000 mm). It is evident that high rainfall variability is not confined to the low-rainfall areas of the region. The large amplitude of the variations is typical for the region and makes it more vulnerable to drought.

The patterns of drought in the region are highly variable in their spatial and temporal dimensions. This variability of drought patterns at local level is demonstrated by the cumulative rainfall deviations for Aleppo, Damascus, and Palmyra in Syria (Fig. 7–4).

Although these locations are <200 km apart, the patterns of longer-term positive or negative anomalies are very different. This figure also confirms that,

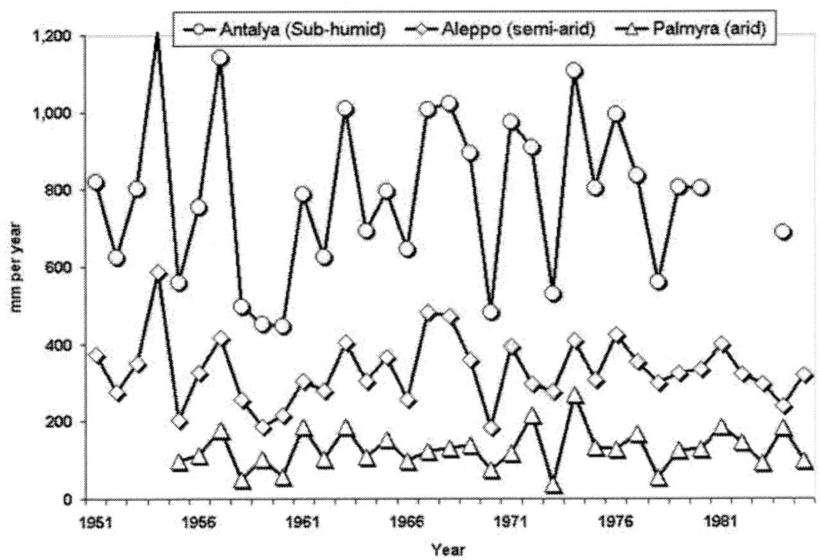

Fig. 7–3. Precipitation fluctuations for three stations under different moisture regimes (1951–1985).

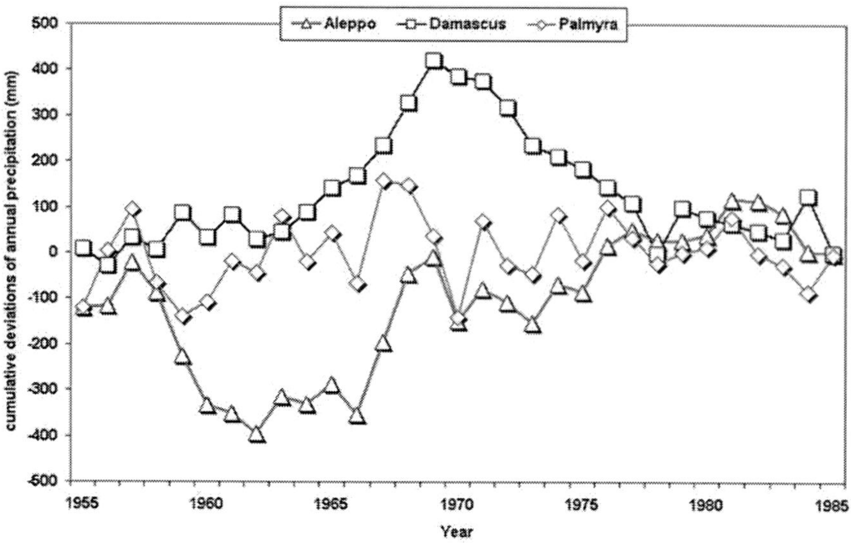

Fig. 7–4. Patterns of positive and negative precipitation anomalies for three locations in Syria.

notwithstanding the regional scale some droughts may take, the impact is usually rather location-specific. For example, in Syria, the drought of 1999 initiated a severe decline in the productivity of the rangelands and barley areas at the steppe margins, which continued for 3 yr. However, the drought had comparatively little effect on the production of wheat (*Triticum aestivum* L.) and tree crops in the higher rainfall areas, <100 km away, which recovered from 2000 onwards.

Throughout the region, drought can strike any time of the growing season. Early-season, mid-season and late-season droughts are all possible. Figure 7–5 shows the occurrences of severe and mild drought within the growing season at Tel Hadya, Syria, for the period 1978 to 2001. For this particular example a severe drought is defined as a precipitation total of <50% of the long-term average, and a mild drought as <70% of the long-term average. The months November to December represent the early season, January to February the mid-season, and March to April the late season.

Water shortage, already a problem in many countries of this arid region, may be amplified by climate change. The Intergovernmental Panel on Climate Change projects for the region small increases in precipitation, but these increases are likely to be countered by increased temperature and evaporation (Watson et al., 1997). According to the Panel, drought is therefore likely to increase and will have the greatest impact on the grasslands, livestock, and water resources of the marginal areas.

Causes of Drought

The causes of drought in the Near East are complex, as they are in most parts of the world. This is to be expected given the geographical extent and exposure to

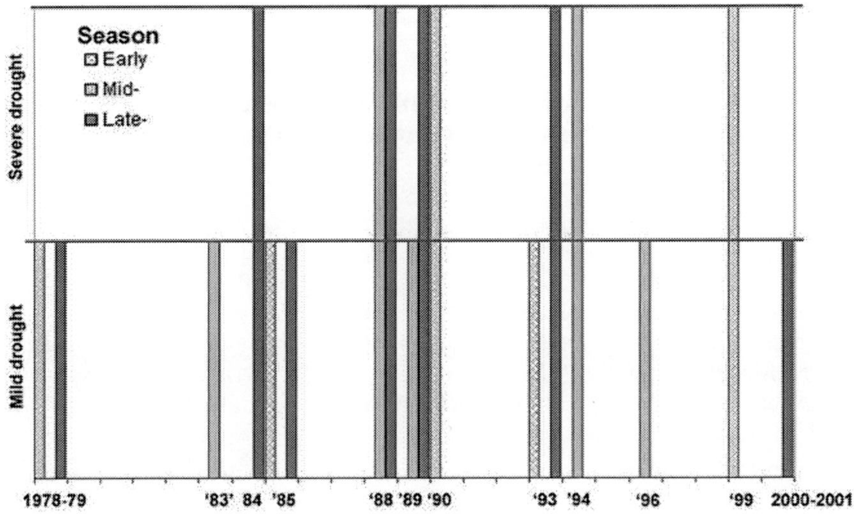

Fig. 7–5. Intra-seasonal droughts at Tel Hadya, Syria.

different oceanic/continental influences and wind systems in the western vs. eastern parts of the region. In North Africa most of the precipitation is generated as a result of depressions that are steered southward from the North Atlantic during blocking episodes by mid-latitude high-pressure cells (Ward et al., 1999). The strength of the blocking is, to a large extent, linked with the North Atlantic Oscillation (NAO) (Lamb and Peppler, 1987), a large-scale mode of climatic variability in the northern hemisphere at monthly, seasonal, inter-annual, and decadal timescales.

A simple index of the NAO is the sea-level air pressure difference between the Azores and Iceland; a positive phase is characterized by strengthened westerly winds across the mid-latitude North Atlantic, leading to mild and relatively wet winters in northern Europe (van Loon and Rogers, 1978), but anomalously dry conditions in the Iberian Peninsula (Zorita et al., 1992) and the Maghreb countries (Lamb and Peppler, 1987). A negative NAO tends to be linked with moist air in the Mediterranean and cold air to northern Europe. The NAO index varies from year to year, but tends to remain in one phase for intervals lasting several years. Since the late 1970s, there has been a tendency for the winter NAO index to be positive (Hurrell, 1995), and for precipitation in subtropical northwestern Africa and the Iberian Peninsula to decline (Ward et al., 1999).

In the Eastern Mediterranean both the Atlantic Ocean and the Mediterranean Sea are the primary source regions for the formation of winter precipitation in the form of mid-latitude cyclones (Turkes, 1996). These migratory low-pressure systems have four primary cyclonic centers near to Crete, Cyprus, southern Italy, and the Gulf of Genoa. Cullen and deMenocal (2000) showed a physical link between precipitation in the Eastern Mediterranean and the NAO and its impact on the stream flow of the Tigris and Euphrates. Their analysis indicates that during positive NAO years, Turkey and—to a lesser extent—northern Syria, become cooler and drier, whereas during negative NAO years anomalous warmer and wetter conditions prevail.

Both in the Maghreb countries, particularly Morocco, and the Eastern Mediterranean a link thus appears to exist with North Atlantic sources of climatic variability. The interactions with other sources of climatic variability are not yet properly established and need further research.

Impact of Drought on Agriculture

Agricultural production in the region is strongly influenced by precipitation fluctuations. This is evidenced by Fig. 7–6, which shows for the period 1961 to 2001 the percentage deviation from the trend production of barley (*Hordeum vulgare* L.) in the Maghreb countries (Morocco, Algeria, and Tunisia). Barley is a very suitable indicator crop because its production nearly always relies on rainfall. If rainfall is inadequate for grain formation, barley is not even harvested. Instead, its biomass is used for sheep (*Ovis aries*) grazing, thereby accentuating the swings in national production statistics. Severe production fluctuations are the rule, and the pattern closely follows precipitation fluctuations. An exception is Turkey, where the production fluctuations are more attenuated because the higher precipitation levels ensure that even in drought years crop failure of barley is unlikely.

Reports of severe drought are common in the region. Cullen et al. (2000) reported droughts in Turkey in 1973, 1984, 1989, and 1990. Prolonged drought periods were also reported in Morocco in 1979 to 1984 (Cullen and deMenocal, 2000) and 1994 to 1995 (Zakaria, 2001). In Tunisia, droughts occurred recently in 1988 to 1989 and 1994 to1995 (Louati et al., 1999).

Starting from 1998 and continuing into 2000, both North Africa and West Asia experienced the worst regional drought in decades. In West Asia, it severely reduced

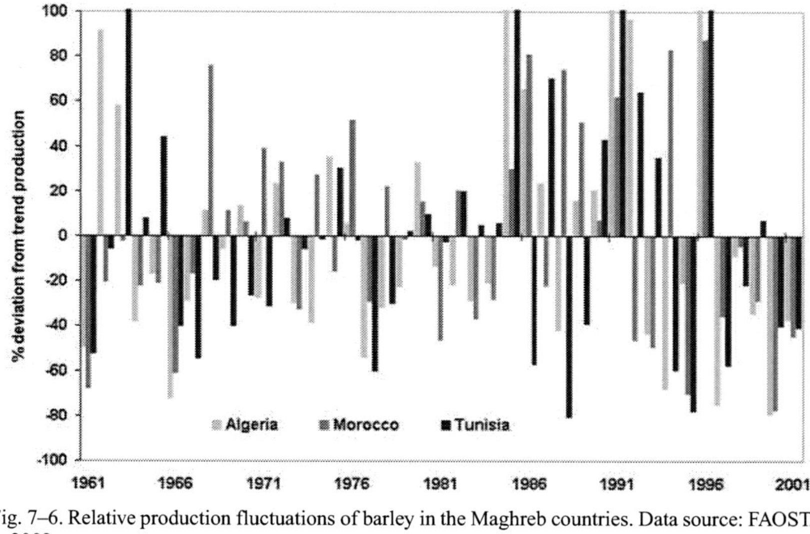

Fig. 7–6. Relative production fluctuations of barley in the Maghreb countries. Data source: FAOSTAT, 2002.

food output in Jordan, Iran, Iraq, and Syria. In 1999, aggregate cereal output in the subregion was 16% lower than in the previous year and 12% lower than the average over the previous 5 yr (1994–1998). In Turkey, which normally contributes approximately 50% of subregional grain production, output fell by 6% as compared to the 5-yr average. As Turkey is also the subregion's main cereal exporter, the volume of exports from the subregion, normally around 5 Tg, declined by about 50% (GIEWS, 1999a). In 2000, the drought forced Iran to import 7 Tg of wheat, making it the world's larger importer (United Nations Interagency Assessment Mission, 2001).

In the Maghreb countries, aggregate cereal production in the period 1990 to 1999 ranged from 4 to 8 Tg in 5 drought years, and 10 to 18 Tg in 5 good years (GIEWS, 2000a). The 1999 cereal crop was also affected by drought, with output estimated at 8 Tg, which was 31% below the previous year's harvest; 2000 was the second consecutive year of reduced harvests in the subregion, particularly in Morocco and Algeria, and lead to further increases in cereal imports, putting more pressure on national budgets (GIEWS, 2000a). While droughts appear already prominently in national-level agricultural statistics, the effects at subnational and local levels can be devastating. The drought episode of 1979 to 1984 in Morocco reduced the small ruminant population by 40 to 50% (Berkat, 2001).

In Iran, the drought, which entered its third consecutive year in 2001, completely destroyed rainfed agriculture in most areas visited by the United Nations Assessment Team (United Nations Interagency Assessment Mission, 2001). More than 200 000 livestock owners lost their only source of livelihood. More than 500 villages in Kerman Province had no drinking water. Many of these villages were abandoned and the population moved to the edges of bigger cities, adding pressure on the urban water supplies (Siadat and Shariati, 2001; United Nations Interagency Assessment Mission, 2001). The same drought reduced the flow in the Tigris and Euphrates rivers in Iraq to about 20% of their average flow, seriously constraining irrigated production, which constitutes more than 70% of cultivated area (GIEWS, 2000b). In Syria, approximately 47 000 nomadic households (329 000 people) had to liquidate their livestock assets, became vulnerable to food shortages, and required urgent food assistance (GIEWS, 1999b). These cases amply illustrate the extent of the social problems induced by drought, and the fact that these problems are certainly not exceptional in the region.

DROUGHT MONITORING SYSTEMS

Drought planning

Comprehensive drought planning requires the integration of functional and institutionalized capabilities related to drought monitoring and early warning, risk assessment, and mitigation and response (Fig. 7–7; Wilhite and Svoboda, 2000). The key principles of drought management are:

- The approach is multidisciplinary and integrated.
- Drought planning must cover different time scales, and include, apart from short-term solutions to mitigate drought impact, also long-term land-use strategies.

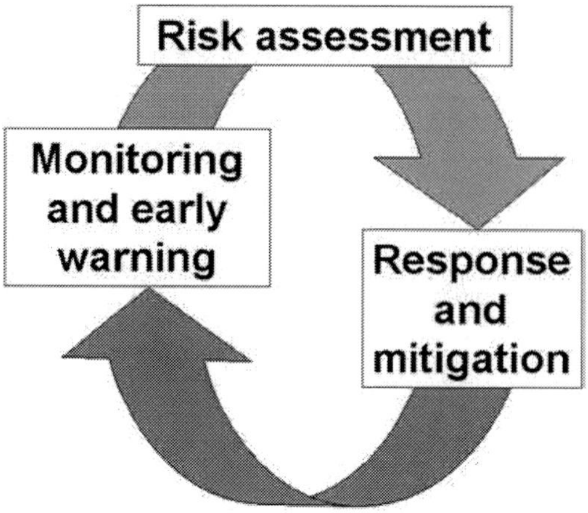

Fig. 7–7. Functional capabilities of comprehensive drought planning.

- Consideration is given to both the biophysical and socioeconomic dimensions of drought.
- Drought management is a key component of sustainable land use in drought-prone areas.

The main strategies identified by ICARDA (1998) for effective short-term and long-term drought mitigation are:

- Seasonal climate forecasts and early warning systems.
- Better targeting of crops and cultivars to specific agroecological environments.
- Natural resource management adapted to the limitations of drought-prone areas, in particular crop breeding and crop management to increase water-use efficiency, soil, and water conservation, including water harvesting, and sustainable use of irrigation water.
- Policy and institutional measures to facilitate implementation of drought mitigation practices, in particular, the conservation and harvesting of water, shifts to more adapted crops etc.

Challenges

The main challenges for comprehensive drought planning in the Near East are:

- Lack of permanent and institutionalized coordination of information from multi-sectoral data collection systems on extent and impact of drought.
- Inadequate drought information products.
- Inadequate characterization of drought through appropriate drought indicators.
- Lack of spatially referenced drought vulnerability profiles.

Institutional Needs

The common approach to drought in the Near East is one of *crisis management* rather than *long-term drought planning*. When a drought becomes apparent, a governmental drought mitigation program is set up, usually steered by an intergovernmental committee, headed by a lead ministry. When the drought subsides, the program is terminated and the committee disbands.

Given this ad-hoc and crisis-oriented approach to drought, there is little room for permanent and integrated drought-monitoring systems. After the drought has been recognized as such, these intergovernmental committees may collect and interpret information about the extent and impact of drought from water supply or irrigation authorities, agricultural extension services, meteorological services, or nongovernmental organizations (NGOs). In most cases, it is very difficult to interpret the patchy and often outdated information emerging from these different sources and parts of the countries and decide on the most appropriate actions for alleviating drought.

For this reason, the countries of the region often appeal for assistance to the specialized agencies of the United Nations, such as the Food and Agriculture Organization (FAO) and the World Food Program (WFP), who have more experience in dealing with these emergencies. This assistance usually takes the form of rapid assessment missions to evaluate the severity and extent of droughts and estimate the food import and emergency assistance requirements.

Experience of well-functioning early warning systems throughout the world, whether for drought or food security, indicates as a common characteristic that they are usually small but tightly integrated units, capable of pulling multidisciplinary information together. Of critical importance for the success of all early warning units is the free flow of information between different governmental and NGOs. To ensure this, a strong coordinating unit, which acts as an interministerial steering committee for the drought monitoring activities in the country, is essential.

The three main functions needed for comprehensive drought planning (Fig. 7–7) can be exercised by a coordinating body in a lead ministry, such as ministries dealing with agriculture or water supply. In other cases, coordination might be more effective if housed in a coordinating ministry, such as a Prime Minister's or President's Office, or a planning ministry, rather than a line Ministry, such as a Ministry of Agriculture, Water Supply, etc. In addition to the coordinating committee, a technical drought early warning unit needs to be established. This unit would compile and interpret all relevant data sources to monitor drought extent and impact, and report through regular or special bulletins to the central drought management unit.

Meteorological data and analytical tools to transform these into relevant drought indicators are critical for the adequate monitoring of droughts. Most meteorological services of the region still define drought as a negative anomaly from normal precipitation, in terms of absolute or percentile deviations (Fig. 7–8).

This information rather serves to characterize the climate during the ongoing year, month, part of the year, or agricultural year, and is not well adapted for defining and monitoring droughts. Well-established drought indicators such as the Palmer Drought Severity Index (Palmer, 1965) or the Standardized Precipitation Index (Guttman, 1999) may be used at research level but are not yet incorporated

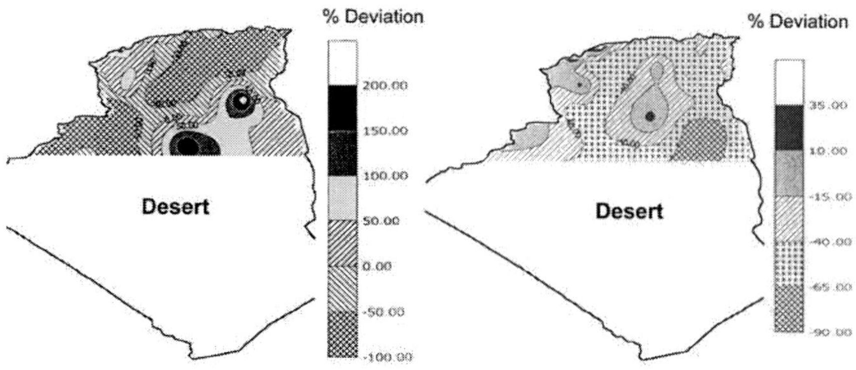

Fig. 7–8. Percentile deviations of precipitation in Algeria, season 2001 to 2002. Left: May 2002; right: cumulative percentile deviations September 2001 to May 2002. Source: ONM, 2002.

in operational drought monitoring. There are no regular bulletins that target the agricultural user community or other stakeholders in drought mitigation through interpretation of the available raw data and communication in terms of impact and seasonal outlook.

Particularly missing in this region are integrated spatial frameworks, such as maps of agroecological and production system zones. Given the tremendous diversity in agroecological conditions and livelihood systems in the region, such information is vital for assessing the vulnerability to and potential impact of drought.

DROUGHT RESEARCH NEEDS

Drought Forecasting

Weather patterns in many parts of the world appear to be related to different phases of the El Niño-Southern Oscillation (ENSO) cycle. Seasonal forecasts are based on this coupling mechanism between sea surface temperatures and global atmospheric circulation. The existence of such linkages is now being used in operational early warning systems, such as the USAID Famine Early Warning System (FEWS), to forecast rainfall patterns for the coming crop growing season in several parts of the world, especially in Sub-Sahara Africa. The seasonal forecasts are produced by global meteorological centers, such as the U.S. Center for Climate Prediction or the U.K. Hadley Meteorological Center, or regional centers, such as the African Center for Meteorological Applications for Development (ACMAD), and transmitted to national meteorological services. Especially in tropical areas, such as Northeast Brazil, South Africa, West Africa, India, and others, it has been established that a significant relationship exists between ENSO and the weather for the coming growing season. However, it has been demonstrated that even in such optimal conditions ENSO is a far from perfect predictor, leading to considerable uncertainty in ENSO-based seasonal climate predictions (Phillips et al., 2002; Podesta et al., 2002).

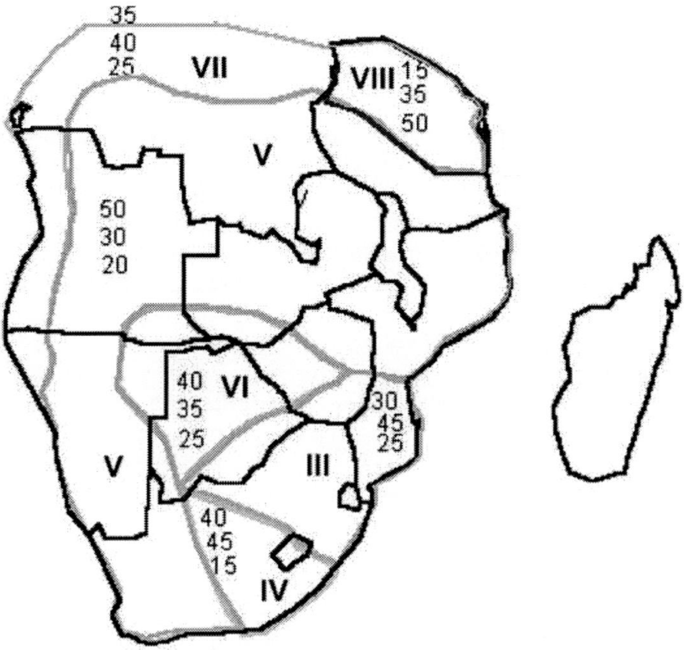

Fig. 7–9. Probabilities of seasonal rainfall patterns for Southern Africa (SADCC, 1999).

In the Near East the relationship between droughts and ENSO events is not firmly established. Ward et al. (1999) see sufficient evidence of an ENSO influence on sea surface temperatures in the Northern Atlantic and on late-season (March–April) precipitation levels in northwestern Africa. Particularly in Morocco, west and north of the Atlas, late-season precipitation would tend to be below normal in warm-event ENSO years. Nevertheless, the consensus is that the ENSO signal in this part of the world is too weak to serve as a reliable predictor of drought. A likely reason is that the effects of this global ocean-weather linkage are substantially modified by more localized weather phenomena, such as the NAO, but also by topography and the nearness of large desert land masses in the Sahara and the Arabian Peninsula, and of enclosed water bodies (Black Sea and Caspian Sea).

A first priority is therefore to improve the reliability or 'skill' of the forecasts in the Near East. In view of the importance of Atlantic sources of rainfall, forecasting skills would depend on whether the NAO, like the ENSO, is a coupled ocean-atmosphere phenomenon, as opposed to a random atmospheric phenomenon. If the former were the case, it would enhance predictability and allow the use of a NAO index in drought early warning systems in the Near East (Cullen and deMenocal, 2000).

In the application of forecasts a cautionary approach is appropriate. For a start, forecasts are statements of possible seasonal outcomes in terms of a set of probabilities. The problem with this kind of information is that it is not unequivocal and requires risk taking.

Figure 7–9 represents, for different climatic regions in Southern Africa, the outlook for rainfall during the period January to March 2000, the most important months of the growing season, on the basis of a forecast made in September 1999 (SADCC, 1999). The top figure for each climatic region represents the probability of above-normal rainfall, the middle figure the probability of normal rainfall, and the bottom figure the probability of below-normal rainfall. Especially in climatic Region III, with probability set 30-45-25, and climatic Region VII, with probability set 35-40-25, it is very difficult to make use of the forecast.

An important issue is therefore how probabilistic forecast information can be incorporated into farmers' decision-making processes. Ingram et al. (2002) give an example of potential response strategies that farmers could implement in response to the receipt of rainfall forecasts and their timing.

Irrespective of the outcome of this research, it is expected that the contribution of seasonal forecasts to the stabilization of agricultural production in the Near East will be relatively modest, as compared to other management strategies adapted to highly variable climates. In fact, as Phillips et al. (2002) point out, instead of dampening them, seasonal forecasts could make production swings even more pronounced.

Drought Characterization

Drought has meteorological, agricultural, hydrological, and socioeconomic dimensions. Different kinds of drought require different indicators. Some indicators are more suited to monitor agricultural drought, others to assess hydrological or meteorological drought. The assessment of socioeconomic drought requires socioeconomic and nutrition-based indicators.

The Standardized Precipitation Index (Guttman, 1999) is an example of a much-used index of meteorological drought. It is simple, can be applied to assess severity of drought at different time scales (Fig. 7–10), and is comparable across different climatic zones. Since it is based on fitting meteorological parameters to probability distributions, it is very suitable for assessing the exceptionality of drought events and beginning and end of a drought, and is thus useful as a trigger for emergency assistance, crop insurance etc.

To monitor agricultural drought, the most suitable indicators are those that are responsive to soil moisture status, and are therefore based on the soil water balance. The reason is that many rainfed crops are sensitive to the timing of soil moisture deficits in relation to crop water requirements and that soil moisture deficits do not necessarily coincide with precipitation deficits. An example of a typical indicator that can be used for agricultural drought monitoring is given in Fig. 7–11. It shows for three growing seasons with different rainfall pattern the cumulative deviations of the actual evapotranspiration (AET) from normal. This indicator is simple, and can be derived from climatic data, a generic water balance model or remote sensing, and allows rapid visualization of the quality of a growing season. The cumulative deviations of AET are a measure of drought intensity, and to some extent drought impact. In addition, they can be carried over from one year to another, indicating patterns of water deficit or surplus that may occur over longer time spans. The AET-based indicators are useful for rapid identification of areas with

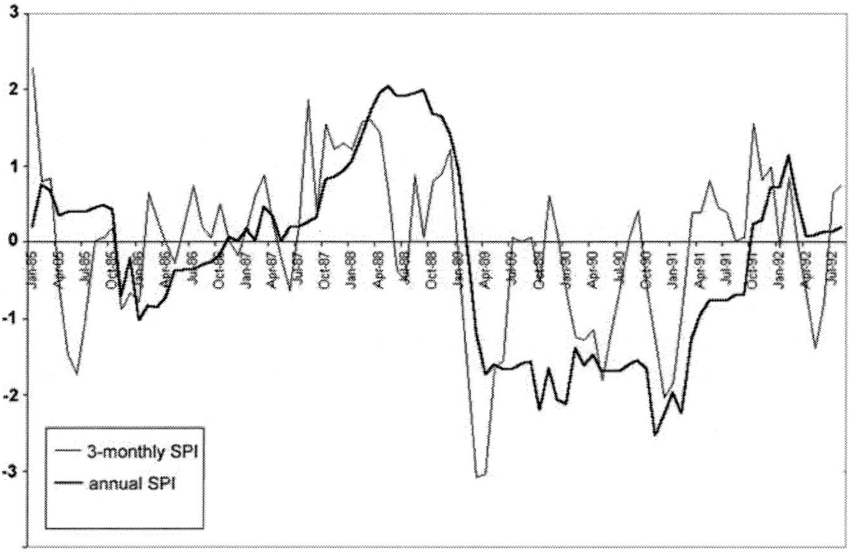

Fig. 7–10. Moving Standardized Precipitation Indices (SPI) for Tel Hadya, Syria, on a 3-monthly and an annual basis, from 1985 to 1991.

severe drought problems. Where meteorological networks are adequate they can be easily spatialized. In addition, in view of the strong correlations between vegetation biomass density and AET, there have been successful attempts to derive AET directly through remote sensing (e.g., Bastiaanssen, 1995; Rosema, 1993).

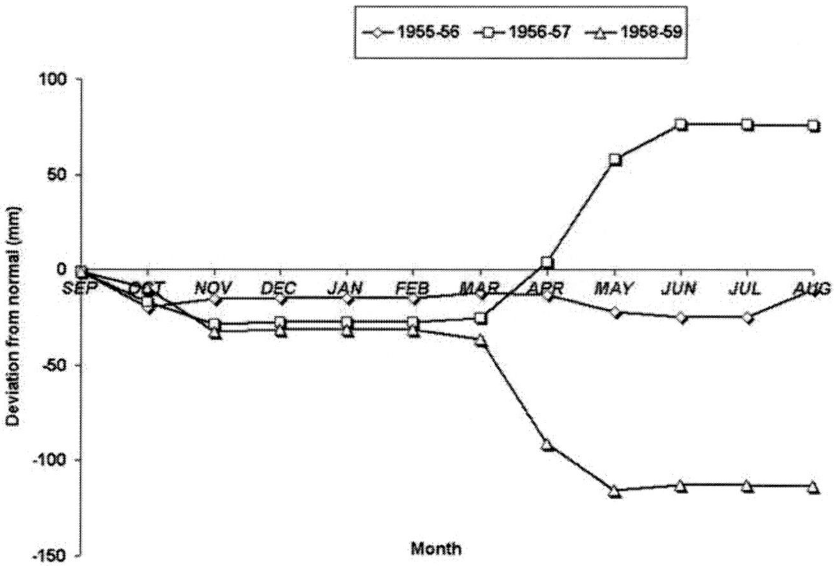

Fig. 7–11. Cumulative deviations of actual evapotranspiration for three growing seasons, Aleppo, Syria

Spatialization of Drought in Data-scarce Areas

The Near East is a region with relatively sparse and poorly distributed data coverage. With the exception of Turkey, which has a good and homogenous coverage throughout the country, most climatic stations in the region are concentrated in coastal and agriculturally important areas. Rangeland (arid) areas and deserts are very poorly covered. However, owing to the pressure of population increase, these areas, which are marginal from an agricultural perspective, are becoming increasingly important from an ecosystem function perspective. As the region is also very diverse in terms of landscapes and topography, temperature regimes are not uniform and, in view of their effect on crop water demand, have to be taken into consideration when using water balance-based drought indicators.

In such environments, characterized by high spatial variations in moisture and temperature regimes, the delineation of drought can be considerably improved by advanced methods of 'topography-guided' spatial interpolation. Several statistical techniques are now available that make use of digital elevation models (DEM) to improve the spatialization of climatic parameters (De Pauw et al., 2000). In view of the strong linkages between climatic variables (especially temperature, but also rainfall, humidity, and sunshine) and topography, the most promising techniques for spatialization in climatology are multivariate approaches that permit the use of terrain variables as auxiliary variables in the interpolation process.

In contrast to the climatic target variables themselves, which are only known for a limited number of sample points, terrain variables have the advantage to be known for all locations in-between, which increases the precision of the interpolated climatic variables significantly. Co-kriging (e.g., Bogaert et al., 1995) and co-splining (e.g., Hutchinson, 1995) are methods that in most cases lead to excellent interpolations. The co-splining approach of Hutchinson in combination with the GTOPO30 digital elevation model (Gesch and Larson, 1996) has been used for regional-level mapping of various basic and derived climatic variables at 1 km resolution (De Pauw, 2002). Where time series of climatic data exist, climate 'surfaces' can be derived from point data and, using appropriate drought indicators, converted into maps of drought severity.

Another important tool for spatialization is remote sensing. Remote sensing has become a standard tool in several food security early warning systems, such as the Famine Early Warning System (FEWS), the Global Information and Early Warning System on Food and Agriculture (GIEWS), and the Monitoring Agriculture with Remote Sensing Project (MARS). This development has been promoted by the decreasing costs of satellite data products and image analysis tools, the difficulty to obtain timely climatic data, and significant correlations between soil moisture status or biomass productivity and some vegetation indices derived from spectral analysis (e.g., the Normalized Difference Vegetation Index, NDVI).

Radiances measured by the National Oceanic and Atmospheric Administration (NOAA) and Meteosat meteorological satellites respond closely to changes in leaf chlorophyll, moisture content, and thermal conditions. By its synoptic view and rapid refresh capability, the information stream from the Advanced Very High Resolution Radiometer (AVHRR) on the NOAA polar-orbiting satellites information stream offers a unique ability to integrate the effects of changing weather, vegetation, soil, and land use. A global monitoring system of drought, based on the combination of three spectral channels of the AVHRR (visible, near-infrared, and infrared) into moisture, thermal, and vegetation health indices, has been developed by the National Environmental Satellite Date and Information Services of the National Oceanic and Atmospheric Administration (NOAA/NESDIS) and generates weekly map products of drought indicators with 16 km^2 resolution (Kogan, 2000). While the AVHRR sensors are expected to remain operational for the near future, a successor of AVHRR, the MODIS instrument on the Terra satellite launched in December 1999, is now producing imagery with NDVI at 250 m resolution and other vegetation indices at 500 m or 1 km resolution.

Meteosat-derived products, which include estimates of precipitation, global, and net radiation, actual and potential evapotranspiration and sensible heat flux, with 5-km spatial resolution and 10-daily or monthly time steps, are now commercially available. These basic products can be combined into models that generate drought indices for crop forecasting purposes.

Drought Risk and Vulnerability Mapping

A first task for an early warning system is to understand the spatial variations of drought risk. If good time series exist of spatially well-distributed climatic stations, drought risk can be spatialized from the probability surfaces of selected drought indicators, such as the Standardized Precipitation Index (SPI) or simulated crop yields.

Figure 7–12 shows an example of an *ex-ante* risk drought risk assessment by a simulation of 200 growing seasons using the SIMTAG wheat model (Stapper, 1993), assuming planting of the variety Potam on or after 15 November each year. The contrast between the two maps shows the extreme variability of crop yields from year to year with which farmers have to cope, as a result of highly variable weather during the growing season (W. Göbel, personal communication, 2000).

By analyzing the inter-annual variations in vegetation indices derived from remote sensing it is also possible to identify areas with particularly high risk of

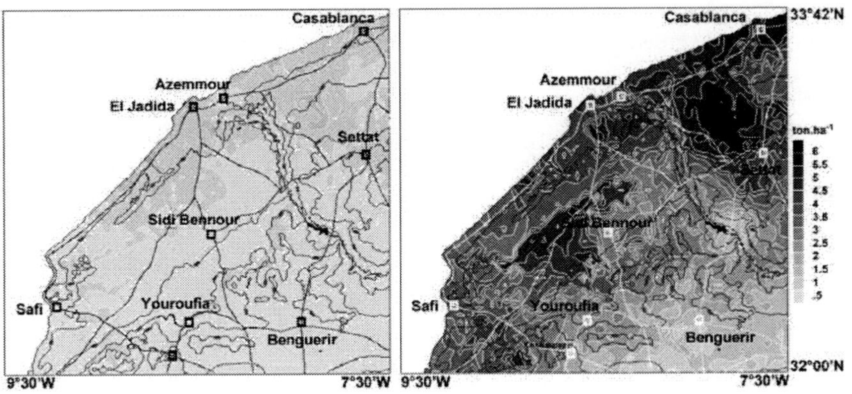

Fig. 7–12. Doukkala, Chaouia, and Abda Regions (Morocco): first (left) and ninth decile (right) of potential wheat grain yields.

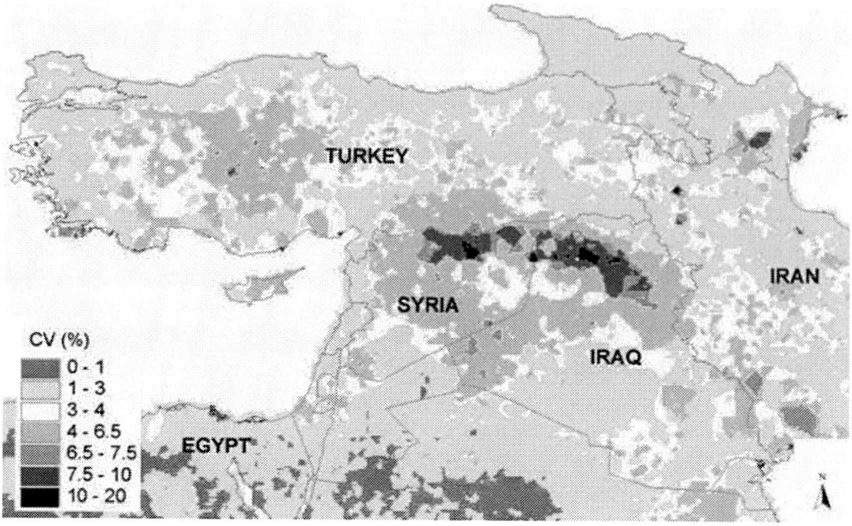

Fig. 7–13. 'Hot spots' of drought vulnerability in West Asia. Source: ICARDA Geographical Systems Laboratory.

drought. Figure 7–13 shows the coefficient of variation of the NDVI derived from a 8-km resolution AVHRR time series for West Asia for the period 1982 to 2000, after correction for changes in land cover.

However, for drought planning it is also essential to go beyond the symptoms of drought, as they appear from the meteorological or hydrological records. As experience in the region shows, access to and the stability of the natural resource capital—particularly natural vegetation, climate, soil, and irrigation water—are major determinants of the resilience of rural livelihood systems against drought. Understanding the underlying causes of vulnerability and anticipating the impact of drought thus requires an integrated approach, which considers both the differences in agroecological and socioeconomic characteristics between different areas.

The basis for mapping vulnerability to drought could be a spatial framework of combined agroecological zones and production systems zones (Fig. 7–14). The agroecological zones can be established by integration of available climatic, soil, terrain, and land cover digital datasets. The production system zones can be derived from remote sensing in combination with farming systems information. By integrating the spatial agroecological and socioeconomic data, 'agro-ecozones' can be established, which have unique characteristics in terms of climate, soil, and water resources, population characteristics and livelihood systems. The agroecozones offer a very useful framework for selecting sampling areas as part of a regular drought monitoring program.

CONCLUSIONS

To combat drought effectively, there must first be recognition that it is an entirely natural phenomenon of dryland environments. The most effective buffering

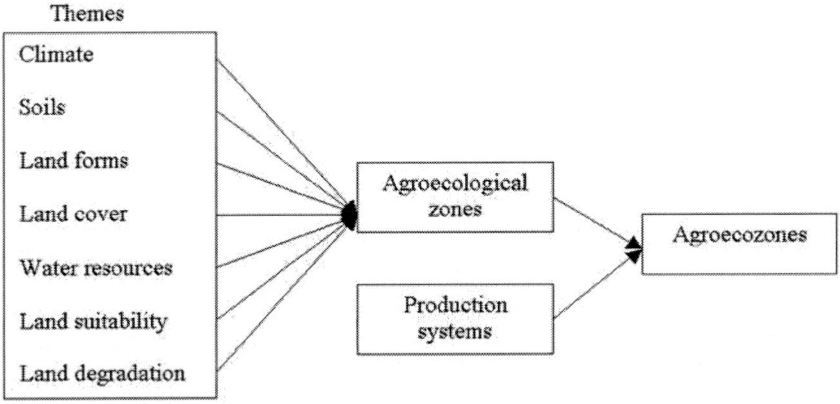

Fig. 7–14. Spatial framework for mapping vulnerability to drought.

against drought is therefore in applying proven dryland management principles for their sustainable use. This would translate, for example, into not growing commodity crops in marginal environments, into reducing water consumption, into grazing rangelands according to their carrying capacity. Drought early warning systems play a crucial role in any long-term management strategy for drylands in the Near East. A major responsibility is to facilitate the free flow of information between institutional and nongovernmental agents in drought monitoring and management. There is a need to make optimal use of multi-thematic, but scarce, information. In this respect, there is much scope, also from the research vantage point, to enhance the explanatory power of representative, but scattered, meteorological stations by extrapolation using remote-sensing platforms. Particularly for this region, drought forecasting skill needs to be improved to be useful for farmer management decisions. Finally, there is a major need to understand the agroecological and livelihood factors that affect the vulnerability to drought of specific population segments.

REFERENCES

Bastiaanssen, W.G.M. 1995. Regionalization of surface flux densities and moisture indicators in composite terrain: A remote sensing approach under clear skies in Mediterranean climates. Ph.D. thesis, Landbouwuniversiteit, Wageningen, The Netherlands.

Berkat, O. 2001. Evaluation de l'impact de la sécheresse sur les parcours et systèmes d'élevage. Lecture module for the Advanced Course on Management Strategies to Mitigate Drought in the Mediterranean: Monitoring, Risk Analysis and Contingency Planning, Rabat, Morocco. 21–26 May 2001. CIHEAM-IAMZ, Zaragoza, Spain, and IAV Hassan II, Rabat, Morocco.

Bogaert, P., P. Mahau, and F. Beckers. 1995. The spatial interpolation of agro-climatic data: Co-kriging software and source data. User's manual. Food and Ariculture Organization Agrometeorology Series 12. FAO of the United Nations, Rome, Italy.

Cullen H.M., and P.B. deMenocal. 2000. North Atlantic influence on Tigris-Euphrates streamflow. Int. J. Climatol. 20:853–863

De Pauw, E. 2002. An agroecological exploration of the Arabian Peninsula. International Center for Agricultural Research in the Dry Areas, Aleppo, Syria.

De Pauw, E., W. Goebel, and H. Adam. 2000. Agrometeorological aspects of agriculture and forestry in the arid zones. Agric. Forest Meteorol. 103:43–58

Food and Agriculture Organization STAT. 2002. FAO statistical databases. FAO of the United Nations, Rome, Italy [Online]. Available at http://apps.fao.org (verified 20 Feb. 2004).

Gesch, D.B., and K.S. Larson. 1996. Techniques for development of global 1-kilometer digital elevation models [Online]. Available at http://edcdaac.usgs.gov/gtopo30/README.html (verified 13 Dec. 2003).

Global Information and Early Warning System. 1999a. Special report: Drought causes extensive crop damage in the Near East raising concerns for food supply difficulties in some parts. 16 July 1999 [Online]. Food and Agriculture Organization GIEWS on food and agriculture, FAO of the United Nations, Rome, Italy. Available at http://www.fao.org/giews/english/alertes/1999/SRNEA997.htm (verified 13 Dec. 2003).

Global Information and Early Warning System. 1999b. Special report: FAO/WFP crop and food supply assessment mission to the Syrian Arab Republic. 23 Aug. 1999 [Online]. Food and Agriculture Organization GIEWS on food and agriculture. FAO of the United Nations, Rome, Italy. Available at http://www.fao.org/giews/english/alertes/1999/SRSYR99.htm

Global Information and Early Warning System. 2000a. Special Alert 304. 5 April 2000 [Online]. Food and Agriculture Organization GIEWS on food and agriculture. FAO of the United Nations, Rome, Italy. Available at http://www.fao.org/giews/english/alertes/2000/SA304NAF.htm (verified 13 Dec. 2003).

Global Information and Early Warning System. 2000b. Special alert 308. 11 May 2000 [Online]. Food and Agriculture Organization GIEWS on food and agriculture. FAO of the United Nations, Rome, Italy. Avalable at http://www.fao.org/giews/english/alertes/2000/SSSA308m.htm

Guttman, N.B. 1999. Accepting the standardized precipitation index: A calculation algorithm. J. Am.Water Res. Assoc. 35 (2):311–322

Hurrell. J., 1995. Decadal trends in the North Atlantic Oscillation: Regional temperatures and precipitation. Science (Washington DC) 269:676–679.

Hutchinson, M.F. 1995. Interpolating mean rainfall using thin plate smoothing splines. Int. J. Geogr. Info. Syst. 9:385–403.

International Center for Agricultural Research in the Dry Areas. 1998. Drought preparedness and mitigation of drought effects. Paper presented at the International Expert Group Meeting for the Preparation of a Sub-Regional Action Programme on Combating Desertification and Drought in Western Asia. Muscat, Oman. 14–16 Sept. 1998. ICARDA, Aleppo, Syria

Ingram, K.T., M.C. Roncoli, and P.H. Kirshen. 2002. Opportunities and constraints for farmers of west Africa to use seasonal precipitation forecasts with Burkina Faso as a case study. Agric.Syst. 74:331–349

Knutson C., M. Hayes, and T. Phillips. 1998. How to reduce drought [Online]. Western Drought Coordination Council, Denver, CO. Available at http://www.drought.unl.edu/plan/handbook/risk.pdf (verified 13 Dec. 2003).

Kogan, F.N. 2000. Contribution of remote sensing to drought early warning. p. 75–87. *In* D.A. Wilhite et al. (ed.) Early warning systems for drought preparedness and drought management. WMO/TD No. 1037. World Meteorological Organization, Geneva, Switzerland.

Lamb, P.J., and R.A. Peppler. 1987. North Atlantic Oscillation: Concept and application. Bull. Am. Met. Soc. 68:1218–1225.

Louati, M. el Hedi, R. Khanfir, A. Al-Ouini, M.L. El Echi, L. Erigui, and A. Marzouk. 1999. Guide pratique de gestion de la sécheresse en Tunisie. Approche méthodologique. Ministère de l'Agriculture, République Tunisienne.

Margat, J., and D. Vallée. 2000. Mediterranean vision on water, population and the environment for the 21st century [Online]. Programme des Nations Unies pour l'Environment. Programme pour l'Alimentation Mondiale. Plan Bleu, Valbonne, France. Available at http://www.planbleu.org/indexa.htm (verified 13 Dec. 2003).

Office National de la Météorologie. 2002. Bulletin mensuel d'informations climatologiques, Mai 2002. ONM, Ministère des Transports, Algers, Algeria.

Palmer, W.C. 1965. Meteorological drought. Res. Paper No. 45. Weather Bureau, Washington, DC.

Phillips, J.G., D. Deane, L.Unganai, and A. Chimeli. 2002. Implications of farm-level response to seasonal climate forecasts for aggregate grain production in Zimbabwe. Agric. Syst. 74:351–369.

Podesta, G., D. Letson, C. Messina, F. Royce, R.A. Ferreyra, J. Jones, J. Hansen, I. Llovet, M. Grondona, and J.J. O'Brien. 2002. Use of ENSO-related climate information in agricultural decision making in Argentina: A pilot experience. Agric. Syst. 74:371–392.

Rosema, A.1993. Using METEOSAT for Operational Evapotranspiration and Biomass Monitoring in the Sahel region. Remote Sens. Environ. 45:1–25.

Siadat, H., and M.R. Shariati. 2001. Drought in Iran: A country report. Internal document. Min. of Agric., Tehran, Islamic Republic of Iran.

South African Development Co-ordination Conference. 1999. Statement from the Southern African Regional Climate Outlook Forum, 13–17 Sept. 1999, Maputo, Mozambique. Southern African Regional Climate Outlook Forum, Harare, Zimbabwe.

Stapper, M. 1993. SIMTAG: Simulation model of wheat genotypes. Model documentation. Univ. of New England, Armidale, Australia and Int. Center for Agric. Res. in the Dry Areas, Aleppo, Syria.

Turkes, M. 1996. Spatial and temporal analysis of annual rainfall variations in Turkey. Int. J. Climatol. 16:1057–1076.

United Nations Educational, Scientific and Cultural Organization. 1979. Map of the world distribution of arid regions. Map at scale 1:25,000,000 with explanatory note. ISBN 92-3-101484-6. UNESCO, Paris.

United Nations Interagency Assessment Mission. 2001. United Nations interagency assessment report on the extreme drought in the Islamic Republic of Iran. Tehran, Islamic Republic of Iran.

van Loon, H., and J.C. Rogers. 1978. The seasaw in winter temperatures between Greenland and northern Europe. Part I: Winter. Monthly Weather Rev.104:365–380.

Ward, M.N., P.J. Lamb, D.H. Portis, M. El Hamly, and R. Sebbari, 1999. Climate variability in Northern Africa: Understanding droughts in the Sahel and the Maghreb. *In* A. Navarra (ed.) Beyond El Niño—Decadal variability in the climate system. Springer Verlag, Berlin.

Watson, R.T., M.C. Zinyawere, R.H. Moss, and D. Dokken (ed.) 1997. The regional impacts of climate change. An assessment of vulnerability. Special report of IPCC Working Group II. Intergovernmental Panel on Climate Change, Washington, DC.

Wilhite, D.A., and M.D. Svoboda. 2000. Drought early warning systems in the context of drought preparedness and mitigation. p.1–16. *In* D.A.Wilhite et al. (ed.) Early warning systems for drought preparedness and drought management. WMO/TD No. 1037. World Meteorol. Org., Geneva, Switzerland

Zakaria, A. 2001. Gestion des effets de la sécheresse au Maroc. Paper presented at the Expert Consultation and Workshop on drought mitigation for the Near East and the Mediterranean. 27–31 May 2001. ICARDA, Aleppo, Syria.

Zorita, E., V. Kharin, and H. von Storch. 1992. The atmospheric circulation and sea surface temperature in the North Atlantic area in winter: The interaction and relevance for Iberian precipitation. J. Clim. 5:1097–1108.

8 Dryland Agriculture on the Canadian Prairies: Current Issues and Future Challenges

Francis J. Larney, H. Henry Janzen, and Elwin G. Smith

Agriculture and Agri-Food Canada, Research Centre[1]
Lethbridge, Alberta

Darwin W. Anderson

University of Saskatchewan
Saskatoon, Saskatchewan

ABSTRACT

During its short existence on the Canadian prairies, dryland agriculture has faced formidable challenges from drought and wind erosion. Conservation tillage practices, widely adopted by farmers in the 1990s, have reduced the reliance on summer fallow and enhanced soil quality. Additionally, simple wheat (*Triticum aestivum* L.)-fallow rotations have evolved into longer more diverse ones which include oilseed and pulse crops. While the old challenges of drought and wind erosion will always be part of dryland farming on the Canadian prairies, new and emerging challenges include protection of water quality and ecosystem health, integration of the livestock sector, reduced reliance on the use of agrochemicals, mitigation of greenhouse gas emissions, climate change adaptation, and farming in a global economy. Tackling these challenges will ensure that dryland prairie agriculture will continue to provide much of Canada's primary and value-added food production in a cost-efficient way to meet future demands of an increasing world population.

INTRODUCTION

Agriculture began on the Canadian prairies only recently. Whereas many of the world's farmlands have been cropped for centuries, even millennia, most of the western Canadian lands were first cultivated within about the last 100 yr. As a result, the soils of the region, as well as the practices imposed on them, have changed markedly in a comparatively short time—from native vegetation about a century ago, to modern farming practices today. Many of the changes in farming practices

[1] Lethbridge Research Centre contribution no. 38703045.

 Challenges and Strategies for Dryland Agriculture. CSSA Special Publication no. 32.

have reflected farmer's (and researcher's) struggles to adapt to challenging conditions, notably the aridity of the climate and the shortness of the growing season. Because agriculture in this region is quite recent, most of the changes in practices have been documented quite well; record-keeping and scientific research were already in place when the plows began to open the prairie sod. Our objective is to review the development of cropping practices on the Canadian prairies. We focus especially on recent developments—since the 1980s—and contemplate briefly how practices might continue to change into the future.

DESCRIPTION OF THE AGRICULTURAL REGION

The Canadian prairie, an area of 500 000 km^2, stretches from the border with the USA (latitude 49°N) to the boreal forest in the north (Fig. 8–1) in the provinces of Alberta, Saskatchewan, and Manitoba. The region represents the northern extent of the Great Plains of North America and comprises >80% of Canada's agricultural land base. Although not contiguous, the arable lands of northwestern Alberta and northeastern British Columbia, associated with the Peace River and its tributaries, are generally included as part of the Canadian prairies. The landscape is predominantly rolling and slopes eastward from ~1200 m above sea level along the Rocky Mountains in Alberta to < 300 m above sea level in the Red River valley of Manitoba (Fig. 8–1).

Land Resources

Soils of the Canadian prairies range in type from Brown to Dark Brown Chernozems in southern Saskatchewan and Alberta, to Black Chernozems in Manitoba, Dark Gray in the central prairies and Gray Luvisolic in the north (Fig. 8–1). The native vegetation of the dry Brown and Dark Brown soils consists mainly of xero-

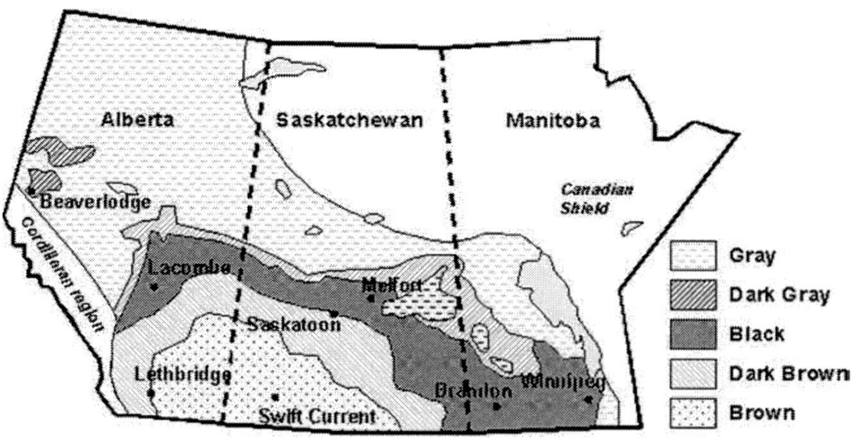

Fig. 8–1. Soil zones of the Canadian prairies.

phytic and mesophytic grasses and forbs. The Black soils occur in the fescue prairie-aspen grove (parkland) and true prairie grasslands, whereas the Dark Gray soils are located in the transitional areas between grasslands and forests. The Gray soils developed under mixed deciduous and evergreen forests, on highly basic, mineral parent materials in subhumid to humid, mild to very cold climates.

Most of the soils developed on glacial deposits from the late Wisconsinan (Wurm) glaciation, mainly on medium-textured glacial till and on silty and clayey materials that were deposited in large glacial lakes. There are local areas of gravelly and sandy sediments, deposited by glacial melt waters. Soils are not strongly weathered and are generally fertile. Organic matter contents of the Ah horizons of uncultivated soils range from 4 to 5% in the Brown soils, 6 to 8% in the Dark Brown soils, and >10% in the Black soils (Anderson et al., 1986). Solonetzic soils occur in lowlands, mainly in south-central Saskatchewan and central and eastern Alberta, where sodium-charged groundwaters influence soil development. A more detailed review of land resources was provided by PFRA (2000).

Some lands in the region have been irrigated since the early 1900s (Ellert, 1995). They are devoted to intensive agriculture and are highly productive, but occupy only a small portion of the agricultural area—about 0.6 million (Mn) ha of the 55 Mn ha of cropland on the Canadian prairies (Statistics Canada, 2002).

Climate

The climate of the prairies is cold and semi-arid to subhumid, and has a strong influence on the distribution of soils (Fig. 8–1; Campbell et al., 1990). The driest area (Brown soil zone of southeastern Alberta and southwestern Saskatchewan) has an annual precipitation of 300 to 350 mm and a water deficit of 400 mm. Dark Brown soils occur in slightly more moist areas. The subhumid regions in the eastern prairies receive up to 500 mm of annual precipitation with water deficits of <200 mm. Roughly 50% of the annual precipitation falls during the growing season (May–September), with 30% falling as snow in the winter months. Growing seasons are short with frost-free periods ranging from ~120 d in southern Manitoba to only 75 to 90 d in the north-central part of the region. The ecoregions of the northern Great Plains (Padbury et al., 2002) broadly reflect these climatic influences.

CROPPING SYSTEMS

Cultivation History

The prairies of western Canada remained uncultivated, apart from small-localized settlements, until the 1880s when the Canadian Pacific Railway was built across the southern prairies. The railway provided access to a large expanse of native prairie that was rapidly cultivated by settlers from Europe, the USA, and eastern Canada. The rapid pace of cultivation can be seen in the area seeded to grain (Mn of ha): <0.1 in 1880, 0.5 in 1890, 1.1 in 1900, 5.0 in 1910, and 11.6 in 1920 (Strange, 1954; area includes wheat [*Triticum aestivum* L.], barley [*Hordeum vulgare* L.], and oat [*Avena sativa* L.]).

In the early years (and still today), the greatest obstacle facing the settlers was the sparse and unpredictable precipitation. A partial solution, widely adopted almost from the outset, was the use of summer fallow, a technique in which the land is left unplanted for an entire year. During that 'fallow' period, weeds were often controlled through intensive tillage, leaving the land black and bare. This practice had numerous advantages: it helped control weeds and made planting easier. But most importantly, it helped replenish soil moisture during the fallow period, so that the next crop was almost certainly spared from severe drought. So beneficial was this practice, that in the drier regions, almost half the land was kept in fallow in a given year; in moister regions, the fraction was almost a third.

Summer fallow may have rescued early farming on the prairies; but soon, it threatened its demise (Janzen, 2001). The intensive tillage and absence of crop growth quickly diminished surface residues, resulting in depleted fertility and severe wind erosion. While the soil-depleting effects of fallow were recognized very early (Shutt, 1910, 1925), there were few alternatives. As Bracken (1915) put it: "*If the fallow dissipates organic matter and nitrogen – and we regret that it does to a very serious degree – then we shall dissipate organic matter and nitrogen until we find a better way because we must have water in the soil and the fallow is the best way to get it there.*" (Cited by Janzen, 2001).

Drought, wind erosion, and depressed markets had a devastating effect on prairie agriculture in the 1930s. Conditions were most severe in the drier Brown and Dark Brown soil zones, where consecutive crop failures and incessant dust storms resulted in the abandonment of many farms. An estimated 10 000 farm families left this area and moved to the moister region on the northern fringe of agriculture, where they began new farms on forested lands (MacEwan, 1986). Others simply stopped farming and moved to cities. With better farming practices, increased precipitation, de-population of the most drought-prone lands, and enhanced markets, the productivity and prosperity of farmlands improved again after the precarious 1930s.

Conservation Tillage

Soon after severe erosion started to deplete and devastate the lands in the early 1900s, farmers and researchers recognized the importance of keeping crop residues on the soil surface. This effort, begun many decades ago, culminated in the development of conservation tillage. This umbrella term includes *zero tillage* (ZT) or *no-till* with low disturbance seeders (also referred to as 'direct seeding' in recent years) as well as *minimum* or *reduced tillage* which allows some tillage (mostly after fall harvest) but maintains surface residue for erosion protection. Conservation tillage on the Canadian prairies was rarely an abrupt change from previous intensive tillage ('conventional tillage'); usually the land had previously been managed by infrequent chisel plowing, wide-blade cultivation or disking, often with deliberate effort to retain residues on the surface.

In recent years, there has been widespread adoption of no-till direct seeding on the Canadian prairies (Larney et al., 1994b; Lindwall et al., 1994). Within 5 yr, its area has increased from 20.9% in 1996 to 31.4% in 2001 (Statistics Canada, 1997, 2002). This trend can be attributed to a yield advantage; the low cost of glyphosate (a herbicide used for weed control during fallow or preseeding), reduced labor, fuel, and equipment requirements, and to a lesser extent improved soil moisture and soil quality conservation (Gray et al., 1996; Zentner et al., 1996c). Moreover, farmer organizations in each province have actively promoted direct seeding through annual workshops and field days. These include the Alberta Conservation Tillage Society, the Saskatchewan Soil Conservation Association, and the Manitoba-North Dakota Zero-Till Farmers Association (Domitruk and Crabtree, 1997). Since conservation tillage has now been widely adopted on the Canadian prairies, most research in the last decade has focused on 'fine-tuning' this management system rather than comparing it with conventional tillage. This 'fine-tuning' aims to enhance the response of traditional prairie crops (e.g., wheat) as well as incorporate new crops into rotations.

Lafond et al. (1996) identified a need to develop crop-specific conservation production practices for each agroecological zone, advocating further research on optimizing stubble height and row spacing to enhance water conservation and determine the impact of such changes on crop growth and development, weeds, and plant disease. Xie et al. (1998) found the best crop performance with paired-row seed/fertilizer placement coupled with 25- to 38-cm row spacings for ZT spring wheat and canola (*Brassica* spp.). However, there was no yield difference between wide (30 cm) and narrow (10 cm) row spacing for ZT spring and winter wheat in the thin Black soil zone in southeast Saskatchewan (Lafond, 1994; Lafond and Gan, 1999). In the semi-arid Brown soil zone of southwestern Saskatchewan, wheat yield, and water-use efficiency was greater with 20 cm rather than 30-cm row spacing (McConkey and Miller, 1999).

Effective fertilizer management is a critical component of any cropping system and aims to maintain economic production while protecting long-term environmental quality. Grant et al. (2001a) found that for soils with high yield potential, enhanced immobilization and/or volatilization of surface-applied N may reduce grain yield of durum wheat (*T. durum* L.) by lowering available N under reduced tillage. They suggested that source, timing, and placement combinations to optimize crop yield may be more important under reduced tillage than conventional tillage.

A point often stressed in direct seeding literature is that "Residue management in direct seeding systems begins at harvest". Improper residue management can reduce the yield of subsequent crops through problems at seeding, such as 'hairpinning', plugging, nutrient tie-up, and toxicity. Also, standing stubble, and its height, alters the microclimate (Cutforth and McConkey, 1997; Cutforth et al., 2002). Seeding wheat into 30-cm tall stubble increased grain yield and water-use efficiency by about 12% compared to wheat seeded into cultivated stubble. Changes in the microenvironment associated with increased surface residue of direct seeding systems may also result in changes to populations of disease-causing agents and other microorganisms. However, Bailey et al. (1998) found that wider row spacings and higher seeding rates with ZT did not lead to increased root diseases in wheat and barley.

Further recent 'fine-tuning' research with direct seeding on the Canadian prairies includes studies on water-use efficiency (Walley et al., 1999), water conservation (Larney and Lindwall, 1995), crop sequence (Arshad et al., 1999, 2002),

Table 8–1. Tillage of seeded area by soil zone, 2001.

Region	Total	Most residue: Incorporate		Most residue: Retain		No-till	
	10^3 ha	10^3 ha	%	10^3 ha	%	10^3 ha	%
Brown	4 526	1 501	33.2	1 164	25.7	1 860	41.1
Dark Brown	7 174	1 992	27.8	2 219	30.9	2 964	41.3
Black and Gray	13 913	6 606	43.6	4 642	33.4	3 211	23.1
Total	25 613	9 551	37.3	8 025	31.3	8 036	31.4

Source: Statistics Canada (2002).

soil organic matter pools (Campbell et al., 1996, 1998a, 1998b; Franzluebbers and Arshad, 1996; Larney et al., 1997; Liang et al., 2003; Malhi et al., 2003), soil nutrient levels and crop response (Nyborg et al., 1995; Selles et al., 1999b; Campbell et al., 2001; McKenzie et al., 2001b; Grant et al., 2002a; McConkey et al., 2002; Soon and Clayton, 2002), soil physical properties (Miller et al., 1998a, 1998b, 1999; Elliott and Efetha, 1999; Larney et al., 2003b), and soil microbial biomass and diversity (Lupwayi et al., 1998, 1999, 2001).

Conservation tillage now accounts for 62.7% of the seeded area of the Canadian prairies (includes most residue retained plus no-till, Table 8–1). The benefits of conservation tillage systems are greater in the moist Black and Gray soil zones, than in the semi-arid Brown soil zone because of a yield advantage and greater savings in land preparation costs (Zentner et al., 2002b). Conservation tillage is more profitable when fallow is minimal because chemical weed control on fallow is more costly than tillage, and the short-term yield advantages of conservation tillage are small (Smith et al., 1996; Zentner et al., 1996c, 2002a). Still, no-till in the Black and Gray soil zones lags behind that in the drier Dark Brown and Brown soil zones (Table 8–1).

Production costs for conservation systems have declined, improving the returns relative to conventional tillage systems. More herbicides are available for weed control and some have declined in cost. Conservation tillage provides significant energy savings in fuel and machinery, particularly for fallow-wheat systems, though these may be offset by increases in the energy input for herbicides, and higher rates of N fertilizer required for conservation tillage (Zentner et al., 1998). Precision seeding equipment for direct seeding is now readily available, further reducing costs by allowing fertilizer application during seeding, allowing seeding despite high levels of crop residue, reducing operational and repair costs, and increasing field time efficiency.

Reducing Summer Fallow

As adoption of conservation tillage has progressed, particularly in the 1990s, the area devoted to summer fallow has diminished (Table 8–2). The additional moisture stored with conservation tillage has made recropping more feasible, reducing the need for moisture replenishment under summer fallow. Fallow continued to be important in the Brown soil zone (40%), but was reduced by half (compared to the 1970s) in the Dark Brown soil zone, and was reduced to 12% of cropland in the

Table 8–2. Cropping percentages by soil zone and decade.

	1970s					1990s				
Soil zone	Fallow	Wheat	Small grains†	Canola	Other crops	Fallow	Wheat	Small grains†	Canola	Other crops
Brown	48	46	5	0	1	40	47	6	2	5
Dark Brown	44	40	15	0	1	22	40	15	15	8
Black and Gray	32	30	29	5	4	12	30	26	24	8
Total	40	39	19	1	1	20	35	20	18	7

†Other than wheat.

Black and Gray Wooded soil zones. (However, this balance may be influenced by prevailing weather conditions; successive drought years could conceivably reverse the trend of declining summerfallow, especially in the drier soil zones.)

Most of the area once under summer fallow has been planted, not to wheat and small grains, but to crops that had only small areas before: canola (*Brassica rapa* L., B. *napus* L.), pulse crops, especially lentil (*Lens culinaris Medik.*) and field pea (*Pisum sativum* L.), mustard (*B. juncea* L.), and flax (*Linum usitatissimum* L.). Crop rotations with oilseeds and pulses had higher and less risky returns than monoculture cereal rotations over a 12-yr period on a Dark Brown Chernozemic soil (Zentner et al., 1996a).

Rates of decline in fallow area, for 1976 to 1998, have been estimated at 1.26% yr^{-1} in the Brown soil zone, 7.5% yr^{-1} in the Dark Brown, and 14.3% yr^{-1} in the combined Black, Gray, and Dark Gray soil zones (Campbell et al., 2002). Fallow remains higher in the semi-arid prairies because historically it contributed to higher and more stable farm income and had lower direct costs for purchased inputs such as fertilizer (Brown, 1987; Zentner and Campbell, 1988). The decline of fallow in the Canadian prairies roughly parallels that reported for the Northern Plains states in the USA (Smith and Young, 2001).

Not only has the area of summer fallow changed, but the method of fallow has changed as well, notably the widespread acceptance of chemical fallow, where some or all tillage operations are replaced with herbicides for weed control (Lindwall and Anderson, 1981; Zentner and Lindwall, 1982). Compared to conventional summer fallow, chemical fallow leaves more of the protective residue cover on the soil surface and lowers risk of soil erosion. By 2001, 17.8% of fallow was no-till chemical and 36.1% was a combination of tillage and chemical (Statistics Canada, 2002).

Fallow Alternatives

Almost from the outset of cultivated agriculture, researchers in western Canada have sought alternatives to summer fallow—crops that would allay the deficiencies of traditional fallow, while preserving its benefits. For example, MacKay in 1899 described a study whose intent was to evaluate the benefits of '*leguminous plants for ploughing under ... in place of the usual fallow*' (Janzen, 2001). Such 'green manure' crops, however, were not widely successful at the time.

The search for fallow alternatives continues today. Single-year legume green manures have been used as a partial- or complete replacements for fallow to increase

N availability in the crop rotation (Rice et al., 1993; Brandt 1996, 1999; Biederbeck et al., 1996, 1998; Bullied et al., 2002; Curtin et al., 2000b). Crops included: black lentil, Tangier flatpea (*Lathyrus tingitanus* L.), chickling vetch (*Lathyrus sativus* L.), and feedpea (*P. sativum* L.); and single-year alfalfa (*Medicago sativa* L.), berseem (*Trifolium alexandrinum* L.), and red clovers (*Trifolium pretense* L.). Green manure crops, grown to trap snow and enhance overwinter soil water recharge, have also been studied (Brandt, 1999). However, in the drier regions, green manure crops may deplete soil moisture, thereby reducing yields of subsequent crops (Campbell et al., 1990; Zentner et al., 1996b). Early incorporation of green manure legumes helps to alleviate water shortages (Biederbeck and Bouman, 1994). For use in drier regions, annual legume green manures should be seeded early (late April–early May) and growth terminated by the first week of July, even if this means foregoing some nitrogen (N_2) fixation (Zentner et al., 1996b).

Another proposed alternative is the use of cover crops, especially to control erosion or protect fall-seeded crops (Thiessen Martens et al., 2001). In southern Alberta, Moyer et al. (2000) determined the effect of short-term, fall rye (*Secale cereale* L.), winter wheat, and annual rye cover crops in the fallow year on weed growth and subsequent wheat yield. Sweet clover underseeded to the crop previous to fallow reduced the companion crop yield, but helped control weeds during the fallow year and increased yield of the following crop (Blackshaw et al., 2001b). Cover crops increase surface residues, but subsequent yields depend on the cover crop and its impact on soil moisture.

Seeder Technology

Seeder technology (opener design, packer wheels, packing force) has developed alongside adoption of direct seeding and producer demand (Johnston et al., 2003). Most producers adopting no-till favor a 'one-pass' seeding and fertilizer operation, so that fertilizer rates or seeding openers may need to be modified to safely apply fertilizer with or away from the seed. Compared with broadcast application, fertilizer placement under the surface residue and below the seed can effectively reduce N immobilization and increase N uptake, recovery, and efficiency in no-till systems (Carter and Rennie, 1984; Malhi and Nyborg, 1992).

Air seeders, using a pneumatic seed and fertilizer delivery system on a cultivator frame, increase the efficiency of seeding operations because of their ease of filling and higher capacity for seed and fertilizer (Read and Dyck, 1990). They also help maintain good seed-soil contact when working through accumulated surface crop residue (Johnston et al., 2001). Seed placement configurations commonly used with air seeders include: (i) hoe opener with a narrow (2 cm) seed band at a 23 or 30-cm row spacing; (ii) an opener 10 to 15 cm wide that splits the seed into two runs and fertilizer can be placed between the seed runs; (iii) spreading seed 15 to 23-cm wide under a 28-cm sweep at a 23 to 30 cm row spacing; and (iv) disc openers with various seed and fertilizer placement configurations. The hoe and disc configurations produce distinct crop rows; the sweep gives more dispersed rows. Higher seeding rates may improve cereal yield when seed is spread rather than placed in distinct rows because of reduced inter-plant competition (Deibert, 1993), but Johnston and Stevenson (2001) found that grain yield did not differ between the sweep and distinct rows, regardless of the seeding rate.

Alternative Seeding Dates

Canola production in most regions of the Canadian prairies is often limited by a short frost-free growing season; hot, dry periods in July during flowering and seed set; and cool wet conditions at harvest in September (Kirkland and Johnson, 2000). Seeding in fall or early spring may allow better use of spring moisture from snowmelt and avoid environmental stress. In Saskatchewan, canola seed yield was 38% greater when seeded in October or mid- to late-April rather than mid-May (Kirkland and Johnson, 2000). The yield advantage from alternative seeding dates was greater and more consistent on stubble than on fallow likely because of reduced soil crusting and protection from wind and extreme soil temperatures. Fertilizer N placement strategies for traditional mid-May canola seeding may be transferable to fall or April seeding (Johnston et al., 2002b).

Crop Diversification

There has been extensive study, in recent years, on crop diversification in the northern Great Plains (Brandt and Zentner, 1995; Entz et al., 2002; Johnston et al., 2002c; Miller et al., 2002a; Zentner et al., 2002b; Gan et al., 2003b) and its relationship with integrated pest management (Derksen et al., 2002; Krupinsky et al., 2002), biodiversity (Olfert et al., 2002), and soil fertility (Grant et al., 2002b). The shift in research emphasis reflects the increasing prominence of noncereal crops in rotations. For example, canola has become central to rotations in the Black and Gray soil zones, and is commonly grown also in drier regions.

Pulse crops, notably field pea and lentil, once rare, are now widely-grown, partly because of N benefits to subsequent crops (Beckie et al., 1997). The area of field pea and lentil production in the region has expanded from <0.25 Mn ha in the early 1900s to about 3 Mn ha by 2001 (Statistics Canada, 2002). This expansion has been aided by new pea and lentil cultivars with improved harvestability, disease tolerance, and yield potential, and by practices that reduce costs and disease incidence (Morrall, 1997; Johnston et al., 2002a).

Progress has also been made in developing other pulse crops. Cultivars of dry bean (*Phaseolus vulgaris* L.) adapted to the Black soil zone of Saskatchewan are now available (Shirtliffe and Johnston, 2002). New cultivars have upright growth, short growing season, and improved pod clearance, allowing them to be grown in narrow rows (15–40 cm) without inter-row tillage, rather than in wide-row (60–75 cm) systems that are prone to erosion. Another legume, chickpea (*Cicer arietinum* L.), has been rapidly adopted by farmers, especially in Saskatchewan where 445 000 ha were seeded in 2001 (Statistics Canada, 2002). Because of its deep rooting and reliance on water stress to hasten maturity, the crop is adaptable to the drier regions of the Canadian prairies (Gan et al., 2002, 2003a, 2003c).

Pulse crops may have significant N fertility benefits in cropping systems; Miller et al. (2002b) reported the average yield of wheat grown after four pulse crops (desi chickpea, dry bean, field pea, and lentil) was 21% higher than after wheat. *Rhi-*

zobia inoculation of pulse crops ensures maximum yields, though in fields that have regularly grown a pea crop the benefits of inoculation are likely to be low (McKenzie et al., 2001a). Dry bean is a poor N-fixer relative to other legumes and the response to inoculant application is cultivar specific, ranging from 51 to 78% of N required, according to Nleya et al. (2001).

Sunflower (*Helianthus annuus* L.) is extensively grown in southern Manitoba, where heat units are adequate (Angadi and Entz, 2002). The recent development of early-maturing, short-stature cultivars has renewed interest in producing sunflower elsewhere on the prairies. Short-stature sunflower cultivars are classified as semidwarf (1.2–1.5 m height) and dwarf (0.8–1.2 m height) types (Schneiter, 1992). Dwarf cultivars, referred to as sunola, emerged in the early 1990s as a drought- and heat-tolerant oilseed (Beckie and Brandt, 1996) that can be produced in short growing seasons and with field equipment used for cereals. However, Miller et al. (1998c) found that sunola had lower seed yield potential and water-use efficiency than canola and mustard in dry conditions.

Miller et al. (2001) examined the adaptation of chickpea, dry bean, field pea, lentil, mustard, safflower (*Carthamus tinctorius* L.), and sunflower under no-till and minimum tillage in southwestern Saskatchewan. Tillage system had no consistent effect on plant densities. Chickpea, lentil, and field pea grown on stubble, yielded 76 to 90% of these crops grown on fallow (compared to 61 to 66% for wheat and mustard), suggesting these pulses may replace fallow in the cropping systems of the dry semi-arid prairie.

There has been a rethinking of forage production as part of diversified prairie farming. While alfalfa remains the most common legume forage, fenugreek (*Trigonella foenum-graecum* L.), is a promising high-protein legume. Perennial cereal rye promotes effective use of spring moisture, produces significant regrowth for silage or grazing, and can protect the soil from erosion (Acharya et al., 2003; Reimann-Philipp, 1995). The role of forages in restoring soil quality characteristics has also been highlighted in recent crop diversification research (Curtin et al., 2000a; Bremer et al., 2002; Entz et al., 2002), especially for increased storage of soil carbon.

The increased diversification on the Canadian prairies has been driven partly by economic factors, notably the increase in oilseed and pulse crop prices relative to wheat (Smith et al., 2001a). Crop selection became more responsive to price changes in the 1990s, an indication of increased flexibility and incorporation of price information in cropping decisions. Economic policy, mostly through crop insurance, may have also influenced cropping decisions (Zentner and Campbell, 1988). Policy is less important in Canada than in comparable production regions of the northern USA where "set-aside" programs directly impact cropping (Smith and Young, 2003). There is no evidence that government policies directly affected the adoption of conservation tillage, though, extension activities likely contributed to increased adoption.

Cereal-based rotations that include oilseeds and pulses have higher and more stable net income, despite higher input costs (Zentner et al., 2002b). Crop diversification itself can have economic benefits by reducing income variability, especially if there is little correlation among commodity prices. However, most annual crop prices are highly correlated, the exception being lentil (Zentner et al., 2002b).

Other economic benefits of diversification include higher revenue because of higher yield, spreading of labor and equipment requirements for crops with different maturity dates, lower pesticide inputs because of reduced disease and weed pressures, and lower fertilizer inputs because of N_2 fixation. Technologies for seeding, harvesting, and weed control have especially benefited oilseeds and pulses by reducing unit production costs.

ISSUES AND CHALLENGES

Cropping practices on the Canadian prairies continue to evolve in response to influences from within and without. In this section, we consider some of the issues and questions that may shape dryland agriculture in this region in coming years and decades. Some of these issues are not new—they represent ongoing struggles with problems known long ago. Others are just emerging, and responses to them are still in their infancy. We look at some of these, roughly in the order in which they emerged.

Soil Erosion

Soil erosion and degradation has been a major environmental issue since the native Canadian prairie was first cultivated by settlers in the late 19th century (Jones, 1986). Burial of native sod by moldboard plows and the rapid depletion of organic matter exposed the soil to severe wind- and sometimes water erosion. In some areas, much of the surface soil was lost; wrote Seager Wheeler (1920), an early farmer: "*In many fields nearly all the top soil has blown away, leaving only the bare subsoil*" (cited by Janzen 2001).

The severe losses, early in the 20th century, soon prompted significant improvements in management practice: less intensive tillage, use of wide-blade cultivators that retained residues on the surface, and strip-farming, among others. More recently, the widespread adoption of direct seeding has further reduced the incidence of wind (and water) erosion by providing higher levels of surface crop residues on dryland fields during times of peak risk (late fall and early spring).

But some risks of erosion remain. One area particularly susceptible to erosion is the "Chinook" belt in southern Alberta on the leeside of the Canadian Rockies, dominated by strong 'Chinook' winds often gusting to more than 100 km hr^{-1}. The word 'Chinook' is derived from the indigenous Blackfoot language and translates to "snow-eater" describing the rapid snowmelt that accompanies the warm Chinook winds during fall, winter, and early spring. Associated freeze/thaw cycles enhance the breakdown of aggregates at the soil surface and increases wind erosion risk. Irrigated soils, which occupy a significant portion of this region, are especially prone to erosion, mainly because of poor soil cover from low-residue crops such as potatoes (*Solanum tuberosum* L.), sugar beet (*Beta vulgaris* L.), and pulses, as well as post-harvest and pre-seeding tillage practices.

Recent wind erosion research on the Canadian prairies has focused on the processes of overwinter change in soil aggregate size distribution as affected by fallow management (Larney et al., 1994a), the impact of freeze-thaw cycles on soil aggregate breakdown (Bullock et al., 2001), freeze-drying effects on wind erodibility (Bullock et al., 1999), the effect of wind erosion on redistribution of soil nutrients and crop yield (Larney et al., 1998) as well as the phenomenon of herbicide transport on wind-eroded sediment (Larney et al., 1999).

Although the risk of soil erosion is now lower than it was decades ago, the potential for wind erosion is always present, given the right combination of factors, for example, several drought years in succession that can reduce crop biomass and residue. Prevention of soil erosion is the best long-term economic strategy as it is not profitable to restore the productivity of eroded soils with inorganic fertilizers (Smith et al., 2000b).

Soil Quality

The rapid deterioration of soil quality alarmed early scientists, but the rate of organic matter loss slowed noticeably soon after cultivation began. Shutt (1925) wrote: "*There is undoubtedly a slowing down in this process of depletion, towards a point of constancy or equilibrium*". Now, after about a century of cultivation, the losses of organic matter in most soils have probably abated (Janzen et al., 1998b). But the risk of further deterioration of soil quality remains in localized areas. Threats to soil quality that remain, in addition to organic matter depletion and erosion, include deteriorating soil structure, salinization, contamination of agricultural soils, and agrochemical entry into groundwater (Acton and Gregorich, 1995). The value of soil quality has been recognized, recently, in efforts to quantify its economic impact (e.g., Smith et al., 2000a).

Landscape-Scale Variation

Landscape-scale variation is being increasingly recognized in agricultural and environmental studies (Jowkin and Schoenau, 1998). Pennock et al. (2001) investigated the effect of topographical position on the yield response of canola. A hummocky, glacial till research site in Saskatchewan was stratified into three topographically defined landform complexes (convex, linear, and concave). Canola seed yields (averaged across years and treatments) increased from 0.53 Mg ha^{-1} in convex complexes to 0.95 Mg ha^{-1} in linear, and 1.42 Mg ha^{-1} in concave landform complexes. Pennock and Corre (2001) found that significant transfers of soil organic carbon and surface soil from convex shoulder units to lower slope positions occurred over 90 yr, consistent with tillage translocation. Models that encompass and interpret this type of variation in fields and across contrasting management practices have the potential to improve the landscape management of Canadian prairie agroecosystems (Manning et al., 2001; Stevenson et al., 2001).

Landscape farming, also known as *precision* agriculture, *variable-rate farming,* or *site-specific agriculture*, is a recent phenomenon on the semi-arid Canadian prairies. Its basis is the knowledge that soils across an agricultural field are variable in terms of their properties and their ability to support crop production. Instead of treating whole fields exactly the same, fertilizer, irrigation water, and pesticides are varied according to existing soil properties and past productivity.

Much research on the Canadian prairies has focused on variable-rate N fertilization (Selles et al., 1999a; Nolan et al., 1999; Pennock et al., 1999; Moulin et al., 1999). Walley et al. (2001) reported little economic rationale for using variable rate fertilizer strategies in a 3-yr study in Saskatchewan, mainly as a consequence of the unpredictable nature of the varied response of wheat to N fertilizer additions. Nevertheless, in the long-term, they believed that variable rate N application strategies could be employed to manage N inputs from the perspective of managing and replacing harvested N.

Landscape variability may also affect the extrapolation of findings from small-scale research plots to farm fields. Bergstrom et al. (2001) found that the effect of tillage practice on organic C mass was contingent on slope position and sampling depth, that is, scale-dependent. Consequently, the effect of tillage practices cannot be wholly understood without some understanding of field-scale variability.

Closely linked with the precision farming concept is the use of remote sensing technology. Although it was heralded as a new technology for agriculture back in the 1960s, it has not yet reached its full potential (Moran et al., 1997). Smith (2003) outlined the potential benefits of remote sensing technology for delineation of soil management zones in a field situation, irrigation and N management, and weed mapping. Multi-spectral remote sensing has the potential for weed/crop discrimination, allowing farmers to selectively apply herbicides to reduce costs and environmental effects.

Integration of the Livestock Sector

From the early days of farming on the Canadian prairies, livestock were seen as necessary to maintain soil quality (Janzen, 2001). The traditional (perhaps idealized) 'mixed farm' is pictured as having diverse crops (cereal grains, pastures for hay, legumes, potatoes, and other root crops) and livestock species (cattle [*Bos taurus*], hogs [*Sus scrofa*], sheep [*Ovis aries*], horses [*Equus caballus*], poultry [*Gallus gallus domesticus*]). But specialization has made individual farms less diverse. The traditional link between crop and livestock production has been broken in many cases. Farms that are 100% cropping rely on inorganic fertilizers as crop nutrient sources unless they procure livestock manure from neighbors. At the other end of the scale are intensive livestock operations or confined animal feeding operations, whereby cattle, pigs, and poultry are raised in large confined houses or open feedlots, often on land bases that are too small to accommodate the large volumes of manure produced. These situations can lead to overapplication of manure nutrients and subsequent detrimental effects of soil, water, and air quality (Hao and Chang, 2003).

In central Saskatchewan, Stevenson et al. (1998) reported that feedlot cattle manure can be applied as a nutrient source when surface applied in a no-till system without significant yield reductions compared with soil incorporation in a conventional tillage system. In southern Alberta, Larney and Janzen (1996, 1997) and Larney et al. (2000a, 2000c) showed livestock manure was an excellent amendment for restoring productivity to eroded soils since reclamation with inorganic fertilizer was not economic (Smith et al., 2000b). The main drawback is that eroded soils are often too far away from manure sources for economical haulage. Composting of livestock manure is a further way of integrating the cropping and livestock sectors. Along with agronomic benefits such as viable weed seed elimination (Larney and Blackshaw, 2003), composting, by reducing mass and water content, compared to raw manure, lowers the haulage requirements of manure nutrients from areas of high intensity livestock production to soils that may be nutrient deficient (Larney et al., 2000b). Both manure and compost also play a role in soil reclamation associated with wellsite drilling activities by the oil and gas industry (Larney et al., 2003a), which, alongside agriculture, is a mainstay of the Canadian prairie economy. If the intensification of livestock operations continues, further efforts may be needed to keep the land and livestock connected, ensuring that the animals remain a benefit to agricultural productivity, rather than a potential hindrance.

Water Quality

Water has always been the scarcest input in dryland farming systems. Population growth and drought as well as agricultural and industrial development are increasing pressure on water supplies in parts of the Canadian prairies. A safe, secure drinking water supply, adequate water quantity for economic production and healthy aquatic ecosystems are part of current water strategies on the prairies (Alberta Environment, 2003). All aspects of water management—water quality, water quantity, aquatic ecosystems, water use, water allocation, water storage, groundwater, wetlands, and economic and social benefits—are integrated within a watershed. Water management is now focused on watersheds rather than individual farms and several Watershed Protection Groups or Watershed Advisory Councils have been established.

Agriculture may generate contaminants with detrimental effects on surface or groundwater quality. Possible contaminants include: sediment, nutrients (N and P), and heavy metals from overapplication of fertilizers and livestock manures, pathogens (e.g., *Escherichia coli*) and parasites (*Giardia, Cryptosporidium*) from livestock manure, and pesticides. Degraded water quality restricts its use for stock watering and irrigation, drinking water supplies, aquatic life, and recreational activities.

Movement of pesticides from agricultural fields to nontarget areas represents a threat to ecosystems adjacent to agricultural land (e.g., wetlands, streams, rivers) as well as irrigation canals and farm dugouts. Waite et al. (2002) reported that 2, 4-D (2, 4-dichlorophenoxyacetic acid) and triallate [*S*-2, 3, 3-trichloroallyl di-isopropyl (thiocarbamate)] concentrations in bulk atmospheric deposition, dugout water, and ambient air in Saskatchewan were related to use of these herbicides on cropland. Hill et al. (2002) detected phenoxy herbicides in Alberta rainfall at levels as high as 17 to 53 $\mu g\ L^{-1}$ or 60 to 149 $\mu g\ m^{-2}$.

Riparian areas are transitional ecosystems between land and water environments. These natural corridors are characterized by lush vegetation bordering rivers, streams, and wetlands. Although riparian areas account for <5% of the prairie landscape, they provide essential wildlife habitat, act as sediment and nutrient filters and are important sources of biodiversity. Across the Canadian prairies, initiatives are underway to ensure the protection of riparian health (PFRA, 2000).

Although current dryland agricultural practices on the Canadian prairie are less intensive than those in many other regions, agriculture has the potential to influence water quality. Furthermore, Canada's agricultural sector markets its products worldwide based on the concept of quality food produced in a clean environment, an image that encourages producers to protect their natural resources.

Integrated Pest Management

Conservation tillage adoption changes the interactions of weeds, insects, and diseases mainly due to the increased amounts of surface residue. Integrated pest management (IPM) describes an evolving philosophy where cultural, biological, and chemical controls are included in a holistic approach to weed, insect, and disease control for crops. Key components of IPM include: crop monitoring, cultural control, resistant cultivars, biological control, and chemical control. Use of competitive crops in diversified cropping systems is also a cornerstone of successful IPM.

Weed management is an integral component of conservation tillage systems (Moyer et al., 1994; Gill and Arshad, 1995; Derksen et al., 2002). Identifying agronomic practices that complement weed management may help reduce dependence on herbicides (Harker et al., 2003). Derksen et al. (1996) reported that changes in weed communities after adopting conservation tillage were not necessarily those expected and were not consistent by species, location, or year. O'Donovan (1996) proposed computerized decision-support systems as a way of achieving economical and environmentally-safe weed management.

In southern Alberta, weed community dynamics were most affected by year-to-year differences in environmental conditions, followed by crop rotation, and then tillage intensity (Blackshaw et al., 2001a). Russian thistle (*Salsola iberica* Sennen & Pau) and kochia [*Kochia scoparia* (L.) Schrad.] densities increased in drier and warmer years. Winter annual weeds such as downy brome (*Bromus tectorum* L.) and flixweed [*Descurainia sophia* (L.) Webb ex Prantl], as well as the perennial dandelion (*Taraxacum officinale* Weber in Wiggers), increased in years with higher than average rainfall in fall or early spring. Russian thistle, downy brome, kochia, and redroot pigweed (*Amaranthus retroflexus* L.) were associated with no-till while wild buckwheat (*Polygonum convolvulus* L.), lamb's-quarters (*Chenopodium album* L.), flixweed, and wild mustard (*Sinapis arvensis* L.) were associated with conventional tillage.

Some weed species on the Canadian prairies have recently become resistant to herbicides, after prolonged use of the same herbicide, or herbicides from the same group with a similar mode of action. Growers are now urged to rotate their herbicides as well as their crops to preclude the buildup of herbicide-resistant weed populations. Field and grain elevator surveys conducted across the prairies indicated that resistance in wild oat (*Avena fatua* L.) to ACCase inhibitors (Group 1 herbicides) occurred most frequently relative to other herbicide groups (Beckie et al., 2002). Group 1-herbicide-resistant wild oat occurred in about 50% of fields surveyed. A relatively high incidence of multiple-group resistance in wild oat in Saskatchewan and Manitoba is of particular concern (Beckie et al., 1999). In

Saskatchewan, 20% of Group 1 herbicide-resistant populations were also resistant to ALS inhibitors (imidazolinones), even though the ALS inhibitors were used infrequently. In Manitoba, 27% of cereal fields surveyed had wild oat resistant to herbicides from more than one group. Four wild oat populations were resistant to all herbicide modes of action for use in wheat. Depending on the nature of resistance in wild oat, alternative herbicides may substantially increase costs. The cost of managing herbicide-resistant wild oat in Saskatchewan and Manitoba using alternative herbicides was estimated at more than $4 Mn annually (Beckie et al., 1999).

A relatively new tool in the area of weed control is the development of herbicide-tolerant crops. In 1999, approximately 76% of the 5.6 Mn ha of canola in western Canada were seeded to herbicide-tolerant cultivars (Harker et al., 2000). Canola tolerant to glufosinate was the first transgenic crop to be registered in Canada (Oelck et al., 1995). Canola tolerant to glyphosate (transgenic) and imidazolinone herbicides (selected via cell culture) followed shortly thereafter. Harker et al. (2000) found that herbicide-tolerant canola provided new opportunities for the control of difficult weeds along with the option to employ in-crop herbicides with new modes of action.

Conservation tillage lowers the impact of some root diseases, but increases the impact of foliar diseases in cereals (Bailey, 1996). Diseases that were less economically important under higher tillage regimes might become more important with reduced tillage, but location and local environment largely influence disease outbreaks. Crop rotation is a key factor in residue management for disease control. Bailey (1996) recommended a well-balanced rotation should be at least 4-yr duration with 50% of the interval devoted to cereals and the remainder divided among pulses, flax, other oilseed crops, or forages.

Organic Agriculture

Organic food is that grown under a production system that promotes soil health, biodiversity, and low stress management of animals without the use of synthetic fertilizers, pesticides, growth regulators, and livestock feed additives (Vladicka and Cunningham, 2002). The use of genetically modified crops, irradiation, and sewage sludge is also prohibited. Although accounting for roughly 2% of the food retail industry in Canada, organic produce has become the fastest growing segment at 15 to 20% annually. The drivers for this increase have been higher agrochemical costs and low commodity prices on the producer side, and increased emphasis on health and nutrition, an aging population, and concerns about food safety, environmental protection, and sustainable agriculture on the consumer side.

Before the introduction of chemical fertilizers and pesticides in the 1940s and 1950s, all agricultural production on the prairies was in fact 'organic'. From the 1960s to the 1980s, interest in organic farming was renewed (Rutherford and Gimby, 1990). In 2000, there were approximately 350 certified organic producers in Alberta with about 180 000 ha in production (Vladicka and Cunningham, 2002). Certified animals included more than 18 000 poultry and almost 7000 cattle. Entz et al. (2001) surveyed cropping records from 13 organic farms in the eastern Canadian prairies (1991–1996) to determine crop rotation pattern, yields, and soil nutrient status. Major crops included cereal grains, forages, and green manure legumes.

Organic grain and forage yields averaged from one-half to almost double conventional yields. Soil N, K, and S levels on organic farms were generally sufficient; however, levels of available soil P were deficient in several instances.

A hybrid of conventional agriculture and organic agriculture is pesticide-free production (PFPC, 2003). In a pesticide-free production (PFP) system, fertilizers can be used, and nonresidual pesticides are permitted in non-PFP years. Some pesticides may be applied to the field in the year of PFP crop production, provided they are not applied directly to the crop and do not have any residual soil activity during the crop year. For example, pre-seeding weed control with a nonresidual herbicide such as glyphosate is within the guidelines, but a preseeding application of a residual herbicide (e.g., triallate for wild oat control) is not. Preseeding glyphosate applications are a critical practice to ensure optimal weed control and crop productivity in direct seeding systems (Johnson et al., 2002).

Greenhouse Gas Emissions and Carbon Storage

The concentrations of greenhouse gases in the atmosphere have increased abruptly in recent decades, prompting fears of possible changes to the world's climate (IPCC, 2001). Canada recently ratified the Kyoto Protocol, committing to reduce its greenhouse gas emissions to 6% below 1990 levels, if the protocol formally takes effect (Olsen et al., 2002). Efforts to curtail emissions of greenhouse gases influence agricultural practices on the Canadian prairies in several ways. First, it provides further incentive to increase C reserves in agricultural soils and quantify any changes (IPCC, 2000). Past research, often from the perspective of soil quality, has provided extensive estimates of the long-term effects on soil C of crop rotation, tillage practices, nutrient application, and other practices (e.g., Campbell et al., 1998b, 2000a, 2000b, 2001; Curtin et al., 2000b, 2002; Ellert and Janzen, 1999; Franzluebbers and Arshad, 1996; Janzen et al., 1997a, 1997b, 1998a, 1998b). Such studies, along with simulation models (e.g., Grant et al., 2001b; Smith et al., 2001b), demonstrate some opportunity for mitigating atmospheric CO_2 increases, at least until the soils again approach a new steady-state C content.

Further research, however, is still needed to quantify the effects of recently-adopted crops and cropping practices. In addition, methods of measuring soil C, traditionally used to monitor soil quality, may need to be adapted for measuring changes in soil C mass (Ellert et al., 2002). But changes in soil C alone do not indicate the net effect on the atmosphere; we need to consider also the other greenhouse gases. For example, some of the benefits of increased C storage could conceivably be offset by increased emissions of nitrous oxide (N_2O), a very potent greenhouse gas. The effects of cropping practices on N_2O emissions have not yet been fully quantified. Some studies in western Canada, however, indicate that adoption of no-till does not increase N_2O emissions and may even reduce emissions (Lemke et al., 1998a, 1998b, 1999). Efforts to reduce greenhouse gas emissions might also influence prairie agriculture by effects on livestock management (ruminants and manures are important sources). In addition, more widespread production of bio-fuels and proposed mitigation strategy, might alter management practices and C returns to soils (Stumborg et al., 1996).

Climate Change

Although appreciable uncertainty remains, there is growing agreement among climatologists that some global warming may occur, even with ongoing mitigation efforts (Karl et al., 1997). The IPCC (2001), in their Third Assessment Report, estimated that global mean temperatures might increase by 1.4 to 5.8°C from 1990 to 2100. Concerns about future climate change have led to studies of historical trends (Cutforth, 2000). Gan (1998) found that the Canadian Prairies have become warmer and somewhat drier in the past four to five decades, but the evidence was insufficient to conclude that a warmer climate would lead to more severe prairie droughts. From 1950 to 1998, winter and spring maximum and minimum temperatures have warmed, snowfall amounts have decreased, and spring runoff has started earlier in southwestern Saskatchewan (Cutforth et al., 1999). As temperatures have warmed, the percentage of precipitation as snow has decreased, but the date of last spring frost has not changed. Analysis of precipitation events at 37 weather stations with 75 yr of records across the Canadian prairies revealed a significant increase of 20 events yr^{-1} mainly due to increased low-intensity events (Akinremi et al., 1999). Precipitation amounts increased significantly by 0.62 mm yr^{-1}. From 1961 to 1995, snowfall declined significantly by 0.95 mm yr^{-1}. These historical studies do not necessarily provide insights into future climates, but they underline the susceptibility of prairie agriculture to changes, especially in precipitation.

Other Global Changes

The burgeoning human population may continue to induce global changes (Ernst, 2000), some of which will affect agricultural practices on the Canadian prairies. Some of these effects are already in view. For example, atmospheric CO_2 concentrations will almost certainly keep increasing (IPCC, 2001), perhaps inducing some CO_2 fertilization. Continued increases in human numbers will increase the demand for food (Tillman et al., 2002), with unpredictable effects on agricultural production on the prairies. Further changes might unfold from the new capacity to alter crop genotypes.

In the end, however, many of the changes coming to the agricultural landscapes of the Canadian prairies are beyond our view. Just as we would not have foreseen, 50 yr ago, that our soils might be considered sinks for atmospheric CO_2, so we are likely to be faced, 50 yr from now, by issues still unseen. Meeting these challenges may depend most on leaving our successors a robust understanding of our agroecosystems.

ACKNOWLEDGMENTS

We thank Leslie Cramer and Yvonne Bruinsma for assistance in editing the manuscript.

REFERENCES

Acharya, S.N., Z. Mir, J.R. Moyer, B.R. Orshinsky, and J.E. Thomas. 2003. Effect of row spacing, and seeding rate on forage yield, and quality of perennial cereal rye (*Secale cereale* L.). Can. J. Plant Sci. 83:363–369.

Acton, D.F., and L.J. Gregorich. 1995. The health of our soils–toward sustainable agriculture in Canada. Publ. 1906/E. Agriculture, and Agri-Food Canada, Ottawa, ON.

Akinremi, O.O., S.M. McGinn, and H. Cutforth. 1999. Precipitation trends on the Canadian prairies. Can. J. Soil Sci. 79:235.

Alberta Environment. 2003. Water for life: Alberta's strategy for sustainability. Publ. 1/935. Alberta Environ., Edmonton, AB.

Anderson, D.W., E. de Jong, G.E. Verity, and E.G. Gregorich. 1986. The effects of cultivation on soils of the Canadian prairies. Trans. 13th Congr. Int. Soc. Soil Sci. 4:1344–1345.

Angadi, S.V., and M.H. Entz. 2002. Agronomic performance of different stature sunflower cultivars under different levels of interplant competition. Can. J. Plant Sci. 82:43–52

Arshad, M.A., A.J. Franzluebbers, and K.S. Gill. 1999. Improving barley yield on an acidic Boralf with crop rotation, lime, and zero tillage. Soil Tillage Res. 50:47–53.

Arshad, M.A., Y.K. Soon, and R.H. Azooz. 2002. Modified no-till, and crop sequence effects on spring wheat production in northern Alberta, Canada. Soil Tillage Res. 65:29–36.

Bailey, K.L. 1996. Diseases under conservation tillage systems. Can. J. Plant Sci. 76:635–639.

Bailey, K.L., G.P. Lafond, and D. Domitruk. 1998. Effects of row spacing, seeding rate, and seed-placed phosphorus on root diseases of spring wheat, and barley under zero tillage. Can. J. Plant Sci. 78:145–150.

Beckie, H.J., and S.A. Brandt. 1996. Sunola response to nitrogen fertilization. Can. J. Plant Sci. 76:783–789.

Beckie, H.J., S.A. Brandt, J.J. Schoenau, C.A. Campbell, J.L. Henry, and H.H. Janzen. 1997. Nitrogen contribution of field pea in annual cropping systems. 2. Total nitrogen benefit. Can. J. Plant Sci. 77:323–331.

Beckie, H.J., A.G. Thomas, A. Légère, D.J. Kelner, R.C. Van Acker, and S. Meers. 1999. Nature, occurrence, and cost of herbicide–resistant wild oat (*Avena fatua*) in small-grain production areas. Weed Technol. 13:612–625.

Beckie, H.J., A.G. Thomas, and F.C. Stevenson. 2002. Survey of herbicide-resistant wild oat (*Avena fatua*) in two townships in Saskatchewan. Can. J. Plant Sci. 82:463-471.

Bergstrom, D.W., C.M. Monreal, and E. St. Jacques. 2001. Influence of tillage practice on carbon sequestration is scale dependent. Can. J. Soil Sci. 81:63–70.

Biederbeck, V.O., and O.T. Bouman. 1994. Water use by annual green manure legumes in dryland cropping systems. Agron. J. 86:543–549.

Biederbeck, V.O., O.T. Bouman, C.A. Campbell, L.D. Bailey, and G.E. Winkleman. 1996. Nitrogen benefits from four green-manure legumes in dryland cropping systems. Can. J. Plant Sci. 76:307–315.

Biederbeck, V.O., C.A. Campbell, V. Rasiah, R.P. Zentner, and G. Wen. 1998. Soil quality attributes as influenced by annual legumes used as green manure. Soil Biol. Biochem. 30:1177–1185.

Blackshaw, R.E., F.J. Larney, C.W. Lindwall, P.R. Watson, and D.A. Derksen. 2001a. Tillage intensity, and crop rotation affect weed community dynamics in a winter wheat cropping system. Can. J. Plant Sci. 81:805–813.

Blackshaw R.E., J.R. Moyer, R.C. Doram, A.L. Boswall, and E.G. Smith. 2001b. Suitability of undersown Sweetclover (*Melilotus officinalis*) as a fallow replacement in semiarid cropping systems. Agron. J. 93:863–868.

Brandt, S.A. 1996. Alternatives to summer-fallow, and subsequent wheat, and barley yield on a Dark Brown soil. Can. J. Plant Sci. 76:223–228.

Brandt, S.A. 1999. Management practices for black lentil green manure for the semi-arid Canadian prairies. Can. J. Plant Sci. 79:11–17.

Brandt, S.A., and R.P. Zentner, 1995. Crop production under alternate rotations on a Dark Brown Chernozemic soil at Scott, Saskatchewan. Can. J. Plant Sci. 75:789–794.

Bremer, E., H.H. Janzen, and R.H. McKenzie. 2002. Short-term impact of fallow frequency, and perennial grass on soil organic carbon in a Brown Chernozem in southern Alberta. Can. J. Soil Sci. 82:481–488.

Brown, W.J. 1987. A risk efficiency analysis of crop rotations in Saskatchewan. Can. J. Agric. Econ. 35:333–355.

Bullied, W.J., M.H. Entz, S.R. Smith, and K.C. Bamford. 2002. Grain yield, and N benefits to sequential wheat, and barley crops from single-year alfalfa, berseem, and red clover, chickling vetch, and lentil. Can. J. Plant Sci. 82:53–65.

Bullock, M.S., F.J. Larney, R.C. Izaurralde, and Y. Feng. 2001. Overwinter changes in wind erodibility of clay loam soils in southern Alberta. Soil Sci. Soc. Am. J. 65:423-430.

Bullock, M.S., F.J. Larney, S.M. McGinn, and R.C. Izaurralde. 1999. Freeze-drying processes, and wind erodibility of a clay loam soil in southern Alberta. Can. J. Soil Sci. 79:127–135.

Campbell, C.A., V.O. Biederbeck, B.G. McConkey, D. Curtin, and R.P. Zentner. 1998a. Soil quality—Effect of tillage, and fallow frequency. Soil organic matter quality as influenced by tillage, and fallow frequency in a silt loam in southwestern Saskatchewan. Soil Biol. Biochem. 31:1–7.

Campbell, C.A., B.G. McConkey, V.O. Biederbeck, R.P. Zentner, D. Curtin, and M.R. Peru. 1998b. Long-term effects of tillage, and fallow-frequency on soil quality attributes in a clay soil in semiarid southwestern Saskatchewan. Soil Tillage Res. 46:135–144.

Campbell, C.A., B.G. McConkey, R.P. Zentner, F. Selles, and D. Curtin. 1996. Tillage and crop rotation effects on soil organic C, and N in a coarse-textured Typic Haploboroll in southwestern Saskatchewan. Soil Tillage Res. 37:3–14.

Campbell, C.A., F. Selles, G.P. Lafond, and R.P. Zentner. 2001. Adopting zero tillage management: Impact on soil C, and N under long-term crop rotations in a thin Black Chernozem. Can. J. Soil Sci. 81:139–148.

Campbell, C.A., R.P. Zentner, S. Gameda, B. Blomert, and D.D. Wall. 2002. Production of annual crops on the Canadian prairies: Trends during 1976–1998. Can. J. Soil Sci. 82:45–57.

Campbell, C.A., R.P. Zentner, H.H. Janzen, and K.E. Bowren. 1990. Crop rotation studies on the Canadian prairies. Publ. 1841/E. Agriculture Canada, Ottawa, ON.

Campbell, C.A., R.P. Zentner, B.C. Liang, G. Roloff, E.C. Gregorich, and B. Blomert. 2000a. Organic C accumulation in soil over 30 years in semiarid southwestern Saskatchewan: Effect of crop rotations, and fertilizers. Can. J. Soil Sci. 80:179–192.

Campbell, C.A., R.P. Zentner, F. Selles, V.O. Biederbeck, B.G. McConkey, B. Blomert, and P.G. Jefferson. 2000b. Quantifying short-term effects of crop rotations on soil organic carbon in southwestern Saskatchewan. Can. J. Soil Sci. 80:193–202.

Carter, M.R., and D.A. Rennie. 1984. Crop utilization of placed, and broadcast ^{15}N-urea fertilizer under zero, and conventional tillage. Can. J. Soil Sci. 64:563–570.

Curtin, D., F. Selles, H. Wang, R.P. Zentner, and C.A. Campbell. 2000a. Restoring organic matter in a cultivated, semiarid soil using crested Wheatgrass. Can. J. Soil Sci. 80:429–435.

Curtin, D., H. Wang, F. Selles, C.A. Campbell, and R.P. Zentner. 2002. Soil fertility effects on carbon fluxes under two spring wheat rotations in a semiarid agroecosystem. Can. J. Soil Sci. 82:155–163.

Curtin, D., H. Wang, F. Selles, R.P. Zentner, V.O. Biederbeck, and C.A. Campbell. 2000b. Legume green manure as partial fallow replacement in semiarid Saskatchewan: Effect on carbon fluxes. Can. J. Soil Sci. 80:499–505.

Cutforth, H. W. 2000. Climate change in the semiarid prairie of southwestern Saskatchewan: Temperature, precipitation, wind, and incoming solar energy. Can. J. Soil Sci. 80:375–385.

Cutforth, H.W., and B.G. McConkey. 1997. Stubble height effects on microclimate, yield, and water use efficiency of spring wheat grown in a semiarid climate on the Canadian prairies. Can. J. Plant Sci. 77:359–366.

Cutforth, H.W., B.G. McConkey, D. Ulrich, P.R. Miller, and S.V. Angadi. 2002. Yield and water use efficiency of pulses seeded directly into standing stubble in the semiarid Canadian prairie. Can. J. Plant Sci. 82:681–686.

Cutforth, H.W., B.G. McConkey, R.J. Woodbine, D.G. Smith, P.G. Jefferson, and O.O. Akinremi. 1999. Climate change in the semiarid prairie of southwestern Saskatchewan: Late winter–early spring. Can. J. Plant Sci. 79:343–350.

Deibert, E.J. 1993. Spring wheat response to seed spreader patterns, and deep band fertilizer placement with an air seeder. J. Prod. Agric. 6:557–563.

Derksen, D.A., R.L., Anderson, R.E. Blackshaw, and B. Maxwell. 2002. Weed dynamics, and management strategies for cropping systems in the northern Great Plains. Agron. J. 94:174–185.

Derksen, D.A., R.E. Blackshaw, and S.M. Boyetchko. 1996. Sustainability, conservation tillage, and weeds in Canada. Can. J. Plant Sci. 76:651–659.

Domitruk, D., and B. Crabtree. 1997. Zero tillage–advancing the art. Manitoba-North Dakota Zero Tillage Farmers Assoc., Minot, ND.

Ellert, B.H., and H.H. Janzen. 1999. Short-term influence of tillage on CO_2 fluxes from a semi-arid soil on the Canadian prairies. Soil Tillage Res. 50:21–32.

Ellert, B.H., H.H. Janzen, and T. Entz. 2002. Assessment of a method to measure temporal change in soil carbon storage. Soil Sci. Soc. Am. J. 66:1687–1695.

Ellert, K.M. 1995. Cropping systems for a sustainable future—Long-term studies in Alberta. Canada-Alberta Environmentally Sustainable Agriculture Agreement, Agriculture, and Agri-Food Canada, Lethbridge, AB.

Elliott, J.A., and A.A. Efetha. 1999. Influence of tillage, and cropping system on soil organic matter, structure, and infiltration in a rolling landscape. Can. J. Soil Sci. 79:457–463.

Entz, M.H., V.S. Baron, P.M. Carr, D.W. Meyer, S.R. Smith, and W.P. McCaughey. 2002. Potential of forages to diversify cropping systems in the northern Great Plains. Agron. J. 94:240–250.

Entz, M.H., R. Guilford, and R. Gulden. 2001. Crop yield, and soil nutrient status on 14 organic farms in the eastern portion of the northern Great Plains. Can. J. Plant Sci. 81:351–354.

Ernst, W.G. (ed.) 2000. Earth systems: Processes, and issues. Cambridge Univ. Press, Cambridge, UK.

Franzluebbers, A.J., and M.A. Arshad. 1996. Soil organic matter pools with conventional, and zero tillage in a cold, semiarid climate. Soil Tillage Res. 39:1–11.

Gan, T.Y. 1998. Hydroclimatic trends, and possible climatic warming in the Canadian prairies. Water Resour. Res. 34:3009–3015.

Gan, Y.T., P.R. Miller, P.H. Liu, F.C. Stevenson, and C.L. McDonald. 2002. Seedling emergence, pod development, and seed yields of chickpea, and dry pea in a semiarid environment. Can. J. Plant Sci. 82:531–537.

Gan, Y.T., P.R. Miller, B.G. McConkey, R.P. Zentner, P.H. Liu, and C.L. McDonald. 2003a. Optimum plant population density for chickpea, and dry pea in a semiarid environment. Can. J. Plant Sci. 83:1–9.

Gan, Y.T., P.R. Miller, B.G. McConkey, R.P. Zentner, F.C. Stevenson, and C.L. McDonald. 2003b. Influence of diverse cropping sequences on durum wheat yield, and protein in the semiarid Northern Great Plains. Agron. J. 95:245–252.

Gan, Y.T., P.R. Miller, and C.L. McDonald. 2003c. Response of Kabuli chickpea to seed size, and planting depth. Can J. Plant Sci. 83:39–46.

Gill, K.S., and M.A. Arshad. 1995. Weed flora in the early growth period of spring crops under conventional, reduced, and zero tillage systems on a clay soil in northern Alberta, Canada. Soil Tillage Res. 33:65–79.

Grant, C.A., K.R. Brown, G.J. Racz, and L.D. Bailey. 2001a. Influence of source, timing, and placement of nitrogen on grain yield, and nitrogen removal of durum wheat under reduced-, and conventional-tillage management. Can. J. Plant Sci. 81:17–27.

Grant, C.A., K.R. Brown, G.J. Racz, and L.D. Bailey. 2002a. Influence of source, timing, and placement of nitrogen fertilization on seed yield, and nitrogen accumulation in the seed of canola under reduced-, and conventional-tillage management. Can. J. Plant Sci. 82:629–638.

Grant, C.A., G.A. Peterson, and C.A. Campbell. 2002b. Nutrient considerations for diversified cropping systems in the northern Great Plains. Agron. J. 94:186–198.

Grant, R.F., N.G. Juma, J.A. Robertson, R.C. Izaurralde, and W.B. McGill. 2001b. Long-term changes in soil carbon under different fertilizer, manure, and rotation: Testing the mathematical model *ecosys* with data from the Breton plots. Soil Sci. Soc. Am. J. 65:205–214.

Gray, R.S., J.S. Taylor, and W.J. Brown. 1996. Economic factors contributing to the adoption of reduced tillage technologies in central Saskatchewan. Can. J. Plant Sci. 76:661–668.

Hao, X., and C. Chang. 2003. Does long-term heavy cattle manure application increase salinity of a clay loam soil in semi-arid southern Alberta? Agric. Ecosyst. Environ. 94:89–103.

Harker, K.N., R.E. Blackshaw, K.J. Kirkland, D.A. Derksen, and D. Wall. 2000. Herbicide-tolerant canola: Weed control, and yield comparisons in western Canada. Can. J. Plant Sci. 80:647–654.

Harker, K.N., G.W. Clayton, R.E. Blackshaw, J.T. O'Donovan, and F.C. Stevenson. 2003. Seeding rate, herbicide timing, and competitive hybrids contribute to integrated weed management in canola (*Brassica napus*). Can. J. Plant Sci. 83:433–440.

Hill, B.D., K.N. Harker, P. Hasselback, J.R. Moyer, D.J. Inaba, and S.D. Byers. 2002. Phenoxy herbicides in Alberta rainfall: Potential effects on sensitive crops. Can. J. Plant Sci. 82:481–484.

Intergovernmental Panel on Climate Change. 2000. Land use, land-use change, and forestry. (R.T. Watson et al. [ed.].) Cambridge Univ. Press, Cambridge, UK.

Intergovernmental Panel on Climate Change. 2001. Climate change 2001: The scientific basis. (J.T. Houghton et al. [ed.]) Cambridge Univ. Press, Cambridge, UK.

Janzen, H.H. 2001. Soil science on the Canadian prairies—Peering into the future from a century ago. Can. J. Soil Sci. 81:489–503.

Janzen, H.H., C.A. Campbell, B.H. Ellert, and E. Bremer. 1997a. Soil organic matter dynamics, and their relationship to soil quality. p. 277–291. *In* E.G. Gregorich, and M.R. Carter (ed.) Soil quality for crop production, and ecosystem health. Elsevier Science Publ. B.V., Amsterdam, The Netherlands.

Janzen, H.H., C.A. Campbell, E.G., Gregorich, and B.H. Ellert. 1998a. Soil carbon dynamics in Canadian agroecosystems. p. 57–80. *In* R. Lal et al. (ed.) Soil processes, and the carbon cycle. CRC Press, Boca Raton, FL.

Janzen, H.H., C.A. Campbell, R.C. Izaurralde, B.H. Ellert, N. Juma, W.B. McGill, and R.P. Zentner. 1998b. Management effects on soil C storage on the Canadian prairies. Soil Tillage Res. 47:181–195.

Janzen, H.H., A.M. Johnston, J.M. Carefoot, and C.W. Lindwall. 1997b. Soil organic matter dynamics in long-term experiments in southern Alberta. *In* E.A. Paul et al. (ed.) Soil organic matter in temperate agroecosystems. CRC Press, Boca Raton, FL.

Johnson, E.N., K.J. Kirkland, and F.C. Stevenson. 2002. Timing of pre-seeding glyphosate application in direct-seeding systems. Can. J. Plant Sci. 82:611–615.

Johnston, A.M., G.W. Clayton, G.P. Lafond, K.N. Harker, T.J. Hogg, E.N. Johnson, W.E. May, and J.T. McConnell. 2002a. Field pea seeding management. Can. J. Plant Sci. 82:639–644.

Johnston A.M., E.N. Johnson, K.J. Kirkland, and F.C. Stevenson. 2002b. Nitrogen fertilizer placement for fall, and spring seeded *Brassica napus* canola. Can. J. Plant Sci. 82:15–20.

Johnston, A.M., G.P. Lafond, G.E. Hultgreen, and G.L. Hnatowich. 2001. Spring wheat, and canola response to nitrogen placement with no-till side band openers. Can. J. Plant Sci. 81:191–198.

Johnston, A.M., G.P. Lafond, W.E. May, G.L. Hnatowich, and G.E. Hultgreen. 2003. Opener, packer wheel, and packing force effects on crop emergence, and yield of direct seeded wheat, canola, and field peas. Can. J. Plant Sci. 83:129–139.

Johnston A.M., and F.C. Stevenson. 2001. Wheat seeding rate for spread, and distinct row seed placements with air seeders. Can. J. Plant Sci. 81:885–890.

Johnston, A.M., D.L. Tanaka, P.R. Miller, S.A. Brandt, D.C. Nielsen, G.P. Lafond, and N.R. Riveland. 2002c. Oilseed crops for semiarid cropping systems in the northern Great Plains. Agron. J. 94:231–240.

Jones, D.C. 1986. "We'll all be buried down here": The Prairie Dryland Disaster 1917–1926. Historical Soc. of Alberta, Calgary, AB.

Jowkin, V., and J.J. Schoenau. 1998. Impact of tillage, and landscape position on nitrogen availability, and yield of spring wheat in the Brown soil zone in southwestern Saskatchewan. Can. J. Soil Sci. 78:563–572.

Karl, T.R., N. Nicholls, and J. Gregory. 1997. The coming climate. Sci. Am. 276:78–83.

Kirkland, K.J., and E.N. Johnson. 2000. Alternative seeding dates (fall, and April) affect *Brassica napus* canola yield, and quality. Can. J. Plant Sci. 80:713–719.

Krupinsky, J.M., K.L. Bailey, M.P. McMullen, B.D. Gossen, and T.K. Turkington. 2002. Managing plant disease risk in diversified cropping systems. Agron. J. 94:198–209.

Lafond, G.P. 1994. Effects of row spacing, seeding rate, and nitrogen on yield of barley, and wheat under zero-till management. Can. J. Plant Sci. 74:703–711.

Lafond, G.P., S.M. Boyetchko, S.A. Brandt, G.W. Clayton, and M.H. Entz. 1996. Influence of changing tillage practices on crop production. Can. J. Plant Sci. 76:641–649.

Lafond, G.P., and Y. Gan. 1999. Row spacing, and seeding rate studies in no-till winter wheat for the Northern Great Plains. J. Prod. Agric. 12:624–629.

Larney, F.J., O.O. Akinremi, R.L. Lemke, V.E. Klaassen, and H.H. Janzen. 2003a. Crop response to topsoil replacement depth and organic amendment on abandoned natural gas wellsites. Can. J. Soil Sci. 83:415–423.

Larney, F.J., and R.E. Blackshaw. 2003. Weed seed viability in composted beef cattle feedlot manure. J. Environ. Qual. 32:1105–1113.

Larney, F.J., E. Bremer, H.H. Janzen, A.M. Johnston, and C.W. Lindwall. 1997. Changes in total, mineralizable, and light fraction soil organic matter with cropping, and tillage intensities in semiarid southern Alberta, Canada. Soil Tillage Res. 42:229–240.

Larney, F.J., M.S. Bullock, H.H. Janzen, B.H. Ellert, and E.C.S. Olson. 1998. Wind erosion effects on nutrient redistribution, and soil productivity. J. Soil Water Conserv. 53:133–140.

Larney, F.J., A.J. Cessna, and M.S. Bullock. 1999. Herbicide transport on wind-eroded sediment. J. Environ. Qual. 28:1412–1421.

Larney, F.J., and H.H. Janzen. 1996. Restoration of productivity to a desurfaced soil with livestock manure, crop residue, and fertilizer amendments. Agron. J. 88:921–927.

Larney, F.J., and H.H. Janzen. 1997. A simulated erosion approach to assess rates of cattle manure, and phosphorus fertilizer for restoring productivity to eroded soils. Agric. Ecosyst. Environ. 65:113–126.

Larney, F.J., H.H. Janzen, B.M. Olson, and C.W. Lindwall. 2000a. Soil quality, and productivity responses to simulated erosion, and restorative amendments. Can. J. Soil Sci. 80:515–522.

Larney, F.J., and C.W. Lindwall. 1995. Rotation, and tillage effects on available soil water for winter wheat in a semi-arid environment. Soil Tillage Res. 36:111–127.

Larney, F.J., C.W. Lindwall, and M.S. Bullock. 1994a. Fallow management, and overwinter effects on wind erodibility in southern Alberta. Soil Sci. Soc. Am. J. 58:1788–1794.

Larney F.J., C.W. Lindwall, R.C. Izaurralde, and A.P. Moulin. 1994b. Tillage systems for soil, and water conservation on the Canadian prairie. p. 305–328. *In* M.R. Carter (ed.) Conservation tillage in temperate agroecosystems. CRC Press, Boca Raton, FL.

Larney, F.J., A.F. Olson, A.A. Carcamo, and C. Chang. 2000b. Physical changes during active, and passive composting of beef feedlot manure in winter, and summer. Bioresour. Technol. 75:139–148.

Larney, F.J., B.M. Olson, H.H. Janzen, and C.W. Lindwall. 2000c. Early impact of topsoil removal, and soil amendments on crop productivity. Agron. J. 92:948–956.

Larney, F.J., T. Ren, S.M. McGinn, C.W. Lindwall, and R.C. Izaurralde. 2003b. The influence of rotation, tillage, and row spacing on near-surface soil temperature for winter wheat in southern Alberta. Can. J. Soil Sci. 83:89–98.

Lemke, R.L., R.C. Izaurralde, S.S. Malhi, M.A Arshad, and M Nyborg. 1998a. Nitrous oxide emissions from agricultural soils of the boreal, and Parkland regions of Alberta. Soil Sci. Soc. Am. J. 62:1096–1102.

Lemke, R.L., R.C. Izaurralde, and M. Nyborg. 1998b. Seasonal distribution of nitrous oxide emissions from soils in the Parkland region. Soil Sci. Soc. Am. J. 62:1320–1326.

Lemke, R.L., R.C. Izaurralde, M. Nyborg, and E.D. Solberg. 1999. Tillage, and N source influence soil-emitted nitrous oxide in the Alberta Parkland region. Can. J. Soil Sci. 79:15–24.

Liang, B.C., B.G. McConkey, J. Schoenau, D. Curtin, C.A. Campbell, A.P. Moulin, G.P. Lafond, S.A. Brandt, and H. Wang. 2003. Effect of tillage, and crop rotations on the light fraction organic carbon, and carbon mineralization in Chernozemic soils of Saskatchewan. Can J. Soil Sci. 83:65–72.

Lindwall, C.W., and D.T. Anderson. 1981. Agronomic evaluation of minimum tillage systems for summer fallow in southern Alberta. Can. J. Plant Sci. 61:247–253.

Lindwall C.W., F.J. Larney, A.M. Johnston, and J.M. Moyer. 1994. Crop management in conservation tillage systems. p. 185–209. *In* P. W. Unger (ed.) Managing agricultural residues. CRC Press, Boca Raton, FL.

Lupwayi, N.Z., M.A. Monreal, G.W. Clayton, C.A. Grant, A.M. Johnston, and W.A. Rice. 2001. Soil microbial biomass, and diversity respond to tillage, and sulphur fertilizers. Can. J. Soil Sci. 81:577–589.

Lupwayi, N.Z., W.A. Rice, and G.W. Clayton. 1998. Soil microbial diversity, and community structure under wheat as influenced by tillage, and crop rotation. Soil Biol. Biochem. 30:1733–1741.

Lupwayi, N.Z., W.A. Rice, and G.W. Clayton. 1999. Soil microbial biomass, and carbon dioxide flux under wheat as influenced by tillage, and crop rotation. Can. J. Soil Sci. 79:273–280.

MacEwan, G. 1986. Entrusted to my care. Western Producer Prairie Books, Saskatoon, SK, Canada.

Malhi, S.S., S.A. Brandt, and K.S. Gill, 2003. Cultivation, and grassland type effects on light fraction, and total organic C, and N in a Dark Brown Chernozemic soil. Can. J. Soil Sci. 83:145–153.

Malhi, S.S., and M. Nyborg. 1992. Placement of urea fertilizer under zero, and conventional tillage for barley. Soil Tillage Res. 23:193–197.

Manning, G., L.G. Fuller, R.G. Eilers, and I. Florinsky. 2001. Topographic influence on the variability of soil properties within an undulating Manitoba landscape. Can. J. Soil Sci. 81:439–447.

McConkey, B.G., D. Curtin, C.A. Campbell, S.A. Brandt, and F. Selles. 2002. Crop, and soil nitrogen status of tilled, and no–tillage systems in semiarid regions of Saskatchewan. Can. J. Soil Sci. 82:489–498.

McConkey, B.G., and P.R. Miller. 1999. Row spacing effects on plant populations, canopy closure, water use, and grain yields in the Brown soil zone. p. 176–197. *In* Proc. Soils and Crops 1999, Univ. of Saskatchewan, Saskatoon, SK. 18–19 Feb. 1999. Ext. Div., Univ. of Saskatchewan, Saskatoon, SK.

McKenzie, R.H., A.B. Middleton, E.D. Solberg, J. DeMulder, N. Flore, G.W. Clayton, and E. Bremer. 2001a. Response of pea to rhizobia inoculation, and starter nitrogen in Alberta. Can. J. Plant Sci. 81:637–643.

McKenzie, R.H., A.B. Middleton, and M. Zhang. 2001b. Optimal time, and placement of nitrogen fertilizer with direct, and conventionally seeded winter wheat. Can. J. Soil Sci. 81:613–622.

Miller, J.J., E.G., Kokko, and G.C. Kozub. 1998a. Comparison of porosity in a Chernozemic clay loam soil under long-term conventional tillage, and no-till. Can. J. Soil Sci. 78:619–629.

Miller, J.J., F.J. Larney, and C.W. Lindwall. 1999. Physical properties of a Chernozemic clay loam soil under long-term conventional tillage, and no-till. Can. J. Soil Sci. 79:325–331.

Miller, J.J., N.J. Sweetland, F.J. Larney, and K.M. Volkmar. 1998b. Unsaturated hydraulic conductivity of conventional, and conservation tillage soils in southern Alberta. Can. J. Soil Sci. 78:643–648.

Miller, P.R., A.M. Johnston, S.A. Brandt, C.L. McDonald, D.A. Derksen, and J. Waddington. 1998c. Comparing the adaptation of sunola, canola, and mustard in three soil climatic zones of the Canadian prairies. Can. J. Plant Sci. 78:565–570.

Miller, P.R., B.G. McConkey, G.W. Clayton, S.A. Brandt, J.A. Staricka, A.M. Johnston, G.P. Lafond, B.G. Schatz, D.D. Baltensperger, and K.E. Neill. 2002a. Pulse crop adaptation in the northern Great Plains. Agron. J. 94:261–272.

Miller, P.R., C.L. McDonald, D.A. Derksen, and J. Waddington. 2001. The adaptation of seven broadleaf crops to the dry semiarid prairie. Can. J. Plant Sci. 81:29–43.

Miller, P.R., J. Waddington, C.L. McDonald, and D.A. Derksen. 2002b. Cropping sequence affects wheat productivity on the semiarid northern Great Plains. Can. J. Plant Sci. 82:307–318.

Moran, M.S., Y. Inoue, and E.M. Barnes. 1997. Opportunities, and limitations for image-based remote sensing in precision crop management. Remote Sens. Environ. 61:319–346.

Morrall, R.A.A. 1997. Evolution of lentil diseases over 25 years in western Canada. Can. J. Plant Pathol. 19:197–207.

Moulin, A.P., H.J. Beckie, and D.J. Pennock. 1999. Strategies for variable rate nitrogen fertilization in hummocky terrain. p. 839–846. *In* P.C. Robert et al. (ed.) Proc. 4th Int. Precision Agric. Conf., St. Paul, MN. 19–22 July 1998. ASA, CSSA, and SSSA, Madison, WI.

Moyer, J.R., R.E. Blackshaw, E.G., Smith, and S.M. McGinn. 2000. Cereal cover crops for weed suppression in a summer fallow-wheat cropping sequence. Can. J. Plant Sci. 80:441–449.

Moyer, J.R., E.S. Roman, C.W. Lindwall, and R.E. Blackshaw. 1994. Weed management in conservation tillage systems for wheat production in North, and South America. Crop Prot. 13:243–259.

Nleya, T., F. Walley, and A. Vandenberg. 2001. Response of four common bean cultivars to granular inoculant in a short-season dryland production system. Can. J. Plant Sci. 81:385–390.

Nolan, S.C., T.W. Goddard, D.C. Penney, and F.M. Green. 1999. Yield response to nitrogen with landscape classes. p. 479–488. *In* P.C. Robert et al. (ed.) Proc. 4th Int. Conf. on Precision Agriculture, St. Paul, MN. 19–22 July 1998. ASA, CSSA, and SSSA, Madison, WI.

Nyborg, M., E.D. Solberg, R.C. Izaurralde, S.S. Malhi, and M. Molina-Ayala. 1995. Influence of long-term tillage, straw, and N fertilizer on barley yield, plant-N uptake, and soil-N balance. Soil Tillage Res. 36:165–174.

O'Donovan, J.T. 1996. Computerized decision support systems: Aids to rational and sustainable weed management. Can. J. Plant Sci. 76:3–7.

Oelck, M.M., R. MacDonald, M. Belyk, V. Ripley, B. Weston, C. Bennett, M. Dormann, P. Eckes, A. Schulz, R. Schneider, and M. Gadsby. 1995. Registration, safety assessment, and agronomic performance of transgenic canola cv. "Innovator" in Canada. p. 1430–1432. *In* Vol. 4, Proc. 9th Int. Rapeseed Congr., Cambridge, UK. 4–7 July 1995.

Olfert, O., G.D. Johnson, S.A. Brandt, and A.G. Thomas. 2002. Use of arthropod diversity, and abundance to evaluate cropping systems. Agron. J. 94:210–216.

Olsen, K., P. Collas, P. Boileau, D. Blain, C. Ha, L. Henderson, C. Liang, S. McKibbon, and L. Morel-à-l'Huissier. 2002. Canada's greenhouse gas inventory, 1990–2000. Environment Canada, Ottawa, ON.

Padbury, G., S. Waltman, J. Caprio, G. Coen, S. McGinn, D. Mortensen, G. Nielsen, and R. Sinclair. 2002. Agroecosystems, and land resources of the northern Great Plains. Agron. J. 94:251–261.

Pennock, D.J., and M.D. Corre. 2001. Development and application of landform segmentation procedures. Soil Tillage Res. 58:151–162.

Pennock, D.J., F.L. Walley, M.P. Solohub, and G. Hnatowich. 1999. Yield response of wheat, and canola to a topographically based variable rate fertilization program in Saskatchewan. p. 797–806. *In* P.C. Robert et al. (ed.) Proc. 4th Int. Precision Agric. Conf., St. Paul, MN. 19–22 July 1998. ASA, CSSA, and SSSA, Madison, WI.

Pennock, D., F. Walley, M. Solohub, B. Si, and G. Hnatowich. 2001. Topographically controlled yield response of canola to nitrogen fertilizer. Soil Sci. Soc. Am. J. 65:1838–1845.

Pesticide Free Production Canada. 2003. 2001 Research report. PFPC [Online]. Available at http://www.pfpcanada.com (verified 13 June 2003).

Prairie Farm Rehabilitation Administration. 2000. Prairie agricultural landscapes—A land resource review. PFRA, Agriculture, and Agri-Food Canada, Regina, SK.

Read, W.B., and B.F. Dyck. 1990. Design considerations for seed, and fertilizer openers. p. 83–103. *In* Proc. Air seeding '90. Regina, SK. 19–21 June 1990. Ext. Div., Univ. of Saskatchewan, Saskatoon, SK.

Reimann-Philipp, R. 1995. Breeding perennial rye. Plant Breed. Rev. 13:265–292.

Rice, W.A., P.E. Olsen, L.D. Bailey, V.O. Biederbeck, and A.E. Slinkard. 1993. The use of annual legume green manure crops as a substitute for summer-fallow in the Peace River region. Can. J. Soil Sci. 73:243–252.

Rutherford, A., and M. Gimby. 1990. The viability of organic farm practice. p. 176–182. *In* Transition to Organic Agriculture Conf., Saskatoon. 31 Oct.–1 Nov. 1990. Ext. Div., Univ. of Saskatchewan, Saskatoon, SK.

Schneiter, A.A. 1992. Production of semi-dwarf, and dwarf sunflower in the northern Great Plains of the United States. Field Crops Res. 30:391–401.

Selles, F., S.A. Brandt, C.A. Campbell, B.G. McConkey, and D. Messer. 1999a. Spatial distribution of soil nitrogen supplying power: A tool for precision farming. p. 407–416. *In* P.C. Robert et al. (ed.) Proc. 4th Int. Precision Agric. Conf., St. Paul, MN. 19–22 July 1998. ASA, CSSA, and SSSA, Madison, WI.

Selles, F., B.G. McConkey, and C.A. Campbell. 1999b. Distribution, and forms of P under cultivator-, and zero-tillage for continuous-, and fallow-wheat cropping systems in the semi-arid Canadian prairies. Soil Tillage Res. 51:47–59.

Shirtliffe, S.J., and A.M. Johnston. 2002. Yield-density relationships, and optimum plant populations in two cultivars of solid seeded dry bean (*Phaseolus vulgaris* L.) grown in Saskatchewan. Can. J. Plant Sci. 82:521–529.

Shutt, F.T. 1910. Western prairie soils: Their nature, and composition. Bull. no. 6, Second Series. Dep. of Agriculture, Central Exp. Farm, Ottawa, ON.

Shutt, F.T. 1925. The influence of grain growing on the nitrogen, and organic matter content of the western prairie soils of Canada. J. Agric. Sci. 15(2):162–177.

Smith, A.M. 2003. Remote sensing: What can it offer agronomy? p. 29–33. *In* Proc. 2003 Alberta Agronomy Update Conf., Lethbridge. 13–14 Jan. 2003. Alberta Agric. Food Rural Dev., Lethbridge, AB.

Smith, E.G., M. Lerohl, T. Messele, and H.H. Janzen. 2000a. Soil quality attribute time paths: Optimal levels, and values. J. Agric. Resour. Econ. 25:307–324.

Smith, E.G., Y. Peng, M. Lerohl, and F.J. Larney. 2000b. Economics of fertilizing with inorganic fertilizers to restore wheat yields on three eroded sites in southern Alberta. Can. J. Soil Sci. 80:165–169.

Smith, E.G., T.L. Peters, R.E. Blackshaw, C.W. Lindwall, and F.J. Larney. 1996. Economics of reduced tillage fallow-crop systems in the Dark Brown soil zone of Alberta. Can. J. Soil Sci. 76:411–416.

Smith, E.G., and D.L. Young. 2001. The economic, and environmental revolution in semi-arid cropping in North America. Ann. Arid Zone 39:347–361.

Smith, E.G., and D.L. Young. 2003. Cropping diversity along the U.S.-Canada border. Rev. Agric. Econ. 25:154–167.

Smith, E.G., D.L. Young, and R.P. Zentner. 2001a. Prairie crop diversification. Current Agric., Food & Resource Issues 2/2001:37–47 [Online]. Available at http://www.CAFRI.org (verified 5 June 2003).

Smith, W.N., R.L. Desjardins, and B. Grant. 2001b. Estimated changes in soil carbon associated with agricultural practices in Canada. Can. J. Soil Sci. 81:221–227.

Soon, Y.K., and G.W. Clayton. 2002. Eight years of crop rotation, and tillage effects on crop production, and N fertilizer use. Can. J. Soil Sci. 82:165–172.

Statistics Canada. 1997. 1996 Census of agriculture, provincial profiles. Ministry of Industry, Gov. of Canada, Ottawa, ON.

Statistics Canada. 2002. 2001 Census of agriculture [Online]. Available at http://www.statcan.ca:80/english/freepub/95F0301XIE/tables/htm (verified 5 June 2003).

Stevenson, F.C., A.M. Johnston, H.J. Becky, S.A. Brandt, and L. Townley-Smith. 1998. Cattle manure as a nutrient source for barley, and oilseed crops in zero, and conventional tillage systems. Can. J. Plant Sci. 78:409–416.

Stevenson, F.C., J.D. Knight, O. Wendroth, C. van Kessel, and D.R. Nielsen. 2001. A comparison of two methods to predict the landscape-scale variation of crop yield. Soil Tillage Res. 58:163–181.

Strange, H.G.L. 1954. A short history of prairie agriculture. Searle Grain Company Limited, Winnipeg, MB.

Stumborg, M., L. Townley-Smith, and E. Coxworth. 1996. Sustainability, and economic issues for cereal crop residue export. Can. J. Plant Sci. 76:669–673.

Thiessen Martens, J.R., J.W. Heppner, and M.H. Entz. 2001. Legume cover crops with winter cereals in southern Manitoba: Establishment, productivity, and microclimate effects. Agron. J. 93:1086–1096.

Tillman, D., K.G. Cassman, P.A. Matson, R. Naylor, and S. Polasky. 2002. Agricultural sustainability, and intensive production practices. Nature (London) 418:671–677.

Vladicka, B., and R. Cunningham. 2002. Snapshot: Organics, a profile of the organic industry, and its issues. Strategic Information Serv. Unit, Alberta Agric. Food Rural Dev., Edmonton, AB.

Waite, D.T., A.J. Cessna, R. Grover, L.A. Kerr, and A.D. Snihura. 2002. Environmental concentrations of agricultural herbicides: 2, 4-D, and triallate. J. Environ. Qual. 31:129–144.

Walley, F.L., G.P. Lafond, A. Matus, and C. van Kessel. 1999. Water-use efficiency, and carbon isotopic composition in reduced tillage systems. Soil Sci. Soc. Am. J. 63:356–361.

Walley, F., D. Pennock, M. Solohub, and G. Hnatowich. 2001. Spring wheat (*Triticum aestivum*) yield, and grain protein responses to N fertilizer in topographically defined landscape positions. Can. J. Soil Sci. 81:505–514.

Xie, H.S., D.R.S. Rourke, and A.P. Hargrave. 1998. Effect of row spacing, and seed/fertilizer placement on agronomic performance of wheat, and canola in zero tillage systems. Can. J. Plant Sci. 78:389–394.

Zentner, R.P., S.A. Brandt, and C.A. Campbell. 1996a. Economics of monoculture cereal, and mixed oilseed-cereal rotations in west-central Saskatchewan. Can. J. Plant Sci. 76:393–400.

Zentner, R.P., and C.A. Campbell. 1988. First 18 years of a long-term crop rotation study in southwestern Saskatchewan—Yields, grain protein, and economic performance. Can. J. Plant Sci. 68:1–21.

Zentner, R.P., C.A. Campbell, V.O. Biederbeck, and F. Selles. 1996b. Indianhead black lentil as green manure for wheat rotations in the Brown soil zone. Can. J. Plant Sci. 76:417–422.

Zentner, R.P., G.P. Lafond, D.A. Derksen, and C.A. Campbell. 2002a. Tillage method, and crop diversification: Effect on economic returns, and riskiness of cropping systems in a thin Black Chernozem of the Canadian prairies. Soil Tillage Res. 67:9-21.

Zentner, R.P., and C.W. Lindwall. 1982. Economic evaluation of minimum tillage systems for summer fallow in southern Alberta. Can. J. Plant Sci. 62:631–638.

Zentner, R.P., B.G. McConkey, C.A. Campbell, F.B. Dyck, and F. Selles. 1996c. Economics of conservation tillage in the semiarid prairie. Can. J. Plant Sci. 76:697–705.

Zentner, R.P., B.G. McConkey, M.A. Stumborg, C.A. Campbell, and F. Selles. 1998. Energy performance of conservation tillage management for spring wheat production in the Brown soil zone. Can. J. Plant Sci. 78:553–563.

Zentner, R.P., D.D. Wall, C.N. Nagy, E.G., Smith, D.L. Young, P.R. Miller, C.A. Campbell, B.G. McConkey, S.A. Brandt, G.P. Lafond, A.M. Johnston, and D.A. Derksen. 2002b. Economics of crop diversification, and soil tillage opportunities in the Canadian prairies. Agron. J. 94:216–230.

9 Crop Diversification for Dryland Agriculture in Central Asia

Raj. S. Paroda and Mekhlis Suleimenov
ICARDA
Tashkent, Uzbekistan

Hasan Yusupov
Gallaaral Grain Research Institute
Uzbekistan

Aitkalym Kireyev
Crop Husbandry Research Institute
Kazakhstan

Rahim Medeubayev
Krasniy Vodopad Station
Kazakhstan

Lyudmila Martynova
Cropping Systems Research Institute
Kyrgyzstan

Khasan Yusupov
Galla-Aral Grain Research Institute
Uzbekistan

ABSTRACT

Dryland agriculture is the major production system in Kazakhstan. In the rest of Central Asia, irrigated agriculture is predominant. In the northern Kazakhstan, spring wheat (*Triticum aestivum* L.) is the major crop grown traditionally in summer fallow, which occupies 20% of the cropped area. Studies were conducted to find out the possibilities to reduce the area under fallow, since this practice mainly leads to soil erosion and is not efficient for moisture conservation. On the contrary, continuous growing of small grains proved to be efficient both in terms of production and economy at different levels of inputs. In the rest of Central Asia, winter wheat is traditionally grown after summer fallow. In the arid areas, fallow-wheat rotation the best farming practice in case if the wheat price is remunerative. In semiarid areas, the replacement of summer fallow with food legumes or safflower (*Carthamus tinctorius* L.) has been found to be economical especially due to prevailing low wheat prices. Alternative crops identified both for the introduction and replacement of wheat are: field pea (*Pisum sativum* L.), lentil (*Lens culinaris* Medik.), chickpea (*Cicer arietinum* L.), mustard (*Brassica junceus* L.), oat (*Avena sativa* L.), buckwheat (*Fagopyrum esculentum* Moench) and millet (*Panicum miliaceum* L.) in northern Kazakhstan; whereas chickpea, field pea and safflower in southern Kazakhstan, Kyrgyzstan, and Uzbekistan.

INTRODUCTION

Central Asia represents three major climatic zones. First, the northern Kazakhstan steppes with sharp continental semiarid climate characterized by long and cold winters and short, dry and hot summers. Spring wheat is the major crop, which is grown as rainfed in rotation with summer fallow practiced once in 3 to 5 yr. Second, the Southern part of Central Asia including Uzbekistan, Tajikistan, and Turkmenistan, representing the northern dry subtropics and characterized by cool winters and very dry hot summers. In this area, till recently, mainly cotton (*Gossypium hirsutum* L.) was grown under irrigation in rotation with alfalfa (*Medicago sativa* L.), now getting replaced by winter wheat. Third, the middle part of the region, occupied by Kyrgyzstan and southern Kazakhstan, is closer climatically to the south of Central Asia but with lower temperatures throughout the year. Winter wheat is the main crop grown both under irrigated and rainfed conditions.

Dryland agriculture in Central Asia is mainly practiced in Kazakhstan. Although the dryland agriculture is also practiced in four other countries of the region, their major production is based on irrigated agriculture. In Kazakhstan, however, the total dryland area accounts for more than 15 million ha, whereas it is hardly around half million hectares in the rest of Central Asia. Hence, mostly this chapter considers the issues of dryland agriculture in Kazakhstan, with the emphasis on crop diversification.

The basic component of dryland farming is a summer fallow. The summer fallow occupies a significant part of dryland agriculture in many countries, such as USA (Black and Bauer, 1983), Canada (Johnson, 1983), and Turkey (Braun, 1999). In fact, wheat-fallow system is widespread. Also, some successful attempts have been made to introduce alternative crops such as: rapeseed in Canada (Johnson, 1983), and sorghum [*Sorghum bicolor* (L.) Moench] in USA (Jones and Johnson, 1983).

During 1950s, a spring wheat-fallow system in Kazakhstan was adopted on the area over 25 million ha based on Canadian experience and considering the fact that soil and weather conditions in northern Kazakhstan were to a great extent similar. After years of experimentation, it became an accepted practice to keep 20 to 25% of cropped area as fallow (Barayev, 1960). Thus, around 5 million hectares of cropland was left fallow every year, being a major source of soil erosion both by wind and water. Later, it was recognized to further reduce a share of fallow to 15 to 20% (Barayev, 1985). Three years later, it was realized that continuous wheat growing not only gave more grain yield but also solved the problem of soil erosion (Suleimenov, 1988).

Subsequently, on the breakdown of the Former Soviet Union, more than 10 million ha of marginal lands being used earlier for wheat production have been abandoned, resulting in significant changes in the farming structure. The main change was the transition from the centrally planned production system to a market oriented one with liberalization of both input and output prices. In the process, also the restriction on fallow land relaxed so the farmers could decide which cropping pattern to adopt.

The scientific studies concerning the possibility of reducing the summer fallow started in 1983, resulting in almost 20-yr data on this aspect. In this chapter, we present the findings from 1998 to 2002 since the Consultative Group on International Agricultural Research (CGIAR) Program for Central Asia and the Caucasus (CAC) got started in Central Asia and the Caucasus.

In southern Kazakhstan, Kyrgyzstan, and Uzbekistan, some dryland area is used for crop production using 3 to 4 yr crop rotation cycle with 1 yr under summer fallow. Winter wheat is the major crop in these rotations. Some area is also occupied by barley (*Hordeum vulgare* L.) but no other crops are grown presently. The major emphasis of the CGIAR Program was on crop diversification and increased cropping intensity.

EXPERIMENTAL DETAILS

Shortandy, Akmola province has been used as a research site in northern Kazakhstan. It is located at 51.5° N, 71° E. The average annual precipitation rate is 350 mm distributed throughout the year with the maximum rainfall in July. One third of the precipitation is from snow, which plays an important role in moisture accumulation prior to sowing of spring wheat. Climate is sharp continental with average annual temperature of 1.2°C: the summer is short, dry and hot, whereas winter is long and very cold. Duration of vegetation period is around 100 d, which is just enough for early-maturing varieties of spring wheat and barley.

Studies on crop diversification at Shortandy were conducted during 2000 to 2002. A total of 10 crops were planted on the fallow. Small grains included bread wheat (*Triticum aestivum* L.) and durum wheat (*T. durum* Desf.), millet, and buckwheat (*Fagopirum esculentum*). Oilseed crops included rapeseed (*B. napus* L.), mustard, and false flax (*Camerlina sativa*). Three food legumes were planted: field pea, chickpea, and lentil (*Lens culinaris* Medik.). All crops were planted in the second half of May. The bread wheat was used as control. In the following year, spring wheat was planted after these 10 crops to select the best preceding crop.

In southeastern Kazakhstan, the site Almalyback is located in Almaty province at 43.5° N, 77° E. A total annual precipitation rate in semiarid steppe is 350 mm, distributed unevenly throughout the year with more rainfall during winter-spring period. Climate is continental but much warmer than in the north: annual average temperature being 7.5°C. Winter wheat is the main crop usually sown after fallow. Different small grains, food grains, and oilseed crops were studied in the arid (annual precipitation 220 mm) and semiarid (350 mm) conditions. Irrigated farming system is predominant in this region.

The Krasniy Vodopad site in southern Kazakhstan is located at 42° N, 70° E. It is semiarid steppe with an average annual precipitation rate of 420 mm, distributed unevenly with the maximum in winter-early spring season. The climate is continental, but much warmer than in the southeast with an average annual temperature of 14.1°C. The main crop is winter wheat usually sown after fallow. Safflower, being an alternative, was compared with wheat and barley. In a crop rotation experiment, winter wheat sown during 2 yr after summer fallow and after alfalfa was compared with wheat sown after chickpea and continuous wheat.

Zhany pahta site in Kyrgyzstan is located at 43° N, 74° E. The annual average precipitation rate is 300 mm, whereas average annual temperature is 8.5°C. Winter wheat sown after fallow was compared with wheat grown after field pea and safflower. Crops were studied using fertilizers at the rates of 60 kg ha^{-1} (P) applied during fallow or before planting crops to replace summer fallow, and 45 kg (N) that was applied to wheat. Two more treatments of improved fallow were studied as well. These were: (i) manure applied at the rate of 30 t ha^{-1} and (ii) in corporation of straw of the harvested crop.

Gallaaral, Uzbekistan is located at 40° N, 68.5° E. Average annual precipitation is 350 mm, with an average annual temperature of 15°C. Fallow-wheat was compared with wheat planted after chickpea, field pea, lentil, and safflower and with continuous cropping of wheat.

RESULTS

Northern Kazakhstan

During 2000 to 2002, being quite favorable for wheat production, Kazakhstan produced a surplus of grain that it was unable to export. This forced the policymakers to diversify the production system, replacing part of the wheat-sown areas. It was during this period when the International Center for Agricultural Research in the Dryland Areas (ICARDA) in partnership with the National Agricultural Research System (NARS) conducted studies on crop diversification in partnership with Kazakh scientists.

Comparative studies on various crops with spring bread wheat revealed the possibility of growing alternative crops that was associated with increased income to the farmers (Table 9–1).

Bread wheat is the major crop occupying up to 80% of the sown area in Kazakhstan. The yields obtained were higher than average because of favorable weather conditions, especially owing to rainfalls. Though the yields were good, the profit margin was rather low (22%) because of low wheat prices.

Durum wheat is not grown in the region, mainly because the prices have not been remunerative until recently. But it provided, on an average, comparable grain yield to that of bread wheat and also provided better profit margin (61%) in view of currently prevailing high prices.

Proso-millet, on an average, gave significantly lower grain yields than bread wheat. It fared well only in relatively dry years, being drought resistant and more suitable for light-textured soils of Pavlodar and Aktobe provinces. Also, it proved to be more profitable than bread wheat with profit margin as high as 78%.

Table 9–1. Yield of different crops planted after fallow at Shortandy, Northern Kazakhstan.

Crops	Year 2000	2001	2002	Mean	Bread wheat
	Grain yield (t ha^{-1})				%
Bread wheat	1.93	2.81	2.61	2.45	100
Durum wheat	1.74	2.90	2.57	2.40	98
Millet	2.15	1.28	1.58	1.67	68
Buckwheat	0.84	1.98	1.58	1.47	60
False flax	0.74	1.23	0.86	0.94	38
Mustard	0.12	1.74	1.57	1.14	46
Rapeseed	0.13	1.09	1.42	0.88	36
Chickpea	1.71	--	1.28	--	66
Field pea	1.96	2.17	1.94	2.02	66
Lentil	1.00	2.14	1.16	1.43	58
LSD(0.05)	0.15	0.22	0.18		

Buckwheat produced only 60% of bread wheat, and yet gave better profit margin (103%). Buckwheat also performed well under favorable weather conditions, revealing that the crop is to be grown in small areas with better moisture storage before planting.

Oilseeds mostly failed because of the damage caused by insect pests, despite treatments applied to control them. The best was mustard, which did well under favorable rainfall conditions. On an average, the profit margin reached 48% even with crop failure in 1 yr.

In general, food legumes have demonstrated good results. The best was field pea followed by lentil with comparable profit margin of 170 and 176%, respectively. Potentially, lentil may overcome field pea as market prices get stabilized. Chickpea failed in one of the 3 yr, since the Russian variety used was susceptible to *Ascochyta blight*. New chickpea varieties from ICARDA have been identified as resistant to *Ascochyta blight*, but they are not released yet. Based on 2 yr data, chickpea was comparable to field pea and better than lentil.

Experiment on spring wheat, planted after different winter crops enabled to identify suitable preceding crops (Table 9–2).

As is evident, there had been no significant difference between small grains as preceding crops compared to bread wheat, while mustard and food legumes (especially chickpea and field pea) demonstrated some advantages. This revealed that food legumes were useful in crop rotations to break a continuous wheat cycle that has been practiced for a long time. Especially, chickpea yielded 66% of bread wheat but had the advantage of higher prices. Being a legume crop, chickpea is also good from the point of view of sustainability.

Also, lentil proved to be quite competitive and demonstrated comparative advantage, although its large-scale adoption may take time since lentil is not a popular food yet and being short statured, it would need manual harvesting, which may be difficult in view of a labor shortage.

Canola (rapeseed) also was found as a good preceding crop and could be used as an alternative crop. Yet the farmers are not familiar with its cultivation and cer-

Table 9–2. Grain yield of spring wheat planted after different crops at Shortandy, northern Kazakhstan

Preceding crops	Year 2000	2001	2002	Mean	Bread wheat
	Grain yield (t ha^{-1})				%
Bread wheat	1.28	1.76	1.94	1.66	100
Durum wheat	1.12	1.80	1.90	1.61	97
Millet (proso)	0.99	2.01	1.86	1.62	98
Buckwheat	1.00	1.71	1.91	1.54	93
Mustard	1.31	1.95	2.90	2.05	123
False flax	1.17	1.90	1.89	1.65	99
Rape	1.42	2.03	2.02	1.82	110
Chickpea	--	2.00	2.28	2.14	116
Field pea	1.39	2.40	2.28	2.02	122
Lentil	1.33	1.76	2.14	1.74	105
LSD(0.05)	0.19	0.15	0.14		

tainly it requires better crop management practices (including fertilizers and pesticides).

Southeastern Kazakhstan

Studies in southeastern Kazakhstan were conducted in arid and semiarid rainfed conditions and also under irrigation. In the arid steppe conditions, crops failed on the stubble in dry year. In the first year, the trial also included sowing on stubble land, because in the past often marginal lands were used for grain production. During the second year, the crops were sown only after fallow (Table 9–3).

For the arid dryland area, there can be no other crop-growing practice except alternate summer fallow and a crop. In dry years, crops can be obtained only on fallow. Even on fallow, many crops failed due to dry spell in 2001. Spring wheat produced 58% as compared to winter wheat. Hence, spring wheat should not be recommended for growing in this region. Farmers still sometimes grow spring wheat

Table 9–3. Comparative crop yields on fallow in arid rainfed area of southeastern Kazakhstan

Crop	Year 2001	2002	Mean
	Grain yield (t ha^{-1})		
Winter wheat	0.63	1.79	1.21
Spring wheat	0.40	1.00	0.70
Spring barley	0.68	2.27	1.48
Oats	0.38	2.40	1.39
Millet (proso)	0.37	1.39	0.88
Grasspea	0.10	1.06	0.58
Lentil	0.29	0.89	0.59
Chickpea	0.32	0.99	0.65
Field pea	0.13	--	--
Safflower	0.77	0.85	0.81
LSD(0.05)	0.11	0.13	

Table 9–4. Comparative grain yields on stubble land in semiarid rainfed area in southeastern Kazakhstan.

Crop	Year 2001	Year 2002	Mean
	Grain yield (t ha^{-1})		
Winter wheat	--	2.87	--
Spring wheat	0.37	2.10	1.24
Spring barley	0.85	2.36	1.61
Oats	0.63	3.35	2.00
Millet (proso)	0.97	1.14	1.06
Chickpea	0.78	0.95	0.87
Grasspea	0.72	2.32	1.52
Lentil	--	0.77	--
Safflower	0.84	1.25	1.05
LSD(0.05)	0.28	0.35	

when for some reasons they fail to plant winter wheat or the crop fails during winter. Barley is a suitable crop for planting in spring ensuring grain yields equal to winter wheat in a dry year and the higher yield in a favorable year. Oats provided variable grain yields: higher than barley in a wet year and lower in a dry year. Proso-millet was found less promising as a spring crop.

Food legumes were found to be low yielding (half the yield of winter wheat), similar to the spring wheat, but they appeared more profitable, taking into consideration that market prices exceed wheat price by almost three to four times. Thus, food legumes could be good alternative crops replacing part of wheat area. Safflower is another alternative oilseed crop with good potential. The yield of safflower was about 67% of wheat. In 2002, wheat prices went down while safflower prices went up, thus making it highly profitable to the farmers. The rotations found promising in arid area were also repeated in the area with more reliable rainfall, when grown on stubble land (Table 9–4).

Spring barley was a more reliable crop than spring wheat especially in the dry year. Oats gave less stable yields compared to barley: higher in wet year and lower in dry year. Proso-millet failed and may not be a good alternative. Food legumes, though lower in yield than winter wheat, could still be good alternatives under a low price scenario for wheat. Grasspea (*Lathyrus sativum* L.) demonstrated rather high grain yield especially in the wet year.

Southern Kazakhstan

Studies in southern Kazakhstan were conducted under rainfed conditions at the Krasniy Vodopad station for consecutive 2 yr. In this region, rainfed winter wheat is a major crop. Spring barley also plays an important role especially when winter wheat fails or has not been planted. In recent years, safflower as an oil seed crop has also become popular and it currently occupies significant area in the region.

Wheat outyielded barley by 4 to 7%. Safflower produced only half of wheat yield. However, safflower has specific advantages being an oil crop with high price in the market. Also, the crop can be grown on marginal lands. Safflower also provides oil cake to feed the animals. In other words the crop is a good cash crop even

Table 9–5. Economic returns from three crops under varying price scenario in southern Kazakhstan

Crops	July 2002	September 2002
	Net profit (US$ per 1 ha)	
Wheat	149.80	78.60
Barley	98.20	66.20
Safflower	43.70	130.20

when processing companies give low net prices. Therefore, there are good prospects for safflower and the area may increase including some areas under irrigated conditions where supplemental irrigation could give good yields.

Market in Kazakhstan had been changing dramatically since the fall of 2002, adversely affecting the economic assessment and once again favoring the crop diversification. Because of three consecutive favorable years in Kazakhstan, the wheat production went up resulting in fall in prices (Table 9–5).

Wheat price on the market in city Shymkent dropped from US$71.40 in July to $45.40 in September, while safflower prices went up from $64.90 to $142.90. This has happened because of considerable wheat and barley yields in the northern plains and a shortage of oil seeds in the country. Thus, wheat was three times more profitable than safflower in July and became less profitable by 51% in September.

During 3 yr of research at the Krasniy Vodopad station, in addition to recommended winter wheat growing during 2 yr after summer fallow and 2 yr after alfalfa, it was winter wheat sown after chickpea. Out of 3 yr, 2000 and 2001 were typically dry, and 2002 was rainy (60% more rainfall than usual rate). The data obtained very distinctly showed that chickpea is a very good alternative to produce food legume and improve grain yield of wheat (Table 9–6).

Continuous winter wheat produced a double crop in the wet year, but on an average, its yield was rather low. In both dry and favorable years, summer fallow provided reliable wheat yield, increased on an average by 76%. However, in the second year after summer fallow, the wheat yield fell down dramatically and gave only 23% advantage over the continuous wheat. Alfalfa grown over 4 yr proved to be a very good preceding crop, providing 63% higher wheat yield compared to continuous cropping, and 93% of wheat sown on summer fallow. Remarkably, the second wheat crop after alfalfa gave higher yield than wheat sown for a second year after fallow. Interestingly, chickpea proved to be as good as alfalfa for a consecu-

Table 9–6. Wheat grain yield as affected by crop sequence in southern Kazakhstan.

Fallow or preceding crop	Year 2000	2001	2002	Mean	Percent of continuous wheat
	Grain yield (t ha^{-1})				
Continuous wheat	0.70	0.96	1.78	1.15	100
Summer fallow	1.62	1.42	3.01	2.02	175
Wheat after fallow	0.93	1.10	2.22	1.42	123
Alfalfa	1.47	1.32	2.86	1.88	163
Wheat after alfalfa	1.27	1.40	2.17	1.61	140
Chickpea	1.53	1.38	2.74	1.88	163

tive wheat crop. This is explained by the fact that legumes improved N availability for the crop. The fallow increased nitrate (NO_3^-) content in an arable land at sowing and heading time compared to continuous wheat by 30 to 34%, alfalfa by 41 to 46%, and chickpea by 34 to 42%. Very important to note that carry over of alfalfa was noticeable during 2 yr, while during the second year after fallow, the content of NO_3^- was only 17 to 25% higher than in continuous wheat. Thus, in the conditions of southern Kazakhstan the best cropping system proved to be growing of wheat during 1 yr after fallow and chickpea, and 2 yr after alfalfa.

Chickpea itself gave very low grain yields in dry years: 0.29 to 0.45 t ha^{-1} and good yield (1.24 t ha^{-1}) in the wet year. On an average, chickpea grain yield was almost half of wheat sown continuously. Yet, the crop could provide good income to the farmers, because the market prices were four to fivefold higher than for wheat.

Kyrgyzstan and Uzbekistan

In Kyrgyzstan, commonly adopted crop rotation in dryland farming is the summer fallow followed by growing grains 2 to 3 yr continuously. In the experiment, three types of fallow were studied with an attempt to improve yields by application of manure or straw. It was also tested against replacement of fallow by sowing crops: chickpea, dry pea, and safflower. The most important advantage of summer fallow is believed to be moisture accumulation. In both years, there was no significant difference between fallow treatments in soil moisture content in spring. In a dry year, application of manure did not improve moisture accumulation in fallow. Growing of crops reduced soil moisture by 12 to 14%. The wheat grain yield was not significantly affected by improved fallow practices and was even reduced by replacement of fallow with crops such as safflower, but not so with dry pea (Table 9–7).

Types of fallow did not affect significantly the wheat yield. Application of manure during fallow period compared with commercial fertilizer gave some positive results. The interesting result was obtained when dry pea was sown instead of summer fallow. It was seen that the wheat yield was only slightly lower than the one after summer fallow. At the same time, dry pea produced 1.34 t ha^{-1} of grain without fertilizers and 1.58 t ha^{-1} with fertilizers. Chickpea was tested only in 2002

Table 9–7. Winter wheat grain yield as affected by summer fallow or preceding crops in dryland conditions of Kyrgyzstan.

Fallow or preceding crop	No fertilizer			Fertilizer		
	2001	2002	Mean	2001	2002	Mean
	Grain yield (t ha^{-1})					
Fallow	1.36	2.66	2.01	1.56	3.24	2.40
Fallow + manure	1.43	2.39	1.91	1.70	3.36	2.53
Fallow + straw	1.28	2.64	1.96	1.48	3.10	2.29
Dry pea	1.29	2.50	1.90	1.48	3.08	2.28
Safflower	0.93	2.19	1.56	1.14	2.51	1.82
LSD(0.05)	0.06	0.37		0.08	0.16	

Table 9–8. Crop grain yields as affected by preceding fallow or crop in Gallaaral, Uzbekistan.

Crop	Year 2000	2001	2002	Mean	Wheat
	Grain yield (t ha^{-1})				%
Continuous wheat	0.78	0.92	1.23	0.98	100
Wheat after fallow	1.03	1.08	1.99	1.37	140
Wheat, 2nd after fallow	0.58	1.18	1.80	1.26	128
Wheat after chickpea	0.82	1.12	1.48	1.14	116
Chickpea on stubble	0.22	0.20	0.95	0.48	49
Safflower on stubble	0.42	0.38	0.78	0.53	54

and provided grain yield comparable to field pea. Safflower did not seem to be a good crop for winter wheat as it reduced wheat yield by 24% as compared to fallow. On the contrary, safflower crop produced 1.46 to 1.95 t ha^{-1} of seeds without fertilizer and with fertilizer, respectively. This obviously compensated the reduction in wheat yield.

An economic assessment demonstrated that the existing practice of fallow-wheat rotation was the worst possible scenario for farm profitability. The most important factor were the market prices, which were not in favor of wheat. During 2002, market prices for 1 metric tonne of different crops were: wheat, \$130; dry pea, \$425; and chickpea, \$930. Hence, the net profit from 1 ha of wheat, dry pea, and chickpea amounted to \$87, \$717, and \$1355, respectively. Even if chickpea prices were somewhat lower and farmers harvested chickpea by hand, it would still be more profitable than wheat. Accordingly, in a favorable year, the farmer can absorb losses incurred during the dry year even if chickpea fails. Remarkably, fertilizers were not profitable for wheat reducing the net profit per hectare from \$87 to \$27, whereas application of fertilizer for field pea and chickpea was found profitable, increasing the net profit by \$28 and \$105, respectively.

In dryland of Uzbekistan, winter wheat is the only crop, and it is planted 2 to 3 yr continuously running after summer fallow. These studies conducted during 3 yr, in general, favored crop diversification (Table 9–8).

The best wheat yield was obtained after fallow as usual. This was obviously the basis for an earlier recommendation to practice fallow once every 3 yr. But the total production in this kind of rotation would not be more than from continuous wheat. On an average, each year this rotation provided as much as 0.88 t ha^{-1} while continuous wheat gave 0.98 t ha^{-1} every year. We do not conclude that the fallow should be eliminated, but it is obvious that continuous winter wheat did not fail even in very dry years. On the other hand, fallow may cause both wind and water erosion.

Replacement of fallow with chickpea was another alternative studied. In both the dry years, with very low rainfall chickpea failed. However, in a wet spring it gave a yield almost half that of wheat. In a dry year, lentil, grasspea, and vetch failed, while in 2002 lentil and buckwheat gave yields comparable to chickpea. Buckwheat was tested only in 2002 and gave 0.85 t ha^{-1}. The wheat yield planted after chickpea was 16% higher than in continuous wheat. One may conclude from this fact that a farmer can use flexible strategy by which he can decide about the cropping system, considering the three scenarios. If weather conditions in the fall are not favorable for planting winter wheat on stubble, plant only on fallow. If weather conditions in spring are dry, leave the rest of the cropland fallow. If soil moisture conditions allow to grow spring crops, planting chickpea, lentil, buckwheat, or safflower will be successful.

DISCUSSION AND CONCLUSION

The data obtained in northern Kazakhstan from studies on alternate crops have indicated that numerous crops could successfully be used for diversification. Considering the production of durum wheat, there appears no problem and the area could be increased quickly provided its price is 20 to 30% higher than for bread wheat. Proso millet is another alternative crop familiar to the producers in the area with light-textured soils (Pavlodar, Aktobe) and its area could also be increased provided good prices prevail. Local people traditionally like proso porridge and hence it is a promising crop for diversification. Oat is another crop with yield potential equal to barley giving about 25 to 30% higher yield than wheat.

It may, however, be appreciated that alternative crops will not occupy large areas and hence cannot replace wheat. They may be grown by small farmers to ensure diversification for increased income, whereas large-scale farmers may continue to grow wheat.

In the context of diversification, food legumes appear to be the best alternatives. Kazakhstan farmers are familiar with field pea cultivation. In the 1960s, when former leader of the Soviet Union Mr. N. Khrushchov came up with an idea to eliminate summer fallow and diversify production, he emphasized the role of maize (*Zea mays* L.) and also the food legumes, which resulted in introduction of crops like maize, field pea, and faba bean in the northern Kazakhstan. As a result, the area under field pea touched 300 000 ha and for faba bean about 20 000 to 30 000 ha in 1964. Subsequently, only maize remained as an important crop due to its use as silage, whereas areas under food legumes dropped, since policies of Mr. Khrushchov could no longer be pursued. At the same time, the breeding program on food legumes was terminated, despite good performance of both field pea and chickpea during the 1950s in northern Kazakhstan. Based on a 9-yr average, yield of wheat at Shortandy was 1.59 t ha^{-1}, whereas chickpea yielded 2.08 t ha^{-1} and field pea gave 2.63 t ha^{-1} (Barayev, 1960). Unfortunately, this data had never been taken note of for crop diversification, as the Government policy only favored growing of wheat crop to meet the food security.

It has been observed that there are many options for crop diversification. The food legumes under favorable weather conditions produce grain yields comparable to that of spring wheat. Even in dry years, despite the yield being low, chickpea, lentil, and field pea had better profit margin. Unfortunately, so far there is no good market potential for the food legumes in Kazakhstan, unless these are marketed to the other countries of Central Asia where prices are much higher. The limiting factor for promotion of food legumes presently is the release of high-yielding disease resistant varieties and their seed production. The other important activity for the region is quality of food legumes and their processing, as these crops are unknown in the region. Even in southern Kazakhstan, food legumes are not consumed as food, whereas just across the border in Uzbekistan they are quite popular and consumed regularly.

The data obtained in the southern and southeastern Kazakhstan, Kyrgyzstan, and Uzbekistan revealed that there are good alternatives to the widespread practice of sowing winter wheat for 2 yr after summer fallow. On the contrary, food legumes, such as chickpea in southern Kazakhstan and field pea in Kyrgyzstan have been found to be good preceding crops for winter wheat. Hence, the best strategy for growing these crops would be to take advantage of moisture conservation practices prior to the sowing. Yet despite low yields, food legumes always proved to be more profitable than wheat.

Chickpea is currently grown on large areas in northern Kazakhstan. At the Dvurechniy farm, Akmola region chickpea was planted on 1145 ha on stubble land and produced almost 1 t ha^{-1} grain yield. In many farms, chickpea is now grown on smaller areas. On the contrary, in southern Kazakhstan, safflower was planted in an area of 70 000 ha during 2002 and is slowly gaining popularity.

In conclusion, there are opportunities for crop diversification in the existing predominant rainfed fallow-spring wheat system followed in northern Kazakhstan. Chickpea, lentil, buckwheat, and dry pea are the potential crops for wider adoption both for the increased income and sustainability of the production systems. For the rainfed agriculture of southern Kazakhstan and Kyrgyzstan, crops like safflower, dry pea, and chickpea appear to be good alternatives to wheat crop with possibilities for higher income to the farmers. Under rainfed conditions of Kazakhstan and Kyrgyzstan, research results have revealed possibilities to reduce summer fallow area partially by growing legume crops for required diversification and long-term sustainability.

REFERENCES

Barayev, A. 1960. Cropping structure and crop rotations in grain farming of northern Kazakhstan. Zemledeliye (Moscow) 7:17–25.

Barayev, A. 1968. Virgin land and science. J. Vestnik selskokhozyastvennoy nauki (Moscow) 1:24–30.

Barayev, A. 1985. To take care of future five year crop production. p.162–166. *In* Barayev. Pochvozashchitnoye zempledeliye, Novosibirsk, 1988. Akademia Nauk RK.

Black, A.L. and A. Bauer. 1983. Soil conservation: Northern Great Plains. p. 247–256. *In* H.E. Dregne and W.O. Willis (ed.) Dryland agriculture. Agron. Monogr. 23. ASA, CSSA, and SSSA, Madison, WI.

Braun, H.-J. 1999. Prospects of Turkey's wheat industry, breeding and biotechnology. p. 1–13. *In* H. Ekiz (ed.) Proc. of the Wheat Symp., Konya, Turkey. 8–11 June 1999. Bahri Dagdash Int. Winter Grains Res. Center. Gurjan ofset, Konya, Turkey.

Johnson, W.E. 1983. Cropping practices: Canadian prairies. p. 407–417. *In* H.E. Dregne and W.O. Willis (ed.) Dryland agriculture. Agron. Monogr. 23. ASA, CSSA, and SSSA, Madison, WI.

Jones, O.R., and W.C. Johnson. 1983. Cropping practices: Southern Great Plains. p. 365–383. *In* H.E. Dregne and W.O. Willis (ed.) Dryland agriculture. Agron. Monogr. 23. ASA, CSSA, and SSSA, Madison, WI.

Suleimenov, M. 1988. About theory and practice of crop rotations in northern Kazakhstan. Zemledeliye (Moscow) 9:25–31.

10 Dryland Cropping in Australia

John F. Angus
CSIRO Plant Industry
Canberra, Australia

Anthony J. Good
Conimbla Road
Cowra, Australia

ABSTRACT

Until the 1980s, almost all Australian dryland crops were cereals grown in rotation with annual pastures and fallows. Rotations are changing rapidly with intensification of cropping and diversification from cereals to pulses and oilseeds. For the past 140 yr, the area of dryland crops has increased at an annual rate of 3.2% and shows no sign of reaching a plateau. Wheat (*Triticum aestivum* L.) yields rose 30% during the 1990s after several decades of stagnation. The trigger for the increase was reduced root disease of wheat grown after broadleaf break crops, followed by increased N fertilizer applied to the healthier crops. Australian farmers and agronomists are optimistic that dryland farming systems can become more profitable and sustainable.

INTRODUCTION

The area of potentially arable land in Australia is about 100 million hectares, equivalent to 1.4% of the continental land area. About half of this potentially arable land is still covered by the original vegetation, mostly woodland in northern Australia, and is used for extensive grazing. Of the area that is both arable and cleared, about 20 million hectares is used annually for crop production at the beginning of the 21st century, the remaining 30 million hectares being used for sheep (*Ovis aries*) and cattle (*Bos taurus*) grazing on sown, volunteer, or native pastures. The enterprise mix on most dryland farms consists of crops grown in phased rotations with pastures that are grazed by sheep to produce wool and meat, and by cattle to produce beef. Most cropping farms contain some areas of nonarable land that is used only for grazing. No ruminant animals are housed. On this mixed farming land, about 40 000 farmers produced, in 2000, about 40 Tg of grain, 0.3 Tg of wool and 1.0 Tg of red meat, with a combined annual value of US$9 billion at the farm gate. Exports make up 66% of Australian grain, 100% of the wool, and 40% of the red meat.

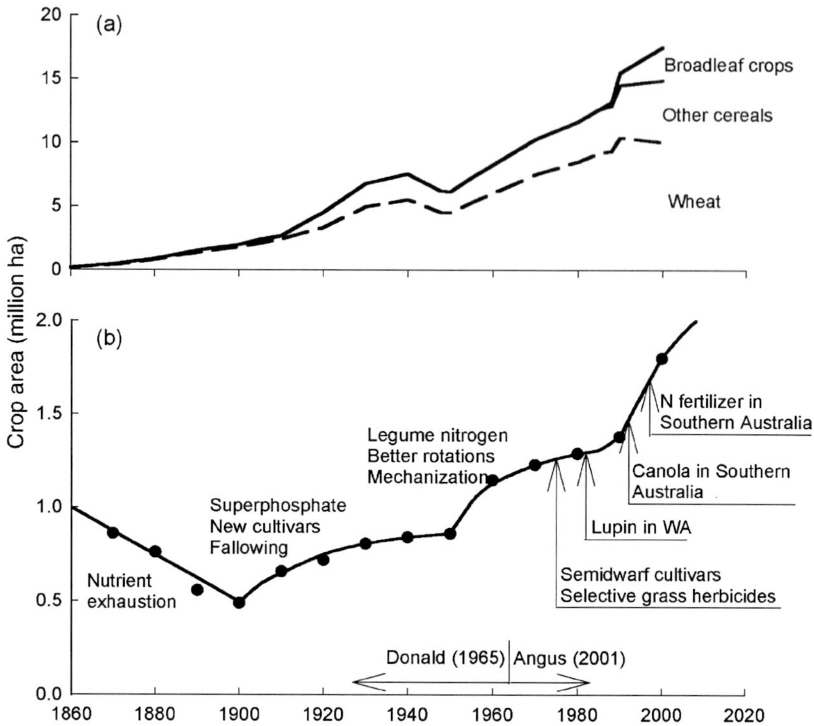

Fig. 10–1. Changes in (a) area of dryland crops and (b) mean wheat yield in Australia.

In addition to dryland crops, Australia exports the products of dairy, sugarcane (*Saccharum officinarum* L.), cotton (*Gossypium hirsutum* L.), wine grapes (*Vitis vinifera* L.), and rice (*Oryza sativa* L.), from high-rainfall and irrigated regions. Together these industries occupy 1.2 million ha with an annual farm-gate value of US$2 billion. Wool and red meat produced on 300 million ha of extensive rangelands and 3 million ha of permanent pastures in high-rainfall regions have an annual farm-gate value of US$3 billion. These data are rounded from surveys of the Australian Bureau of Agricultural and Resources Research (www.abare.gov.au).

Crops have been grown in southern Australia since the early 19th century. Over the 140 yr of records, the average annual increase in the cropped area was 3.2%, with no sign of a plateau (Fig. 10–1a). The only interruptions to the steady increase were a surge during the economic depression of the 1930s and a retraction during and after the Second World War. The increase mostly reflects replacement of pasture with crops, while at the same time pastures were increasingly cleared of trees. The rate of tree-clearing decreased during the 1990s with concern about greenhouse gases. While some dryland farms are now 100% cropping, most still retain at least 20% of the land under pasture.

The trend in national wheat yields from 1860 to 1960 consisted of an initial decrease during an exploitative phase, followed by two phases when new technol-

Table 10–1. Profile of mixed Australian dryland farms 1995/1996 to 1998/1999.

Area			Proportion of receipts from:		
	Land	1684 ha		Crops	65%
	Crop	563 ha		Wool	13%
				Sheep meat	8%
				Beef	7%
Labor					
	Full-time	2.4			
			Farm		
Livestock				Capital value	US$790 000
	Sheep	1733		Cash income	US$49 200
	Beef	101		Business profit	US$14 300

Source: (Knopke et al., 2000).

ogy raised yields to new plateaus (Donald, 1965). After 1965, wheat yields reached another plateau despite adoption of semi-dwarf varieties and selective herbicides (Fig. 10–1b). Then in the decade after 1990, wheat yield rose by 30%, in response to the increased area of break crops, particularly canola (*Brassica napus* L.), which reduced the level of root disease in the following wheat crops. These healthier crops then gave profitable responses to N fertilizer, leading to a sudden rise in the use of N fertilizer on dryland crops (Angus, 2001).

Most Australian grain is produced on large, export-oriented and unsubsidized family farms, with relatively little share-farming or leasing. Until the 1990s, farm families did almost all on-farm and off-farm operations. Since then contracting has become widespread, particularly harvest and haulage. Table 10–1 presents a profile of mixed crop-livestock farms and specialist crop farms. The number of farms is decreasing by 2.5% per year, mostly due to aggregation by neighbors. The combination of increasing cropping intensity, farm consolidation, and increased yield means that the crop production per farm is increasing at an annual rate of about 8%. The result is that the largest one third of farms produce about three quarters of the grain.

This chapter describes the present state of Australian dryland cropping systems, where they are unique and how they have evolved, particularly in relation to the main crop, wheat (*Triticum aestivum* L.) and the integration of crops with livestock. It speculates where the cropping systems are heading and the future research challenges. Generalizations are hazardous for systems that stretch 3000 km from southwest to northeast, and while the chapter attempts to convey the variety of environments and management systems, inevitably this brief description fails to capture their full diversity.

ENVIRONMENTS

Dryland crops are grown in a crescent in eastern Australia, separated by the arid Nullabor Plain from the triangle in Western Australia (Fig. 10–2). The inland margin runs close to the 300 mm isohyet in the south, but closer to the 500 mm isohyet in the northeast. The margin on the coastal side runs between the 500 and 600 mm isohyets in the south and greater in the north, but is defined by unsuitable ter-

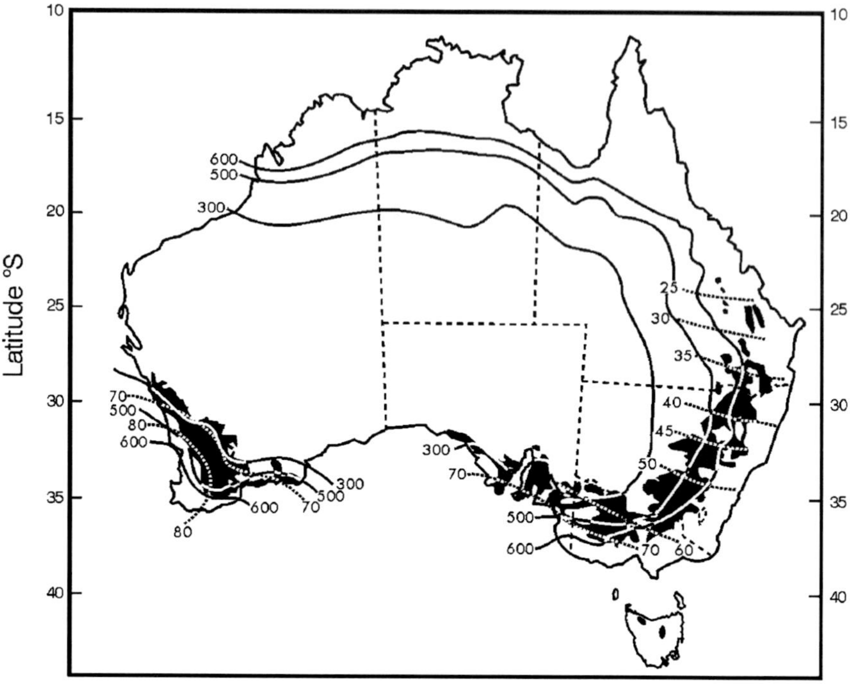

Fig. 10–2. Location of the Australian dryland cropping region in relation to (———) annual average rainfall (mm) and (--------) the percentage of rainfall from May to October.

rain and soils as well as excess rainfall. Before being cleared for farming, most of this region was open grassy woodland, typically with a density of 20 to 100 trees ha^{-1}, mostly of the *Eucalyptus* and *Acacia* genera. There are still isolated trees in cropped fields, most of which are remnants of the original woodland. The species of trees are usually a good indication of the potential for crop production. The woodland origin of much of the Australian croplands contrasts with the grassland of other dryland cropping regions in the North American prairie, South American pampas, and Eastern European steppe.

The proportion of annual rainfall during the growing season of winter crops (May–October) decreases from 80% in the southwest to 20% in the northeast. The variability of rainfall between years reflects the summer–winter distribution, with rain most reliable in the southwest and least reliable in the northeast (Nix, 1975). The extreme variability of weather in the northeast is due to the El Niño Southern Oscillation (ENSO).

Farm decisions are dominated by climate factors (www.longpaddock.dpi.qld.gov.au) and by short and medium-term weather forecasts (www.bom.gov.au/silo). The variability in rainfall leads to annual wheat yields of some districts having a coefficient of variation >75%. However, the spread of the Australian grains industry in different climate zones means that national yield is less variable.

Winters are mostly mild and there is no snow; the lowest mean daily temperatures vary from 8°C in the south to 15°C in the northeast. Winter growth of cereals is slow in the south, but never stops altogether, while winter growth of canola and most pulses is slower than the cereals. The mild winter temperatures means that crops can be sown from the 'break' of season, which can occur as early as February, until July, although almost all crops are sown in May–June, with a yield penalty for sowing after May. By the time of wheat flowering, temperatures range from 12°C in the south to 18°C in the northeast, and by maturity from 18 to 23°C. There is a significant risk of frost at flowering for almost all crops and in almost all regions. Evaporative demand is lowest during the winter period from sowing to stem elongation, when potential evapotranspiration is 1 to 2 mm d^{-1}, but rises rapidly to 4 to 5 mm d^{-1} by flowering, and 7 to 8 mm d^{-1} by physiological maturity. The selection of maturity type and sowing date therefore involves a compromise between the risks of frost and terminal drought.

Environmental Biology of Dryland Crops

Most Australian dryland crops grow in winter and spring. Even in the summer-rainfall areas, most crops grow in these seasons, using soil water stored from a summer fallow. These so-called winter crops have a spring habit, with little or no requirement for vernalization. The exceptions are some cereals with mild vernalization requirements that can be sown during late summer and autumn, grazed during winter and still flower at the optimum time during spring. Almost all are long-day plants, that is, their development is hastened in long days. The effect is that even with sowing dates spread from May to July, crops tend to flower within a narrow window. The duration of typical wheat crops varies from 125 d in the warm north to 200 d in the cooler south. Harvest typically starts in early October in the northeast and finishes in the southeast by early January.

Although Australian grain-growing regions have semi-arid or subhumid climates, the yields of commercial crops are not generally limited by water supply, but by management factors such as nutrition, root disease, and late sowing (French and Schultz, 1984). Only in the driest seasons is water supply a clearly limiting factor; in favorable and average seasons, most commercial yields are below the water-limited potential. The variability of rainfall means that flexible crop management is needed to capture the high yield potential in favorable seasons but minimize loss during drought. Cereals have few foliar diseases because breeding programs generally stay ahead of the major pathogens. The only pesticide routinely applied is fungicidal seed dressing. There are few insect pests of cereals. However broadleaf crops suffer from a range of biotic problems, particularly fungal pathogens, and fungicides and insecticides are usually needed.

Plant density of cereals varies from about 200 plants m^{-2} in high-rainfall districts to 70 plants m^{-2} in semi-arid regions. Tillering is complete about 60 to 80 d after sowing and the number of tillers is highly dependent on water, nutrient, and root-disease status. Tiller density usually decreases during the course of stem elongation and reaches a stable value by flowering.

Rooting depth varies with sowing date and soil type, with root zones as shallow as 0.7 m for late-sown crops growing in sodic-clay subsoils, and as deep as 2.5

m for early-sown crops growing on sandy soils. For many crops growing in subsoils with medium texture and no chemical limitations, the rooting depth is 1.0 to 1.5 m.

ROTATIONS AND CROP SEQUENCES

Australian dryland cropping started with wheat, oat (*Avena sativa* L.), and barley (*Hordeum vulgare* L.). Wheat has always been the main crop, comprising more than 90% of the dryland cropping area during the 19th century and decreasing to <60% at the end of the 20th century, with the expansion of broadleaf crops and other cereals. For the first 100 yr, crops were grown in loose rotation with unsown pastures and, starting in the early 20th century, after long fallows (Fig. 10–1b). From the 1950s, a ley-farming system evolved after widespread adoption of improved annual pastures grown in phased rotation with cereals. For the next 20 to 30 yr, typical rotations were 1 to 3 yr of annual pasture alternating with 1 to 3 yr of crop. The pastures are established by undersowing with the last crop and are then self seeding until the next cropping phase. In the acid-soil regions of southeastern and southwestern Australia, the main pasture legume is subterranean clover (*Trifolium subterraneum* L.). In the alkaline-soil regions of South Australia, western Victoria, and western New South Wales, the preferred pasture legumes are species of annual *Medicago*. During the 1990s several promising new species of annual pasture legumes were released (www.clima.uwa.edu.au). Sown pastures are not widely grown in the northeast grains region, where continuous cropping and fallowing is more common and soil organic matter is decreasing.

The grazing animals on the pasture were sheep for wool and meat, and a relatively small proportion of cattle. Cattle predominate in the subtropical north and sheep are concentrated in the winter-rainfall regions of the south. Returns from pasture were excellent during the early 1950s and adequate until the late 1960s, enabling farmers to supply the pastures with high levels of inputs, particularly phosphate. During this period, an important role of cropping was in establishing improved annual pastures. The consequence was that biological fixation by pasture legumes replenished soil N and supported increased cropping after the 1960s. The sheep population decreased from 160 m in 1970 to 100 m in 2000 because of the decreasing profitability of animal products, particularly wool. Despite the low direct profit from livestock, they remain important because of complementarities with crops. They use pastures growing on non-arable areas and crop stubbles. Livestock provide a diversified source of income in regions where cereals are the only well-adapted crops, and generally provide a more reliable income than crops during drought. The ley pastures used for livestock production contribute N to following crops and also provide a means of combating herbicide-resistant weeds (see below).

Mixed farms are normally subdivided into 10 to 20 paddocks, using wire fences. Most crop farms have a boundary fence and some retain internal fences to allow for livestock in future. Paddock position and size were determined more by livestock rather than crop considerations, particularly 'dams' which hold 1000 to 5000 m^3 of water for livestock collected as runoff.

The downturn in prices for wheat and wool in the late 1960s led to interest in alternative crops. The first widely grown alternative was narrow-leaf lupin (*Lupi-*

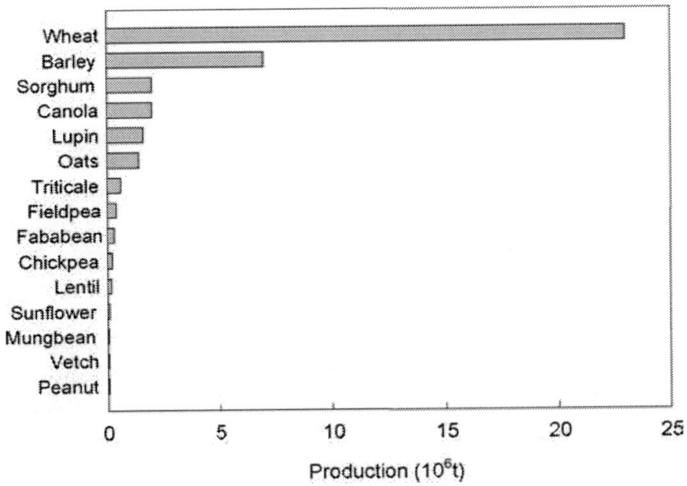

Fig. 10–3. Production of dryland crops in Australia, 2000.

nus angustifolius L.), which was widely adopted in the early 1980s and is a major part of the rotation in Western Australia. Since then an additional nine new crop species have expanded to an area more than 100 000 ha, but only narrow-leaf lupin and canola have expanded to 1 million ha (Fig. 10–3); apart from triticale (*Triticosecale* spp.), all the new crops are broadleaf species, most of which are concentrated in the higher rainfall regions.

Australia produces relatively small amounts of grain specifically for stockfeed. This is partly because of limited demand by the domestic pig, poultry and feedlot industries, but is mostly because the dry harvest conditions favor production of high value food grain, which farmers grow in preference to feed grains. Stockfeed is mostly supplied from grain that fails to meet a quality requirement such as for malting barley or milling wheat.

Financial returns from broadleaf crops are generally less than from cereals and their major contribution is from the additional yield of the following cereal. This result has been shown in many field experiments with grain legumes and canola, summarized in Table 10–2. The 19% yield increase of wheat growing after canola

Table 10–2. Yield of wheat grown after break crops from a survey of published experiments.

Previous crop	Yield as% of wheat after wheat	Number of comparisons
Wheat	100	
Lupin	146	75
Field pea	142	52
Chickpea	153	4
Canola	119	26

Source: www.regional.org.au/au/asa/2001/plenery/4/angus.htm; www.regional.org.au/au/gcirc/2/333.htm

is mostly because of reduced root disease, either because of the passive process or depriving the pathogens of a host for a year, or the active process of biofumigation when isothiocyanates are released from canola roots into the soil. The diseases are members of the fungal root rot complex such as take-all, as well as nematodes (Sarwar et al., 1998). Other possible reasons for the increased yield of wheat growing after canola are increased N mineralization and reduced loss of assimilates to mycorrhizae, because of the reduction in mycorrhizal colonization of the roots of wheat growing after canola (Ryan and Graham, 2002). The reason for the larger increases in yield of wheat growing after pulses is presumably because of additional N as well as reduced root disease. Another possible reason is the benefit to the following crop is from H_2 gas released into the soil by some types of legume rhizobia (Dong and Layzell, 2001).

A recurring pattern for industries based on these broadleaf crops has been an initial production boom followed by a contraction to a relatively small region or a minor part of the rotation. The cause of the decreases has mostly been development of pathogens. The consequence of maintaining a large number of minor crops is high per-unit infrastructure costs of research and marketing.

Another common way to control cereal root disease is by 'winter cleaning' of pasture. This process consists of removing annual grasses from a mixed grass-legume pasture in the year before a cropping phase, using a grass herbicide or a low rate of paraquat. Removing annual pasture grasses is important because many are host of the major root pathogens of cereals. Winter cleaning has the additional benefit of releasing large amounts of mineral N to subsequent crops (Harris et al., 2002).

In the northeast, pastures are not widely grown in rotation with crops, but a wide range of summer and winter crop species are well adapted. However, because of the variable rainfall it is difficult to plan a fixed rotation. During the 1950s, farmers developed a system of 'opportunity cropping' in which an appropriate crop is sown whenever there is sufficient soil water, irrespective of season. For example, when there is a full profile of soil water in spring, farmers may sow a long-season crop such as dryland cotton rather than a short-season crop such as mungbean. Following this system, farmers may grow continuous crops during *la niña* periods or no crops during *el niño* periods. This system of flexible crop sequences instead of fixed rotations has lessons for other regions where the need for flexibility may be management factors such as herbicide residue or N status, rather than soil-water status.

The most recent addition to rotations is an increasing adoption of grazing lucerne (alfalfa, *Medicago sativa* L.). Since the mid-1990s, lucerne has begun to replace annual pastures in phased rotations with dryland crops, with the lucerne phase generally lasting 3 yr. In wetter regions, lucerne is undersown with the last crop, but in drier regions it is sown alone. The advantages of lucerne over annual legumes are its greater forage productivity and N_2 fixation, and its ability to de-water the subsoil, leading to reduced risk of water-logging, soil acidification, and downslope salinity (Peoples et al., 1998). The increased area of lucerne in the acid-soil regions of the southeast and southwest would not have been possible without the lime applied to grow canola. The general improvements in crop management that accompanied the adoption of canola crops also assisted in lucerne establishment.

SOILS AND NUTRIENTS

Soil properties vary enormously from the regional scale to the within-field scale (CSIRO, 1983). As a generalization, in the northeast, most crops are grown on alkaline clays such as the Vertisols, which retain water conserved during a fallow. In the wetter parts of the east and southeast, soils with acid topsoil and with loamy surface texture and an abrupt texture change to clay at 0.1 to 0.3 m. Sandy acid soils are widespread in the wetter parts of the west, in some places with acid subsoils as well as topsoils. Some of the acidity is natural, but much has also been induced by farming, as discussed below. Across the drier regions of the south, many crops are grown on alkaline soils with sandy surface layers, most of which have problem subsoils which are highly sodic and contain salt and boron (B) at increasing concentrations with depth (Rengasamy 2002). In such soils the effective rooting depth of annual crops is 0.6 to 0.8 m. Where the subsoil is not sodic and contains little salt or B, the effective rooting depth is 1.0 to 1.5 m. Even where the subsoils are chemically benign, the bulk density is usually in a range from 1.6 to 1.8 g cm^{-3}, values that could be expected to limit root penetration. The available water-holding capacity within the root zone is also highly variable, with extremes of 250 mm for Vertisols and 60 mm where there are problem subsoils. Medium-textured soils with no chemical limitations in the root zone have an available water-holding capacity of 100 to 130 mm.

Almost all soils are deficient in phosphorus (P) and sulfur (S). Until the 1980s, most of the P requirement was met with single superphosphate, which contains 50% more S than P, so that the S status of soils increased, at least until the replacement of single superphosphate with ammonium phosphates that are banded with the seed. The main requirement for S is now by broadleaf crops, particularly canola. The S is supplied to these crops in gypsum or compound fertilizer, and is generally in excess of the needs of a single crop so that the residual is sufficient to supply cereals and pastures that follow in rotation. The potassium (K) status of soils is generally high, except on sandy soils in parts of Western Australia, and little K fertilizer is applied. Micronutrients are naturally deficient in some regions and the incidence of deficiency appears to be increasing with greater product removal. For example, zinc (Zn) deficiency is widespread in the generally sandy soils across southern Australia and the alkaline clays of the northeast. Many acid soils in the southeast are molybdenum (Mo) deficient, and there are pockets of copper (Cu) and manganese (Mn) deficiency, mostly on alkaline soils. When deficiencies are detected or suspected, they are routinely corrected with micronutrients co-granulated in solid starter fertilizer and in foliar sprays.

The N dynamics of dryland crops are influenced by the historical depletion of soil organic matter and its later replenishment from N_2 fixation by legume pastures (Fig. 10–1b). The outcome is that mineralization of soil organic matter and residue organic matter contributes about 80% of the N requirement of dryland crops, in contrast to about 50% worldwide (Angus, 2001). In both cases, N fertilizer supplies the rest. Until the 1990s, mineralization contributed an even greater proportion to crop N in Australia. It was only with the rapid adoption of canola and other breakcrops in the 1990s that cereal root diseases were reduced to a level where cereal responses to N fertilizer became reliable, as discussed in the Rotation section. Australian dryland crops received about 5 kg N ha^{-1} from fertilizer during the 1980s, but the amount rose abruptly during the mid-1990s to about 30 kg N ha^{-1}, mostly as urea (Angus, 2001).

There are other unusual features of N dynamics in Australian systems. One is that the diverse quality of residues of crops and pastures, combined with the variable rainfall in summer and autumn lead to variable rates of soil N mineralization, so the levels of mineral N in the soil is difficult to predict. At the time when winter crops are sown, the mineral-N content of the top 60 cm of soil in commercial fields averages 80 to 120 kg N ha^{-1}, with a range from 20 to 400 kg N ha^{-1}. Most of the high values are found in paddocks at the end of the pasture phase, when above-average rainfall in summer and autumn can lead to rapid fallow mineralization. In-crop mineralization is more predictable, with relatively high rates soon after sowing and before flowering, the two times when the soil is most likely to be warm and moist. However, the rate of N mineralization is relatively low during the cool conditions of early spring, when crops are at the stem-elongation phase and N-demand is normally at its greatest. The result is that the timing of N supply from the soil is badly synchronized with crop demand.

In situations where the supply of mineral N exceeds demand by the crop, there can be large amounts of residual mineral N available for the following crop. Generally this mineral N remains in the top 30 cm of soil, but can be present as a bulge of nitrate (NO_3^-) at 0.5 to 1.0 m when leached in the wetting front. It is unusual for mineral N to be leached deeper than the root zone, and subsoil bulges of NO_3^- are normally recovered by subsequent crops.

The other unusual feature of the N dynamics is that an excessive amount of N in a season with a terminal drought causes a yield reduction in cereals known as 'haying off'. The cause of haying-off is not simply because there is insufficient soil water for grain-filling, but is mainly because reserves of soluble carbohydrate are depleted at high N levels of crop tissue (Angus and van Herwaarden, 2001). The risk of haying off was a powerful disincentive against applying N fertilizer in the past. The use of the 60-cm soil test for mineral N and simple crop-N budgets have helped grain-growers to select optimum N fertilizer rates. Some of the N fertilizer is incorporated or banded into the soil at the time of sowing, but increasing amounts are top-dressed during the period between stem-elongation and flowering, when the crop demand is greatest. This system of tactical management of N fertilizer takes account of the amount of soil water, tests of crop-N status and forecast rainfall needed to wash fertilizer into the soil. Because of the importance of rain after top-dressing, the system is most effective in the southern regions where spring rain is reliable. The split application system reduces the risk of haying off and leads to greater N-use efficiency than when all fertilizer N is applied at sowing.

TILLAGE, CONSERVATION FARMING, AND WEED CONTROL

Land preparation is a mixture of minimum tillage, no-till, and some conventional cultivation. Weeds are normally controlled with a knockdown herbicide before cultivation. This system of minimum tillage has gradually replaced multiple cultivations after knockdown herbicides became available in the 1970s.

In the summer-rainfall areas and in drier parts in the south, crops are preceded by a fallow of up to 10 mo, maintained by cultivation or herbicide (Fischer, 1987). In the wetter regions of the south, soil is cultivated or sprayed with herbicide in autumn to maintain a short fallow of 1 to 3 mo. In these situations, the soil is cultivated with the same tine implement used for seeding. One cultivation is normally needed to loosen topsoil compacted by grazing animals. No-till is less widely adopted than minimum tillage, even though it has been promoted for 30 yr. Cultivation is also needed in wetter regions to overcome slow seedling growth and reduced yield, due to the activity of inhibitory microbes in undisturbed soils (Simpfendorfer et al., 2002). In drier environments, slow seedling growth of no-till crops leads to less water use early in the life cycle but no yield decrease when the conserved water is used during grain-filling (Fischer, 1987). The realization that tillage helps to overcome soil inhibitory microbes is leading to a revival of no-till, now with deep blade tines and vigorous soil disturbance below the seed, rather than with the system of narrow, shallow tines and minimum soil disturbance when no-till was first introduced. In the summer-rainfall regions, no-till seeding is done with spear points into a drying clay soil, while in the south, seed is sown with about 8 cm of loose soil below the seed.

Stubbles of winter crops are usually grazed by sheep after harvest. Although there is little nutritional value in the straw itself, there is some grazing value for a low density of sheep from weed seedlings and grain dropped at harvest. Grazing by the sheep suppresses but does not kill weeds and trampling by the sheep helps to decompose the stubble. Indigestible weeds growing in the stubble after summer rains are controlled with broad-spectrum herbicides to retain soil water and prolong the period of N-mineralization. Residues are mostly burned as late as possible before sowing in the autumn, generally with a 'cool burn' in the afternoon or evening when the fire is easiest to control. Burning leads to loss of N and S from the field, but most of the other nutrients remain in the ash. The exceptions of stubble burning are increasing examples of no-till into standing or disturbed stubble, particularly those of broadleaf crops.

In-crop weeds are mostly controlled with pre-emergence and post-emergence herbicides. Summer weeds growing in the stubbles after rain are controlled with broad-spectrum herbicides, to retain soil water and prolong the period of N mineralization. Herbicide resistance has developed in several species of in-crop weeds, the worst case being annual ryegrass (*Lolium rigidum* L.), which is a component of annual pastures. It has developed multiple and cross resistance to the major herbicide groups (Powles and Holtum, 1994). Despite the growing problems, graingrowers are generally confident of controlling herbicide-resistant weeds by rotating crops and herbicide groups. If that fails, their backup strategy is to grow pastures and prevent seed set by weeds, using a combination of heavy grazing, hay cutting, and "knockdown" herbicides.

Herbicide-tolerant crops are also being adopted to help control weeds. Canola cultivars with non-genetically modified (GM) herbicide tolerance have been widely used since the mid-1990s and cultivars with GM herbicide tolerance are being cautiously released in 2003. The reluctance to adopt GM in grain crops is partly because of some domestic community opposition but is mostly due to concern that importers will use GM as a reason to restrict grain imports.

Environmental Problems

Soil acidification, erosion, compaction, and secondary salinization are associated with crop production (www.nlwra.gov.au). Of these, acidification of the poorly buffered topsoils in the higher rainfall areas of the southeast and southwest is probably the most severe. The main causes of acidification are product removal and leaching of NO_3^- from the topsoil. The leaching mostly occurs during winter, particularly under annual pasture or the first crop after annual pasture, following high rates of N mineralization from legume residues in autumn. Little or no lime was applied until the 1990s. Before then, the growing problem of acidification was avoided by growing acid-tolerant crops (e.g., lupin, triticale, and tolerant wheat cultivars). Liming was rapidly adopted along with canola during the 1990s because of canola's intolerance of acidity and because the profitability of the canola-wheat sequence easily covered the cost of lime. Without lime, canola production would have been impossible on the extensive areas of acid soil. By 2000, lime was applied to all acid soils used for canola.

Soil erosion by water and wind was a serious problem of crop production when long fallowing was widespread but has been greatly reduced with the adoption of reduced tillage and no-till. However, dust storms during the 2002 to 2003 drought, some of which were due to erosion from cropping fields, may cause a re-evaluation of the extent of plant cover needed to prevent erosion. Soil compaction is another consequence of previous excessive cultivation, particularly on soils with a naturally dense subsurface. The use of tractors weighing >10 t is a growing problem for subsoil compaction.

Secondary salinity is a serious and growing problem of dryland cropping in Western Australia. During the 1990s, environmentalists became concerned that similar problems will develop in eastern Australia. In this case, 'secondary' refers to salinization induced by land management, rather than original salt. The cause is increased groundwater recharge because evapotranspiration by annual crops and pastures is less than the original native vegetation. The problem is greatest in Western Australia because a combination of sandy soils and high winter rainfall causes annual recharge to groundwater that is naturally high in salt. When the groundwater rises to the surface, salt is discharged into streams and low-lying parts of the landscape. Recent research suggests that the increased water use by higher yielding crops and perennial pastures is sufficient to prevent or greatly reduce groundwater recharge [see the special issue of *Australian Journal of Agricultural Research* 52, Vol. 2 (2001)].

The reduced biodiversity of many regions in Australia is being increasingly recognized by grain growers and the community in general. The ecosystems replaced by dryland cropping are poorly represented in national parks, and several species of plants and animals are extinct or endangered in the cropping regions. Activities by farmers to preserve biodiversity are in their infancy, and consist of fencing patches of remnant vegetation on farms, and by establishing corridors to connect existing areas of bush and so facilitate movement of native animals.

Scale of Enterprises and Rate of Operations

From the 1950s until the 1990s much of the productivity gain in the Australian grains industry was from increased scale of operations, particularly the width and speed of equipment used for sowing, spraying, and harvesting. For example, a single operator driving a 400 kW tractor and towing an airseeder with a width of 10 to 15 m can apply seed and fertilizer at a rate of 10 to 15 ha h^{-1}, and therefore sow up to 1500 ha of crop in a 3-wk window. Many grain growers strive to obtain full utilization of this scale of equipment by purchasing or sharefarming additional land.

Harvesting is by machines with 7 to 11 m-wide open fronts and a capacity of up to 30 t h^{-1}. The grain generally contains <11% water because of the generally hot and dry conditions, so harvesting can proceed for 24 h d^{-1}. No artificial grain drying is needed. Contract harvesting is common because harvesters move from the north in October to the south in January. The high rate of grain harvesting, combined with relatively long distances to grain receival depots means that there are frequent bottlenecks at harvest. Many growers set up temporary storages on farm and then test grain quality before sale after harvest.

Block Farming

With the increasing size of farms and machinery during the 1990s, there has been a trend for similar crops to be grown in contiguous fields to save traveling time during farm operations and minimize spray drift to neighboring fields and properties. Block farming is most common where farmers purchase land that is not adjacent to the home farm. A typical large farm contains three to five blocks of crops and pastures. Accompanying this trend is removal of some internal fences to increase field size and allow economies of scale in machinery operations. Large field sizes create difficulties for management of livestock, particularly rotational grazing of lucerne.

Precision Farming and Controlled Traffic

Grain harvesters purchased since the late 1990s contain yield monitors, mostly linked to global positioning systems (GPS) receivers, so that many grain growers have access to yield maps. It is not clear what changes have been implemented as a result of that access, but there is growing evidence from them that there are zones of poor growth within fields because soils are physically or chemically hostile. Soils in these zones often contain high levels of available P and mineral N where fertilizer has not been used by previous crops, so there is increasing interest in variable fertilizer applications to better match inputs with crop demand. There is also interest in using yield maps and remote sensing to identify zones of consistently unprofitable crop. In some cases it is possible to convert these zones from crop to pasture after fencing, or to retire them from all production and re-establish native vegetation where costs consistently exceed returns.

Related to the trend towards precision agriculture is a growing interest in controlled traffic. The initial impetus is improved field efficiency by reducing overlap and extended operating hours, based on satellite guidance systems accurate to 0.2

to 0.4 m. A small proportion of growers have adopted such systems. A future possibility, based on guidance with accuracy down to 0.02 m will be driving all machines in permanent tracks. Many growers are preparing for this possibility with spray booms that are double or triple the width of seeders. Possible benefits of permanent tracks may be side-banding of fertilizer, inter-row spraying of herbicide, and reduced soil compaction, particularly by tractors and seed bins weighing 10 to 20 t.

A variant of controlled traffic is the system of raised beds in southern regions where there is a risk of water-logging. These beds are typically 2-m wide, 10 cm above the furrow, and aligned to so that water drains without causing erosion. Experiments and farmer acceptance suggests that this system is effective.

RESEARCH AND INFORMATION

Grain research is funded by a grower levy of 1% of the gross value of production, supplemented by federal government funds equivalent to 0.7% of production (www.grdc.com.au). Funds are allocated competitively to research providers in state government agencies, universities, CSIRO, and other federal agencies, private companies, grower groups, and international agencies such as CIMMYT. Most of the research funding supports crop breeding by state government agencies, universities, and CSIRO. There is a small and growing private crop breeding industry, starting with hybrid sorghum and wheat, currently focused on hybrid and herbicide-resistant canola and moving towards GM cereals and oilseeds. Most current wheat varieties are derived using germplasm from CIMMYT programs. Pulses are mostly bred by state government agencies.

Extension is provided by about 400 district agronomists employed by state governments, 300 private consultants, and 800 agribusiness agronomists. Growers have a strong network of local, regional, and national groups dealing with production and conservation issues, for example, the Birchip Cropping Group and Landcare, (www.bcg.org.au; www.landcareaustralia.com.au).

As of 2000, about 90% of grain growers have facsimile machines, 70% have personal computers, and 30% have Internet access. The main commercial uses for the Internet are for communication, grain marketing, and weather information, such as short-term rainfall forecasts and radar images to plan operations. There is increasing use of decision support systems for farm management.

GRAIN QUALITY, STORAGE, TRANSPORT, AND MARKETING

Almost all Australian wheat is white-grained, and, because of the generally dry weather at harvest, is not dried artificially. Protein content has generally been highest for grain grown in the northeast and lowest in the south. Until the late 1990s, mean protein was <10%. With the increased use of N fertilizer, particularly late top-dressing of the southeast, the average level of grain protein is increasing.

After sale by farmers, grain is stored by grower-owned companies, formed after privatization of state government agencies. Australia lacks navigable rivers so there is no barge transport and most grain is carted to ports by rail or road. All exported wheat is marketed from a single desk by a grower-controlled company (www.awb.com.au). Most other grains are traded freely on domestic and export markets. Depending on the grain price during the growing season, many grain-growers make forward sales and hedge these sales. Australian grain has the advantages that it is produced in the off-season from the northern hemisphere and has low freight costs to the growing markets of Asia and the Pacific. Most wheat is exported for noodle and flat bread markets in East Asia, West Asia, and North Africa. Most barley is of malting grade and exported to East Asia. Oilseeds and pulses are also mostly exported into Asian markets.

ECONOMIC AND COMMUNITY ASPECTS

Australian farmers receive the world price for grain, with no government subsidies. The long-term trend in world prices is negative, driven by the subsidized grain dumped on world markets by the USA and the European Union. Because Australian grain prices are generally lower than those received by competitors while the input costs are similar, the unfavorable price:cost ratio leads to relatively low input use. Increased productivity and economic efficiency therefore involve better combination and timing of inputs rather than increased amounts.

The increased productivity involves more grain produced from fewer farms, a process common to many industries. Small towns servicing the farms are losing population, along with services such as schools and hospitals. Services are increasingly provided in larger centers, so that farm families have long distances to travel.

CONCLUSIONS

During the last decade, the profitability of Australian dryland cropping has increased through higher yield, improved grain quality, increased scale of operations, and lower cost of marketing and transport, all with little or no government protection. The farming systems are highly flexible and the mix of crops and livestock changes rapidly is response to price, problems such as disease, and opportunities such as new technology. The result has been increased exports despite the corruption of world markets with subsidized grain. At the same time the industry is addressing environmental problems of soil acidification, erosion, compaction and salinity, and there is evidence that they are being reduced.

Australian farmers and agronomists are optimistic about continued improvements in profitability and sustainability of dryland farming based on known technology and new science. There are challenges in continuing to diversify the mix of crop species and maintain pastures in the rotation, particularly perennial pastures. Better integration of cropping and grazing appears to be a key to overcoming the environmental problems. Other major challenges are managing herbicide-resistant weeds, physical and chemical limitations in subsoils, and improving systems to cope with a variable climate.

ACKNOWLEDGMENTS

We are grateful to Bernard Hart, Mark Peoples, and Michael Poole for helpful discussions.

REFERENCES

Angus, J.F. 2001. Nitrogen demand by Australian agriculture. Aust. J. Exp. Agric. 4:277–288.

Angus J.F., and A.F. van Herwaarden. 2001. Increasing water use and water use efficiency in dryland wheat. Agron. J. 93:290–298.

Commonwealth Scientific and Industrial Research Organization. 1983. Soils: An Australian viewpoint. CSIRO, Melbourne.

Donald, C.M. 1965. The progress of Australian agriculture and the role of pastures in environmental change. Aust. J. Sci. 27:187–198.

Dong, Z., and D.B. Layzell. 2001. H_2 oxidation, O_2 uptake and CO_2 fixation in hydrogen treated soils. Plant Soil. 229:1–12.

Fischer, R.A. 1987. Responses of soil and crop water relations to tillage. p. 194–221. *In* P.S. Cornish and J.E. Pratley (ed.) Tillage, new directions in Australian agriculture. Inkata Press, Melbourne.

French, R.J., and J.E. Schultz. 1984. Water use efficiency in wheat in a Mediterranean-type environment. II. Some limitations to efficiency. Aust. J. Agric. Res. 35:765–775.

Harris, R.H., G.J. Scammell, W. Müller, and J.F. Angus. 2002. Wheat productivity after break crops and grass-free annual pasture Aust. J. Agric. Res. 53:1271–1283.

Knopke, P, V. O'Donnell, and A. Shepherd. 2000. Productivity growth in the Australian grains industry. Australian Bureau of Agric. and Resource Econ., Canberra.

Nix, H.A. 1975. The Australian climate and its effects on grain yield and quality. p.183–226. *In* A. Lazenby and E.M. Matheson (ed.) Australian field crops 1. Wheat and other temperate cereals. Angus and Robertson, Sydney.

Peoples, M.B., R.R. Gault, G.J. Scammell, B.S. Dear, J. Virgona, G.A. Sandral, J. Paul, E.C. Wolfe, and J.F. Angus. 1998. The effect of pasture management on the contributions of fixed N to the economy of ley-farming systems. Aust. J. Agric. Res. 49:459-474.

Powles, S.B., and J.A.M. Holtum, (Ed.) 1994. Herbicide resistance in plants: Biology and biochemistry. Lewis Publ., Boca Raton, FL.

Rengasamy, P. 2002. Transient salinity and subsoil constraints to dryland farming in Australian sodic soils: An overview. Aust. J. Exp. Agric. 42:351–361.

Ryan, M.H., and J.H. Graham. 2002. Is there a role for arbuscular mycorrhizal fungi in production agriculture? Plant Soil 244:263–271.

Sarwar, M., J.A. Kirkegaard, P.T.W. Wong, and J.M. Desmarchelier. 1998. Biofumigation of brassicas. III *In vitro* toxicity to soil-borne fungal pathogens. Plant Soil. 201:103–112.

Simpfendorfer, S., J.A. Kirkegaard, D.P. Heenan, and P.T.W. Wong. 2002. Reduced early growth of direct drilled wheat in southern New South Wales—Role of root inhibitory pseudomonads. Aust. J. Agric. Res. 53:323–331.

11 Breeding for Drought Resistance in a Changing Climate

Salvatore Ceccarelli, Stefania Grando, Michael Baum, and Sripada M. Udupa

ICARDA
Aleppo, Syria

ABSTRACT

Drought is one of the major plagues affecting crop production worldwide. Climate changes will increase the frequency of droughts, particularly in Southeast Asia and Central America; by 2050 water shortages are expected to affect 67% of the world's population. Drought resistance has been always a challenge to plant breeders. Physiologists, biochemists, geneticists, and breeders have dissected drought resistance into individual components, with the aim of finding simple associated traits. Examples include proline accumulation, osmotic adjustment, stomatal conductance, canopy temperature, and various root characteristics. Today, most scientists agree that a drought resistance gene does not exist, and that differences in drought resistance are due to the effects of several genes, affecting different characters that interact with each other. This is because drought seldom occurs in isolation and often interacts with other stresses, mostly temperature extremes thus determining many combinations of stresses. Therefore, while drought is a global issue, its effects need to be addressed locally because every dry area has its own type of drought. Future research to increase the level of drought tolerance in crops needs to address the interaction between a number of traits and assemble those combinations of traits that maximize economic yields per unit of water used, possibly with the use of molecular tools.

INTRODUCTION

Drought continues to be a challenge to agricultural scientists in general and plant breeders in particular, despite many decades of research (Blum, 1993). Drought, or more generally, limited water availability is the main factor limiting crop production. Although it reaches the front pages of the media only when it causes famine and death, drought is a permanent constraint to agricultural production in many developing countries, and an occasional cause of crop losses in developed ones.

The development, through breeding, of cultivars with higher harvestable yield under drought conditions would be a major breakthrough (Ceccarelli and Grando, 1996). However, the ability of some plants to give a higher economic yield under drought than others is a very elusive trait from a genetic point of view. This is because occurrence, severity, timing, and duration of drought vary from location to location, and in the same location from year to year, and cultivars successful in one dry year may fail in another, or cultivars resistant to terminal drought may not be resistant to intermittent drought, or to drought occurring early in season (Turner, 2002). To make matters worse, drought seldom occurs in isolation; it often interacts with other abiotic stresses (particularly temperature extremes), and with biotic stress (root diseases and nematodes being the most obvious examples). Also, areas with a high risk of drought (and/or other abiotic stresses) generally have low-input agriculture (Cooper et al., 1987). Thus, as if breeding for drought resistance alone was not complex enough already, it is made even more so by its interactions with other stresses. *The problem of drought is a challenge due to its complexity, as well as its gravity.*

In this chapter, we discuss key problems associated with breeding for stressed environments, analyze possible reasons for the limited success breeding has had in dry areas, and indicate ways to overcome the inherent difficulties in breeding for stressed environments. Most of the examples will be taken from barley (*Hordeum vulgare* L.), one of the most drought-tolerant cereal crops.

Drought: A Lesson from History

Water has always played a central role in human civilizations, being a cause of either life or death. The regions surrounding the "Twin Rivers" of the Tigris and Euphrates, and the mighty Nile and Indus, were the birthplace of civilization (Sardar, 1995). The earliest domestication of crops took place along the arch (today known as the Fertile Crescent) stretching from Palestine and Jordan, up to northern Syria-southern Turkey, and down to what is today Iraq and part of Iran. During Roman times this region provided the Mediterranean civilizations with the bulk of their food (Cooper et al., 1987).

Abundance of water, as either rainfall or rivers, meant life; its scarcity often meant death. Most of the major famines of the last few centuries were due to, or associated with, drought (Plucknett, 1991). Even the most recent ones, such as those in Africa, although aggravated by civil wars, have been fundamentally associated with severe droughts and have affected 150 million people in 22 countries (McWilliam, 1986). Droughts are thought to be the major cause of collapse of empires and societies, such as the Akkadian Empire around 2200 B.C. (Weiss et al., 1993) and the early bronze society in the southern part of the Fertile Crescent (Rosen, 1990).

Today, water continues to be the major factor limiting crop production in most dryland areas and is one of the major causes of poverty and underdevelopment. Today, more than one billion people (18% of the world population) have no access to clean water, and in developing countries women, on average, walk several kilometers daily to collect water in portable containers.

Seldom is precipitation sufficient and adequately distributed to enable crops to achieve full yield potential (Papendick and Campbell, 1990). A classification of soils in the USA indicated that almost half of the land surface is affected by low water availability, and that drought alone affects 25% of the U.S. soils, accounting

Table 11–1. Water use in irrigated and rainfed agriculture.

Losses (or used) through:	Irrigated (% of available water)	Rainfed (% of rainfall)
Storage and conveyancing	30	0
Runoff and drainage	44	40–50
Evaporation	8–13	30–35
Transpiration	13–18	15–30

Source: Wallace (2000).

for 41% of the insurance indemnities paid to U.S. farmers between 1939 and 1978 (Boyer, 1982).

Irrigation has often been seen as the way to alleviate drought. And indeed, agricultural production in some areas is only possible with irrigation. However, only 16% of the total cropland in the world is irrigated, with an increase of only 2% (from 14%) during the 10 yr between 1977 to 1979 and 1987 to 1989. Boyer (1982) estimated that because of competition with municipal and industrial uses and possible soil degradation no more than 10% of U.S. agricultural land would ever be irrigated. The major problem with irrigation is salinization; this affects one third of all irrigated land (Ashraf, 1994) and is estimated to endanger half of the irrigated cropland (Strauss, 1993). Soil salinity played an important part in the breakup of the Sumerian civilization in Mesopotamia around 1700 B.C. (Jacobsen and Adams, 1958) and it is an ever-present problem in semi-arid environments with high evaporation (Boyer, 1982).

In addition of being a limited resource, water is used in both rainfed and irrigated agriculture, with a low efficiency of 13 to 18% (irrigated) and of 15 to 30% (rainfed) (Table 11–1). There are two main ways of increasing water-use efficiency in agriculture: one is to convert more water into transpiration, and this is the domain of physical engineering, hydrology and agronomy, and the second is to increase the transpiration/water use ratio, and this is the domain of pant breeding, physiology, and genetics.

Limitation and/or competition for resources, and the need to avoid further environmental degradation, suggest to many scientists that fitting crops to the environment is a more sustainable strategy than modifying the environment to fit the crops, particularly for drought-affected areas. Therefore, the central questions are whether crops and varieties within crops differ genetically in yield when water is a limiting factor and whether we can exploit these differences.

Drought: Not by Rain Alone

Drought seldom occurs in isolation from other environmental stresses. Using different crops and different environments, the percentage of variance in grain yield explained by total rainfall varies between 6 and 79% (Table 11–2). These studies, conducted in the Near East and North Africa, show that the relationship between rainfall and yield improves through temperature and/or rainfall distribution, not just one of the two and that their effect on yield varies widely between crops (Ceccarelli and Grando, 1996). In the case of barley, the range was between 6% (Oudina and Bouzerzour, 1993) to 68% in our own work.

Table 11–2. Percentage of the variance for grain yield explained by total annual rainfall in different crops.

Crop	Variance %	Source
Lentil	33–45	Erskine and El Ashkar (1993)
Wheat	75	Blum and Pnuel (1990)
Wheat	31–79	Hadjichristodoulou (1982)
Barley	5–31	Hadjichristodoulou (1982)
Barley	52–63	van Oosterom et al. (1993a)
Barley	6	Oudina and Bouzerzour (1993)

At a given location, grain yield of barley, estimated by the mean of several breeding lines, can be very different in different years, even with nearly the same amount of rainfall (Fig. 11–1). Therefore, in the real world, crops exposed to drought will, at the same time, face numerous other yield-limiting factors, both biotic and abiotic, which modify the response to available water (Passioura, 2002). One of the major challenges for the plant breeder is to select for the ability to tolerate various combinations of all these stresses. Only such ability will eventually contribute to making people living in areas affected by drought less vulnerable.

BREEDING FOR DROUGHT RESISTANCE

Breeding Philosophies

Two contrasting breeding philosophies have been used to breed drought-tolerant cultivars: the first uses selection under optimum growing conditions and is based on the assumption that an increased yield potential will have a carry-over (or "spillover") effect even when the improved cultivars are grown under less favorable conditions. The second, on the contrary, uses direct selection in the presence of drought, that is, in the target environment, and can take two forms: (i) selection for physiological or developmental traits (also called analytical breeding), and (ii) direct selection for grain yield (also called empirical or pragmatic breeding).

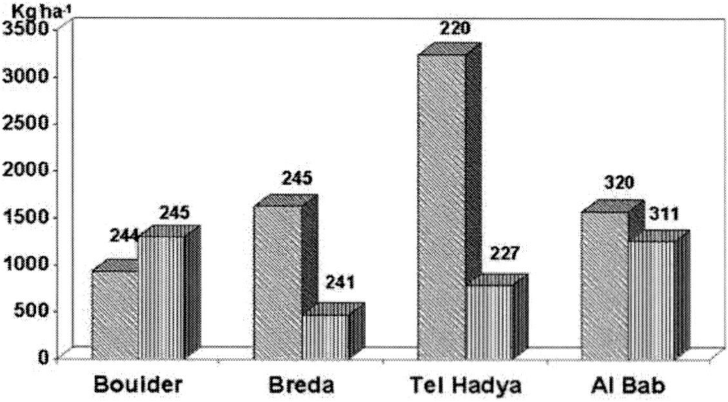

Fig. 11–1. Not by rain alone: similar amount of total rainfall (in mm above bars) may result in different grain yields because of other factors such as rainfall distribution, air temperature, and air humidity.

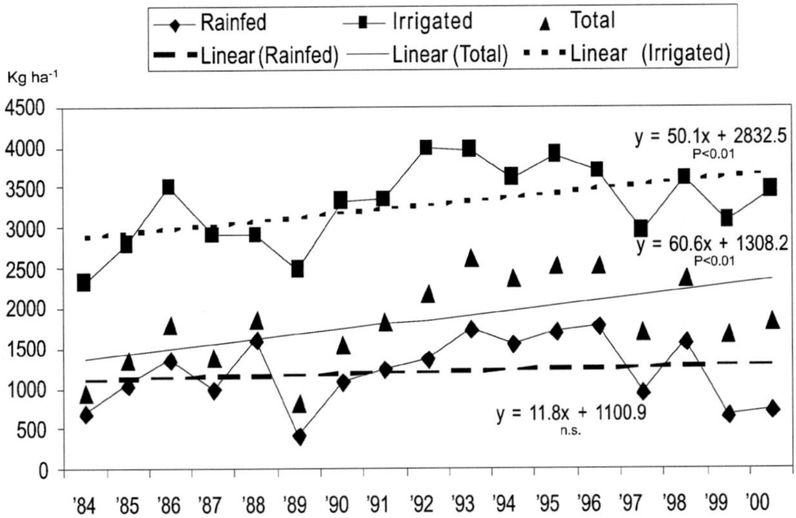

Fig. 11–2. An example of no spillover: wheat in Syria (kg ha^{-1} from 1984–2000). (Source: Official Statistics of the Ministry of Agriculture, Syria.)

The first philosophy has failed to produce convincing results about the existence of the "spillover" effect. In barley, we have consistently found a negative relationship between yield potential and yield under stress conditions (Ceccarelli et al., 1992; van Oosterom et al., 1993b; van Oosterom and Ceccarelli, 1993). The circumstantial evidence of the absence of the spillover effect is the fact that drought continues to negatively affect agricultural productions worldwide, despite the spectacular increases in crop yields obtained through breeding under optimum conditions. A specific example of the absence of the spillover effect is offered by wheat (*Triticum* spp.) in Syria (Fig. 11–2), where yields, in the period 1984 to 2000 (the first improved cultivar was released in 1983), increased by a staggering 60 kg ha^{-1} per year ($P < 0.01$), thus transforming Syria from an importing to an exporting country. However, this large yield increase was mostly due to an increase of 50 kg ha^{-1} per year ($P < 0.01$) under irrigated conditions, while the yield increase under rainfed conditions (11.8 kg ha^{-1} yr) was not significantly different from zero, even though the adoption of improved cultivars reached nearly 90%.

Breeding for drought resistance based on putative traits (defined as traits associated with drought resistance, but easier to select for than grain yield) has been, and still is, very popular (Richards et al., 2002). Traits, which have been investigated, include physiological/biochemical traits (such as osmotic adjustment, proline content, stomatal conductance, epidermal conductance, cell membrane stability, cell wall rheology, canopy temperature, relative water content, leaf turgor, abscisic acid content, transpiration efficiency, water-use efficiency, carbon isotope discrimination, and retranslocation), and developmental/morphological (such as leaf emergence, early growth vigor, leaf area index, leaf waxiness, stomatal density, tiller development, flowering time, maturity date, vernalization, and root characteristics). Several papers have been published on the putative role of these traits, for exam-

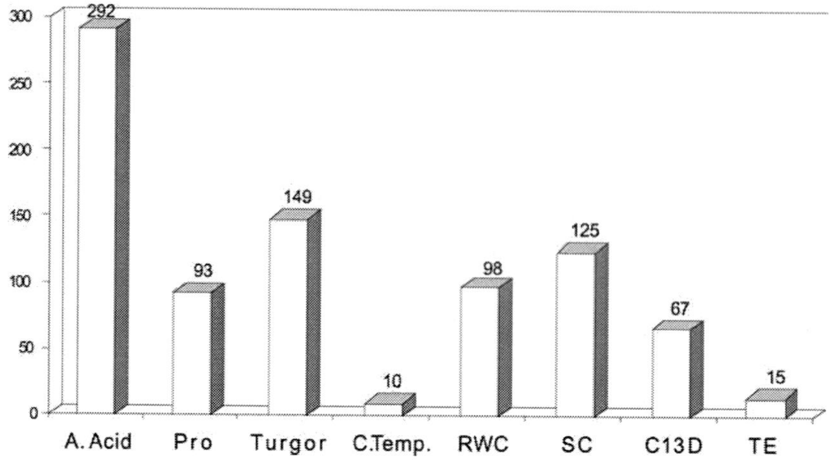

Fig. 11–3. Number of papers published on some putative traits (A. Acid = Abscissic Acid; Pro = Proline, C. Temp. = Canopy Temperature, RWC = Relative Water Content, SC = Stomatal Conductance, C13D, C13 discrimination, TE = Transpiration Efficiency) associated with drought resistance in plants from 1992 to 2002. (Source: Agricola data base.)

ple, nearly 850 papers were published between 1992 and 2002 on the eight traits shown in Fig 11–3.

In the case of barley, we have found that the traits more consistently associated with higher grain yield under drought are *growth habit, early growth vigor, earliness, plant height under drought, long peduncle and a short-grain filling duration* (Acevedo and Ceccarelli, 1989). Three traits that deserve a special mention are leaf epidermal conductance (Sinclair, 2000), osmotic adjustment (Serraj and Sinclair, 2002), and desiccation tolerance (Ramanjulu and Bartels, 2002) that appear related to survival under severe stress conditions. Even though the low yields resulting from survival traits may look irrelevant from the perspective of high-input agriculture, they are crucial to the livelihood of farmers in some of the driest regions of the world.

While the analytical approach has been very useful in understanding which traits are associated with drought tolerance and why, it has been less useful in actually developing new cultivars showing an improved drought resistance under field conditions. Under such conditions, drought varies in timing, intensity and duration, and therefore it is the interaction among traits to determine the overall crop response to the variable nature of the drought stress rather than the expression of any specific trait (Ceccarelli et al., 1991). A typical example is offered by early growth vigor, a trait that is unanimously considered important in reducing the amount of water lost by evaporation from the soil surface, and therefore in increasing water-use efficiency (Richards et al., 2002) in crops grown on current rainfall (absence of stored moisture). The study of barley landraces from Syria has revealed that also genotypes with a modest early vigor can successfully achieve the same result with a prostrate growth habit (Ceccarelli and Grando, 1999).

Recently, a similar level of complexity has been revealed by the analysis of drought responses at the molecular level (Fig. 11–4). Cold, drought, and salinity

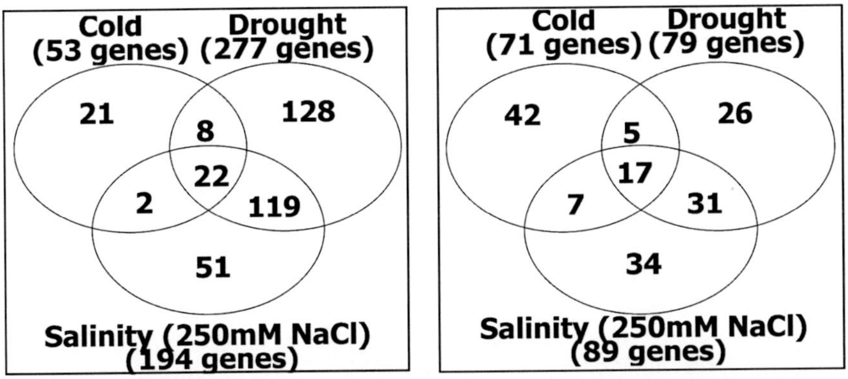

Fig. 11–4. Up- (left) and down- (right) regulated genes by cold, drought, and salinity stress in *Arabidopsis thaliana*.

stress up-regulate or down-regulate several genes in *Arabidopsis thaliana* (L.) Heynh. (Seki et al., 2001; Seki, personal communication, 2002). Several genes are affected by drought and salinity (119 up-regulated and 31 down-regulated), a smaller number by drought and cold and by cold and salinity. A relatively large number of genes (22 up-regulated and 17 down-regulated) are affected by all the three abiotic stresses (Seki et al., 2001; Seki, personal communication, 2002). If we consider that drought and cold can occur at different developmental stages of the crop, for a variable length of time and with a variable intensity, the level of complexity of the molecular responses is likely to be even greater than what is shown in Fig. 11–4. In fact, recent data on barley show that the genes up-regulated and down-regulated at the given stress treatment are different from those up-regulated and down-regulated at a different stress treatment (Öztürk et al., 2002).

Breeding for drought resistance based on direct selection for grain yield in the target environment (empirical or pragmatic breeding) appears intuitively as the most obvious solution. However, it has faced the major criticism that since field-drought is such a "moving target", the chances of progress appear slow at best and possibly remote. One major consequence of this attitude has been to study a "less mobile" target by simulating drought in laboratory (or green house) conditions, which result in generally irrelevant shocks (Passioura, 2002). Several studies have been and are being conducted addressing "laboratory drought" with the main justification of discovering mechanisms and genes (Yamaguchi-Shinozaki and Shinozaki, 1994; Kasuga et al., 1999; Nakashima et al., 2000; Seki et al., 2001).

We recognized that ultimately it is the drought resistance under field conditions that needs to be improved, but major problems with selection in rainfed conditions in general, and in stress environments in particular, are the *precision of selection* and the *existence of several target environments,* each characterized by its own specific type of drought, or more generally, a combination of stress.

The issue of the precision of selection under moisture stress conditions has been addressed by several papers (Simmonds, 1991, Singh et al., 2003), which have focused attention on the magnitude of heritability (the ratio of genetic to phenotypic variance, or, in simpler terms, a measure of the accuracy with which the phenotype reflects the genotype). A literature review (Ceccarelli, 1994, 1996a) showed the absence of a consistent relationship between grain yield (as a measure of stress intensity) and the magnitude of heritability. Recently, vom Brocke et al. (2002) showed that it is possible to detect genetic differences in pearl millet even under severe moisture stress conditions, while Al Yassin et al. (2003) found no relationships between magnitude of heritability and grain yield of barley for a wide range of yield levels. Furthermore, considerable progresses have been made both in experimental designs and statistical analysis that considerably improve the estimate of experimental error (e.g., Singh et al., 2003). Therefore, today it is possible to combine precision and relevance by conducting trials in the target environment even when this is a stressed environment.

The second issue, which has been frequently debated in relation to empirical breeding, is how to deal with the multitude of target environments (Cooper, 1999). This issue is intimately associated with that of *broad* and *specific adaptation*, which has been debated in plant breeding since the early 1920s (Hayes, 1923; Engledow, 1925). One of the causes of controversy is the definition of stress environment, which is well illustrated by the distribution of the crops in different agro-climatic environments. For example, in a country such as Syria, with a large spatial variability of rainfall within short distances (van Oosterom et al., 1993c), bread wheat (*Triticum aestivum* L.), durum wheat (*T. turgidum* var. *durum* L.) and barley among the cereals, and faba bean (*Vicia faba* L.), chickpea (*Cicer arietinum* L.), and lentil (*Lens culinaris* L.) among the food legumes, are grown in progressively drier environments with some overlapping. Therefore, a stress environment for faba bean or bread wheat is moderately favorable for durum wheat and chickpea, and a stress environment for durum wheat and chickpea is moderately favorable for barley and lentil. At the drier end, barley and lentil are the only rainfed crops, and the other cereals or legumes are only grown under supplementary or full irrigation. The situation described for Syria applies to most countries of the Mediterranean basin and West Asia, and is only altered by irrigation.

Another cause of controversy is the confusion between *adaptation over time* and *adaptation across space*, even though the distinction is of fundamental importance (Cleveland, 2001). Adaptation over time (also called stability or dependability) refers to the performance of a cultivar in a given location across several years; if the cultivar performs consistently better than a reference cultivar it is said to be *stable*. Adaptation over space refers to the performance of a cultivar in several locations; if the cultivar performs consistently better than a reference cultivar, it is said to be *widely adapted* (Ceccarelli and Grando, 2002). One recurrent problem in the literature on this topic is the use of the term 'environments' to indicate locations and years, and, more often, combinations of locations and years. This makes it difficult to understand whether wide or narrow adaptation refers to time or space or both (e.g., Cooper, 1999).

An additional problem has been the widespread use of the term "wide" adaptation for those varieties grown on large areas. The term has confused geographically-wide adaptation, often associated with lack of vernalization requirement and of photoperiod sensitivity, with environmentally-wide adaptation, that is, the ability to perform consistently well in locations or years which differ in soil characteristics (type, depth, fertility, etc.) water availability, biotic stresses, etc. (Cleveland et al., 2000). Unfortunately, for many, wide adaptation still means the ability to produce well both under stress and nonstress conditions.

It can be argued that wide adaptation over time (also defined as stability) is much more important to farmers than wide adaptation across space. The latter is, for obvious reasons, the major concern of seed producers. Two other causes of the controversy are the range of environmental variation sampled and the type of genetic (or breeding) material being used; most studies comparing the two strategies are biased either because they fail to note the different ranges of environments considered in each (Cleveland et al., 2000), or because breeding materials selected for specific adaptation to stress environments are not included.

Two of the most recent examples of using a narrow range of environments is on two-row barley in Canada (Atlin et al., 2000), where the ratio between the highest yielding environment and the lowest (nearly 3 t ha^{-1}) was 1.8, and where, not surprisingly, no crossover interaction was found, and winter wheat in France, where the average yields of the two most extreme environments were 5.9 and 9.5 t ha^{-1} (Brancourt-Hulmel et al., 2003). On the other hand, most studies supporting the concept of breeding for wide adaptation, counting on the "spillover effect" in marginal environments from selection conducted in optimal or suboptimal environments, are based on comparisons between modern varieties and farmer varieties, and are not correct. The right comparison should be between *varieties selected under optimum conditions* and *varieties selected under stress*.

Few breeding programs have conducted enough breeding cycles under marginal conditions to achieve measurable gains. Thus, studies comparing modern varieties (MVs) selected for favorable environments and MVs selected for unfavorable environments are limited. The few studies that do exist invariably show repeatable crossover interactions (Calhoun et al., 1994; Ceccarelli, 1996a). Most arguments relating repeatability of crossover interactions with breeding for specific or wide adaptation, or a combination of the two (Cooper, 1999; Yan et al., 2000), do not take into account that in plant breeding programs material has to be selected and discarded at each cycle.

Several international breeding programs implicitly recognize the existence of repeatable crossover interactions by targeting the development of germplasm for discrete target areas (Ceccarelli and Grando, 2002). For example, the Centro Internacional de Meyoramiento de Mais y Trigo (CIMMYT) divides the global target environments for bread wheat into 12 mega-environments (Braun et al., 1996), which are further subdivided. For example, mega-environments 4, 5, 8, and 11 are divided into 4a, 4b, 4c, 5a, 5b, 8a, 8b, 11a, and 11b based on other biotic and/or abiotic stresses. Similarly, the International Rice Research Institute (IRRI) targets nurseries to seven irrigated, three rainfed, one deepwater, and two upland ecosystems (Chaudhary and Ahn, 1996).

The basic philosophy of these programs is to generate and distribute widely adapted germplasm and to add traits that confer specific adaptation at a later stage (Braun et al., 1996). In both cases, the subdivision between target environments is an implicit recognition of specific adaptation, and in the case of rice (*Oryza sativa* L.), in breeding for individual target environments with independent breeding programs. In the case of wheat, all the breeding material, regardless of the mega-environment for which is targeted, undergoes heavy selection pressure for yield under optimum conditions and for disease resistance, and therefore is specifically adapted to all those locations and countries sharing reliable rainfall (or where the crop is irrigated) and that are affected by biotic stresses.

Breeding for specific adaptation is particularly important in the case of crops predominantly grown in unfavorable conditions because unfavorable environments are more diverse than favorable environments (Ceccarelli and Grando, 1997). Furthermore, unfavorable environments tend to produce repeatable crossover interactions (Ceccarelli, 1989). Breeding for specific adaptation to unfavorable conditions is often considered an undesirable breeding objective because it is usually associated with a reduction of potential yield under favorable conditions (Ceccarelli, 1996a). This issue has to be considered in its social dimension and in relation to the difference between adaptation over space vs. adaptation over time; for example, Australian farmers prefer maximizing yield in favorable years, while for North African and Near Eastern farmers yield in very poor years is more important.

Choice of Germplasm

The choice of the selection environment has important consequences on the assessment of the potential value of different types of germplasm in relation to breeding for drought tolerance. In many developing countries, and for crops grown in low-input conditions and in stress environments, landraces are still the backbone of agricultural production (Ceccarelli, 1984; Grando et al., 2001). The reasons why farmers prefer to grow only landraces or continue to grow landraces even after partial adoption of modern cultivars are not well documented, but include quality attributes such as food and feed quality, and seed storability (Brush, 1999). Landraces are often able to produce some yield even in difficult conditions where modern varieties are less reliable. For example, where farmers have adopted modern cultivars they have kept the landraces in the most unfavorable areas of the farm (Cleveland et al. 2000).

The value of landraces as sources of drought tolerance is well documented in the case of barley in Syria (Ceccarelli et al., 1995; Grando et al., 2001) and in several other crops elsewhere (Brush, 1999). The comparisons between barley landraces and modern cultivars under a range of conditions from severe stress (low input and low rainfall) to moderately favorable conditions (high inputs and high rainfall) have consistently indicated that:

1. landraces yield more than modern cultivars under low-input and stress conditions (Table 11–3);
2. the superiority of landraces is not associated with mechanisms to escape drought stress, as shown by their heading date;
3. within landraces there is considerable variation for grain yield under low-input and stress conditions, but all the landraces derived lines yield something whereas some modern cultivars fail;
4. landraces are responsive to both inputs and rainfall and the yield potential of some lines is high, though not as high as modern cultivars;
5. it is possible to find modern cultivars which under low-input and stress conditions yield almost as well as landraces, but their frequency is very low.

Table 11–3. Grain yields (kg ha^{-1}) of pure lines derived from Syrian landraces and modern cultivars at three levels of stress† in northern Syria (Ceccarelli, 1996b).

Environment	Landraces (N = 44)	Modern (N = 206)	Difference	P‡
Stress	1038	591	447	<0.01
Intermediate	3105	3291	−186	NS
Nonstressed	4506	6153	−1647	<0.01

† As defined by average precipitation and soil fertility (Ceccarelli, 1996b).
‡ Based on t-tests for groups of unequal size; NS = Nonsignificant.

The data in Table 11–3 also suggest that selection conducted only in high-input conditions is likely to miss most of the lines that would have performed well under low-input conditions.

The assumption of most breeding programs that landraces are genetically inferior is based on work conducted in research stations. Even those breeding programs addressing target environments that have low yield potential because of the combination of biotic and abiotic stresses have rarely challenged this assumption.

Decentralized Selection

Selection for specific adaptation in barley at ICARDA has been implemented initially by expanding the evaluation of early segregating populations in dry sites (such as Breda and Bouider in Syria), and by modifying the breeding method from a classical pedigree to a modified bulk method (Ceccarelli, 1996b). The most important results of this choice have been the re-evaluation of the role of landraces in breeding for drought tolerance described earlier, and the discovery of the importance of the wild progenitor of cultivated barley, *Hordeum spontaneum* (C. Koch), as a source of resistance to extreme levels of drought. Gas exchange observations made at anthesis in a wet site showed that *H. spontaneum* had widely open stomata, higher net photosynthesis, and lower pre-dawn leaf water potential at this stage of development than cultivated barley (Table 11–4). The ability of some accessions of *H. spontaneum* to tolerate extreme levels of drought stress was evident during the severe drought of 1987, when two lines of *H. spontaneum* were the only survivors in the breeding nurseries grown at Bouider which received only 176 mm rainfall (Grando et al., 2001).

These lines had some photosynthetic activity early in the morning, even though six times less than in absence of stress, stomata were open and the predawn leaf water potential was negative. At the same time, the stomata of the black-

Table 11–4. Gas exchange parameters of *H. spontaneum* accessions (means of 12 accessions) and ratio *H. spontaneum/H. vulgare* at Tel Hadya, Syria.

Parameter	Units	*H.spontaneum*	*H. spontaneum/H. vulgare*
Net photosynthesis	µmol m^{-2} s^{-1}	16.90 ± 0.80	1.49
Leaf conductance	mol m^{-2} s^{-1}	0.40 ± 0.03	3.57
Transpiration efficiency		4.35 ± 0.09	0.63
Leaf temperature	C°	25.50 ± 0.36	0.85
Predawn leaf water potential	Mpa	0.89 ± 0.25	1.35

seeded local landrace Arabi Aswad, considered by farmers to be very resistant to drought, were closed even though the pre-dawn leaf water potential was slightly higher than in *H. spontaneum*. By midday, the stomatal conductance of *H. spontaneum* decreased, the net photosynthesis became negative, while, the stomatal conductance of Arabi Aswad was zero.

Decentralized selection, namely selection in the target environment, is mostly based on grain yield which is considered to be the integrated response of numerous physiological, morphological, and developmental traits that allow a particular genotype or population to perform better than others to the particular type of drought encountered in a specific site and a specific year, and to the combination of that particular type of drought with other stresses (Ceccarelli et al., 1991). Genotypes or populations selected under this specific set of conditions are then tested for a second season in numerous sites and, if again successful, for a third season. The process makes use of the large year-to-year variability that characterizes stress environments, so that in a short period of time, such as 3 yr, and in 7 to 10 sites, there is a high probability that a genotype or a population is exposed to a number of different types of drought in terms of occurrence, timing, duration, and severity (Ceccarelli and Grando, 1996). Therefore, the process progressively accumulates favorable alleles for performance under various conditions.

Since the method relies heavily on grain yield, the classical pedigree method has been replaced by a bulk-pedigree method in which the segregating populations (F_3, F_4, and F_5) are yield tested in numerous locations, and single heads are selected within the best populations, and eventually tested for yield and used as parents in crosses (Ceccarelli, 1996b). Although the process relies heavily on grain yield, we also consider a number of other attributes such as early vigor, habit of growth, flowering time, plant height, peduncle length, peduncle extrusion leading to complete extrusion of the spike above the flag leaf, tillering, and when extreme stress conditions occur, leaf rolling and leaf senescence. Genotypes and populations with the desirable expression of these traits are also selected irrespective of their grain yield.

Therefore, the process combines the characteristics of the empirical and the analytical approach, and results in a slow but steady increase in yield under a combination of stresses. Figure 11–5 shows an example of this steady increase using numerous lines developed between 1985 and 2000: yield increased from <0.96 t ha^{-1} in Harmal to 1.4 t ha^{-1} in Arta. The latter, however, has the major handicap of becoming very short under drought stress. A reduction of plant height is undesirable because it forces the farmers to harvest by hand which is much more expensive than combine harvesting. The use of *H. spontaneum* and of one of the tallest selections from the black-seeded landrace, Zanbaka, resulted in yields of nearly 1.5 t ha^{-1}, with a plant height only marginally lower than that of Zanbaka.

In adopting decentralized selections as their breeding strategy, breeders have to be prepared to face crop failures in the case of extremely dry years. We believe that these extreme dry conditions are useful to assess the level of drought resistance available in the breeding program, and to identify genotypes that possess survival traits. Unfortunately, in these circumstances, most breeders would apply irrigation "to save the breeding material", thus missing out on collecting important information.

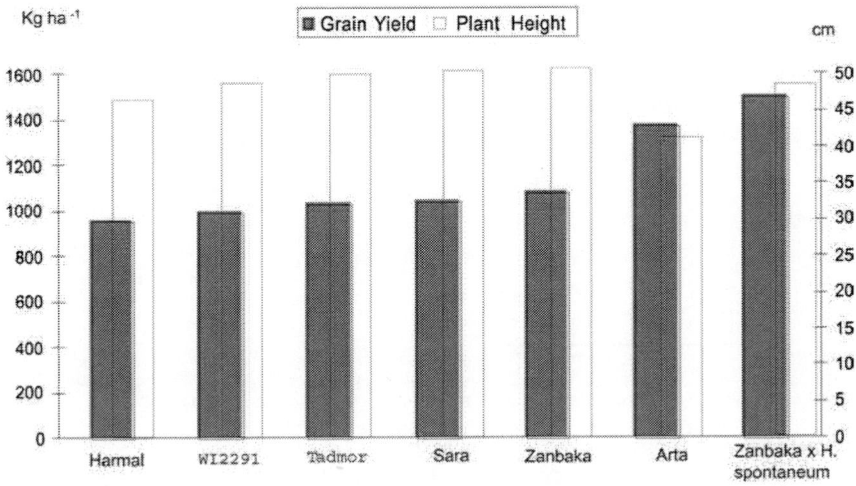

Fig. 11–5. Grain yield and plant height of barley lines developed during the last 15 yr (rainfall 197 mm).

PARTICIPATORY PLANT BREEDING

Selecting for specific adaptation has the advantage of adapting cultivars to the physical environment, and hence is more sustainable than other strategies that rely on changing the environment to fit new cultivars adapted to more favorable conditions. However, one of the most serious limitations of decentralized selection for specific adaptation to unfavorable environments is in the large number of potential target environments. Moreover, the number of target environments increases if we consider that environment is not only climate, soil, agronomic practices, farming system, etc., but also people living in that physical environment, their perception of risk associated with yield variation over time, their uses of the crop, and the consequent importance of quality traits even if neutral in terms of adaptation to the physical environment. Clearly, selection for specific adaptation to unfavorable conditions needs a larger sample of selection environments than selection for favorable environments.

The participation of farmers in the very early stages of selection offers a solution to the problem of fitting the crop to a multitude of both target environments and users' preferences (Ceccarelli et al., 1996, 2000). Although decentralized selection and farmers' participation are unrelated concepts, the acceptance of the former as a breeding strategy almost inevitably leads to the acceptance of the latter as a tactical necessity. It is worth mentioning that, although farmer participation is often advocated on the basis of equity, there are sound scientific and practical reasons for farmer involvement to increase the efficiency and the effectiveness of the breeding program. Decentralized-participatory plant breeding could also be particularly effective in those situations where seed is supplied by the informal seed system, as is the case for several crops in marginal environments.

The essence of decentralized-participatory plant breeding is depicted in Fig. 11–6, where A represents a typical centralized-nonparticipatory plant breeding pro-

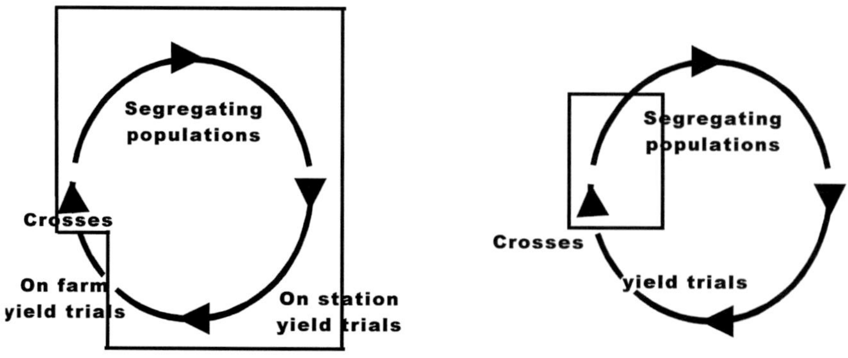

Fig. 11–6. The stages of a possible type of decentralized-participatory plant breeding program (B) are the same as in a centralized-nonparticipatory plant-breeding program (A), but only the generation of variability (crosses) and part of the selection (segregating populations) take place on station (in this hypothetical example). The line represents the fence of a research station.

gram, with most of the process taking place in one or more often, in several research stations, and only a small fraction taking place in farmers' fields. A model of a decentralized-participatory plant breeding program (B) has the same steps are the same as in A, but only the generation of variability (crosses) and part of the selection (segregating populations) takes place within the research stations.

The implementation of truly decentralized-participatory plant breeding requires not only the transfer to farmers' fields of part of the breeding materials that are usually grown on station, and the transfer to farmers of part of the decisions that are usually taken by the breeder alone or together with a team of other scientists. Therefore, decentralized-participatory plant breeding must, of necessity, involve several farmers or farmers' communities, each of which represents different target environments not only from a physical but also from a social point of view.

Participatory plant breeding has now been implemented on barley in numerous countries, and even though it was not specifically designed to breed for drought tolerance, the breeding material with the highest level of drought tolerance in Syria originated from this activity. In 2000, the total rainfall in most areas of Syria was below average (ICARDA, 2001) and crop yields were severely affected. In some areas, the rainfall was so low that the crop did not even germinate, in many others the crop failed to produce grain.

The trials, planted in eight farmers' fields in Syria as part of the participatory barley breeding program, were affected by different intensities of drought: one extreme was Melabya (Hassakeh province) with only about 50 mm rainfall in the entire season and no germination, and the other extreme was Suran with 252 mm rainfall and an average grain yield of 1.8 t ha^{-1} (ranging from 1.0–3.2 t ha^{-1}). The driest sites, where some new barley entries were able to produce some grain and/or some biomass were Bylounan and Jurn El-Aswad (Raqqa province) with 87 and 121 mm, respectively, and Bari Sharky (Hama Province) with 130 mm. Average grain and biomass yield were very low but some lines were able to produce between

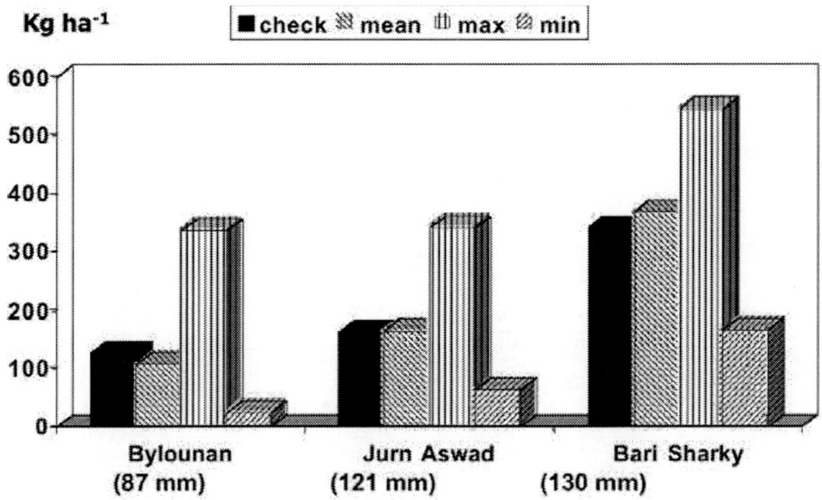

Fig. 11–7. Grain yield (kg ha^{-1}) under severe drought stress in three locations in Syria in 2000.

300 and 500 kg ha^{-1} of grain (Fig. 11–7) and between 500 and 3000 kg ha^{-1} of biomass yield.

Drought tolerance was assessed in the field when the plots where close to maturity with a score from 1 (most plants with a spike and seed and no symptoms of leaf rolling or wilting) to 5 (absence of spikes, leaf desiccation and/or wilting). In the four locations, the crosses between *H. spontaneum-41* (one of the *H. spontaneum* lines identified in 1987 and described earlier) and landraces were the most drought tolerant (Fig. 11–8), while the modern germplasm was nearly always the least tolerant. Such crosses were also the germplasm most frequently selected by

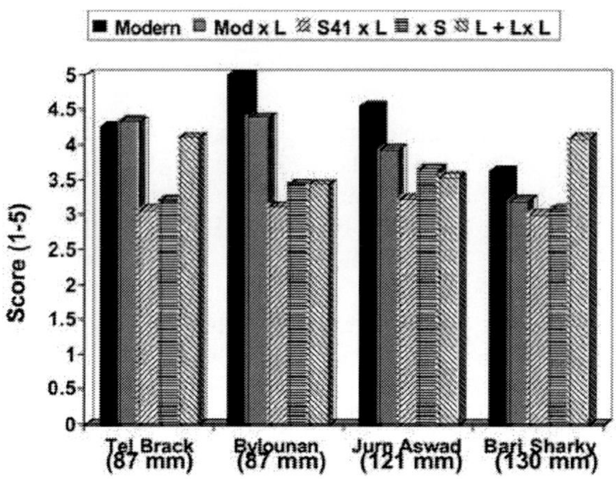

Fig. 11–8. Drought susceptibility (1 = resistant 5 = susceptible) of five types of germplasm grown under severe stress in four locations in Syria in 2000.

farmers in three of four locations, while the modern germplasm was always selected the least. In one location, the farmers more often selected the landraces and the crosses with landraces, which in that location did not differ greatly with the crosses between *H. spontaneum 41* and landraces in the drought score. This is a further indication that, if carefully selected, some *H. spontaneum* lines can contribute significantly to enhance the drought resistance of cultivated barley.

MOLECULAR APPROACHES

The use of DNA molecular-marker techniques is now sufficiently well developed to be exploited in breeding programs to improve resistance/tolerance to both abiotic and biotic stresses, and can considerably reduce the complexity of combining in the same cultivar numerous desirable traits (Graner et al., 1991; Forster et al., 2000; Ramsay et al., 2000). These techniques make it feasible to develop linkage maps, which are used to locate and estimate phenotypic effects of quantitative trait loci (QTL). In fact, one of the first molecular approaches has been the identification and the localization of those traits known to be associated with resistance to drought and with other stresses occurring in various combinations with drought (Yin et al., 1999, Foster et al., 2000, Teulat, 1998, 2001, 2002).

Today, the chromosomal location of most of the traits described earlier in this chapter are known: for example: vernalization requirement in barley has been identified by comparative studies with wheat (group-2 and group-5), and located as Vrn-H1 or Sgh2 on 5H, Vrn-H2 (originally Sgh1) on 4H and Sgh3 on 1H (Laurie et al., 1995). In addition, Pan et al. (1994) reported a frost tolerance QTL on 5H, around the same location. This locus on 5H has been revealed with a DHn probe and named Dhn9 (Choi et al., 2000). Recently, Choi et al., (2002) mapped the barley HvCbf3 gene, ortholog of the *Arabidopsis* CBF/DREB1 gene to a position more proximal to the Dhn1/Dhn2 – DHn9 interval. A major QTL affecting drought, induced abscisic acid accumulation, has been mapped on chromosome 5H, tightly linked to one of the dehydrin (see below) loci, the Dhn1/Dhn2 locus, suggesting a genetic linkage between abscisic acid accumulation and stress tolerance (Quarrie et al., 1994). These results emphasize the important role of chromosome 5H for abiotic stress resistance (Cattivelli et al., 2002).

The major genes for photoperiodic response are the Eam1 or Ppd-H1 location on chromosome 2H (Laurie et al., 1995), and Ppd-H2 on chromosome 1H (Laurie et al., 1995). Additionally, earliness per se loci have been identified as eps2S on 2H, eps3L on 3HL, eps4L on 4HL, eps5L on 5HL, eps6L.1 and eps6L.2 on 6HL, eps7S on 7HS and eps7L on 7HL (Laurie et al., 1995, Davis et al., 1997).

The chromosomal location of some other traits related to drought tolerance is known. Of particular interest are those traits that have been evaluated under Mediterranean environments. For example, Teulat et al. (1997), using a barley population developed at ICARDA, found that the centromeric region and the long arm of chromosome 7H are involved in the control of relative water content, and 10 QTLs for osmotic adjustment were identified (Teulat et al., 1998, 2001). In the same population, two QTL for carbon isotope discrimination were identified on chromosomes

2H and 4H (Teulat et al., 2002). These were co-localized with QTL's for osmotic adjustment and 1000-kernel weight.

Quantitative trait loci for agronomic traits related to drought resistance were localized with the help of a genetic linkage map in recombinant inbred lines (RILs) of the cross Arta/*Hordeum spontaneum*41-1 (Grando et al., 2000, Baum et al., 2003). For one of the most important characters, plant height under drought stress, QTLs on chromosomes 2H, 3H, and 7H were detected in this population. The "plant height" QTLs, especially the one on 3H, showed pleiotropic effects on traits such as days to heading, grain yield, and biological yield. The QTLs were also identified for other traits associated with adaptation to Mediterranean environment such as cold tolerance (5H), days-to-heading (7H, 3H), and tiller number (3H).

In another population adapted to Mediterranean environment, Tadmor/Sel160 (Baum et al., 1996) a number of QTLs related to yield under stress conditions such as QTLs for days to heading, cold damage, grain yield and early growth vigor, clustered on chromosome 5H. A QTL for plant height was located on chromosome 3H, probably in the vicinity of the location of the *denso* gene.

Chromosome group 5 has the highest concentration of QTLs and of major loci controlling plant adaptation to the environment, particularly those controlling heading date, and frost and salt tolerance. Although the molecular responses to cold and drought stresses share a common set of genes (see below), the loci describing the genetic bases of cold and drought tolerance are different. Nevertheless, multiple-stress QTLs and linked markers have also been detected, suggesting the existence of common mechanisms for different stresses, or of clusters of genes controlling different stress tolerance processes.

A large number of genes have been isolated whose expression is affected by the application of drought stress or by more than one abiotic stress. One of the best known is the late embryogenesis-abundant (LEA) proteins gene family, induced by dehydration, cold, abscisic acid treatment, salt, and osmotic stress. One class of LEA, the dehydrins, are the main group of stress-related proteins, and some are expressed during dehydration, others during cold treatment (Choi et al., 1999, 2000; Zhu et al., 2000).

Regulation of Stress Responsive Genes

The crop plants have evolved complex molecular mechanisms by which they tolerate and adapt to adverse growth conditions. When they encounter drought stress conditions, the cells reprogram their cellular processes by triggering a network of signaling events that start with stress perception and end with a cellular response, such as gene expression in the nucleus. Many genes that normally are silent, are activated (up-regulated) under stress conditions. Other genes are silenced or their activity is reduced (down-regulated) under stress conditions. Abscisic acid (ABA), a phytohormone, has been shown to play a critical role in plant stress responses, because many stress signals increase the level of ABA (Shinozaki and Yamaguchi-Shinozaki, 2000).

Recent studies in *A. thaliana* suggest that osmotic stress imposed by drought or high salt is transmitted through at least two pathways, one ABA-dependent and the other ABA-independent. Cold, on the other hand, exerts its effect on gene expression largely through an ABA-independent pathway (Shinozaki and Yamaguchi-Shinozaki, 2000). This paradigm has been established largely by molecular studies that identify the cis- and trans-acting elements responsible for gene induction by stress signals and ABA (Shinozaki and Yamaguchi-Shinozaki, 2000). For example, some cold- and drought-inducible genes contain a cis-acting element called DRE/CRT (for drought/cold-responsive element) that is critical for drought/cold-induced gene expression, but not for ABA induction. Conversely, ABA-induced expression often relies on the presence of a different cis-acting element called ABA-responsive element (ABRE). Although drought and salinity trigger a sustained increase in endogenous ABA levels and are sufficient to induce the expression of ABRE-containing genes, cold induces only a transient increase in ABA level and often is insufficient to induce ABRE-containing genes (Shinozaki and Yamaguchi-Shinozaki, 2000).

Recently, microarray technology has become a powerful tool for the analysis of gene expression at genomic level. This DNA chip-based technology arrays cDNA sequences on a glass slide at a density of up to 1000 genes cm^{-2}. This technology has been used in *A. thaliana* to identify various genes involved in major abiotic stress tolerance such as drought, salinity, and cold (Seki et al., 2001; Seki, personal communication, 2002). They isolated mRNAs from drought, cold, or high-salinity stress-treated *A. thaliana* plants, and wild-type unstressed plants were used for preparation of Cy3- and Cy5-labelled cDNA probes, respectively.

These cDNA probes were mixed and hybridized with the cDNA microarray containing more than 7000 genes. Based on the hybridization pattern, the stress genes are grouped into induced (up-regulated) and silenced (down-regulated) genes (Fig. 11–4). Analysis of overlapping on the Venn diagram showed that 22 genes were induced under all three stresses and many of them function as regulatory genes. Furthermore, the diagram also shows differences and cross-talk of gene expression among drought, cold, and high-salinity stress responses.

As shown in Fig. 11–4, 277, 53, and 194 genes were identified as drought-, cold-, and high-salinity-induced genes, respectively; 141 genes were induced by both drought and high salinity, whereas only 30 genes were induced by both drought and cold stress; and 24 genes were designated as cold and high-salinity-inducible genes.

A similar trend was also observed with regards to down-regulated genes of drought, cold, or salinity stress.

These results demonstrate the existence of greater cross talk between drought and salinity stress signaling processes than those between cold and high-salinity stress signaling processes. This observation at molecular level also agrees with breeders' observations that some of the drought-tolerant varieties also show some tolerance to salinity.

Some of the induced genes includes functional proteins such as late embryogenesis abundant proteins; heat-shock proteins; cold-inducible proteins; osmoprotectant biosynthesis-related proteins; carbohydrate metabolism-related proteins; water-channel proteins; sugar transporters; potassium transporters; proteases; senescence-related genes; protease inhibitors etc., involved in cell protection from osmotic stress and regulatory proteins like transcription factors (e.g., DREB1A), protein kinases, protein phosphatases, etc., involved in signal transduction and regulation. Many of the drought, cold, or high-salinity stress down-regulated genes include photosynthesis-related genes and genes involved in carbohydrate metabolism, reduce the rate of these two key cellular metabolism under these stress conditions.

Knowledge on molecular biology of drought tolerance and other abiotic stresses in model plants enabled us to determine the genes involved in various stress tolerance more accurately than before. The stress genes identified and their expression profiles in the model plants and crop varieties are extremely useful for designing better tolerant crops varieties with multiple stress tolerance both through conventional breeding and through transgenic approaches. Notable example in this direction is the use of DREB1 genes of *A. thaliana* for engineering of tolerant varieties of tomato (Lycopersicon esculentum Mill.) and tobacco (*Nicotiana tabacum* L.) to drought, cold, and salinity (Kasuga et al., 1999; Hsieh et al., 2002a, 2002b). The identified new genes can also be used for isolating homologs and further mapping (Choi et al., 2002) and marker-assisted selection.

CONCLUSIONS

The objective of this chapter was to discuss what plant breeders can do when the target environment of their breeding program is characterized by chronic low yields due to numerous factors such as climatic, nutritional, and biotic stresses. The discussion is biased as it concentrates mostly on one crop (barley) and one type of stress environment (dry Mediterranean with cold winters and hot summers, and crops grown on current rainfall). While this is inevitable, given the authors' familiarity with that crop and that environment, there are perhaps some general concepts that, with some modifications, could be useful in other crops and in other types of stress environments.

The first concept is that in these environments, climatic, nutritional, and biotic stresses usually occur together (though not necessarily all of them all the time); and, so far, there is little substitute for actually exposing the breeding material to a real field situation. Although little practiced, the idea is not new. More than a quarter of a century ago Hurd (1971) published a paper with the title Can We Breed for Drought Resistance? The first sentence of the paper was "My answer to the above very pertinent question is a confident and optimistic 'Yes'". He concluded: "One method is to grow large populations in early generation under typical dry growing conditions." More recently, Bramel-Cox et al. (1991) recognized that the key to increased production with fewer external inputs would be through a reevaluation of the identification and use of selection and testing environments.

Although this concept is obvious to many and not new, selection for stress environments is still seldom done in the target environments. This may not always be necessarily a deliberate choice of one breeding strategy or another, but is simply due to the distance of suitable selection sites from main cities and all the associated inconveniences. We hypothesize that in these cases an interesting solution may be offered by farmers' participation in breeding (Ceccarelli et al., 1996, 1996b, 2000). Conducting selection in farmer's fields has the advantage of exploiting genetic differences under farmers' condition and the additional advantage of making use of the farmers' knowledge of the crop.

The second concept is the use of germplasm usually ignored by most plant breeders, such as landraces and wild relatives. This approach is a direct consequence of choosing to work in the target environment. It has led, over nearly 20 yr, to the development of three barley cultivars, now grown in a number of farmers' fields in central and northern Syria and in environments considered too difficult and therefore beyond the plant breeder's domain.

In stress environments, every effort should be made to control environmental variability in trial and nurseries evaluation. When working at stress sites, the breeder should forget the typical research-station style of work. The methodology, experiment designs, and plot techniques used in the very homogeneous environment of the experiment station are not suitable; in fact, when the conclusion is reached that progress cannot be made in a stress environment, it is probably for that reason.

The main conclusion of the chapter is that breeding for stress environments is possible provided it is conducted with strategies and methodologies that little have in common with those used in breeding for favorable environments. Adaptation over time can be improved by breeding for specific adaptation to a given type of stress environment. This can be achieved by taking advantage of the temporal variability of stress environments, which permits exposure of the same breeding material to variable combinations of stresses over a (relatively) short period. We are aware that this is fundamentally different from the modern trend of plant breeding towards broad adaptation over space. The difference represents the contrasting interests of farmers and seed companies. Farmers are interested in cultivars that are consistently superior on their farm, regardless of how they perform at other locations or in other countries. Seed companies however, want to market as much seed of as few cultivars as possible. Breeders have been breeding, perhaps unconsciously, more for seed companies and for their personal prestige than for farmers. The two objectives coincide when selection and target environments are similar, but this approach has bypassed millions of small farmers in difficult environments.

Recent advances in plant genomics have enabled to dissect various molecular mechanisms (signal transduction pathways) involved in drought, cold, and salt stress tolerance and in identifying various genes involved in these stress tolerance. Information generated in genomics should be integrated in the practical plant breeding. Various genes identified both in model plants and crop plants could be used for developing stress tolerant plants through either marker-assisted selection or direct gene transfer.

REFERENCES

Acevedo, E., and S. Ceccarelli. 1989. Role of physiologist-breeder in a breeding program for drought resistance conditions. p.117–139. *In* F.W.G. Baker (ed.) Drought resistance in cereals. CAB Int., Wallingford, UK.

Al Yassin, A., S. Grando, and S. Ceccarelli. 2003. Heritability, environment and type of germplasm. Euphytica (submitted).

Ashraf, M. 1994. Breeding for salinity tolerance in plants. Crit. Rev. Plant Sci. 13:17–42.

Atlin, G.N., K.B. McRae, and X. Lu. 2000. Genotype x region interaction for two row barley yield in Canada. Crop Sci. 40:1–6.

Baum, M., S. Grando, G. Backes, A. Jahoor, A. Sabbagh, and S. Ceccarelli. 2003. QTLs for agronomic traits in the Mediterranean environment identified in recombinant inbred lines of the cross 'Arta' x H. spontaneum 41-1. Theor. Appl. Genet. 107:1215–1225.

Baum, M., H. Sayed, J.L. Araus, S. Grando, S. Ceccarelli, G. Backes, V. Mohler, A. Jahoor, and G. Fischbeck. 1996. QTL analysis of agronomic important characters for dryland conditions in barley by using molecular markers. p. 241–243. *In* A. Slinkard et al. (ed.) Proc. of the V Int. Oat Conf. and of the VII Int. Barley Genetics Symp., Saskatoon. Brewing Malting Barley Res. Inst., Winnipeg, ON.

Boyer, J.S. 1982. Plant productivity and environment. Science (Washington, DC) 218:443–448.

Blum, A. 1993. Selection for sustained production in water-deficit environments. p. 343–347. *In* International crop science I. CSSA, Madison, WI.

Blum, A., and Y. Pnuel. 1990. Physiological attributes associated with drought resistance of wheat cultivars in a Mediterranean environment. Aust. J. Agric. Res. 41:799–810.

Bramel-Cox, P.J., T. Barker, F. Zavala-Garcia, and J.D. Eastin. 1991. Selection and testing environments for improved performance under reduced-input conditions. p. 29–56 *In* D.A. Sleeper et al. (ed.) Plant breeding and sustainable agriculture, considerations for objectives and methods. CSSA Spec. Publ.18. CSSA, Madison, WI.

Brancourt-Hulmel, M., G. Doussinault, C. Lecomte, P. Bérard, B., Le Buanec, and M. Trottet. 2003. Genetic improvement of agronomic traits of winter wheat cultivars released in France from 1946 to 1992. Crop Sci. 43:37–45.

Braun, H.J., S. Rajaram, and M. van Ginkel. 1996. CIMMYT's approach to breeding for wide adaptation. Euphytica 92:175–183.

Brush, S.B. (ed.) Genes in the field: On-farm conservation of crop diversity. p. 51–76. IPGRI/IDRC/Lewis Publ., Boca Raton, FL.

Calhoun, D.S., G. Gebeyehu, A. Miranda, S. Rajaram, and M. van Ginkel. 1994. Choosing evaluation environments to increase wheat grain yield under drought conditions. Crop Sci. 34:673–678.

Cattivelli L., P. Baldi, C. Crosatti, N. Di Fonzo, P. Faccioli, M. Grossi, A.M. Mastrangelo, N. Pecchioni, and A.M. Stanca. 2002. Chromosome regions and stress-related sequences involved in resistance to abiotic stress in *Triticeae*. Plant. Mol. Biol. 48:649–665.

Ceccarelli, S. 1984. Utilization of landraces and *Hordeum spontaneum* in barley breeding for dry areas. Rachis 3(2):8–11.

Ceccarelli, S. 1989. Wide adaptation. How wide? Euphytica 40:197–205.

Ceccarelli, S. 1994. Specific adaptation and breeding for marginal conditions. Euphytica 77:205–219.

Ceccarelli, S., 1996a. Adaptation to low/high input cultivation. Euphytica 92:203–214.

Ceccarelli, S., 1996b. Positive interpretation of genotype by environment interactions in relation to sustainability and biodiversity. p. 467–486. *In* M. Cooper and G.L. Hammers (ed.) Plant adaptation and crop improvement. CAB Int., Wallingford. UK, ICRISAT, Andra Pradesh, India; IRRI, Manila, Philippines.

Ceccarelli, S., E. Acevedo, and S. Grando. 1991. Breeding for yield stability in unpredictable environments: Single traits, interaction between traits, and architecture of genotypes. Euphytica 56:169–185.

Ceccarelli, S., and S. Grando. 1996. Drought as a challenge for the plant breeder. Plant Growth Regulation 20:149–155.

Ceccarelli, S., and S. Grando. 1997. Increasing the efficiency of breeding through farmer participation. p. 116–121. *In* Ethics and equity in conservation and use of genetic resources for sustainable food security, proceeding of a workshop to develop guidelines for the CGIAR, 21–25 Apr. 1997. Foz de Iguacu, Brazil. IPGRI, Rome, Italy.

Ceccarelli, S., and S. Grando. 1999. Barley landraces from the fertile crescent: a lesson for plant breeders. p.51–76. *In* S.B. Brush (ed.) Genes in the field: On-farm conservation of crop diversity. IPGRI/IDRC/Lewis Publ., Boca Raton, FL.

Ceccarelli, S., and S. Grando. 2002. Plant breeding with farmers requires testing the assumptions of conventional plant breeding: Lessons from the ICARDA barley program. p. 297–332. *In* D.A. Cleveland and D. Soleri (ed.) Farmers, scientists and plant breeding: Integrating knowledge and practice. CABI Publ. Int., Wallingford, Oxon, UK.

Ceccarelli, S., S. Grando, and R.H. Booth. 1996. International breeding programmes and resource-poor farmers: Crop improvement in difficult environments. p. 99–116. *In* P. Eyzaguirre and M. Iwanaga (ed.) Participatory plant breeding. Proc. of a workshop on participatory plant breeding, Wageningen, The Netherlands. 26–29 July 1995. IPGRI, Rome, Italy.

Ceccarelli, S., S. Grando, and J. Hamblin. 1992. Relationships between barley grain yield measured in low and high yielding environments. Euphytica 64:49–58.

Ceccarelli, S., S. Grando, R. Tutwiler, J. Baha, A.M. Martini, H. Salahieh, A. Goodchild, and M. Michael. 2000. A methodological study on participatory barley breeding. I. Selection phase. Euphytica 111:91–104.

Ceccarelli, S., S. Grando, and J.A.G. van Leur. 1995. Understanding landraces: The fertile crescent's barley provides lesson to plant breeders. Diversity II:112–113.

Chaudhary, R.C., and S.W. Ahn. 1996. International network for genetic evaluation of rice (INGER) and its modus operandi for multi-environment testing. p. 139–164. *In* M. Cooper and G.L. Hammers (ed.) Plant adaptation and crop improvement. CAB Int., Wallingford, UK; ICRISAT, Andra Pradesh, India; IRRI, Manila, Philippines.

Choi, D.W., M.C. Koag, and T.J. Close. 2000. Map location of barley *Dhn* genes determined by gene-specific PCR. Theor. Appl. Genet. 101:350–354.

Choi D.W., E.M. Rodriguez, and T.J. Close. 2002. Barley Cbf3 Gene identification, expression pattern, and map location. Plant Physiol. 129:1781–1787.

Choi D.W., B. Zhu B., and T.J. Close. 1999. The barley (*Hordeum vulgare* L.) dehydrin multigene family: Sequences, allele types, chromosome assignments, and expression characteristics of 11 *Dhn* genes of cv. Dicktoo. Theor. Appl. Genet. 98:1234–1247.

Cleveland, D.A., 2001. Is plant breeding science objective truth or social construction? The case of yield stability. Agric. Human Values 18:251– 270.

Cleveland, D.A., D. Soleri, and S.E. Smith. 2000. A biological framework for understanding farmers plant breeding. Econ. Bot. 54:377–390.

Cooper, M. 1999. Concepts and strategies for plant adaptation research in rainfed lowland rice. Field Crops Res. 64:13–34.

Cooper, P.J.M., P.J. Gregory, D. Tully, and H.C. Harris, 1987. Improving water use efficiency of annual crops in the rainfed farming systems of West Asia and North Africa. Exp. Agric. 23:113–158.

Davis, M.P., J. D. Franckowiak, T. Konishi, and U. Lundqvist (ed.). 1997. Barley Genet. Newsl. 26 (Special Issue):1–517. The Am. Malting Barley Assoc., Milwaukee, WI.

Engledow, F.L. 1925. The economic possibilities of plant breeding. p. 31–40. *In* F.T. Brooks (ed.) Proc. of the Imperial Botanical Conf. F.T. Brooks, London.

Erskine, W., and F. El Ashkar. 1993. Rainfall and temperature effects on lentil (*Lens culinaris*) seed yield in Mediterranean environments. J. Agric. Sci. (Cambridge) 121:347–354.

Forster, B.P., R.P. Ellis, W.T.B. Thomas, A.C. Newton, R. Tuberosa, D. This, R.A. El-Enein, H. Bahri, and M. Ben Salem. 2000. The development and application of molecular markers for abiotic stress tolerance in barley. J. Exp. Bot. 51:19–27

Grando, S., G. Backes, S. Ceccarelli, A. Sabbagh, A. Jahoor, and M. Baum. 2000. QTL Analysis for agronomic traits in recombinant inbred lines of the cross Arta x H. spontaneum 41-1. p. 61–63. *In* Proc. 8th Int. Barley Genetics Symp.,Vol. III, Adelaide, Australia. 22–27 Oct. 2000. Dep. of Plant Sci., Waite Campus, South Australia.

Grando, S., R. von Bothmer, and S. Ceccarelli. 2001. Genetic diversity of barley: Use of locally adapted germplasm to enhance yield and yield stability of barley in dry areas. p. 351–372. *In* H.D. Cooper et al. (ed.) Broadening the genetic base of crop production. CABI, New York/FAO, Rome/IPRI, Rome.

Graner, A., A. Jahoor, J. Schondelmaier, H. Siedler, K. Pillen, G. Fischbeck, G. Wenzel, and R.G. Herrmann. 1991. Construction of an RFLP map of barley. Theor. Appl. Genet. 83:250–256.

Hadjichristodoulou, A. 1982. The effects of annual precipitation and its distribution on grain yield of dryland cereals. J. Agric. Sci. (Cambridge) 99:261–270.

Hayes, H.K. 1923. Controlling experimental error in nursery trials. J. Am. Soc. Agron. 15:177–192.

Hsieh, T.H., J. Lee, Y. Charng, and M.-T. Chan. 2002b. Tomato plants ectopically expressing arabidopsis cbf1 show enhanced resistance to water deficit stress. Plant Physiol. 130(2):618 –626.

Hsieh, T.H, J.T. Lee, P.T. Yang, L.H. Chiu, Y.Y. Charng, Y.C. Wang, and M.T. Chan. 2002a. Heterology expression of the Arabidopsis C-repeat/dehydration response element binding factor 1 gene confers elevated tolerance to chilling and oxidative stresses in transgenic tomato. Plant Physiol. 129:1086–1094.

Hurd, F.A. 1971. Can we breed for drought resistance? p. 77–88. *In* K.L. Larson and J.D. Eastin (ed.) Drought injury and resistance in crops. CSSA Spec. Publ. 2. CSSA and SSSA, Madison, WI.

International Center for Agricultural Research in Dry Areas. 2001. Annual report for 2002. ICARDA, Aleppo, Syria.

Jacobsen, T., and R.M. Adams. 1958. Salt and silt in ancient Mesopotamian agriculture. Science (Washington, DC) 128:1251–1258.

Kasuga, M., Liu, Q., Miura, S., Yamaguchi-Shinozaki, K., and Shinozaki, K. 1999. Improving plant drought, salt, and freezing tolerance by gene transfer of a single stress-inducible transcription factor. Nature Biotechnol. 17:287–291.

Laurie, D.A., N. Pratchett, J.H. Bezant, and J.W. Snape. 1995. RFLP mapping of 5 major genes and 8 quantitative trait loci controlling flowering time in a winter x spring barley (*Hordeum vulgare* L.) cross. Genome 38:575–585.

McWilliam, J.R. 1986. The national and international importance of drought and salinity effects on agricultural production. Aust. J. Plant Physiol. 13:1–13.

Nakashima, K., Z.K. Shinwari, Y. Sakuma, M. Seki, S. Miura, K. Shinozaki, and K. Yamaguchi-Shinozaki. 2000. Organization and expression of two Arabidopsis DREB2 genes encoding DRE-binding proteins involved in dehydration- and high-salinity-responsive gene expression. Plant Mol. Biol. 42:657–665.

Oudina, M., and H. Bouzerzour. 1993. Variabilité du rendement de l'orge (*Hordeum vulgare* L.) sous l' influence du climat des hauts plateaux setifiens. p. 110–120. *In* M. Jones et al. (ed.) The agrometeorology of rainfed barley-based farming systems. ICARDA, Aleppo, Syria.

Öztürk, Z.N., V.Talamé, C.B. Michalowski, N. Gozukirmizu, R. Tuberosa, and H.J. Bohnert. 2002. Monitoring large-scale changes in transcript abundance in drought- and salt-stressed barley. Plant Mol. Biol. 48:551–573.

Pan, A., P.M. Hayes, F. Chen, T.H.H. Blake, T.K.S. Wright, I. Karsai, and Z. Bedö. 1994. Genetic analysis of the components of winter hardiness in barley (*Hordeum vulgare* L.). Theor. Appl. Genet. 89:900–910.

Papendick, R.I., and G.S. Campbell. 1990. Concepts and management strategies for water conservation in dryland farming. p. 119–127. *In* P.W. Unger et al. (ed.) Challenges in dryland agriculture—A global perspective. Proc. Int. Conf. on Dryland Farming. Texas Agric. Exp. Stn., College Station.

Passioura, J.B. 2002. Environmental biology and crop improvement. Funct. Plant Biol. 29:537–546.

Plucknett, D.L. 1991. Saving lives through agricultural research. Issues Agric. (CGIAR) 1:20.

Quarrie, S.A., M. Gulli, C. Calestani, A. Steed, and N. Marmiroli. 1994. Location of a gene regulating drought-induced abscisic acid production on the long arm of chromosome 5A of wheat. Theor. Appl. Genet. 89:794–800.

Ramanjulu, S., and D. Bartels. 2002. Drought- and desiccation-induced modulation of gene expression in plants. Plant Cell Environ. 25:141–151.

Ramsay, L., M. Macaulay, S. Degli Ivanissevich, K. Maclean, L. Carsle, J. Fuller, K. J. Edwards, S. Tuvesson, M. Morgante, A. Massari, E. Maestri, N. Marmiroli, T. Sjakste, M. Ganal, W. Powell, and R. Waugh. 2000. A simple sequence repeat-based linkage map of barley. Genetics 156:1997—2005

Richards, R.A., G.J. Rebetzke, A.G. Condon, and A.F. van Herwaarden. 2002. Breeding opportunities for increasing the efficiency of water use and crop yield in temperate cereals. Crop Sci. 42:111–121.

Rosen, A.M. 1990. Environmental change at the end of early Bronze Age Palestine. p. 247–255. *In* P. De Miroschedji (ed.) L'urbanisation de la Palestine à l'âge du Bronze ancien. BAR Int., Oxford, UK.

Sardar, Z. 1995. Cruising for peace. Nature (London) 373:483–484.

Seki, M., M. Narusaka, H. Abe, M. Kasuga, K. Yamaguchi-Shinozaki, P. Carninci, Y. Hayashizaki, and K. Shinozaki. 2001. Monitoring the expression pattern of 1300 Arabidopsis genes under drought and cold stresses by using a full-length cDNA microarray. Plant Cell 13:61–72.

Shinozaki, K., and K. Yamaguchi-Shinozaki. 2000. Molecular responses to dehydration and low temperature: Differences and cross-talk between two stress signaling pathways. Curr. Opin. Plant Biol. 3:217–223.

Serraj, R., and T.R. Sinclair. 2002. Osmolyte accumulation: Can it really help increase crop yield under drought conditions? Plant Cell Environ. 25:333–341.

Simmonds, N.W. 1991. Selection for local adaptation in a plant breeding programme. Theor. Appl. Genet. 82:363–367.

Sinclair, T.R. 2000. Model analysis of plant traits leading to prolonged survival during severe drought. Field Crops Res. 68:211–217.

Singh, M., R.S. Malhotra, S. Ceccarelli, A. Sarker, Grando, S., and W. Erskine. 2003. Spatial variability models to improve dryland field trials. Exp. Agric. 39:1–10.

Strauss, M.S. 1993. The challenge ahead. p. 401–405. *In* International crop science I. CSSA, Madison, WI.

Teulat B., C. Borries, and D. This. 2001. New QTLs identified for plant water status, water-soluble carbohydrates, osmotic adjustment in a barley population grown in a growth-chamber under two water regimes. Theor. Appl. Genet. 103:161–170.

Teulat, B, O. Merah, X. Sirault, C. Borries, R. Waugh, and D. This. 2002. QTLs for grain carbon isotope discrimination in field-grown barley. Theor. Appl. Genet. 106:118–126.

Teulat, B., P. Monneveux, J. Wery, C. Borries, I. Souyris, A. Charrier, and D. This. 1997. Relationship between relative water content and growth parameters under water stress in barley: A QTL study. New Phytol. 137:99–107.

Teulat, B., D. This, M. Khairallah, C. Borries, C. Ragot, P. Sourdille, P. Leroy, P. Monneveux, and A. Charrier. 1998. Several QTLs involved in osmotic-adjustment trait variation in barley. Theor. Appl. Genet. 96:688–698.

Turner, N.C. 2002. Optimizing water use. p. 119–135. *In* J. Nösberger et al. (ed.) Crop science: Progress and prospects. CAB Int., Wallingford, UK.

van Oosterom, E.J., and S. Ceccarelli. 1993. Indirect selection for grain yield of barley in harsh Mediterranean environments. Crop Sci. 33:1127–1131.

van Oosterom, E.J., S. Ceccarelli, and J.M. Peacock. 1993a. Yield response of barley to rainfall and temperature in Mediterranean environments. J. Agric. Sci. (Cambridge) 121:307–314.

van Oosterom, E.J., S. Ceccarelli, and J.M. Peacock. 1993b. Yield response of barley to rainfall and temperature in Mediterranean environments. J. Agric. Sci. (Cambridge) 121:307–314.

van Oosterom, E.J., D. Kleijn, S. Ceccarelli, and M.M. Nachit. 1993c. Genotype x environment interactions of barley in the Mediterranean region. Crop Sci. 33:669–674.

vom Brocke, K., T. Presterl, A. Christinck, E. Weltzien, and H.H. Geiger. 2002. Farmers' seed management practices open up new base populations for pearl millet breeding in a semi-arid zone of India. Plant Breed. 121:36–42.

Wallace, J.S. 2000. Increasing agricultural water use efficiency to meet future food production. Agric. Ecosyst. Environ. 82 (1–3):105–119.

Yamaguchi-Shinozaki, K., and K. Shinozaki. 1994. A novel cis-acting element in an Arabidopsis gene is involved in responsiveness to drought, low-temperature, or high-salt stress. Plant Cell 6:251–264.

Yan, W., L.A. Hunt, Q. Sheng, and Z. Szlavnics. 2000. Cultivar evaluation and mega-environment investigation based on the GGE biplot. Crop Sci. 40:597–605.

Weiss, H., M. A. Courty, W. Wetterstrom, F. Guichard, L. Senior, R. Meadow, and A. Curnow. 1993. The genesis and collapse of third millennium North Mesopotamian civilization. Nature (London) 261:995–1004.

Yin, X., P. Stam, J.C. Dourleijn, and M.J. Kropff. 1999. AFLP mapping of quantitative trait loci for yield-determining physiological characters in spring barley. Theor. Appl. Genet. 99:244–253.

Zhu, B., D.W. Choi, R. Fenton, and T.J. Close. 2000. Expression of the barley dehydrins multigene family and the development of freezing tolerance. Mol. Gen. Genet. 264:145–153.

12 Localization of Quantitative Trait Loci for Dryland Characters in Barley by Linkage Mapping

Michael Baum, Stefania Grando, and Salvatore Ceccarelli

ICARDA
Aleppo, Syria

Gunther Backes and Ahmed Jahoor

Risø National Laboratory
Roskilde, Denmark

ABSTRACT

The success of breeding for yield stability in stressful environments has been limited due to the high variability, timing, duration and severity of a number of climatic stresses. In the northwest of Syria, these stresses include cold, terminal drought, and heat occurring during the rainfed barley (*Hordeum vulgare* L.)-growing season. With the advent of DNA-marker techniques, it has become feasible to develop linkage maps for agricultural crops. Together with statistical procedures, these linkage maps can be used to locate and estimate phenotypic effects of quantitative trait loci (QTL). A genetic linkage map has been developed for recombinant inbred lines of the cross 'Arta'/*Hordeum spontaneum* 41-1. A total of 194 recombinant inbred lines randomly chosen from a population of 494 RILs were mapped with 189 markers including one morphological trait (brittle rachis locus). The linkage map extended to 890 cM. Agronomic traits such as grain yield, biological yield, days-to-heading, plant height, cold tolerance, and others were evaluated at the ICARDA's research stations at Tel Hadya and Breda in the years 1996/1997 and 1997/1998 and QTLs for agronomic traits related to drought resistance were localized. For the most important character "plant height under drought stress", QTLs on 2H, 3H, and 7H were detected. The "plant height" QTLs, especially the one on 3H, showed pleiotropic effects on traits such as days-to heading, grain yield, and biological yield. Cold tolerance on 5H, days to heading on 7H and 2H, kernel weight on 3H were QTLs identified and localized at similar locations as in other QTL studies. However, for a number of traits QTLs and locations have been described for the first time such as for tillering capacity on 3H, growth habit on 6H and 1H, early growth vigor on 6H. This should allow us to exploit the information for marker-assisted selection of lines better adapted to the Mediterranean drylands.

INTRODUCTION

Barley is an important cereal crop in the Near East, Asia, Central Africa, Latin and North America, and Europe. It covers more than 40 million ha in developing countries where often is the only possible rainfed crop farmers can grow (Blum, 1988; Ceccarelli, 1994). Barley grain is used as feed for animals, malt, and human food. Barley straw is used as animal feed in West Asia, North Africa, Ethiopia, Eritrea, Yemen, in the Andean region, and the Far East. Barley straw is also used for animal bedding and as cover material for hut roofs. After combine harvesting, barley stubble is grazed in summer in large areas of West Asia and North Africa. Barley is also used as an animal feed at the vegetative stage (green grazing) or is cut before maturity and either directly fed to the animal or used for silage.

In the highlands of Tibet, Nepal, Ethiopia, Eritrea, Yemen, in the Andean Region, in North Africa, Afghanistan, and Central Asian States, barley is used as human food and it is the main source of calories for some of the poorest people in the world. Its importance derives from the ability to grow and produce in marginal environments, which are often characterized by drought and low temperature conditions. Therefore, the development of lines with increased drought and other abiotic stresses tolerance becomes increasingly important (Ceccarelli and Grando, 1996). Barley was domesticated some 10 000 yr ago from its wild progenitor *Hordeum vulgare* ssp. *spontaneum* (hereafter referred to as *H. spontaneum*), which can contribute useful genes for several characters. Resistance to powdery mildew (Moseman et al., 1983; Jahoor and Fischbeck, 1987, Gustafsson and Claesson, 1988; Nevo, 1992; Lehman et al., 1998), leaf rust (Moseman et al., 1990; Nevo, 1992), and other diseases (Nevo, 1992) has been identified in *H. spontaneum*, and its use in breeding for disease resistance has been reported by several authors (Backes and Jahoor, 2001). The species showed variation for important agronomic traits such as seminal root morphology (Grando and Ceccarelli, 1995), floral structure (Giles and Bengtsson, 1988), salt tolerance (Nevo et al., 1993), grain size (Giles, 1990), milling energy (Ellis et al., 1993), grain protein content (Jaradat, 1991), earliness, biomass, grain yield, plant height under drought, and drought tolerance (Nevo, 1992; Grando et al., 2001; Ellis, 2002). This vast potential store of genetic resources is as yet largely unexploited. *Hordeum spontaneum* is thus expected to have a potential in contributing useful genes in barley breeding as a donor of adaptive traits to extreme stress conditions, as suggested by its distribution in the driest areas of West Asia.

One of the most useful traits of *H. spontaneum* in relation to stress tolerance is its plant height under drought. This is important because one of the most evident effects of drought is a reduction in plant height, making harvest by combine difficult or impossible (Grando et al., 2001). Therefore, the introgression of this trait into cultivated barley will contribute to stabilize the income of the rural poor. Gene introgression from *H. spontaneum* into cultivated barley is possible because crosses between the two species are easily made, and the progeny is fully fertile. However, *H. spontaneum* includes numerous complex, undesirable traits such as brittle rachis, low kernel weight, asynchronous tillering, and rough awns. An additional difficulty is that the improvement of plant height under drought often causes a reduction in tillering, and hence in both grain and straw yield. This makes gene introgression from *H. spontaneum* into cultivated barley a long and difficult process.

It is evident that in order to fully exploit the potential of crosses between cultivated barley and *H. spontaneum,* a large number of recombinant lines derived from each cross has to be evaluated. The use of DNA-molecular marker techniques can considerably reduce the complexity of combining a number of desirable traits in the same line. These techniques make it feasible to develop linkage maps for barley (Graner et al., 1991; Ramsey et al., 2000). Together with statistical techniques, these linkage maps are used to locate and estimate phenotypic effects of QTLs (Tinker et al., 1996; Bezant, 1997; Yin et al., 1999; Zhu et al., 1999; Foster et al., 2000; Teulat, 1998, 2001a, 2001b, 2002). The QTL analysis provides a powerful tool to locate genes on chromosomes. Therefore, the identification of molecular markers closely linked to QTLs of agronomic interest, or to negative traits, will allow the use of marker-assisted selection and thus increase the efficiency in the exploitation of *H. spontaneum*.

The objective of this study was to identify trait-marker linkages in a population of recombinant inbred lines of a cross between 'Arta' and *H. spontaneum* 41-1 using the QTL approach, for traits related to drought tolerance, such as earliness and the ability to maintain plant height under drought, and straw characteristics.

MATERIALS AND METHODS

Plant Material and Agronomic Traits

A population of 494 F_7 RILs derived by single seed descent from the cross 'Arta'/*H. spontaneum* 41-1 was developed at ICARDA. 'Arta', a high-yielding pure line selected from the Syrian white-seeded landrace 'Arabi Abiad', is well adapted to Syrian conditions, and combines a high number of tillers and high kernel weight, but becomes very short under dry conditions. *Hordeum spontaneum* 41-1, a pure line selected for its adaptation to severe drought stress conditions, combines earliness with acceptable cold tolerance, and with the ability to maintain plant height under severe drought. At least six characters are expected to segregate from this cross: rachis brittleness, awn roughness, peduncle extrusion, plant height, tillering, and kernel size. The main objective of this cross was to develop lines combining the grain yield and tillering ability of 'Arta' with the plant height and the adaptation to severe drought stress conditions of *H. spontaneum* 41-1 (Grando et al., 2001).

The 494 lines and the parental lines were planted, in the cropping seasons 1996 to 1997 and 1997 to 1998, at ICARDA's research stations located near Tel Hadya (36°01′ N; 37°20′ E, elevation 300 m above sea level) and near Breda (35°56′ N; 37°10′ E, elevation 354 m above sea level) in Syria. The experimental design was a 20 x 25 m α-lattice, with two replicates. Plot size was 1.8 m wide (eight rows 20 cm apart), 6 m long in 1996 to 1997 and 1.6 m wide (six rows at 20 cm apart), 6 m long in 1997 to 1998. In both years and locations the trials were grown under rainfed conditions. Total rainfall in Tel Hadya was 433.7 and 410.5 mm in 1996 to 1997 and 1997 to 1998, respectively, whereas Breda received 230.8 and 227.4 mm in 1996 to 1997 and 1997 to 1998, respectively.

The following agronomic traits were recorded: Early growth vigor as a visual score at 5 – 6 leaf stage, using a scale from 1 = good vigor to 5 = poor vigor; growth habit as a visual score at 5 – 6 leaf stage, using a scale from 1 = erect to 5 = prostrate; cold damage as a visual score, using a scale from 1 = absence of damage (all leaves completely green) to 5 = leaf blades and sheaths yellow; days-to-heading as number of days from emergence to awns appearance in 50% of the plants in a plot; plant height measured in centimeters from ground level to the base of the spike, at maturity; number of tillers per square meter, calculated from the number of tillers counted on two rows of 1 m each 20 cm apart; biological yield in kg ha^{-1}, measured by hand harvesting the six central rows in 1996 to 1997 and the four central rows in 1997 to 1998 of each plot for the entire plot length; grain yield in kg ha^{-1}, measured after threshing the harvested sample; 1000-kernel weight in grams, measured as the average of three samples of 100 kernels per each plot.

DNA Extraction and Marker Assays

Total genomic DNA was extracted following the procedure described by Saghai-Maroof et al. (1984), with minor modifications. Fresh leaves from 4 to 6 wk-old seedlings were collected for DNA extraction. The quality of the extracted DNA was visually checked on a 1% agarose gel.

Genetic mapping was carried out using Amplified Fragment Length Polymorphic (AFLP) markers and microsatellite-based markers. The protocol for the AFLP assay was carried out as described by Zabeau and Vos (1993) with minor modifications using combination of *Pst*I and *Mse*I or *EcoR*I and *Mse*I restriction enzymes and respective adapters. Pre-amplification was carried out using one base pair extension primers. Selective amplification of restriction fragments was conducted using primers with two, or three selective nucleotides. Polymerase chain reaction amplifications were separated on 6% denaturing polyacrylamid gels and stained with silver nitrate stain while fluorescent dyes were used on an ABI377 sequencer at Risø.

Simple-sequence repeat markers (SSR) were amplified on a PE-9600 or 9700 systems using the published protocols for the respective markers (Ramsay et al., 2000). Products were then electrophoretically separated on 6% polyacrylamid gels using standard sequencing gels with silver staining (Bassam et al., 1991) at ICARDA or with fluorescence dyes on an ABI377 sequencer at Risø National Laboratory, Denmark.

Linkage Mapping and QTL Analysis

Segregation analysis was performed with the MAPMAKER computer software program; the Join Map 2.0 (Stam and Ooijen, 1995) software package was employed for map construction. Recombination fractions were converted to centiMorgans (cM) according to the Kosambi mapping function (Kosambi, 1944). The QTL analysis was performed using PLABQTL v. 1.1 (Utz and Melchinger, 1995).

RESULTS AND DISCUSSION

The linkage map based on the 'Arta'/*H.spontaneum* 41-1 population originally contained 189 marker loci (Baum et al., 2003) including 1 morphological marker locus (btr = brittle rachis), 158 AFLP loci and 30 SSR loci, the latter with mostly known location (Ramsey et al., 2000) that served as anchor markers. For QTL mapping, closely mapped markers were deleted to allow a better interval mapping. The map spans over 890 cM and has an average interval length of 7.3 (Fig.12–1). Significant distortion segregation was observed with 10 of the 129 loci. For the traits cold damage, growth habit, and growth vigor a nonparametric QTL analysis was carried out, as those traits were characterized by an extremely skewed distribution or were determined by categories instead of measurements. Table 12–1 reports the QTLs detected in the 'Arta'/*H. spontaneum* 41-1 population, ordered by chromosomal position.

The 'Arta'/*H. spontaneum* population is specifically adapted to the Mediterranean environment. It is interesting to examine whether the major QTLs identified in this study are unique or common to other barley populations. With the ex-

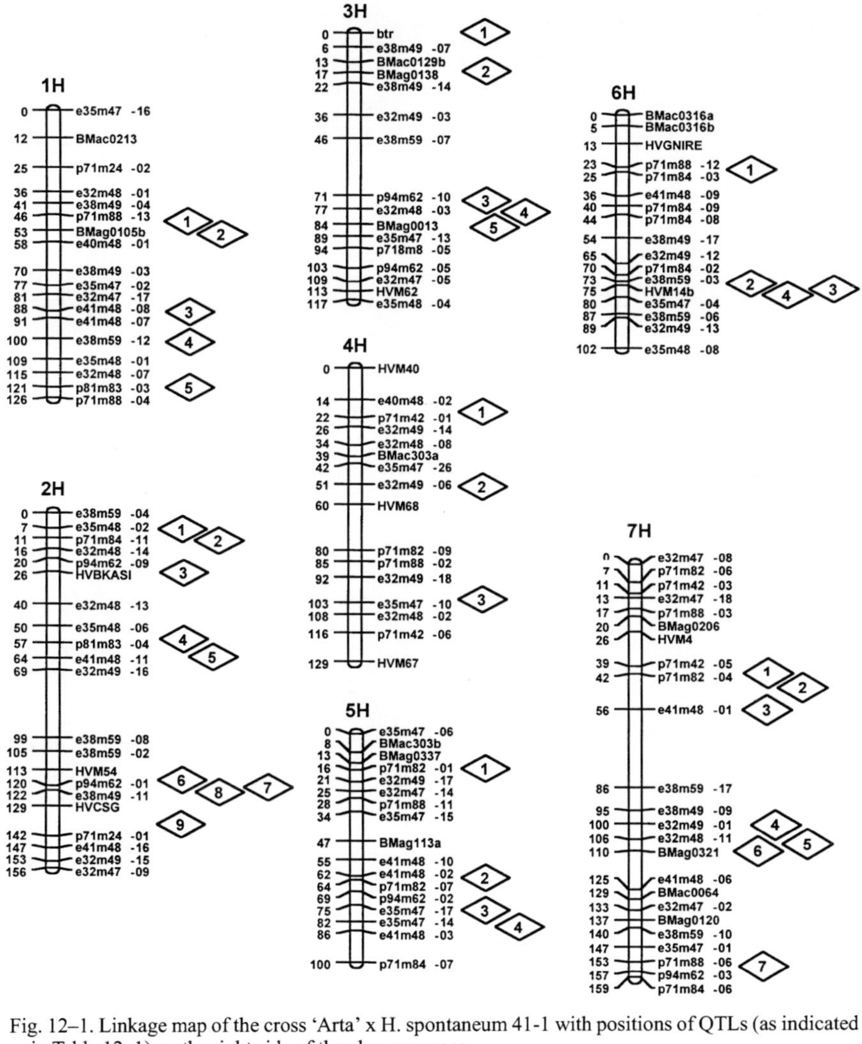

Fig. 12–1. Linkage map of the cross 'Arta' x H. spontaneum 41-1 with positions of QTLs (as indicated in Table 12–1) on the right side of the chromosomes.

Table 12–1. Position, statistics, effects, and explained phenotypic variance of the quantitative trait loci (QTLs) detected.

Trait†	Env‡	LOD/*K*§	Effect¶	$Var._{expl.}$#
1H-1: 47 (C.I.: 45–50)				
PH	Br97	3.0	1.39 [Hs]	1.4%
1H-2: 53 (C.I.: 45–56)				
GrH	Th97	*27.4*	0.65 [Hs]	10.2%
GrH	Th98	*13.3*	0.21 [Hs]	6.7%
KW	Th97	3.3	1.08 [Ar]	7.4%
1H-3: 88 (C.I.: 82–92)				
TN	Br97	2.6	36.6 [Ar]	6.3%
1H-4: 100 (C.I.: 81–109)				
GrH	Th97	*15.9*	0.49 [Hs]	7.0%
1H-5: 120 (C.I.: 115–123)				
GY	Br97	4.5	108 [Ar]	5.6%
GY	Br98	3.6	91 [Ar]	5.5%
BY	Br98	4.0	220 [Ar]	7.6%
2H-1: 7 (C.I.: 0-11)				
KW	Th97	9.0	1.79 [Ar]	17.7%
BY	Br98	2.6	219 [Ar]	7.1%
TN	Br97	4.0	36.8 [Ar]	6.6%
2H-2: 12 (C.I.: 9-16)				
DH	Th97	6.7	1.9 [Hs]	8.5%
2H-3: 26 (C.I.: 20–38)				
CD	Th97	*15.8*	0.41 [Ar]	2.0%
2H-4: 55 (C.I.: 52–62)				
KW	Th97	6.9	1.69 [Ar]	14.8%
2H-5: 63 (C.I.: 58–62)				
KW	Br97	*10.7*	2.52 [Ar]	24.3%
KW	Br98	7.4	1.68 [Ar]	15.4%
2H-6: 115 (C.I.: 109–119)				
PH	Br98	5.0	3.84 [Hs]	8.9%
2H-7: 119 (C.I.: 114–122)				
KW	Th98	4.8	1.60 [Ar]	9.8%
2H-8: 122 (C.I.: 120–126)				
DH	Th97	5.6	2.12 [Ar]	11.0%
2H-9: 136 (C.I.: 125–143)				
KW	Th97	4.8	1.36 [Ar]	9.4%
3H-1: 0 (C.I.: 0–3)				
BY	Br97	5.1	406 [Ar]	8.3%
GY	Br97	8.6	186 [Ar]	16.2%
KW	Br97	2.5	0.86 [Ar]	3.6%
GY	Br98	8.1	234 [Ar]	29.8%
BY	Th97	54.9	3284 [Ar]	72.6%
GY	Th97	76.7	2545 [Ar]	84.8%
BY	Th98	25.1	1916 [Ar]	44.1%
GY	Th98	49.4	1565 [Ar]	67.5%
3H-2: 16 (C.I.: 13–21)				
TN	Br97	5.1	49.3 [Ar]	49.3%
BY	Br98	4.0	223 [Ar]	8.1%
DH	Th97	3.3	1.79 [Hs]	7.6%
KW	Th97	5.7	1.20 [Ar]	9.2%
3H-4: 77 (C.I.: 71–83)				
PH	Br97	8.4	5.16 [Hs]	11.8%
PH	Br98	16.7	7.4 [Hs]	29.1%

(continued on next page)

Table 12–1. Continued.

Trait†	Env‡	LOD/*K*§	Effect¶	Var.$_{expl.}$#
GY	Br98	2.7	97 [Ar]	6.1%
PH	Th97	4.4	3.92 [Hs]	4.8%
DH	Th97	9.8	3.14 [Ar]	20.0%
GY	Th97	3.6	245 [Ar]	4.9%
3H-5: 85 (C.I.: 80–90)				
KW	Th97	5.9	1.30 [Ar]	10.5%
4H-1: 16 (C.I.: 6–25)				
PH	Br97	4.4	3.97 [Hs]	11.5%
PH	Br98	2.6	3.12 [Hs]	6.3%
4H-2: 52 (C.I.: 43–59)				
KW	Th97	4.9	1.18 [Ar]	8.9%
KW	Th98	5.7	1.79 [Ar]	11.2%
CD	Th98	*14.0*	0.14 [Ar]	8.0%
4H-3: 103 (C.I.: 94–107)				
DH	Br97	2.6	1.02 [Ar]	5.1%
KW	Br97	8.2	2.14 [Ar]	17.1%
TN	Br97	4.8	35.0 [Hs]	5.5%
KW	Br98	8.0	1.65 [Ar]	14.9%
KW	Th97	2.6	1.06 [Ar]	7.3%
GY	Th98	3.0	170 [Ar]	2.4%
CD	Th98	*14.0*	0.14 [Ar]	2.8%
5H-1: 15 (C.I.: 12–20)				
DH	Br97	3.6	1.23 [Hs]	5.8%
DH	Th97	5.2	2.24 [Hs]	10.7%
KW	Th98	3.1	1.24 [Ar]	6.0%
5H-2: 52 (C.I.: 48–69)				
CD	Th97	*53.3*	0.72 [Hs]	8.4%
CD	Th98	*34.4*	0.23 [Hs]	4.0%
5H-3: 78 (C.I.: 64–82)				
KW	Br97	3.3	0.48 [Hs]	7.1%
CD	Th97	*65.0*	0.77 [Hs]	
PH	Th97	4.4	2.90 [Hs]	5.5%
5H-4: 84 (C.I.: 76–95)				
KW	Th97	2.6	1.03 [Ar]	7.1%
GY	Th97	2.6	244 [Ar]	5.0%
CD	Th98	44.6	0.26 [Hs]	11.4%
6H-1: 23 (C.I.: 18–29)				
KW	Th97	3.9	1.30 [Ar]	11.0%
KW	Th97	4.6	1.17 [Ar]	8.7%
KW	Th98	4.8	1.31 [Ar]	7.2%
6H-2: 72 (C.I.: 69–75)				
KW	Br98	3.1	1.12 [Hs]	7.8%
KW	Th98	4.4	1.15 [Hs]	5.8%

† Abbreviations: GV = early growth vigor; GH = growth habit; CD = cold damage; DH = days-to-heading, d ; PH = plant height, cm; TN = number of tillers per m^2 ; BY = biological yield, kg ha^{-1}, GY = grain yield, kg ha^{-1}; KW = 1000-kernel weight, g.

‡ Env = Means of environment.

§ LOD/*K*: For the case of marker-QTL-marker (MQM) analysis, the logarithm of the odds (LOD)-score is given, for the case of nonparametric analysis, the K-value from the Kruskal-Wallis ANOVA (K-value *italic* in the table)

¶ Effect: Difference of the genotype with two alleles of the one parent compared to the genotype with two alleles from the other parent (in parenthesis: the parent with the higher trait expression, [Ar] = Arta, [Hs] = *Hordeum spontaneum*.

Var.$_{expl}$: Phenotypic variance explained for the trait.

ception of the studies of Teulat et al. (1998, 2001b), barley populations or lines were tested in more favorable environments (Hayes et al., 1993; Pan et al., 1994; Backes et al., 1995; Thomas et al., 1995; Tinker et al., 1996; Bezant et al., 1997; Yin et al., 1999; Forster et al., 2000; Marquez-Cedillo et al., 2001; Ivandic et al., 2002).

Four QTLs were detected in total for biological yield, three of them were detected in one environment. The QTL on 3H-1, at the same position of the *btr* locus, was detected in Br97, Th97, and Th98 and showed the highest effect. 'Arta' was always contributing the allele with the higher biological yield. Six QTLs were detected for grain yield; three of them were environment-specific and three were common for more than one environment (1H-5, 3H-1, and 3H-4). For all detected QTLs, the higher grain yield was linked with the allele from 'Arta'. Nineteen QTLs were identified for kernel weight, 14 linked to single environments and five to more than one environment (2H-5, 4H-2, 4H-3, 6H-1, and 6H-2). In five QTLs (5H-3, 6H-2, 6H-3, 7H-1, and 7H-6), identified for kernel weight, the allele increasing the kernel size originated from the *H. spontaneum* line. In general, the level of explained phenotypic variance was high for the QTLs identified for this trait. The major QTL for kernel weight on 3H-2 could correspond to the grain yield QTL identified by Marquez-Cedillo et al. (2001) on 3H. The major QTL for kernel weight on 2H-5 might coincide with the grain yield QTLs and other traits that were mapped at the *vrs* locus on 2H (Tanno et al., 2002; Marquez-Cedillo et al., 2001; Hayes et al., 1993). The kernel weight QTL on 4H-3 might match the kernel weight QTL identified by Bezant et al. (1997).

When QTLs were detected for plant height, the allele for taller plants originated always from the *H. spontaneum* line. One QTL was detected in all three environments (3H-4), but showed a much higher effect in Breda than in Tel Hadya, one QTL in Breda in both years (4H-1) and five QTLs in individual environments. The major QTL for plant height on 3H-4 is located proximal to the Bmac0013 SSR marker and likely positioned at the *sdw1* locus (Ellis et al., 2002) or *denso* locus (Laurie et al., 1993), one of the dwarfing genes with commercial significance in barley breeding. The *H. spontaneum* allele at this position has a positive effect on stem elongation and increased plant height under drought stress. Similarly to our findings for the increased plant height alleles, the *sdw1* dwarfing gene has shown linkage to date of heading, a number of agronomic traits (Barua et al., 1993; Thomas et al., 1995; Ivandic et al., 2002), and a number of physiological traits (Yin et al., 1999). The other plant height QTLs on 2H-6, 4H-1, and 6H-3 (favorable allele from *H. spontaneum*) could correspond to the QTLs found in the Mediterranean environment by Teulat et al. (2001b).

Cold damage was evaluated in Th97 and Th98, resulting in the detection of eight QTLs. Two of them were detected in Th97 only, four in Th98, and two in both environments (5H-2, 7H-5). Only for the QTLs with minor effects (2H-3, 4H-3, 6H-4), the allele from the *H. spontaneum* line showed the better protection. For all other QTLs, the allele of 'Arta' showed better ability to protect the plant from cold damage. Three of the four alleles with high effect on cold damage (5H-2, 5H-3, 5H-5) were localized on chromosome 5H. Verbalization requirement in barley has been identified by comparative studies with wheat (*Triticum aestivum* L.), and located as *Sh2* on chromosome 5H (Laurie et al., 1995; Ivandic et al., 2002; Pan et al., 1994). A locus for vernalization requirement (sgh1) has been identified on 4H (Laurie et al 1995). This is likely to correspond to the cold damage QTL on 4H-3. The cold damage QTLs on 7H-5 have not, to our knowledge, been reported before.

Seven QTLs for 'days-to-heading' were recorded in Th97 and Br97. Five QTLs were detected in Th97, two in Br97, and one in both environments (5H-1). 'Arta' contributed the allele with the late heading in four of the eight QTLs (2H-8, 3H-4, 4H-3, 7H-3). The QTL on 7H-3 showed by far the largest effect (3.5 d difference and 36% explained variance). The location on 7H-3, with the allele for the earlier heading originating from the *H. spontaneum* line, may correspond to the one found by Backes et al. (1995), Hayes et al. (1993), and Teulat et al. (2001b). The QTL on 3H-4 coincides with that of Bezant et al. (1997). An additional QTL for days-to heading was identified on 2H-2. This might correspond to the Eam1 or Ppd-H1 location on 2H (Cattivelli et al., 2002; Laurie et al., 1995; Backes et al., 1995; Hayes et al., 1993), but the one on 2H-8 seems to be unique to this populations. The QTL on 5H-1 might match the QTL identified by Marquez-Cedillo et al (2001).

The 'brittle rachis' is possibly caused by complementary genes at two tightly linked loci btr1 and btr2 as suggested by Komatsuda and Mano (2002). They mapped the brittle rachis trait to 3H but to a more distal location on 3HS, which is in agreement with the position suggested by Franckowiak et al. (1997). Kandemir et al. (2000) mapped Btr2 of *H. spontaneum* in the interval MWG798B – MWG014 on 3H. In the A'H population, brittle rachis was treated as a qualitative trait and was mapped to 3H at the proximal end of the linkage group.

All four QTLs for tiller number were detected in Br97 (1H-3, 2H-1, 3H-2, 4H-3). The *H. spontaneum* line contributed the allele for higher number of tillers for the QTL on 4H-3, 'Arta' for the other three QTLs. Proximal to the btr location at 3H-2 a QTL for tiller number was mapped (49% explained variance). Increased plant height, asynchronous tillering and reduced tillering are co-segregating traits introgressed from *H. spontaneum*. High tillering is an important character in relation to phenotypic plasticity in response to drought. The higher tillering capacity originated from 'Arta'. Additionally, QTLs for tillering capacity were identified on 4H-3 originating from *H. spontaneum* and on 1H-3 and 2H-1 from 'Arta'.

Growth habit was observed in Th97 and Th98, three QTLs were found, two common in both years (1H-2 and 6H-4) and one only in Th97 (1H-4). For the two common QTLs, the largest effect was found in Th97. For the QTL on 6H-4, the 'Arta'-allele caused the more prostrate growth type, for the two other QTLs, it was the allele from the *H. spontaneum* line. The QTLs for growth habit, with the alleles for prostrate habit originating from *H. spontaneum* were located on 1H (1H-2, 1H-4). Prostrate growth habit results in good ground cover in winter, and hence reduces losses of water by soil evaporation. Ceccarelli et al. (1991) reported high ground cover in winter associated with higher yield under drought conditions. The QTL with the strongest effect was found on 6H (6H-4) and the allele for the prostrate growth originated from 'Arta'.

Only one QTL was found for the early growth vigor at 6H-4 together with the QTL for growth habit mentioned before. The 'Arta'-allele provided the more vigorous growth. Interesting to note is that also a QTL for growth vigor was located on 6H at the same position with the allele for more vigorous growth coming from 'Arta'. Early growth vigor, which describes the ability to grow at low temperature, is considered to be a beneficial trait in Mediterranean type drought-prone environments (Passioura, 1986). The 1H and 6H locations seem to be important for plant phenology traits.

In conclusion, a number of QTLs were identified and localized and some of the locations correspond to locations identified already in other QTL studies such as the plant height on 3H, cold tolerance on 5H, days-to-heading on 7H and 2H, kernel weight on 3H. However, for a number of other traits QTLs and locations have been described for the first time, such as for tillering capacity on 3H, growth habit on 6H and 1H, and early growth vigor on 6H. This should allow us to exploit the information for marker-assisted selection of lines better adapted to the Mediterranean drylands.

ACKNOWLEDGMENT

This research was jointly funded by BMZ Grant No. 96.7860.8-001.00, by the Arab Fund for Economic and Social Development and by the Danish Research Council for Developmental Research (Proj.-No. 90978). We are grateful to Dr. W. Powell, SCRI, for providing barley microsatellite primer sequences, even prior to publication.

REFERENCES

Bassam, B.J., G. Caetano-Anolles, and P.M. Gresshoff. 1991. Fast and sensitive silver staining of DNA in polyacrylamide gels. Anal. Biochem. 196:80–83.

Barua, U.M., K.J. Chalmers, W.T.B. Thomas, C.A. Hackett, V. Lea, P. Jack, B.P. Forster, R. Waugh, and W. Powell. 1993. Molecular mapping of genes determining height, time to heading, and growth habit in barley (*Hordeum vulgare*). Genome 36:1080–1087.

Baum, M., S. Grando, G. Backes, A. Jahoor, A. Sabbagh, and S. Ceccarelli. 2003. QTLs for agronomic traits in the Mediterranean environment identified in recombinant inbred lines of the cross 'Arta' x *H. spontaneum* 41–1. Theor. Appl. Genet. 107:1215–1225.

Bezant, J., D. Laurie, N. Pratchett, J. Chojecki, and M. Kearsey. 1997. Mapping QTLs controlling yield and yield components in a spring barley (*Hordeum vulgare* L.) cross using marker regression. Mol. Breed. 3:29–38.

Backes, G., and A. Jahoor. 2001. Wege zur Nutzung des genetischen Potentials der Wildgerste – mit molekularen Markern zu neuen Resistenzgenen. Bericht der Arbeitstagung der 52. Tagung der Vereinigung der Pflanzenzûchter und Saatgutkaufleute Österreich, Gumpenstein:81–86.

Blum, A. 1988. Drought resistance. p. 43–76. *In* Plant breeding for stress environments. CRC Press, Boca Raton, FL.

Cattivelli, L., P. Baldi, C. Crosatti, N. Di Fonzo, P. Faccioli, M. Grossi, A.M. Mastrangelo, N. Pecchioni, and A.M. Stanca. 2002. Chromosome regions and stress-related sequences involved in resistance to abiotic stress in *Triticeae*. Plant Mol. Biol. 48:649–665.

Ceccarelli, S. 1994. Specific adaptation and breeding for marginal conditions. Euphytica 77:205–219.

Ceccarelli, S., E. Acevedo, and S. Grando. 1991. Breeding for yield stability in unpredictable environments: Single traits, interaction between traits, and architecture of genotypes. Euphytica 56:169–185.

Ceccarelli, S., and S. Grando. 1996. Drought as a challenge for the plant breeder. Plant Growth Regulation 20:149–155.

Ellis, R.P., 2002. Wild barley as a source of genes for crop improvement. p. 65–83. *In* G.A. Slafer et al. (ed.) Barley science: Recent advances from molecular biology to agronomy of yield and quality. Food Products Press, The Haworth Press Inc., Binghamton, NY.

Ellis, R.P., E. Nevo, and A. Beiles. 1993. Milling energy polymorphism in *Hordeum spontaneum* Koch in Israel and its potential utilization in breeding for malting quality. Plant Breed 111:78–81.

Forster, B.P., R.P. Ellis, W.T.B. Thomas, A.C. Newton, R. Tuberosa, D. This, R.A. El-Enein, H. Bahri, and M. Ben Salem. 2000. The development and application of molecular markers for abiotic stress tolerance in barley. J. Exp. Bot. 51:19–27.

Franckowiak, J.D., U. Lundqvist, and T. Konishi. 1997. New and revised names for barley genes. Barley Genet. Newsl. 26:4–8.

Giles, B.E. 1990. The effects of variation in seed size on growth and reproduction in the wild barley *Hordeum vulgare* ssp. *spontaneum*. Heredity 64:239–250.

Giles, B.E., and B.O. Bengtsson. 1988. Variation in anther size in wild barley (*Hordeum vulgare* ssp. *spontaneum*) Hereditas 108:199–205.

Grando, S., and S. Ceccarelli. 1995. Seminal root morphology and coleoptile length in wild (*Hordeum vulgare* ssp. *spontaneum*) and cultivated (*Hordeum vulgare* ssp. *vulgare*) barley. Euphytica 86:73–80.

Grando, S., R. von Bothmer, and S. Ceccarelli. 2001. Genetic diversity of barley: Use of locally adapted germplasm to enhance yield and yield stability of barley in dry areas. p. 351–372. *In* H.D. Cooper et al. (ed.) Broadening the genetic base of crop production. CABI, New York/FAO, Rome/IPRI, Rome.

Graner, A., A. Jahoor, J. Schondelmaier, H. Siedler, K. Pillen, G. Fischbeck, G. Wenzel, and R.G. Herrmann. 1991. Construction of an RFLP map of barley. Theor. Appl. Genet. 83:250–256.

Gustafsson, M., and L. Claesson. 1988. Resistance to powdery mildew in wild species of barley. Hereditas 108:231–237.

Hayes, P.M., B.H. Liu, S.J. Knapp, F. Chen, B. Jones, T. Blake, J. Franckowiak, D. Rasmusson, M. Sorells, S.E. Ullrich, D. Wesenberg, and A. Kleinhofs. 1993. Quantitative trait locus effects and environmental interaction in a sample of North American barley germplasm. Theor. Appl. Genet. 87:392–401.

Ivandic, V., B.P. Forster, C.A. Hackett, N. Nevo, R. Keith, and T.B.W. Thomas. 2002. Analysis of simple sequence repeats (SSRs) in wild barley from the Fertile Crescent: Associations with ecology, geography and flowering time. Plant Mol. Biol. 48:511–527.

Jahoor, A., and G. Fischbeck. 1987. Sources of resistance to powdery mildew in barley lines derived from *Hordeum spontaneum* collected in Israel. Plant Breed. 99:274–281.

Jaradat, A.A. 1991. Grain protein variability among populations of wild barley (*Hordeum spontaneum* C Koch) from Jordan. Theor. Appl. Genet. 83:164–168.

Kandemir, N., D.A. Kudrna, S.E. Ullrich, and A. Kleinhofs. 2000. Molecular marker assisted genetic analysis of head shattering in six-rowed barley. Theor. Appl. Genet. 101:203–210.

Komatsuda, T., and Y. Mano. 2002. Molecular mapping of the intermedium spike-c (*int-c*) and non-brittle rachis 1 (*btr1*) loci in barley (*Hordeum vulgare* L.). Theor. Appl. Genet. 105:85–90.

Kosambi, D.D. 1944. The estimation of map distance from recombination values. Ann. Eugen. 12:172–175.

Laurie, D.A., N. Pratchett, C. Romero, E. Simpson, and J.W. Snape. 1993. Assignment of the *denso* dwarfing gene to the long arm of chromosome 3 (3H) of barley by the use of RFLP markers. Plant Breed. 111:198–203.

Laurie, D.A, N. Pratchett, J.H. Bezant, and J.W. Snape. 1995. RFLP mapping of 5 major genes and 8 quantitative trait loci controlling flowering time in a winter x spring barley (*Hordeum vulgare* L.) cross. Genome 38:575–585.

Lehmann, L.C., R. Jönsson, and M. Gustafsson. 1998. Identification of resistance genes to powdery mildew isolated from *Hordeum vulgare* ssp. *spontaneum* and landraces of barley. J. Swedish Seed Assoc. 108:94–101.

Marquez-Cedillo, L.A., P.M. Hayes, A. Kleinhofs, W.G. Legge, B.G. Rossnagel K. Sato, S.E. Ullrich, and D.M. Wesenberg. 2001. The North American Barley Genome Mapping Projec: QTL analysis of agronomic traits in barley based on the doubled haploid progeny of two elite North American varieties representing different germplasm groups. Theor. Appl. Genet. 103:625–637

Moseman, J.G., E. Nevo, and D. Zohary. 1983. Resistance of *Hordeum spontaneum* collected in Israel to infection with *Erysiphe graminis hordei*. Crop Sci. 23:1115–1119.

Moseman, J.G., E. Nevo, and M.A. El-Morsidy. 1990. Reactions of *Hordeum spontaneum* to infection with two cultures of *Puccinia hordei* from Israel and United States. Euphytica 49:169–175.

Nevo, E. 1992. Origin, evolution, population genetics and resources for breeding of wild barley, *Hordeum spontaneum*, in the Fertile Crescent. p. 19–43. *In* P.R. Shewry (ed.) Barley: Genetics, biochemistry, molecular biology and biotechnology, CAB Int., Wallingford, UK.

Nevo, E., T. Krugman, and A. Beiles. 1993. Genetic resources for salt tolerance in the wild progenitors of wheat (*Triticum dicoccoides*) and barley (*Hordeum spontaneum*) in Israel. Plant Breed. 110:338–341.

Pan, A., P.M. Hayes, F. Chen, T.H.H. Blake, T.K.S. Wright, I. Karsai, and Z. Bedö. 1994. Genetic analysis of the components of winter hardiness in barley (*Hordeum vulgare* L.). Theor. Appl. Genet. 89:900–910.

Passioura, J.B. 1986. Resistance to drought and salinity: Avenues for improvement. Aust. J. Plant Physiol. 13:191–201.

Ramsay, L., M. Macaulay, S. Degli Ivanissevich, K. Maclean, L. Carsle, J. Fuller, K.J. Edwards, S. Tuvesson, M. Morgante et al. 2000. A simple sequence repeat-based linkage map of barley. Genetics 156:1997–2005.

Saghai-Maroof, M.A., K.M. Soliman, R.A. Gorgensen, and R.W. Allard. 1984. Ribosomal DNA spacer length polymorphism in barley: Mendelian inheritance, chromosomal location and population dynamics. Proc. Natl. Acad. Sci. USA 81:8014–8018.

Stam, P., and J.W. van Ooijen. 1995. JoinMap® version 2.0: Software for the calculation of genetic linkage maps. CPRO-DLO, Wageningen, The Netherlands.

Tanno, K., S. Taketa, K. Takeda, and T. Komatsuda. 2002. A DNA marker closely linked to the *vrs1* locus (row-type gene) indicates multiple origins of six-rowed cultivated barley (*Hordeum vulgare* L.) Theor. Appl. Genet. 104:54–60.

Teulat, B., D. This, M. Khairallah, C. Borries, C. Ragot, P. Sourdille, P. Leroy, P. Monneveux, and A. Charrier. 1998. Several QTLs involved in osmotic adjustment trait variation in barley (*Hordeum vulgare* L.). Theor. Appl. Genet. 96:688–698.

Teulat, B., C. Borries, and D. This. 2001a. New QTLs identified for plant water status, water-soluble carbohydrates, osmotic adjustment in a barley population grown in a growth-chamber under two water regimes. Theor. Appl. Genet. 103:161–170.

Teulat, B., O. Merah, I. Souyris, and D. This. 2001b. QTLs for agronomic traits from Mediterranean barley progeny grown in several environments. Theor. Appl. Genet. 103:774–787.

Teulat, B., O. Merah, X. Sirault, C. Borries, R. Waugh, and D. This. 2002. QTL for grain carbon isotope discrimination in field-grown barley. Theor. Appl. Genet. 106:118–126.

Thomas, W.T.B., W. Powell, R. Waugh, K.J. Chalmers, U.M. Barua, P. Jack, V. Lea, B.P. Forster, J.S. Swanston et al. 1995. Detection of quantitative trait loci for agronomic, yield, grain, and disease characters in spring barley (*Hordeum vulgare* L.) Theor. Appl. Genet. 91:1037–1047

Tinker, N.A., D.E. Mather, B.G. Rossnagel, K.J. Kasha, A. Kleinhofs, P.M. Hayes, D.E. Falk, T. Ferguson, L.P. Shugar et al. 1996. Regions of the genome that affect agronomic performance in two-row barley. Crop Sci. 36:1053–1062.

Utz, H.F., and A.E. Melchinger. 1995. PLABQTL: A computer program to map QTL version 1.0. Inst. of Plant Breed., Seed Sci., and Population Genetics, Univ. of Hohenheim, Stuttgart, Germany.

Yin, X., P. Stam, C. Johan Dourleijn, and M.J. Kropff. 1999. AFLP mapping of quantitative trait loci for yield-determining physiological characters in spring barley. Theor. Appl. Genet. 99:244–253.

Zabeau, M., and P. Vos. 1993. Selective restriction fragment amplification: A general method for DNA fingerprinting. Eur. Pat. App. 92402629.7 (Publ. no. 0 534858 A1).

Zhu, H., G. Briceño, R. Dovel, P.M. Hayes, B.H. Liu, C.T. Liu, T. Toojinda, and S.E. Ullrich. 1999. Molecular breeding for grain yield in barley: An evaluation of QTL effects in a spring barley cross. Theor. Appl. Genet. 98:772–779.

13 Durum Wheat Adaptation in the Mediterranean Dryland: Breeding, Stress Physiology, and Molecular Markers

Miloudi M. Nachit and Ismahane Elouafi

ICARDA
Aleppo, Syria

ABSTRACT

From antiquity, food from wheat (*Triticum aestivum* L.) has been the staff of life for much of the world's population. Wheat is the word's staple food crop, along with rice (*Oryza sativa* L.). Two main categories of wheat are based on growth requirements and grain quality, and thus end use. As wheat has its origin in the Middle East, both bread wheat and durum wheat are grown here. Durum (*T. turgidum* L. var. durum) is cultivated extensively in the dryland of the Mediterranean region under drought prone and environmental variable conditions. Drought tolerance research is complex and requires a large number of testing sites and seasons to determine the genotypic drought resistance. Drought tolerance research requires various tools such as the exploitation of the genetic variation of landraces and *Triticum* wild relatives, use of representative testing environments/sites, stress physiology, and use of molecular markers. Drought resistance is of major interest in International Center for Agricultural Center in Dry Areas (ICARD) dryland durum breeding program; this involves the use of *Triticum* wild relatives in the breeding; representative testing sites, stress physiological traits, and molecular markers are discussed. Crosses between durum and its wild relatives have generated durum genotypes with better performance under environments with abiotic constraints such as drought and temperature extremes (heat and drought). The carbon isotope discrimination technique showed relatively the largest association with grain yield combined with low genotype-environment interactions and was mapped and the chromosomal region controlling it was identified on 4BS chromosome. The molecular markers are considered as potential tools to be used in marker-assisted selection, to improve drought tolerance and productivity of durum in the Mediterranean region. These techniques will be incorporated in developing targeted crosses and selecting durum improved germplasm.

INTRODUCTION

Durum is grown mainly in the dryland of the Mediterranean region under stressed and variable environmental conditions. The Mediterranean climate is characterized by low and highly erratic annual rainfall varying from 200 to 800 mm, with usually poor rainfall distribution, and periods of drought and temperature extremes (cold and heat) that can occur at any plant stage of development (Nachit et al., 1992a, 1992b). These environmental conditions are the main causes of yield reduction and fluctuation in the Mediterranean region. As irrigation on a large scale for durum commercial production is not available to increase production, the only applicable alternative is the improvement of drought tolerance and yield stability through genetics/plant breeding, stress physiology, and use of molecular markers as tools by the breeders (Nachit, 1998b). Consequently, in Syria, during the last 10 yr the contribution made by the stress tolerant and productive durum genotypes developed by the ICARDA is reflected in the spectacular production increase, from <1 to 5.4 Tg, without any significant increase in the cropped area (1.2 m.ha). However, with unrelenting population growth in most countries of the West Asia and North Africa region, and virtually all show food deficit, the major challenge for researchers is to increase food—mainly wheat output.

The ICARDA dryland durum-breeding program was initiated in 1977 in northern Syria. The ICARDA main research station (Tel Hadya) and its related research sites across a rainfall gradient (Lattakia, Terbol, Kfardan, and Breda,) in the Middle East region are located in the heart of the Fertile Crescent. A region where a wealth of wheat landraces is found along their wild relatives in different agro-ecological zones, from lowland plains to highland plateaus; and from favorable to stressed environmental conditions (Nachit and Ouassou, 1988; Nachit, 1992). Durum wheat originates from the Middle East Fertile Crescent region and is cultivated for several millennia: The cereal-based agriculture of the Mediterranean civilizations (Egyptian, Phoenician, Carthagian, Assyrian, Greek, Roman, and Arab) was closely linked to the cultivation of durum. Thousands of landraces that were grown over long periods are still found along its wild relative parents (*T. urartu, T. monococcum*, and *Aegilops* ssp.). In the Mediterranean dryland, abiotic stresses such as drought, cold, and heat are the most important constraints along with multiple biotic stresses (diseases, insects, and viruses). The highest number and virulence of disease races and insect biotypes are also found in this region. Thus, these combinations of abiotic and biotic stresses, makes the breeding work in the Mediterranean dryland both complex and challenging.

The main objective of ICARDA durum breeding is to develop genotypes and genetic stocks that combine yield potential with resistance to drought and other abiotic and biotic stresses; and improved grain quality. The purpose of this chapter is to give broad review of the durum research program, with emphasis on the challenges involved, particularly drought research and conventional and modern strategies to breed drought resistance.

DRYLAND BREEDING METHODOLOGIES

Before developing a selection procedure for abiotic stresses, it is crucial to determine the frequency of occurrence of a particular stress and its timing in relation to crop development in each agro-ecological zone. Results of our earlier selection work under contrasting environments show that genotypes that selected under favorable conditions do not necessarily do so under less favorable conditions and visa versa. It may be difficult to select for genotypes with high yield potential in dryland environments, but it is far more difficult to select for moisture stress tolerance under high-input environments (Nachit, 1992). However, it appears that selection only under extreme environmental conditions (too favorable or too dry) is not an efficient way to identify cultivars for Mediterranean dryland that are characterized by a high year-to-year variability and by an unpredictable alternation between favorable and less favorable seasons. Breeding cultivars that combine yielding ability with stress tolerance and yield stability are therefore a prerequisite to the Mediterranean dryland conditions.

In our selection approach, all early segregating populations are subjected to the stresses encountered in the Mediterranean dryland, for example, drought, cold, heat, rusts, *septoria tritici*, root rot, Hessian fly, stem sawfly, etc., with the aim of identifying the populations that do particularly well in certain environments and are not sensitive to the stresses of other environments (Nachit et al., 1995). The pedigree method of selection is used to select individual plants from the populations that were selected across sites/environments. As for the bulk method, it is more extensively used to select among populations across sites during the winter and summer testing. This method presents the double advantage of testing more crosses at several sites/environments. The promising bulked segregating populations are tested in several agroecological zones of West Asia and North Africa (WANA) region in collaboration with national programs in the Mediterranean region. With this selection procedure, it is possible to identify at early stages the populations that combine productivity and stability with tolerance to biotic and abiotic stresses; and are heterogeneous for a number of important characteristics in the populations reaching the advanced testing stage.

Double Gradient Selection Technique

Dryland productivity can fluctuate from 0 to 6 t ha^{-1}. The fluctuation is accompanied by variations in attacks by biotic stresses; for example, in favorable rainfall season dryland productivity (1–6 t ha^{-1}) attacks by rusts and *septoria tritici* are a threat to the crop; whereas in the dry seasons Hessian fly and wheat stem sawfly are the major biotic constraints. Thus, the basis of our breeding work is to select durum wheat populations and advanced lines with resistance to abiotic and biotic stresses; and to test for yield stability and productivity under Mediterranean dryland conditions (Nachit, 1992). The cornerstone of this strategy is the introgression of resistance genes from landraces and wild relatives to durum advanced genotypes and the utilization of contrasting and representing environments in the Mediterranean region. A Double Gradient Selection Technique (DGST), developed in the 1980s, reflects abiotic and biotic stresses in WANA for the temperature extremes varying from cold to hot and for the water regime varying from severe drought to irrigated conditions.

The DGST employs five environments that are extensively used during the various phases of selection in segregating populations and testing of advanced lines. The environments/sites are:

- Tel Hadya/Syria is the main research station of ICARDA (36°01′ N; 36°56′ E, and at 284 m above sea level) has a Mediterranean continental climate with average annual precipitation of 335 mm;
- Breda station also in Syria (35°56′ N, 37°10′ E, and 300 m above sea level) with clay soil and an average annual precipitation of 260 mm is characterized by drought and harsh continental climatic conditions, and provides a dry site with cold winters and high natural infestation of wheat stem sawfly.
- Lattakia station (35°32′ N, 35°46′ E, 7 m above sea level) with an annual average rainfall of 784 mm has mild winters and severe disease pressure, and is a high rainfall site used to test for resistance to diseases under natural and artificial infestation, particularly to *Septoria tritici* and Barley Yellow Dwarf Virus.
- The Terbol in Lebanon's Beka'a Valley (33°33′ N, 35°59′, at 890 m above sea level) is characterized by cold winters but favorable growing conditions, and it has a fine clay soil and an average annual precipitation of 524 mm. It is used during the winter season to test for yield potential and resistance to yellow rust and cold; and during the summer to screen for resistance to heat, stem and leaf rusts.
- The Kfardan station also in Lebanon's Beka'a Valley, Lebanon (34°01′ N, 36°06′ E, and 1080 m above sea level) is characterized by extreme temperature fluctuations and has a fine clay soil and an average annual precipitation of 402 mm.

Further, the DGST employs within the Tel Hadya main station six environments (Nachit et al., 1995). Realizing that an important part of our work, particularly in the early stage, has to be done at experimental research station, we have developed a stress screening technique using simulated environments at one site (Nachit, 1983). The technique is based on growing the same germplasm using staggered sowing dates (*early planting, rainfed, irrigated, late planting, summer planting, and sowing after rotation with hay-vetch*) at one site. The germplasm is thus subjected to different stresses according to date of sowing (cold, drought, terminal stress, and heat).

Thus at Tel Hadya station durum wheat germplasm is grown under six environmental conditions:

- Early sowing (mid October) with supplemental irrigation (450 mm, including rainfall) to simulate a crop cycle with long duration and favorable growing conditions. The early sowing conditions subject durum wheat plant to cold damage (winterkill) during the tillering stage, to frost during the anthesis stage, and to attacks of yellow rust (*puccinia striiformis*) and *septoria tritici*.
- Normal date of sowing under dryland conditions representing Mediterranean continental dryland conditions.
- Late sowing (early April) to simulate a short growing season, with severe terminal stress (drought and heat). The late sowing conditions subject durum wheat plant to terminal heat and drought stresses, particularly during the grain filling stage. Late sowing conditions increase the attacks by

aphids, Barley Yellow Dwarf Virus, and Hessian fly (*Mayetiola destructor*).

- Summer sowing (early July–mid-October) is used to test for high temperature conditions during all crop growth stages. The summer sowing conditions subject durum wheat plant to extremes high temperatures and sirocco.
- Sowing under irrigated conditions for high yield potential selection.
- Sowing after rotation with vetch to induce favorable dryland conditions and slow release of N, mobilized from vetch root decomposition.

Thus, the testing sites/environments of the Double Gradient Selection Technique provide two interacting selection gradients for rainfall and temperature regimes; and also such environmental conditions encompass the main abiotic and biotic stresses that prevail in the Mediterranean dryland.

Broadening the Genetic Base for Drought Resistance with Landraces

The Mediterranean durum landraces were found to possess desirable traits lacking in other materials, such as resistance to drought and cold, early growth vigor, long peduncle, and high fertile tillering. Using of durum wheat landraces in the hybridization program, we showed that substantial progress can be achieved in developing improved cultivars for dry areas (Nachit, 1992). Furthermore, the knowledge of physiological mechanisms involved in drought tolerance is a prerequisite to increasing durum dryland productivity. High relative water content and high osmotic adjustment were identified as important traits for drought tolerance. We selected for several morph-physiological traits related to drought tolerance in populations arising from crosses between durum and its wild relatives (Nachit et al., 1995). Crosses initiated in the mid-1980s at our program are now generating several advanced lines with better performance under environments with abiotic constraints such as drought and temperature extremes (heat and drought).

Triticum Wild Relatives to Improve Drought Tolerance and Yield Potential

Besides landraces, wheat relatives, for example, *Triticum dicoccoides*, *T. monococcum*, *Aegilops ssp.*, etc. provide valuable sources for widening the genetic base of durum and improving its resistance to abiotic and biotic stresses (Nachit et al., 2000). The wild relatives *Aegilops* species such as *Ae. geniculata*, *Ae. biuncialis*, *Ae. triuncialis,* and *Ae. neglecta*, *Ae. speltoides*, *Ae. longissima*, and *Ae. Searsii*, *T. dicoccoides*, *T. Araraticum*, *T. urartu*, *T. boeoticum*, *T. carthlicum, and T. dicoccum* showed different tolerances to abiotic stresses (drought, cold, and heat). Several advanced abiotic stress-tolerant durum genotypes were generated through hybridization with this material. Further studies (Nachit et al., 1997) showed that yield potential of the genotypes generated from hybridization with *Triticum* wild relatives produce high grain yields under dryland conditions than durum x durum genotypes with no negative effect on yield potential, thus confirming earlier finding that stress tolerance can be combined with yield potential (Nachit and Ouas-

Table 13–1. Grain yield of some durum genotypes derived from crosses with Triticum wild relatives under drought and terminal stress conditions.

Cross/Genotype	Grain yield	Increase
	kg ha^{-1}	%
Rufom-5/*T. Araraticum*500140//Carzio ICD92-0764-WABL-1AP-0TR	2139	201
Sbl1/*T.dicoccoides*600545//Omguer-1 ICD92-0750-WABL-2AP-0TR	1946	183
Haucan/*Ae. Columnaris*400020//Omtel1/3/Omlahn-3 ICD91-0604-WABL-11AP—4AP-0TR	1939	183
Haucan/*Ae.Columnaris*400020//Omtel-1/3/Omlahn-3 ICD91-0604-WABL-13AP-5AP-0TR	1936	182
Checks: Haurani (local)	1062	100
Korifla (improved)	1533	144

Increase (%): Percentage increase over the local check Haurani.

sou, 1988). Furthermore, wild relatives conferring abiotic (drought, cold, heat) and biotic (Hessian fly, Russian wheat aphid, cereal cyst nematodes) stress resistance are crossed with adapted durum genotypes and recombinant lines are being developed.

Crosses with wild relatives such as *T. monococcum, T. dicoccoides, T. dicoccum, Aegilops* species, etc., have shown significant grain yield increase in the dry areas, as well as increasing yield stability. Table 13–1 shows the yield performance of some newly developed advanced lines derived from crosses between durum and wild relatives (*T. araraticum, T.dicoccoides, Ae. columnaris*) under terminal stress conditions. Large yield increases are achieved in comparison to the durum landrace Haurani and the improved stress tolerant and productive durum cv. Korifla (Cham3). The grain yield of these lines, which is more than 1 t than Haurani's yield, shows the progress achieved in tolerance to terminal stress.

Under cold and dryland conditions (Table 13–2), the percentage of selection frequency was highest in the crosses with *T. carthlicum* and *T. dicoccoides* followed

Table 13–2. Selection for resistance to biotic and abiotic stresses in durum x wild relatives crosses for continental and temperate drylands.

Durum × wild relatives	Continental dryland D, C, YR, ST, WSSF†	Temperate dryland D, H, LR, SR, BYDV
	%	%
Aegilops ssp.	38.5	54.0
T. monococcum	42.5	51.7
T. dicoccoides	57.9	38.5
T. dicoccum	41.0	44.0
T. carthlicum	62.1	21.0
T. polonicum	17.0	51.4
T. araraticum	25.5	20.0
T. persicum	39.3	12.5

† D = drought, C = cold, H = heat, YR = yellow rust, ST = *Septoria tritici*, WSSF = heat stem sawfly, LR = leaf rust, SR = stem rust.

by *T. monococcum, T. dicoccum, T. persicum, and Aegilops* species. In contrast, in drought x heat conditions, the highest selection (%) was found in the crosses involving *Aegilops* species, *T. monococcum, T. polonicum*, *T.dicoccum,* and *T. diccoccoides*. Under continental cold conditions, the crosses with *T. polonicum* showed a low percentage of selection, but high percentage under hot conditions. In contrast to *T. carthlicum*, the highest selection (%) was made under cold and the lowest under hot conditions.

Drought Tolerance and Morpho-physiological Traits

For the drier zones, emphasis is placed on resistance to drought, heat, winterkill, frost at anthesis stage, and to pathogens and insects specific to dry areas, for example, root rot, common bunt, wheat stem sawfly, and Hessian fly. Genetic stocks have been developed with combined resistance to drought and cold (winterkill). The number of advanced lines possessing consistent and high yields is steadily increasing as reflected by an increased number of lines released in the Mediterranean region (Nachit, 1998b).

To identify drought tolerant germplasm, the *empirical selection approach* is employed. Evaluation for drought tolerance requires a large number of testing sites and seasons. Empirical selection approaches to genetically improve grain yield in dryland have been effective, but they are time-consuming and require the employment of large amounts of resources (Nachit, 1998b). However, the *analytical selection approach* using the morpho-physiological traits has been slowly adopted for selection in the segregating generations in some dryland breeding programs (Nachit, 1992). The employed trait for dryland selection needs to be simple and rapid to use, and less expensive than the field and design techniques used in the empirical selection approach.

Understanding the drought tolerance basis of the morpho-physiological traits in durum will offer the potential to select germplasm based on key traits linked with grain yield in dryland agriculture. Morpho-physiological traits can be used as indirect selection criteria for grain yield; however, their effectiveness depends on correlations with grain yield under drought and the degree to which each trait is genetically controlled. Usually, a dryland crop deploys a complex set of interactions to grow and survive under moisture stress (Clarke et al., 1992; Nachit et al., 1992a).

The progress achieved in drought resistance breeding can be shown in the example of Omrabi3 and Omrabi17. These lines originated from a cross between a Middle East landrace (Haurani) and a CIMMYT high yielding variety (JoriC69). Because of their high performance under stress and favorable conditions, they were included in the farmers' field verification trials of the low and high rainfall areas of Syria and Lebanon. Omrabi3 was released as Cham5 for the dry areas in Syria. As for thermal conditions, Omrabi17 is adapted to low rainfall areas with continental climates. In contrast, Korifla (released in Algeria, but under the names of Cham3 in Syria, and Petra in Jordan) is adapted to dry areas with mild winters. These results confirm earlier findings in which yield and stress resistance can be successfully combined (Nachit and Ouassou, 1988).

Plant characterization is another tool we use to improve our breeding work (Nachit and Jarrah, 1986). We have found that earliness, fertile tillering, spike fer-

Table 13–3. Comparison of grain yield between ICARDA dryland durum improved genotypes and Middle East durum landraces over years × sites.

Year × Site†	Improved cultivars	Middle East landraces	T-test‡
	kg ha^{-1}		
GY95BR	1776	1322	4.6***
GY95IR	3636	2816	3.8***
GY95RF	3121	2402	3.6**
GY96BR	2720	1972	4.1***
GY96EP	4261	3674	1.6NS
GY96IR	6236	3948	5.6***
GY96RF	4582	3620	5.2***
GY97RF	2778	2225	2.6*
GY01BR	2834	2504	2.8**
GY01RF	4897	3639	3.2**
GY02BR	2743	2224	3.7**

† Years: 1995, 1996, 1997, 2001, and 2002; and the sites: Breda (BR), Tel Hadya Irrigated (IR), Rainfed, and Early Planting (EP).

‡ Significance level: 5, 1, 0.1: 2.09, 2.85, 3.85, respectively. NS = not significant.

tility, peduncle length, and early plant vigor are associated with higher grain yield under drought conditions. Most of our improved lines show desirable values for these traits. The analytical approach is likely to be more useful for areas with severe moisture stress. It relies on the plant's different adaptive mechanisms in a stress environment, with the possibility that breeding and selection for these adaptations will contribute to growth and yield under stress.

Under Mediterranean drylands, fertile tillering ability is the most potent predictor of durum grain yield under moisture stressed conditions (Nachit et al., 1992a). The trait contribution estimates have shown that fertile tillering can account for more than 30% of the total variability in grain yield. In addition, spike fertility and earliness account for 5.3 and 4.1% of the total variability in grain yield. Furthermore, measurements are made for relative water content, osmotic adjustment, isotopic carbon discrimination, canopy temperature, chlorophyll fluorescence, chlorophyll content, osmolyte accumulation (soluble sugars and proline), root parameters (volume, weight, deepness, and number), leaf anatomy and morphology, boron toxicity, zinc deficiency, cold damage at vegetative and reproductive stage, and heat damage at vegetative and reproductive stage (Nachit et al., 1993; Dib et al., 1994).

The Middle East landraces (Haurani Nawawi, Haurani 27, Normal Haurani, Hamari Ahmar, Akbash, Kishk, Baladia Hamra, Senatore Cappelli, Gezira 17) compared with the improved cultivars (Belikh-2, Cham-1, Korifla, Omrabi-5, Omrabi-3, Lahn, Massara-1, Bicre, Deraa, Daki, and Omlahn-3) developed at ICARDA durum program during the last 25 yr. Genetic gain (Table 13–3) made in grain yield in dry areas encompass environments from severe dry (Breda) to irrigated (Tel Hadya-Irrigated) conditions.

Morphophysiological traits were studied, to determine which changes were made through selection in addition to grain yield. The fertile tillering ability and plant height were the most affected traits (Table13– 4). The fertile tillering ability was increased significantly while the plant height and anthesis were reduced. As

Table 13–4. Morphophysiological changes made through breeding in ICARDA dryland durum improved genotypes compared with the Middle East durum landraces.

Trait/Year/Site†	T-test	Trait/Year/Site	T-test‡
Tillering95BR	+4.01***	Chlorophyll96RF	+2.65*
Tillering95IR	+3.97***	Photosynthesis95IR	+3.15**
Tillering95IR	+3.43**	Photosynthesis95BR	+3.00**
Plant height	−4.75***	Photosynthesis96IR	+2.96**
Anthesis96IR	−2.98**	Photosyntheis96BR	+2.73*
Osmotic adjustment95BR	+2.89**	Photosynthesis96IR	+2.69*

*,**,*** Significant levels at 0.05, 0.01, and 0.001, respectively.
† Years: 1995 and 1996 and sites: Breda (BR) and Tel Hadya Irrigated (IR)and Rainfed (RF).
‡ Significance level: 5, 1, 0.1: 2.09, 2.85, 3.85, respectively.

for the physiological traits, chlorophyll content, osmotic adjustment, and photosynthesis were also significantly increased through selection.

Yield Potential and Stability

Although the delimitation of agro-ecological zones in the south Mediterranean region decreases the genotype x environment interactions, it does not necessarily eliminate it because the year-to-year and site-to-site variations within a zone can still be very important and make it imperative to look for cultivars possessing an acceptable degree of consistency of superior performance (commonly called stability) across a series of environments within a particular agro-ecological zone.

The cultivars' performance reflects the interaction of genetic and environmental factors. The relative performance of genotypes or crosses may vary in different environments, in which case the genotypes are said to interact with the environments. Results from this project as well as other works show the overwhelming evidence of genotype x environment interactions. For an efficient breeding program in a given region of interest, it is important to know the causes and nature of genotype x environment interactions (Nachit et al., 1992a, 1993). Multi-location testing provides the data for assessing the consistency of relative cultivar performance. It also enables the identification of cultivars that combine desirable traits such as resistance to various diseases and insects and tolerance to drought, cold, and heat. These cultivars provide good sources of parental material for the hybridization program.

Germplasm developed under the DGST's approach showed far better combination of high yield and yield stability than germplasm developed under high input or low input environments. The variety Waha was released in contrasting environments and in countries of the Mediterranean region under different names in Portugal, Algeria, Cyprus, Syria, Turkey, Jordan, and Sudan. Waha is used in the ICARDA breeding program as a check for yield stability. But through intensive breeding and selection for yield stability during the last decade, new durum cultivars were identified with better yield stability, disease resistance, and grain quality than Waha; for example, Omrabi lines, Massara, Genil5, and Omruf2 (Table 13–5).

Through continuous breeding for gene pyramiding for high yield potential and yield stability, durum genotypes with high yield potential were produced (Table

Table 13–5. Durum genotypes with stable productivity in the Mediterranean dryland, 26 sites.

Entry	Mean grain yield†	Relative stability‡
	kg ha^{-1}	%
Omrabi5	3053	188
Massara	3000	192
Genil3	2956	157
Omrabi3	2925	146
Omruf2	2824	108
Cham1	2803	100
LSD (0.05)	334	

† Sites: 10 from North Africa, 11 from Middle East, 5 from South Europe.
‡ Relative stability = (MDMYL of check entry/MDMYL of test entry) × 100. Cham-1 = 100%. MDMYL = Mean of difference from maximum highest yielder at each location divided by location mean.

Table 13–6. Yield potential of newly developed durum under irrigated conditions, Terbol station, Lebanon, 2002.

Cross/Name	Grain yield
	kg ha^{-1}
Miki1	13905
Ouaserl-1	12861
Ouasloukos	12211
Ouaserl2	12061
Aghrass2	11945
Amedakul1	11905
Checks:	
Haurani (Local check)	5961
Waha (Improved check)	9811
LSD = 670	
CV(%) = 12.7	

13–6). Under favorable and irrigated conditions of Terbol station (650 mm) in Lebanon's Beka'a Valley, the newly developed advanced durum genotype Miki-1 attained a yield level of 13 905 kg ha^{-1} with water-use efficiency for grain yield of 21.4 kg ha^{-1} mm^{-1} a grain yield advantage of over 4 t ha^{-1}. This gain was achieved mainly through continuous crossing and selecting grain yield contributing genes, which derived from the hallmark advanced durum genotypes in the program. This suggests that within the available variable material through crosses and selection, novel genes combination(s) can be developed to boost the improvement of a desirable trait, such as grain yield potential in this case.

Drought Resistance and Molecular Markers

A durum core collection of landraces and improved genotypes, showing a wide range of response to abiotic and biotic stresses and grain quality parameters, was probed with molecular markers and scored for grain yield, grain quality, and traits related to abiotic and biotic stresses (Nachit et al., 2000). Multivariate analyses to determine possible associations between quantitative traits and molecular

Table 13–7. Grain yield and morphophysiological traits that associated with RFLP markers.

Trait	RFLP-Markers
Grain yield **	KSUG48, CDO1090, CDO395, BCD1661
Awns length***	BCD348
Peduncle length***	BCD782, BCD292
Early growth vigor**	BCD758
Productive tillering under stress**	BCD292
Spike fertility under stress**	BCD348
Kernel weight***	BCD342
Leaf rolling index**	BCD348, BCD1355f
Canopy temperature **	CDO669, BCD305
Fluorescence inhibition***	BCD292, BCD758
Osmotic adjustment **	BCD475, BCD1438
Proline content **	BCD758
Carbon isotope discrimination***	CDO1090, CDO1312, KSUG48

, * Significant at 1, and 0.1 level, respectively.

markers were determined. Subsequently, exact linkages were confirmed in the mapping populations. Indirect selection with molecular markers has an advantage over direct selection, as their "heritability" is 1.0. When the correlation between a molecular marker and a trait is greater than the heritability of the trait, marker assisted selection should be advantageous, particularly when a trait is highly affected by environmental variations. Molecular markers in durum were found to be associated with grain yield (Table 13–7) under dry conditions and with some morpho-physiological traits related to drought tolerance (Nachit et al., 1993; Nachit, 1998b). The markers CDO395 and BCD1661 are associated with high grain yields; and CDO1090 and KSUG48 with both grain and carbon isotope discrimination (Table 13–7). Osmotic adjustment, canopy temperature, chlorophyll inhibition, and proline content showed also strong relationships with some molecular markers.

Seven mapping populations were developed at our program in 1991 and 1992 using the single seed method (Nachit et al., 1995) and two of these populations were mapped (Nachit et al., 2001; Elouafi and Nachit, 2003). From the seven mapping populations, three were developed for drought tolerance and yield stability:

- Jennah Khetifa/Cham1 for drought tolerance and biotic stress resistance (Nachit et al., 1995, 2001);
- Omrabi5/*T. dicoccoides*//Omrabi5 for grain quality and drought resistance (Nachit, 1998; Elouafi and Nachit, 2003); and
- Lahn/Cham1: for high yield potential and yield stability.

In contrast, the other populations were developed for adaptation in the three main agro-ecological Mediterranean environments (Nachit et al., 1995): Haurani/Cham1 for continental areas; Cama-di-Abou/Cham1 for temperate dryland; Zenati Bouteille/Cham1 for the Atlas highland; and Kunduru/Cham1 for Anatolia highland.

For drought tolerance and yielding under dryland conditions, carbon isotopic discrimination was found as a useful tool (Farquhar and Richards, 1984; Nachit 1998) and be associated with water-use efficiency in terms of grain weight per mm (Hubick and Farquhar, 1989; Johnson and Bassett, 1991; Nachit, 1998a). It correlates also with transpiration efficiency (ratio of biomass production to water transpired or mol CO_2 fixed per mol H_2O transpired) and high grain yield under favorable and dry conditions, and was found to show large heritability (Ehdaie and Waines, 1994) and low genotype x environment interactions in the Mediterranean dryland (Nachit, 1998b). This result suggests that carbon isotope discrimination can be usefully employed in applied breeding programs to improve productivity under stressed conditions. However, contrasting results for associations with yield and transpiration efficiency were noticed between the Mediterranean group (Araus et al., 1997; Nachit, 1998b) found that this association is positive, and the Australian group (Farquhar and Richards, 1984) reported the opposite. The positive association may reflect the large grain yield dependence on stomata conductance (Morgan et al., 1993), which also relates to the severe terminal stress of the Mediterranean continental dryland (Nachit, 1998b). Consequently, these results have prompted the use of carbon isotope discrimination in the breeding program as promising screening tools for identification of cultivars with high productivity under Mediterranean dryland conditions (Nachit, 1998b).

Molecular markers could assist in identifying traits that are difficult to select through phenotype (Autrique et al., 1996). Drought tolerance in durum wheat had shown that some markers can be associated with grain yield in drylands and with morpho-physiological traits for drought tolerance (Nachit et al., 1993). Positive significant correlations between carbon isotope discrimination and grain yield (0.54***), fertile tillers (0.30***), grains per spike (0.42***), and 1000 kernel weight (0.22**) were reported (Nachit, 1998b). The association with grain yield was similar to that of a number of fertile tillers and a number of spike kernels with grain yield. The testing in different environmental conditions showed also that carbon isotope discrimination was less affected by environment than grain yield. The genotype ranking for this trait in different sites did not change significantly from site-to-site, as it did for grain yield, thus confirming observations for bread wheat (Ehdaie et al., 1991; Condon and Richards, 1992; Matus et al., 1997).

Chromosomal Regions Controlling Grain Quality and Drought Tolerance

Genetic linkage maps were constructed for the durum populations Jennah Khetifa x Cham1 (JKC) (Nachit et al., 2001) and the backcross population Omrabi5/*T. dicoccoides*//Omrabi5 (MDM) (Elouafi and Nachit, 2003) and quantitative trait loci (QTL) were identified at the ICARDA durum-breeding program. Markers linked to resistance to drought and temperature extremes; to improvement in grain quality are being developed. Molecular markers of Restriction Fragments Length Polymorphisms (RFLPs), Amplified Fragments Length Polymorphisms (AFLPs), Single Sequence Repeats (SSRs), drought candidate genes, dehydrin, gliadin, and glutenin genes are used to screen for abiotic stresses and grain quality. These markers are probed on the mapping population to determine their chromosomal location and identify quantitative trait loci (QTL) for different traits of interest. The identified QTLs are validated on the genotypes included in the ICARDA Durum Core Collection. Thus, 144 Durum Genotypes were surveyed with molecular markers (RFLPs, SSRs, AFLPs) and seed storage proteins (Gliadins and Glutenins), and assessed in different environments for grain yield, agronomic traits, and abiotic stress (drought and temperature extremes); biotic stress (diseases, insects, viruses); and grain quality. For drought tolerance test, the Durum Core Collection (DCC) genotypes were also assessed for kernel carbon isotope discrimination.

In the durum mapping populations Jennah Khetifa x Cham-1 (JKC) and Omrabi5/*T. dicoccoides*//Omrabi5 (MDM), QTLs for the grain qualities were determined for gluten strength, protein content, yellow pigment, flour extraction, and

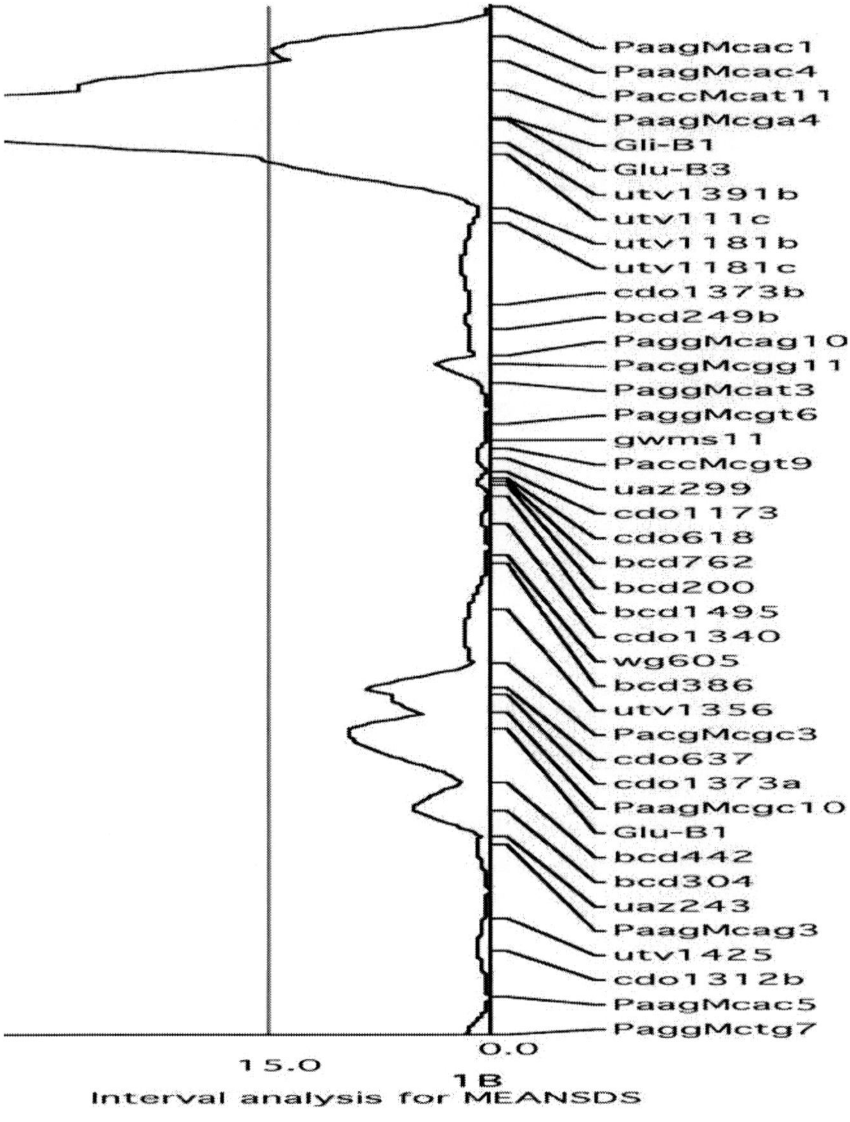

Fig 13–1. *Gluten strength-QTL* on 1Bs chromosome in the durum map Jennah Khetifa × Cham1.

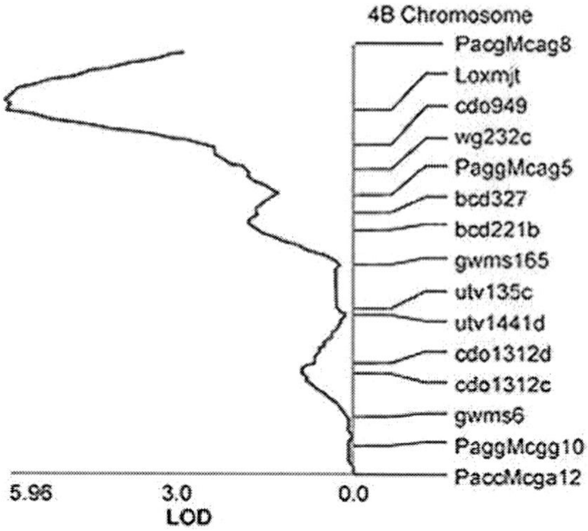

Fig. 13–2. Interval analysis for carbon isotope discrimination on 4B, Breda, Syria, 1999.

test weight (Elouafi et al., 2000, 2001; Elouafi and Nachit, 2002). Figure 13–1 shows the major QTL on the short arm of 1B chromosome.

Genetic markers are particularly useful when attempting to develop drought tolerance, as conventional selection for this trait is both costly and time-consuming, having to be conducted over many sites and seasons. Drought tolerance is an especially important breeding objective for the ICARDA collaborative project on durum wheat with the national breeding programs in the Mediterranean region. Using the same mapping population as above, various markers were found to be linked to some of the morpho-physiological traits associated with drought tolerance. These traits included the temperature of the crop canopy, the chlorophyll inhibition, and the osmotic adjustment. The same markers were also linked with grain yield and some of the components of grain yield (number of fertile tillers and of kernels per spike). The main aim is to identify simple traits that can be quickly and easily used in the field.

For traits of agronomic importance, such as the drought tolerance trait—carbon isotope discrimination, they were found to be positively correlated with grain yield and yield components (Nachit, 1998b), the markers linked to this trait are used to breed durum cultivars with high values for trait. Further, in the Jennah Khetifa x Cham-1 and Omrabi5/*T. dicoccoides*//Omrabi5 populations, different QTLs for carbon isotope discrimination (*CID-QTLs)* were identified on different chromosomal regions (4BS and 3BL) and are used to pyramid the genes for CID in durum. Indeed, the association of carbon isotope discrimination with grain yield was just as strong as that of the number of fertile tillers and the number of kernels per spike with grain yield. Analysis of QTL has revealed the approximate location of the genes coding for the carbon isotope discrimination trait on chromosome 4B (Fig. 13–2).

CONCLUSIONS

Durum wheat is grown on large areas in the Mediterranean dryland and is economically an important crop to the farmers and food processing industry. During the last decade, a spectacular adoption of the stress tolerant and productive durum varieties was made, which has boosted wheat production in the Mediterranean drylands. The results demonstrate the potential of improving dryland productivity through the introgression of stress resistance from Mediterranean landraces and the durum wild relatives. Furthermore, the results also show the usefulness of molecular markers to improve grain quality and drought tolerance in durum wheat. Molecular markers may also speed and facilitate further the pyramiding and transferring of desirable traits into durum wheat. Preliminary use of molecular markers in mapping traits of grain quality and drought tolerance was shown and the potential use of marker-assisted selection for drought tolerance and grain quality discussed. However, more fundamental studies need to be conducted, and results generated, in order to devise sound approach for the use of molecular markers in marker assisted selection and breeding. Durum wheat production is likely to expand in the Mediterranean in the years ahead, with some countries becoming specialized producers.

ACKNOWLEDGMENT

This work was supported by ICARDA. We wish to especially acknowledge the excellent field technical and lab assistance of the durum breeding research technicians and assistants at ICARDA. We are thankful for the helpful discussion and review by Dr. John Ryan, ICARDA soil scientist. The experiments comply with the laws of the countries where they were performed.

REFERENCES

Araus, J.L., T. Amaro, Y. Zuhair, and M.M. Nachit. 1997. Effect of leaf structure and status on carbon isotope discrimination in field-grown durum wheat. Plant Cell Environ. 20:1484–1494.

Autrique, E., M.M. Nachit, P. Monneveux, S.D. Tanksley, and M.E. Sorrells. 1996. Genetic diversity in durum wheat based on RFLPs, morphophysiological traits, and coefficient of parentage. Crop Sci. 36:735–742.

Clarke, J.M., R.M. DePauw, and T.F. Townley-Smith. 1992. Evaluation of methods for quantification of drought tolerance in wheat. Crop Sci. 32:723–728.

Condon, A.G. and R.A. Richards. 1992. Broad sense heritability and genotypes x environment interaction for carbon isotope discrimination in field-grown wheat. Aust. J. Agric. Res. 43:921–934.

Dib, T.A., P. Monneveux, E. Acevedo, and M.M. Nachit. 1994. Evaluation of proline analysis and chlorophyll fluorescence quenching measurements as drought tolerance indicators in durum wheat (*Triticum turgidum* L. var. durum). Euphytica 79:65–73.

Ehdaie, B., A.E. Hall, G.D. Farquhar, H.T. Nguyen, and J.G. Waines. 1991. Water-use efficiency and carbon isotope discrimination in wheat. Crop Sci. 31: 1282–1288.

Ehdaie, B., and J.G. Waines. 1994. Genetic analysis of carbon isotope discrimination and agronomic characters in a bread wheat cross. Theor. Appl. Genet. 88:1023–1028.

Elouafi, I., and M.M. Nachit. 2002. QTL identification of milling characters in *Triticum turgidum* L. var durum". p. 57. *In* Q. Zhang (ed.) First Int. Symp. on Genetics and Crop Genetic Improvement, Wuhan, China. 21–25 Sept. 2002.

Elouafi, I., and M.M. Nachit. 2004. A genetic linkage map of *Durum* x *Triticum dicoccoides* backcross population based on SSRs and AFLP markers and QTL analysis for milling characters. Theor. Appl. Genet. 108(3):401–413. Epub 2003 Dec. 16. PMID:14676946 [PubMed-in process]. Available at http://www.ncbi.nlm.nih.gov/entrez/query.fcgi?cmd=Retrieve&db=PubMed&list uids=14676946&dopt=Abstract (verified 23 Feb. 2004).

Elouafi I., M.M. Nachit, A. Elsaleh, A. Asbati, A. Martin, L.M. Martin, and D.E. Mather. 2000. QTL-mapping of genomic regions controlling gluten strength in durum *(Triticum turgidum* L. durum var.). p. 505–510. *In* Royo, Nachit, DiFonzo, Araus (ed.) Proc. Durum wheat improvement in the Mediterranean region: New challenges (OPTIONS mediterraneennes), Zaragoza, Spain. 12–14 Apr. 2000.

Elouafi I., M.M. Nachit, and L.M. Martin. 2001. Identification of a microsatellite on chromosome 7B showing a strong linkage with yellow pigment in durum wheat (*Triticum turgidum* L. var. durum). Hereditas 135:255–261.

Farquhar, G.D., and R.A. Richards. 1984. Isotope composition of plant carbon correlates with water-use efficiency of wheat genotypes. Aus. J. Plant Physiol. 11:539–552.

Hubick, K.T., and G.D. Farquhar. 1989. Carbon isotope discrimination and the ratio of carbon isotope gained to water lost in barley cultivars. Plant Cell Environ. 12:795–804.

Johnson, R.C., and L.M. Bassett. 1991. Carbon isotope discrimination and water-use efficiency in four cool season grasses. Crop Sci. 31:157–162.

Matus, A., A.E. Slinkard, and C. Van Kessel. 1997.Genotype x environment interaction for carbon isotope discrimination in spring wheat. Crop Sci. 37:97–102.

Morgan, J.A., D.R. LeCain, T.N. McCaig, and J.S. Quick. 1993. Gas exchange, carbon isotope discrimination and productivity in winter wheat. Crop Sci. 33:178–186.

Nachit, M.M. 1983. Use of planting dates to select stress tolerant and yield stable genotypes for the rainfed Mediterranean environment. Rachis 3:15–17.

Nachit, M.M. 1992. Durum wheat breeding for Mediterranean dryland of North Africa and West Asia. p. 14–27. *In* S. Rajaram et al. (ed.) Proc. Durum Wheat Workshop "Discussion on Durum Wheat: Challenges and Opportunity". 23–25 Mar. 1992. CIMMYT, Ciudad Obregon, Mexico.

Nachit M.M. 1998a. Durum breeding research to improve dryland productivity in the Mediterranean region. p. 1–15. *In* M.M. Nachit et al. (ed.) Proc. the SEWANA Durum Research Network, 20–23 Mar. 1995. ICARDA, Aleppo, Syria.

Nachit, M.M. 1998b. Association of grain yield in dryland and carbon isotope discrimination with molecular markers in durum (*Triticum turgidum* L. var. durum). p. 218–223. *In* Proc. 9th Int. Wheat Genetics Symp., Saskatoon, SK, Canada.

Nachit, M.M., A. Asbati, M. Azrak, and I. Elouafi. 1997. Durum breeding. Annual report. ICARDA, Aleppo, Syria.

Nachit M.M., A. Asbati, M. Azrak, N. Rbeiz, and A. El Saleh. 1995. Durum wheat improvement. p. 68–89. In S. Varma (ed.) Cereal Program Annual Report 1995. ICARDA, Aleppo, Syria.

Nachit, M.M., M. Baum, E. Autrique, M.E. Sorrells, T.A. Dib, and P. Monneveux. 1993. Association of morphophysiological traits with RFLP markers in durum wheat. p. 159–171. *In* P. Monneveux and M. Ben Salem (ed.) Proc. Tolerance a la secheresse des cereales en zone mediterraneenne. Diversite genetique et amelioration varietale, Montpellier, France. 15–17 Dec. 1992.

Nachit, M.M., I. Elouafi, M.A. Pagnotta, A. El-Saleh, E. Iacono, M. Labhilili, A. Asbati, M. Azrak, H. Hazzam et al. 2001. Molecular linkage map for an intraspecific recombinant inbred population of durum wheat (*Triticum turgidum* L. var. durum). Theor Appl. Genet. 102:177–186

Nachit, M.M., and M. Jarrah. 1986. Association of some morpho-physiological characters to grain yield in durum wheat under Mediterranean dryland conditions. Rachis 5:33–34

Nachit, M.M., P. Monneveux, J.L. Araus, and M.E. Sorrells. 2000. Relationship of dryland productivity and drought tolerance with some molecular markers for possible MAS in durum (*Triticum turgidum* L. var. durum. p. 203–206. *In* C. Roya et al. (ed.) Options Mediterraneennes No. 40 "Durum Wheat Improvement in the Mediterranean Region: New Challenges" Zaragoza, Spain. 12–14 Apr. 2000.

Nachit, M.M., and A. Ouassou. 1988. Association of yield potential, drought tolerance and stability of yield in *Triticum turgidum* var. *durum*. p. 867–870. *In* T.E. Miller and R.M.D. Koebner (ed.) Proc. of the 7th Int. Wheat Symp. 1, Cambridge, UK. 13–19 July 1988.

Nachit, M.M., M.E Sorrells, R.W. Zobel, H.G. Gauch, R.A. Fischer, and W.R. Coffman. 1992a. Association of morpho-physiological traits with grain yield and genotype-environment interaction in durum wheat. I. J. Genet. Breed. 46:50–55.

Nachit, M.M., M.E Sorrells, R.W. Zobel, H.G. Gauch, R.A. Fischer, and W.R. Coffman. 1992b. Association of environmental variables with sites' mean grain yield and genotype-environment interaction in durum wheat. II. J. Genet. Breed. 46:41–49.

14 Sustainable Barley-Legume Rotations for Semi-Arid Areas of Lebanon

Sui-Kwong Yau

American University of Beirut
Beirut, Lebanon

Mustapha Bounejmate and John Ryan

ICARDA
Aleppo, Syria

Adel Nassar

ICARDA
Terbol, Bekaa, Lebanon

ABSTRACT

In semi-arid areas of West Asia and North Africa with a Mediterranean-type climate merging into a continental one, farmers have been increasingly practicing more continuous barley (*Hordeum vulgare* L.) cultivation, which is likely to be unsustainable in the long run. The objectives of the study were to: (i) determine if barley monoculture is really unsustainable, (ii) ascertain if barley and total dry matter yields can be increased and sustained by including a legume in the rotation, and if so, (iii) determine which barley-legume rotation is more productive in terms of dry matter. A long-term rainfed cropping trial, which was the first of its kind in Lebanon, was conducted at the Agricultural Research and Education Center in Lebanon's Bekaa Valley (512 mm annual precipitation). Eight different two-phase, barley-based rotations were compared. Thus, barley was grown after: barley, lentil (*Lens culinaris* Medik.), common vetch (*Vicia sativa* L.), bitter vetch [*V. ervilia* (L.)Willd.], common vetch for grazing, medics (*Medicago* spp.) for grazing, common vetch for hay, and common vetch with barley for hay. Seed and straw yield under barley monoculture declined after 3 yr to about 50% of the trial mean yield due to severe infestation of wild barley [*Hordeum spontaneum* (C. Koch) Thell.]. In the barley phase, barley-legume rotations yielded more seed and straw, and had higher stability than the barley monoculture. In the legume phase, all the legume treatments, except medics, yielded more dry matter than barley monoculture. For total drymatter per rotation cycle, barley in rotation with common vetch for seed and straw gave the highest yield and the lowest coefficient of variation (CV). In conclusion, the study showed that barley monoculture is unsustainable, and that barley and total drymatter yields could be sustained by including a legume in the rotation. Thus, farmers in the semi-arid areas of Lebanon should adopt a barley-legume rotation, such as barley-common vetch.

 Challenges and Strategies for Dryland Agriculture. CSSA Special Publication no. 32.

INTRODUCTION

Barley is the dominant winter crop in arid and semi-arid areas of West Asia and North Africa (WANA), including the northern Bekaa Valley of Lebanon (Cooper et al., 1987). In such environments, barley usually gives higher grain and drymatter yield than wheat, because it is more tolerant to dryness, poor soils, and salinity (Van Oosterom and Acevodo, 1992). Barley grain, straw, and stubble are the traditional and predominant feed for sheep (*Ovis aries* L.). Relative to wheat, the thinner stems of the two-row barley landraces are more digestible and preferable to the sheep.

Associated with the increase in population and wealth in the region, the demand on meat has increased rapidly. As a consequence, the number of sheep and goats (*Capra hircus* L.) kept in the region has increased rapidly as well, causing a feed shortage problem. In 1985 to 1989, there was a total of 163 million sheep in WANA (Belaid and Morris, 1991). A recent survey of small ruminant production systems in the Bekaa Valley of Lebanon showed that inadequate feed supplies and high prices of feeds were among the top problems ranked by farmers (Hammadeh et al., 1994). The increase in feed demand led farmers to grow barley continuously instead of their customary barley-fallow rotations (Belaid and Morris, 1991).

Growing continuous barley could have brought short-term economical benefits but is expected to be unsustainable (Jones, 1998). Cereal monoculture is known to deplete soil nutrients, and increase pest and weed populations, leading to reduction in yield. A sustainable and productive option to replace barley monoculture is needed. One suggestion is to introduce a legume into the rotation to reduce the intensity of barley cropping (Harris, 1995).

The planting of legumes in rotation with cereals has been demonstrated to be beneficial in semi-arid areas (Papastylianou, 1990; Jones and Singh, 2000). Many legumes are adapted to such environments and have the potential of being used in rotation with barley. Lentil is a traditional food legume crop in many semi-arid WANA areas, and its straw is a valuable ruminant feed. Recent research found that common vetch is a versatile forage legume (Jones and Arous, 1999). Besides harvesting seed and straw, common vetch is well suited to green-stage grazing, hay making, and growing in mixture with barley for hay production. Bitter vetch is another species found to be productive in semi-arid areas. In Australia, where alkaline soils dominate, medics are common forage legumes of the self-regenerating ley system (Cocks et al., 1980), which has contributed to yield increase of small grains and soil fertility since its adoption (Weston et al., 2002).

No research on the effects of barley monoculture and introduction of legumes to increase diversity or sustainability has been conducted in Lebanon. To fill this gap of inadequacy, a long-term rotation trial was set up to find a sustainable, more profitable, and environment-friendly alternative to barley monoculture for farmers in the semi-arid northern Bekaa Valley. The objectives of the study were to: (i) determine if barley monoculture is viable in the long run, (ii) establish if barley and total dry matter yields can be increased and sustained by alternative cropping with a legume and, if promising, (iii) identify which legume in the rotation with barley is more productive in terms of dry matter.

MATERIALS AND METHODS

The long-term rotation trial, which was the first of its kind in Lebanon, was conducted under rainfed conditions at the Agricultural Research and Educational Center (33° 56′ N, 36° 5′ E, 995 m above sea level) in the Bekaa Valley (Fig. 14–1). The long-term annual precipitation and mean temperature of the Center is 513 mm and 13.9°C, respectively. The soil is an alkaline (pH 8.0), clayey, Vertic Xerochrept (Ryan et al., 1980).

Eight rotations were studied: barley-barley monoculture and seven two-course barley-legume rotations: barley-lentil for seed and straw, barley-bitter vetch for seed and straw, barley-common vetch for seed and straw, barley-common vetch for green-stage grazing by lambs, barley-common vetch for hay production, barley-common vetch in mixture with barley for hay production, and barley-medics for green-stage grazing by ewes and their lambs. Varieties used in the trial were: Rihane 03 barley (facultative six-row type); mixture of *Medicago* spp. (75% *M. rigidula* L., line 1919; 25% *M. rotata* Boiss. and *M. noeana* Boiss.) for medics; Talia 2 lentil; ICARDA accession 3030 bitter vetch; and Syrian Local (ICARDA accession 2541) common vetch.

The trial was set up in 1994–95. Results from 1997–98 to 2001–02 were analyzed and reported here. The experiment was in a randomized complete block design with two replicates. The size of the plots was 0.1 ha (10 m by 100 m) except

Fig. 14–1. Map of Lebanon on which the site of the trial is marked.

for barley-medics (100 m by 100 m) and barley-vetch for grazing (25 m by 100 m) rotations. Both phases of each treatment were present each year.

Sowing with a commercial drill in rows spaced 15 cm apart took place in November, which is the optimal sowing time for the northern Bekaa. Selective herbicides were used to control grass weeds in legume crops and broad-leaf weeds in barley. Nitrogen (30 kg N ha^{-1}) was broadcasted by hand in the spring as ammonium nitrate to the barley plots only. Samples of plants were cut at ground level from each plot, collected, oven-dried at 80°C for 48 h, and weighed. Grazing of medics and vetch started around the end of March. The moving-cage technique was followed to measure the dry matter yield of medics. In years in which grazing continued longer, dry matter production was probably over-estimated in the vetch plots. Thus, in the comparison between legumes, yield of vetch under grazing was not included in the analysis. In the plots of vetch for hay, plant materials were collected and dried when the vetch reached early-pod set in late April to early May. Lentil, common vetch for seed, and bitter vetch were hand-harvested at ground level before physiological maturity in May to prevent seed shattering and leaf loss. For barley, samples were hand-harvested at maturity in late May. In 2001–02, the numbers of two-row and six-row heads in each barley plot were counted. Further details of the trial management can be found in Yau et al. (2003).

The ANOVA directive of the GENSTAT package (Genstat 5 Committee, 1993) was used for the analysis of variance. In the combined ANOVA analysis, the random-year and fixed-treatment model was adopted (McIntosh, 1983). The coefficient of variation on treatment yield relative to the year mean (Yau and Hamblin, 1994) was used as an agronomic type of stability measure across years.

The 5 yr of the trial encompassed a diverse spectrum with respect to annual precipitation and temperature. Precipitation ranged from 366 mm in 1999–2000 to 560 mm in 1997–98. Mean temperatures of the 5 yr were above the 41-yr long-term average, with 1998–99 being the warmest year.

RESULTS AND DISCUSSION

Figure 14–2 shows that the grain yield of barley monoculture was close to the average of the different rotations in the first 3 yr, but dropped sharply from 1997–98 onwards to 52% of the trial mean in the last 5 yr. This rapid yield decline clearly showed that barley monoculture was not sustainable under the conditions of the study. One apparent cause of such a decrease of barley grain yield under barley monoculture was infestation by two-row wild barley. There was a buildup of the wild barley over the years (Yau et al., 2003). In 2001–02, the number of two-row wild barley heads was as high as 83% of the six-row Rihane heads (Fig. 14–3). Wild barley increased under barley monoculture because it could not be controlled by herbicide application. Wild barley population also increased under medics and vetch for grazing because no herbicide was used on them. Wild barley still exists widely in the Fertile Crescent. Our study most probably is the first report to highlight the problem of wild barley infestation under barley monoculture in the region (Yau et al., 2003).

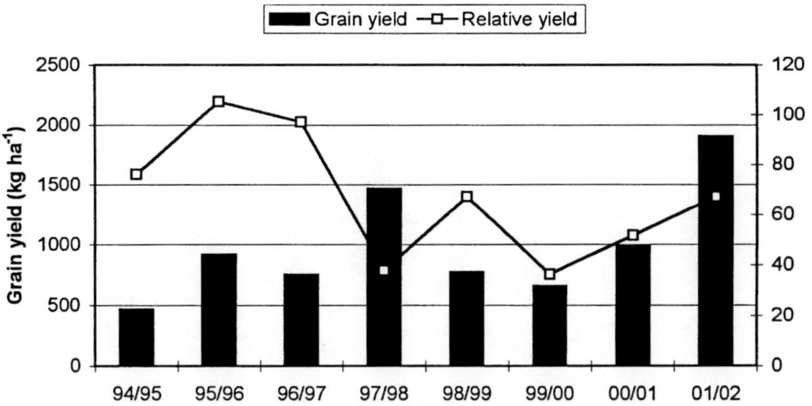

Fig. 14–2. Mean barley grain yield of the trial and grain yield of barley monoculture relative to the trial mean yield over 8 seasons (1994–95 to 2001–02).

There were significant differences in barley grain and straw yield between rotations, but the rotation-by-year interaction was not significant. Mean grain and straw yield of barley monoculture was lower than all the other rotations (Fig. 14–4 and 14–5). Besides, the barley monoculture had the highest CV, showing that it gave unstable yield across the years. Our results on the poor performance of barley monoculture relative to barley-legume rotations supported those obtained in neighboring countries, for example, in Cyprus by Papastylianou (1990), and in Syria by Jones and Singh (2000). In contrast to the poor yield of barley monoculture, the barley-vetch for hay rotation gave the highest barley grain and straw yield. It also had the lowest CV in grain yield. The barley-lentil rotation had the lowest CV in terms of straw yield.

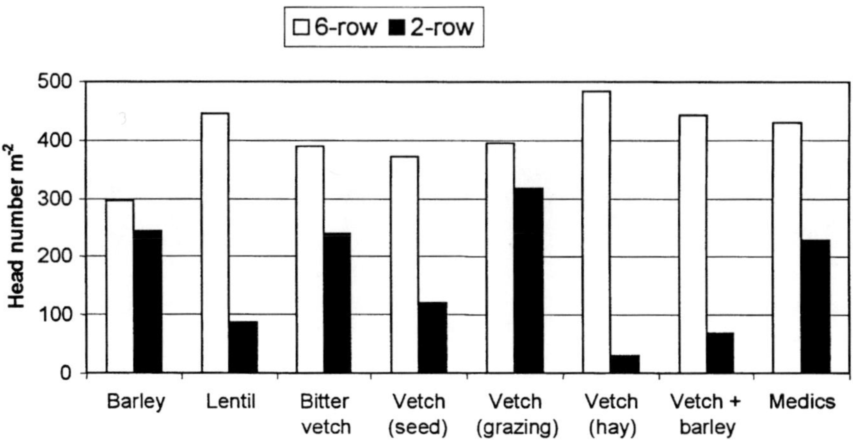

Fig. 14–3. Head numbers of cultivated (Rihane, 6-row) and wild (2 row) barley grown under different rotations in 2002.

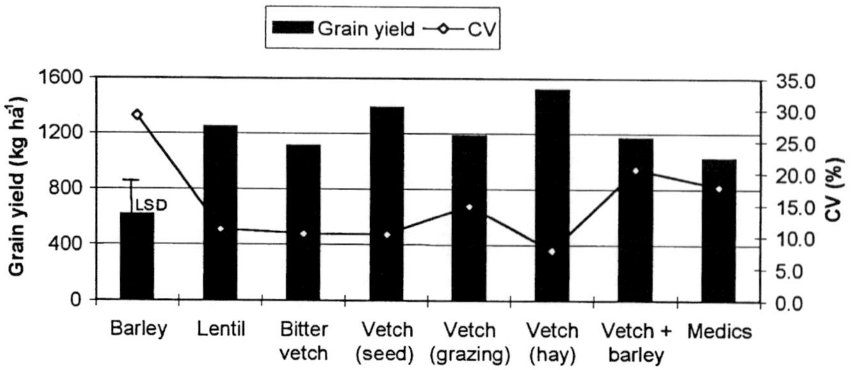

Fig. 14–4. Mean grain yield and its coefficient of variation of barley grown under different rotations, 1997–98 to 2001–02.

In the legume phase, differences in seed yield between rotations were significant, and there was no significant rotation-by-year interaction. Mean seed yields of bitter and common vetch were higher than that of lentil and barley monoculture (Fig. 14–6). Barley monoculture was also lower than lentil, bitter vetch, and common vetch in straw yield. Although the economics of the different rotations has not been studied in detail at this stage, we showed in an earlier report (Yau et al., 2001) that despite the higher cost for hand-harvesting of legume seed, the net income from barley-legume rotations was much higher than that from barley monoculture.

Since two legume treatments were used for grazing and two others were used for hay production, one way to compare the productivity of all the legumes is to compare their dry matter yield. Figure 14–7 shows that all the legume treatments, except medics, yielded higher dry matter than barley monoculture. Barley-bitter vetch and barley-common vetch for seed and straw rotations were the highest yielding, with the latter rotation having the lowest CV.

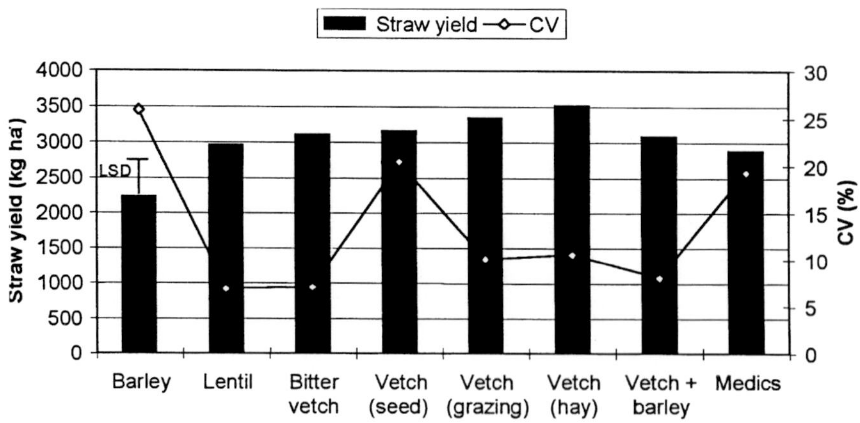

Fig. 14–5. Mean straw yield and its coefficient of variation of barley grown under different rotations, 1997–98 to 2001–02.

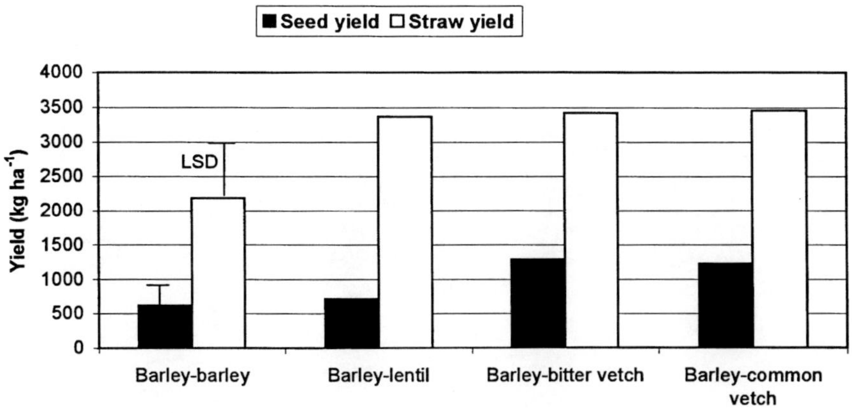

Fig. 14–6. Mean seed and straw yield of lentil, common vetch, and bitter vetch in comparison with barley, 1997–98 to 2001–02.

The rotation main effect on total drymatter yield per cycle was significant, but the rotation-by-year interaction was not. Barley-medic gave poorer yield than the other five rotations (Fig. 14–8), which were not significantly different. The poor performance of the barley-medic rotation was disappointing, which might provide another possible explanation on why projects on introducing ley farming in Syria and other WANA countries failed (Christiansen et al., 2000).

Barley in rotation with common vetch for seed and straw gave the highest yield and the lowest CV (Fig. 14–8). Among the different feed legume crops, we are more inclined to support common vetch. Besides producing the highest and most stable drymatter yield in this study, it is a versatile crop suitable to meet different farmers' demands (Jones and Arous, 1999).

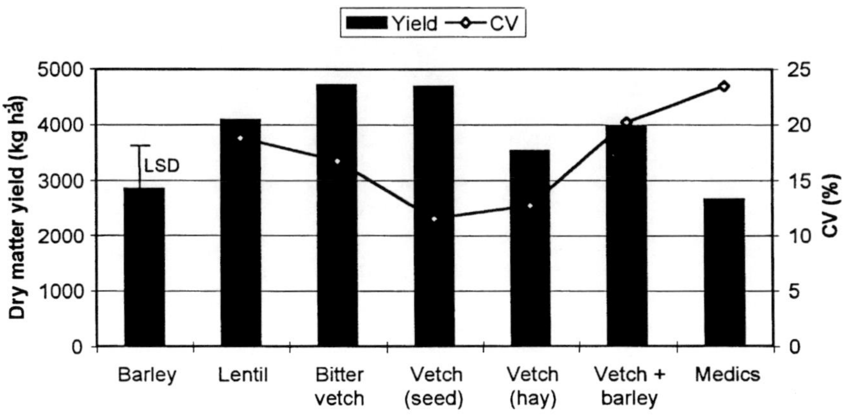

Fig. 14–7. Dry matter yield of the different legumes in comparison with barley, 1997–98 to 2001–02.

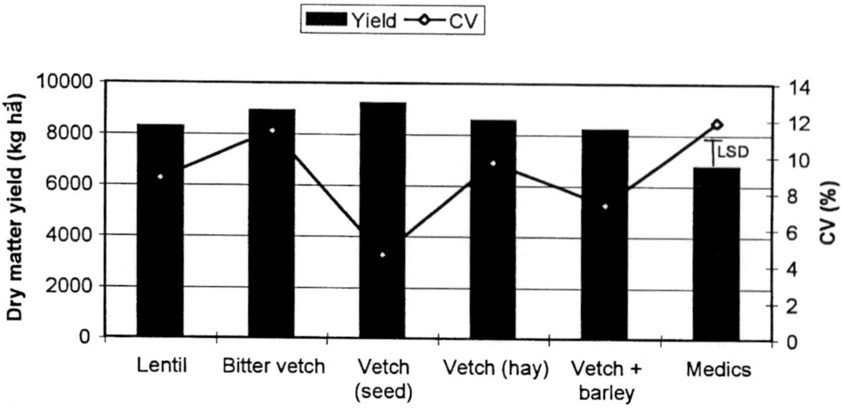

Fig. 14–8. Dry matter yield per cycle in the different barley-legume rotations, 1997–98 to 2001–02.

CONCLUSIONS

The findings of this study from Lebanon regarding continuous cropping of barley coincide with observations elsewhere in the Mediterranean region. While many factors such as disease buildup and reduced fertility contribute to declining barley yields, the infestation of wild barley in this study is an additional factor contributing to the system's unsustainability. The study indicated positive alternatives in that it showed that barley-legume rotations increased barley yield and stability, and yielded higher total drymatter. Thus, farmers in the semi-arid areas of Lebanon should be encouraged to discontinue practicing barley monoculture and adopt a barley-legume rotation, such as barley-vetch. Experience with technology transfer efforts in countries of the region such as Syria suggests a high likelihood of farmers adopting vetch in their cropping systems.

REFERENCES

Belaid, A., and M.L. Morris. 1991. Wheat and barley production in rainfed marginal environments of West Asia and North Africa: Problems and prospects. CIMMYT Economics Working Paper 91/02. CIMMYT, Mexico, DF, Mexico.

Cooper, P.J.M., P.J. Gregory, D. Tully, and H.C. Harris. 1987. Improving water use efficiency in the rainfed farming systems of West Asia and North Africa. Exp. Agric. 23:113–158.

Christiansen, S., M. Bounejmate, H. Sawmy-Edo, B. Mawlawi, F. Shomo, P.S. Cocks, and T.L. Nordblom. 2000. Tah village project in Syria: Another unsuccessful attempt to introduce ley-farming in the Mediterranean basin. Exp. Agric. 36:181–193.

Cocks, P.S., M.J. Matheson, and E.J. Crawford. 1980. From wild plants to pasture cultivars: Annual medics and subterranean clovers in southern Australia. p. 569–596. *In* R.J. Summerfield and A.H. Bunting (ed.) Advances in legume science. Royal Botanic Gardens, Kew, UK.

Genstat 5 Committee. 1993. Genstat 5 Reference Manual. Clarendon Press, Oxford.

Harris, H.C. 1995. Long-term trials on soil and crop management at ICARDA. Adv. Soil Sci 19:447–469.

Hammadeh, S.K., F. Shomo, T. Nordblom, and T. Goodchild. 1994. A rapid survey of small ruminant production in the Bekaa Valley, Lebanon. Small Ruminant Res. 21:173–180.

Jones, M.J. (ed.) 1998. The challenge of production system sustainability. Long-term studies in agronomic research in dry areas. Abstracts of presentations and workshop conclusions. ICARDA, Aleppo, Syria.

Jones, M.J., and Z. Arous. 1999. Effect of time of harvest of vetch (*Vicia sativa* L.) on yields of subsequent barley in a dry Mediterranean environment. J. Agron. Crop Sci. 182:291–294.

Jones, M.J., and M. Singh. 2000. Long-term yield patterns in barley-based cropping systems in northern Syria. 1. Comparison of rotations. J. Agric. Sci. (Cambridge) 135:223–236.

McIntosh, M.S. 1983. Analysis of combined experiments. Agron. J. 75:153–155.

Papastylianou, I. 1990. The role of legumes in the farming systems of Cyprus. p. 39–49. *In* A.E. Osman et al. (ed.) The role of legumes in the farming systems of the Mediterranean areas. Kluwer Academic Publ., Dordrecht, The Netherlands.

Ryan, J., G. Musharrafieh, and A. Barsumian. 1980. Soil fertility characterization at the agricultural research and educational center of the American University of Beirut. Publ. 64. Am. Univ. of Beirut, Beirut.

Van Oosterom, F.J., and A. Acevedo. 1992. Adaptation of barley (*Hordeum vulgare L*) to harsh Mediterranean environments. I. Morphological traits. Euphytica 62:1–14.

Yau, S.K., and J. Hamblin. 1994. Relative yield as a measure of entry performance in variable environments. Crop Sci. 34:813–817.

Yau, S.K., S. Haj Hassan, A. Nassar, and R. Maacaroun. 2001. Grain legumes in rotation with barley: A sustainable cropping system for northern Bekaa, Lebanon. p. 344. *In* Proc. European Conf. on Grain Legumes, 4th, Cracow, Poland. 8–12 July 2001. European Assoc. for Grain Legume Res., Paris.

Yau, S.K., M. Bounejmate, J. Ryan, A. Nassar, R. Baalbaki, and R. Maacaroun. 2003. Barley-legumes rotations for semi-arid areas of Lebanon. Eur. J. Agron. 19:599–610.

15 Cool-Season Grain Legumes Production and Rhizobial Interactions in Australian Dryland Agriculture

Jo Slattery

Rutherglen Research Institute
Rutherglen, Victoria, Australia

Kadambot H. M. Siddique

The University of Western Australia
Crawley, Western Australia, Australia

John Howieson

Murdoch University
Murdoch, Western Australia, Australia

ABSTRACT

In this chapter we review recent advances in cool-season pulse production in Australia and how rhizobial, soil, and environmental factors impact on productivity. Nationally, pulse production has continued to increase to about 2 x 10^6 t yr^{-1}, but in recent years the capacity for nitrogen (N_2) fixation has been limited, especially due to insufficient moisture in 2002 and through the emergence of Ascochyta blight in chickpea *(Cicer arietinum* L.) crops across southern Australia, in addition to abiotic factors such as extremes in soil pH (highly acidic or alkaline soils), temperature, soil moisture, nutrients, and chemical residues have a significant impact on N_2 fixation and pulse production in Australia.

INTRODUCTION

Farming systems in Australia are continually in a state of flux because of the need for diversification, to overcome disease, increasingly higher inputs driven by a demand for increased production and in many situations to cope with specific soil and climate needs. Some of these key changes are reflected in the demand for inclusion of legumes (both grain and pasture legumes) in rotations of varying form and cropping intensity. Legumes are unique in having a special ability to acquire N symbiotically with root nodule bacteria. Since the 1970s, pulse crops have been

Table 15–1. Area and production figures for pulse production in Australia. Source: FAOSTAT, 2003.

Year	Area	Production
	10^3 ha	Tg
1970	51	42 300
1980	189	173 847
1990	1 393	1 353 670
1999	2 229	2 987 600
2000	2 230	2 152 000
2001	2 008	2 507 000
2002	1 842	1 359 000

successfully integrated into Australian farming systems due to their important role in the cereal rotation. In 1970, total pulse production was only 42 300 Tg rising to 2 987 600 Tg in 1999 (FAOSTAT, 2003) (Table 15–1).

Australia's main pulse growing regions are in south Western Australia, South Australia, Victoria, and inland areas west of the Great Dividing Range running from northern Victoria, through New South Wales (NSW) to southern Queensland (Fig. 15–1). The geographic range is vast, with pulses produced under a wide range of environmental conditions. Much of the cropping is under dryland (rainfed) condi-

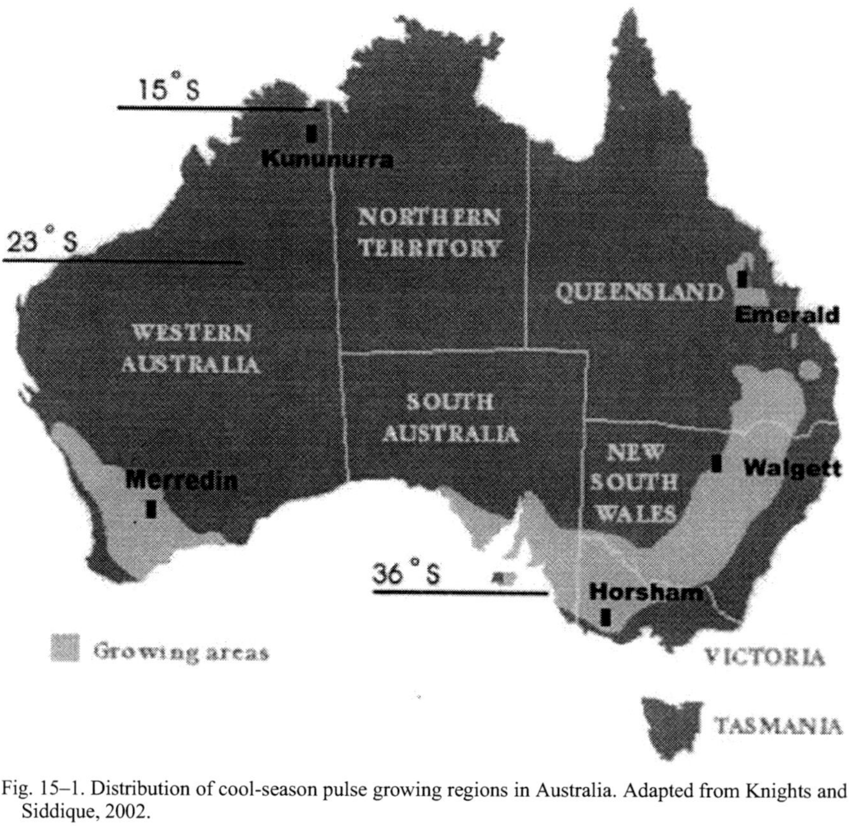

Fig. 15–1. Distribution of cool-season pulse growing regions in Australia. Adapted from Knights and Siddique, 2002.

tions, with the southern region experiencing a Mediterranean-type environment characterized by a predominantly winter rainfall (250–750 mm yr^{-1}) and hot dry summers (Siddique et al., 2000). The soils within these regions are typically low in N and P; rainfall can be low and soil moisture often below optimum for plant growth (Unkovich et al., 1997).

Australian farmers are appreciating the agronomic and financial benefits of including pulse crops in their farming rotations. Some benefits of incorporating legumes into farming systems in Australia were recently summarized by Howieson et al. (2000). The major reasons for the industry's growth can be attributed to the poor returns from traditional continuous cereal crops; recognition of the need to move towards a more sustainable farming system; improvement in soil N reserves; reduction in weeds; and provision of a disease break that rotating cereals with legumes can provide (Reeves et al., 1984; Siddique and Sykes, 1997).

This review examines cool-season pulse production in Australia and the impact of rhizobial, soil, and environmental factors on productivity. We also consider the relationship between the legumes and their symbiotic N_2-fixing root nodule bacteria (the rhizobia) and how the symbiosis in the root nodules function to influence plant productivity and soil fertility.

PULSE SPECIES IN AUSTRALIAN FARMING SYSTEMS

Across Australia, cool-season grain legumes have become increasingly popular in farming systems in the low to medium rainfall (300–500 mm yr^{-1}) cropping regions (Siddique and Sykes, 1997). More recently plant breeding programs have identified species for neutral to alkaline soils in south-western Australia (Siddique et al., 1999) and for the neutral to mildly acidic soils of south-eastern Australia. In Western Australia the research has focused on selecting cool-season grain legumes in soils unsuitable for narrow-leaf lupin (*Lupinus angustifolius* L.). The main pulse species suitable for farming systems identified with a dual role for human consumption are field pea (*Pisum sativum* L.), chickpea, faba bean (*Vicia faba* L.), albus lupin (*Lupinus albus* L.) and lentil (*Lens culinaris* Medik.) (Siddique et al., 1999). Narbon bean (*Vicia narbonensis* L.), *Lathyrus* spp. and vetches (*Vicia* spp.) were also identified to have potential for use in stock feeds.

In Australia, identifying those species with suitable adaptation traits and seed yield is dependent on host environment. Clearly, soil pH, clay content, and rainfall were the soil and environmental factors identified in Western Australia as most crucial in determining seed yields (Siddique et al., 1999) and growing seasonal rainfall often being more important than soil pH in NE Victoria (Table 15–2). Siddique et al. (1999) reported on the performance and adaptation of a number of grain legume species at 36 sites over three growing seasons. Field pea, faba bean, narbon bean, and common vetch were superior in adaptation and grain yield, but profitability of each crop needs also to be taken into account. Producing legume crops for both human consumption and the stockfeed market increased the flexibility and on-farm use of that crop. Lentils and chickpea yields are generally lower, but potential profitability is greater because of the relatively high value of their grain for human consumption export markets.

Table 15–2. The effect of growing seasonal rainfall on crop yield for various pulse crops during 1999 to 2002 when grown on an acid soil (pH_{Ca} 4.7) at Rutherglen, Victoria.

Pulse legume	1999†	2000	2001	2002
			t ha^{-1}	
Lentil	2.95–3.39	0.25–0.50	3.20–3.80	1.41–1.51
Faba bean	3.23–4.38	1.58–2.72	5.33–6.54	1.49–1.66
Field pea	2.87–3.69	1.55–2.36	4.50–5.20	1.45–1.51
Vetch	1.67–2.20	1.93–2.08	Not tested†	Not tested
Narbon bean	1.76–3.16	0.36–0.72	2.50–3.64	Not tested
Growing seasonal rainfall (GSR)	324 mm	423 mm	280 mm	156 mm

† NT = Not tested.

A recent overview of pulse production in Australia (Siddique and Sykes, 1997) has described the importance of pulse crops in the Australian farming systems. Since 1998, improved domestic and export markets, low grain prices for some crops (e.g., lupins) and the recent outbreak of Ascochyta blight (*Ascochyta rabiei*) in chickpea have changed the earlier predicted expansion of chickpea industries. Other grain legume species that have attracted interest among farmers are vetch (*Vicia sativa* L.), narbon bean, and *Lathyrus* spp. Not all species are likely to target the high-value human consumption market. Some *Vicia* and *Lathyrus* species exhibit anti-nutritional factors, accumulate unfavorable low molecular weight peptides in the seed, which have implications for use in human and stockfeed markets. Plant breeding and further selection of these species are required to improve their adaptation and reduce toxin levels before they are widely adopted in Australian farming systems.

Introduction of pulses into a wider range of climatic and soil environments across southeastern Australia are essential for the future expansion of the industry in Australia. Adaptation of pulse species to new regions requires the input of breeders to modify their selection criteria and for agronomists to develop optimum management packages. For example, the soils in southwest and northeast Victoria, southern NSW and Tasmania can be highly acidic with a pH_{CaCl2} within the range of 3.5 and 5.5, and selection of plant species for these soils requires the plant breeding program to specifically select species that tolerate acid soils and higher annual rainfall.

Lupin

Narrow-leaf lupin is well adapted to the deep, coarse-textured, acid to neutral soils, but it is less suited to more alkaline soils (Thompson et al., 1997). Low grain prices relative to other species limits expansion of the lupin industry when lupins are grown on soils that can be used to produce other higher valued grain legume and oilseed crops. The role of lupins in rotations on sand plain soils of Western Australia is being challenged by the development of herbicide-resistant weeds and the emergence of alternatives such as herbicide-tolerant canola and Cadiz yellow Serradella. More options in the rotation are not necessarily detrimental as wider rotation aids weed management and helps ameliorate fungal diseases in lupins. Effective disease and weed management improved regional adaptation and enhanced grain quality, leading to higher returns will help the continued growth of lupin industry in Australia.

Field Pea

Field pea is highly adapted to a wide range of soil types and environments with the ability to produce very high yields when compared to other grain legumes (Siddique et al., 1999). However, fungal disease (black spot), prostrate growth habit, mechanical harvesting difficulties, and the conservative price of human and stock feed markets limit rapid adoption of the crop. Recently, the release of new semi-dwarf, semi-leafless plant types that maintain erectness have been shown to benefit mechanical harvest and improved crop yields. In years where growing seasonal rainfall is limited, field pea is superior in exploiting moisture by exhibiting traits for early flowering and podding characteristics that allow the plant to complete early growth development more efficiently (Thompson et al., 1997; Siddique et al., 1999).

Chickpea

Chickpea production rapidly increased during the 1980s and the 1990s reaching a peak of 309 000 ha in 1998 (Knights and Siddique, 2002). With the emergence of Ascochyta blight in commercial crops, chickpea production has fallen by 90% in southern Australia. However, during the past three seasons production in northern NSW and Queensland has rapidly expanded (Knights and Siddique, 2002) where less foliar fungicide applications are needed to control the disease. Management of Ascochyta blight is costly, with recovery of the industry in the southern states reliant on the breeding program selecting varieties with improved disease resistance. Other biotic stresses that can limit chickpea production are foliar diseases like botrytis grey mold, sclerotina, phytophthora root rot, some virus diseases, nematodes, and pod borer (Knights and Siddique, 2002). A range of available herbicide products can control grass and broadleaf weeds. Terminal drought is the major abiotic constraint limiting chickpea production, with waterlogging causing production loss on the finely textured duplex soils.

Faba Bean

Unlike chickpea and lentils with some tolerance to drought, faba bean has a shallow root system and previously been shown to be sensitive to water stress and high temperatures (Loss and Siddique, 1997). In a recent study, faba bean produced high biomass and seed yields when grown across a range of dryland Mediterranean environments in south Western Australia (Loss and Siddique, 1997) and early sowing is also critical when faba bean is grown in low rainfall regions. In contrast, most other faba bean studies were conducted in areas where average rainfall exceeded 400 mm per annum (Carter et al., 1995). It has also been demonstrated that pulses with hypogeal emergence, for example, chickpea, faba bean, field pea, and lentil can tolerate sowing depths of 5 to 10 cm. Deep sowing can provide benefits to pulse crops resulting in improved crop establishment, growth, nodulation, and seed yield (Siddique and Loss, 1999).

Lentil

Lentil introduction has expanded from a modest crop in 1994 to an industry of 125 000 ha producing 180 000 Tg in 2001. Increased issues relating to agronomy, cultivars, disease, or quality have become evident with this rapid expansion in cropped areas. Improved understanding of the suitability of lentil cultivars to specific regions is based on the phenological responses and interactions with disease (Materne et al., 1999). Sowing rate, disease control, and physical damage to seed caused by mechanical harvest and post-harvest handling have limited lentil adaptation and production in Australia (Siddique et al., 1998). Soil pH limits lentil production with current research now investigating high yielding varieties and specific packages for different regions.

NITROGEN FIXATION AND PULSE PRODUCTION

The concept that legume N_2 fixation can be enhanced through selection and breeding has been recently reviewed (Herridge and Rose, 2000) with three general strategies proposed for increasing legume N_2 fixation through breeding. The strategies of most relevance to the expansion of pulses in Australian dryland agriculture are maximizing legume biomass and seed yield according to the environment and agronomic management and the optimizing of legume nodulation. In a symbiotic relationship the plant is often regarded as the dominant partner in the symbiosis between the legume and the bacteria. No matter how effective, competitive, and persistent a strain of rhizobia might be, it cannot realize its full capacity for N_2 fixation if limiting factors such as plant disease, nutrient deficiency, mineral toxicity (e.g., Al, Mn), salinity, unfavorable pH, weed competition, temperature extremes, insufficient or excessive soil moisture, inadequate photosynthesis, or the influence of grazing and management practices impose limitations on the vigor of the host legumes (Brockwell et al., 1995a; Peoples et al., 1995; Thies et al., 1991).

Soil Moisture

Many studies have shown growth potential and N_2 fixation to be adversely affected by a lack of moisture, or conversely, excess moisture in water-logged soils at critical periods of plant growth. Rainfall can be variable in terms of its timing or abundance during the growing season or at grain filling of crops late in the season. These considerations need to be addressed when plants are bred and developed for specific regions. In Mediterranean climates, higher temperatures with large soil water deficits during spring and early summer are the main contributors to the rapid termination of the growing season, plant production, and N_2 fixation in annual legumes. Low growing seasonal rainfall and extended dry periods across most of Australia during 2002 reduced total pulse production by 34% compared to production in 2001 (Table 15–2). Over a 4-yr period at a site near Rutherglen in Victoria, low rainfall in 2002 and high rainfall in 2000 reduced pulse production compared to the yields obtained in 1999 and 2001, when seasonal rainfall was close to the long-term average.

Temperature

Soil temperature can affect legume nodulation and N_2 fixation of cool-season pulses. Legumes grown at low temperatures experience delays in the formation of nodules and the onset of N_2 fixation (Peltzer et al., 2002). Even if background rhizobial populations are present, this delay in nodulation can impact on N_2 fixation and ultimately affect growth and seed yield. The major symptoms of cold in cool-season pulses are reduced leaf production, area expansion, poor dry matter production, chlorosis, and, finally, necrosis of the older leaves. These symptoms are generally observed in cool-season chickpea in Australia. A daily average temperature between 0 and 10°C is considered a threshold level for cold or chilling stress in cool-season pulses. Chickpea is more sensitive to chilling injury than field pea, faba bean, lentil, or lupin (Siddique, 2000). Increased seedling deaths from root disease are also a major consequence of low soil temperatures (Brockwell et al., 1995b). Frost damage in spring (radiation frost) is a major cause of yield loss in pulses and crops in general in Australia. Flowering, early pod formation, and seed filling are the most sensitive stages in pulses.

Unfavorable pH

The primary soil factors affecting legume production as well as the growth and survival of *Rhizobium* spp include the extremes of soil pH; acidity, alkalinity, and also salinity. Continued expansion and widespread adoption of pulses across Australia is restricted and limited to tolerance of nodulation and N_2 fixation of these species to acid soils (Howieson et al., 2000). With soil acidity affecting 90 million ha of agriculturally productive land (AACM, 1995) in Australia, and approximately one third of this land having a $pH_{CaCl2} < 5.0$ (Chartres et al., 1990), acidity can be described as a major problem affecting the symbiosis and the subsequent productivity of many legumes. Exceptions to this generalization include the acid-adapted *Ornithopus* and *Lupinus* symbioses with bradyrhizobia (Howieson et al., 1998).

It is well recognized that many Australian soils are naturally acidic, but certain agricultural practices are further acidifying the soil, with legume-based agriculture contributing to the process. Previous work has shown that continuous subterranean clover pastures have caused a decline in soil pH of one unit (Williams, 1980; Ridley et al., 1990), while lupin crops have lowered the soil pH by 1.5 units (Dolling, 1995; Slattery et al., 1998).

Legumes and rhizobia are both susceptible to soil pH outside their normal range. A schematic presentation illustrates the pH effect on *Rhizobium* activity including the effects on *Rhizobium* survival on nodulation and on plant growth (Slattery et al., 2001). Specific effects on the symbiosis can be attributed to low rhizobial populations due to excessive H^+, Mn^+, and Al^{+3}; inhibition of root hair growth and reduction in root growth (Unkovich et al., 1997). Acid soils provide an enormous challenge to rhizobial survival (Dilworth et al., 2001) and the nodulation process and establishment and functioning of the legume symbiosis (Howieson et al., 2000). In Australia, we have a selection program for *Rhizobium* germplasm for acidic soils and in conjunction with the pulse breeding programs new breeding lines are evaluated against improved *Rhizobium* germplasm.

Chickpea is an example of a cool-season pulse with a highly specific plant-*Mesorhizobium ciceri* symbiosis that has adapted well to acidity and fertility stresses, even though it has a highly successful history on alkaline soils in the Indian subcontinent, West Asia, and North Africa (Howieson et al., 2000). Rai (1991) demonstrated that only 5% of rhizobial strains examined in this symbiosis were suitable for nodulation and growth in strongly acid environments, but once acid-tolerant strains were identified, these strains proved to be robust. Poor persistence of *Rhizobium leguminosarum* bv *viciae* in acid soils across southern Australia is demonstrated by low nodulation score and poor plant growth (Carter et al., 1995; Slattery et al., 2003) and reflects the need for finding strains with more tolerance to acid conditions.

On alkaline soils, background *Rhizobium leguminosarum* bv *viciae* populations persist well, with pulse production reliant on adequate soil moisture (Slattery et al., 2003). Plant species are more sensitive to high pH or high bicarbonate levels when grown in soils where the pH is above 6.0. Reduced nodulation and N_2 fixation can be evident in alkaline-sensitive legume species. Cool-season pulses are also sensitive to acid soil conditions and require a neutral to alkaline soil pH for their optimum growth and yield. Among the cool-season pulses, lentil is the most sensitive to low pH, followed by chickpea, faba bean, and field pea. A drop of one pH unit below the threshold value (pH_{CaCl2} 5.5) can cause more than an 86% reduction in seed yield of lentil. Cool-season pulses generally appear to be more sensitive to acidity than cereals (Siddique, 2000).

Legume Nutrition

In legumes, mineral nutrients can limit N_2 fixation through direct effects upon plant growth and the rhizobia or through indirect effects upon the symbiosis (O'Hara et al., 1989; Peoples et al., 1995; Slattery et al., 2001; O'Hara et al., 2001). Current evidence shows that of the 18 essential nutrients (macronutrients: C, O, N, H, P, S, K, Ca, Mg, Fe, and micronutrients: Mn, Cu, Zn, Mo, Co, Ni, Se, and B) with the exception of B, all have a requirement for enzyme function and only a few nutrients have a function for energy transfer and signal regulation. Those nutrients described as having a specific role in nodulation and/or N_2 fixation, as summarized by O'Hara (2001), are P, K, Ca, Fe, Cu, Zn, Mo, Co, Ni, Se, and B.

The soil fertility impacts on N_2 fixation in Australian soils are primarily Al and Mn associated with soil acidity, P and Ca deficiencies on the less fertile soils, often those with low clay content and low cation exchange capacity. Phytotoxic concentrations of Al and Mn impact directly by inhibiting root or shoot growth (Helyar, 1987; Slattery and Coventry, 1999). In nutrient-depleted soils in Tanzania, the addition of P fertilizer dramatically increased *Phaseolus vulgaris* nodulation and grain yield (Giller et al., 1998). Similarly, in the Victorian Mallee, the response to nodulation on *Vicia narbonensis* by the addition of 30 kg P ha^{-1} was an increase in nodulation from 0 to 3.2 and plant dry matter increases from 1.1 to 2.4 g $plant^{-1}$. This same response to P can be shown for pasture legumes. Where P was added to pastures in soils deficient in P, marked increases in subterranean clover composition and production occurred (Peoples et al., 1998).

Boron toxicity is recognized as a widespread constraint to crop production, particularly in the low rainfall areas of southern Australia. In these areas, B toxicity is caused by a high B concentration in the subsoil. Field observations and limited glasshouse studies suggest that the majority of the commercial cultivars of chickpea, lentil, field pea, and faba bean are highly sensitive to B toxicity. Faba bean and field pea, however, appear to be relatively more tolerant to B toxicity compared with lentil and chickpea (Siddique, 2000).

Residual Chemicals in the Soil

With the continued expansion of pulse production and the need for higher inputs to produce the crop, highly selective herbicides, fungicides, and pesticides are essential for weed, pest, and disease management in crop rotations. In many cases, farmers have become reliant on seed-dressing chemicals for insect or disease control or pre- and post-emergent application for the control of weeds. Herbicide residues have been known to reduce N_2 fixation of subsequent crops and pastures, especially where edaphic and climatic conditions have prevented rapid chemical breakdown (Ferris et al., 1992; Douka et al., 1995; Slattery et al., 2001). A more recent review investigates residue chemical breakdown with particular reference to Australian alkaline soils (Sarmah et al., 1998).

Studies on the effects of residual herbicides or pesticides on N_2-fixing symbiosis is complex and dependent on a range of factors and interactions including the active constituent, application rate, climate, soil characteristics as well as the legume species studied. Low rainfall, alkaline soils, and drought years especially can contribute towards slow degradation of soil chemical residues (Unkovich et al., 1997). The sulfonylurea (triasulfuron, metsulfuron-methyl, and chlorsulfuron) herbicides are a group of compounds designed to control broad-leaved weeds and some grasses for a variety of crops. Low rates (10–40 g ha^{-1}) and ease of application have made these herbicides popular with farmers. There are numerous factors influencing the breakdown and sorption of the sulfonylureas with soil pH, soil organic matter, temperature, water content, and microbial populations considered the main influences. Evidence from the literature suggests that the degradation process is not a single factor but a combination of several factors with sorption of sulfonylureas having a strong correlation with soil pH (Sarmah et al., 1998).

The impact of chemical residues on N_2 fixation and plant growth is varied. Chemical residues can reduce rhizobial growth, nodulation, N_2 fixation or yield in grain legumes, with the level of inhibition dependent on strain differences (Martinez et al., 1996; Durgesha, 1994). In northwest Victoria, concern is often expressed about the direct effect of chemical residues on the legume, the rhizobia, or the symbiosis. The soils in this region are alkaline soil (soil pH_w between 7 and 9) and paddocks are often continually cropped. In a recent study, lentil plants were assessed for nodulation and plant growth from a site where sulfonylurea chemicals had been applied as treatments over the previous 3 yr. The toxicity of the chemical residue was evident when compared with the nil chemical treatment, as shown by the low nodulation (nodule score and number) and reduced plant growth (Table 15–3). Similar observations have also been recorded with chickpea on alkaline soils of NSW

Table 15–3. Nodulation (nodule number and nodule score) and herbage dry matter (g plant^{-1}) for lentil and medic plants growing in a soil at Horsham, Victoria (1999) following the application of Chlorosulfuron to the soil in the previous years (1996, 1997, and 1998).

Legume	Treatment/Year	Nodule number	Nodule score	Herbage yield
Lentil	Control	39	3.89	4.28
	Chlorosulfuron 1998	12	2.33	0.55
	Chlorosulfuron 1997	25	3.55	1.45
	Chlorosulfuron 1996	40	4.37	2.82
LSD ($P < 0.05$)		14	0.95	1.16
Medic	Control	11	2.22	7.36
	Chlorosulfuron 1998	7	2.28	0.34
	Chlorosulfuron 1997	13	2.58	1.75
	Chlorosulfuron 1996	11	2.28	3.28
LSD ($P < 0.05$)		8	0.96	2.41

and Western Australia. Clearly, when chemicals are used there is a need to consider their long-term effects on the soil and to subsequent crops.

RHIZOBIAL SELECTION AND DEVELOPMENT

Prior to European settlement, Australian soils contained no indigenous rhizobia capable of effective symbiosis with the majority of our introduced pulse legumes. Since then, many strains capable of infecting pulse crops have been introduced to the soil as seed inoculant. Today, populations of naturalized rhizobia in the alkaline soils of the Victorian Mallee, the major Victorian pulse-growing regions are present in sufficient numbers for effective nodulation of the host legume (Slattery et al., 2003). Even so, with the expansion of pulses into a new region with no previous pulse history, inoculation will remain necessary, as few of the soils will contain sufficient background rhizobia. In hostile soil environments, the potential for N_2 fixation is limited, as rhizobia are often absent or those present are ineffective. Low soil pH has been shown to affect the viability of *Sinorhizobium meliloti* (Brockwell et al., 1991) and persistence and distribution of *Rhizobium leguminosarum* bv *trifolii* (Richardson and Simpson, 1988; Slattery and Coventry, 1995). As such, the expansion of pulses into acid soil environments must consider the selection of appropriate *Rhizobium* germplasm and ensure that it is highly compatible with both the legume and target environment. Growing pulses in new and varied soil types has serious implications for *Rhizobium* survival, effective nodulation, and the need for re-inoculation of following pulse crops.

Germplasm Acquisition

Previous rhizobial germplasm collections have been focused in the Mediterranean region, a region with a climatic homology aligned to southern Australia and used as a major consideration when seeking legumes for use in Australian plant improvement programs (Howieson, 1995). More recently, edaphic homology is considered necessary with recent pulse rhizobial germplasm collections occurring in the major pulse regions of the Victorian Mallee where background populations are high.

Selecting Matching Rhizobial Strains

In recent years, considerable effort has been spent matching the root nodule bacteria to both the legumes and the intended soil environment (Howieson et al., 1995; Slattery and Coventry, 1995; Howieson, 1999; Howieson et al., 2000; Slattery et al., 2001; Slattery and Pearce, 2002). Certain characteristics are desirable for strains to be used as an inoculant. These include an ability to colonize the soil environment targeted for the host, competitiveness with the background populations, an ability to form effective symbiosis with the host legumes, a lack of deleterious effects on nontarget hosts, and genetic stability in culture, storage, and the soil (Brockwell et al., 1995a; Howieson 1999).

Strains of rhizobia differ widely in their ability to survive, form nodules, and fix N in soil environments, and for this reason considerable attention has been given to selecting rhizobia with specific symbiotic and competitive attributes suited to a range of soil environments. For example, screening of germplasm collected from lentils growing in mildly acidic soils (pH_w 6.0) in Nepal may prove useful for the future expansion of lentils into more marginal soils of Australia (Slattery and Pearce, 2003).

Authentication and screening of the rhizobial isolates for genetic stability and N_2 fixation in glasshouse conditions allows the processing of multiple isolates under controlled conditions in a sterile environment (Howieson et al., 2000; Slattery and Pearce, 2002). Laboratory screening of strains for their suitability in a wide range of soil pH conditions needs to be undertaken in parallel with plant assays and field experimentation, especially when studying the effects of specific soil toxicities or deficiencies on *Rhizobium* growth and survival. In the past, considerable variation between the sensitivity of rhizobial growth on defined medium and under plant assays has occurred (Slattery et al., 2001). In a more recent study, results obtained from laboratory screening for acid-tolerance, supported the glasshouse study whereby the selected strains formed an effective symbiosis with acid-tolerant lentil lines compared to the symbiosis between these lines and the commercial inoculant strain (Slattery and Pearce, 2003).

In a recent review, Sessitschet et al. (2002) summarized technologies and strategies for selecting quality inoculant strains by examining more closely the complex interaction between the edaphic environment with the genotypes of both the legume and its microsymbiont. This step-by-step process used a flow chart to illustrate the range of decisions required for consideration prior to initiation of a legume or rhizobial selection program. Within this process, improving the competitive ability of the inoculant strain improves successful colonization of plant roots, nodule formation, and then subsequently N_2 fixation.

Legume Inoculation

Inoculation of legumes with *Rhizobium* bacteria is a well-established practice (Herridge et al., 2002). Even with a quality product, the inoculation process is not always successful. Using elite strains as inoculants or selection of strains with specific host requirements with more tolerance to hostile soil environments is one way of establishing well-nodulated pastures (Howieson and Ewing, 1986). More often, responses to inoculation are reliant on background rhizobial populations being absent (when a new legume is first sown in a paddock), or present only in small numbers (Brockwell et al., 1995b). If populations of ineffective rhizobia are present in the soil, then introduction of an inoculant strain can be difficult due to competition. When the inoculant strain is outnumbered by the background populations, inoculant responses and establishment of the inoculant strain rarely occurs in pasture or pulse legumes (Riffkin et al., 1999; Slattery et al., 2003).

CONCLUSIONS

This review highlights the recent developments in the adaptation of cool-season pulse crops to a wider range of soil types and environments in Australia. In most areas, grain legumes crops are being developed, and inclusion of legumes into the farming system provides farmers with rotational benefits to following cereal and oilseed crops. Further selection and breeding programs are developing cultivars more suited to marginal environments and are also investigating the seed toxicity issues associated with some species. Development and promotion of variety and species combinations with specific agronomic packages for various regional environments will further enhance the adoption of cool-season pulses in Australia. Rhizobial inoculants contribute greatly to the N_2-fixation process, whereby appropriate strains selected must effectively nodulate their host and best suit the edaphic environment. The interaction of the rhizobia with the legume is an integral part of the N_2-fixation process and for this reason rhizobial research in Australia aligns closely with the pulse-breeding programs.

REFERENCES

AACM International. 1995. Social and economic feasibility of ameliorating soil acidification, a national review. Land and Water Resources Res. and Develop. Corp., Canberra Australia.

Brockwell, J., P.J. Bottomley, and J.E. Thies. 1995a. Manipulation of rhizobia microflora for improving legume productivity and soil fertility: A critical assessment. Plant Soil 174:143–180.

Brockwell, J., R.R. Gault, M.B. Peoples, G.L. Turner, D.M. Lilley, and F.J. Bergersen. 1995b. Nitrogen fixation in irrigated lucerne grown for hay. Soil Biol. Biochem. 27:589–594.

Brockwell, J., A. Pilka, and R.A. Holliday. 1991. Soil pH is a major determinant of the numbers of naturally occurring *Rhizobium meliloti* in non-cultivated soils in central New South Wales. Austr. J. Exp. Agric. 31:211–219.

Carter, J.M., J.S. Tieman, and A.H. Gibson. 1995. Competitiveness and persistence of strains of rhizobia for faba bean in acid and alkaline soils. Soil Biol. Biochem. 27:617–623.

Chartres, C.J., R.W. Cummings, J.A. Beatie, G.M. Bowman, and J.G Wood. 1990. Acidification of soils on a transect from plains to slopes, south-western NSW. Aust. J. Soil Res. 28:539–549.

Dilworth, M.J., J.G. Howieson, W.G. Reeve, R.P. Tiwari, and A.R. Glenn. 2001. Acid tolerance in legume root nodule bacteria and selecting for it. Austr. J. Exp. Agric. 41:435–446.

Dolling, P.J. 1995. Effect of lupins and location on soil acidification rates. Austr. J. Exp. Agric. 35:753–763.

Douka, C.E., J. Doskaris, L. Protopapadaki, S. Visantinopoulos, and A.C Xenoulis. 1995. Relationship between biological nitrogen fixation and herbicide degradation. p. 679. *In* Proc. of the Int. Nitrogen Fixation Conf.

Durgesha, M.H. 1994. Effect of dinitroaniline herbicides on rhizobia, nodulation and nitrogen (C_2H_2) fixation of four groundnut cultivars. Ann. App. Biol. 124:75–82.

FAOSTAT. 2003. Data bases website: http:/www.fao.org/waicent/portal/statistics_en.asp (verified 3 Feb. 2003).

Ferris, I.G., R.N. Pederson, M.W. Schwinghamer, and B.M. Haigh. 1992. Sulfonylurea herbicides in cereal farming systems-detection, persistence and impact. p.1–11. *In* D.J. Reuter and S.D. Bowen (ed.) Proc., Natl. Workshop on Subsoil Constraints to Root Growth and High Water Use Efficiency by Plants, Barossa Valley, South Australia. August 1992. CSIRO, Adelaide.

Giller, K.E., F. Amijee, S.J. Brodick, and O.T. Edje. 1998. Environmental constraints to nodulation and nitrogen fixation of *Phalseolus vulgaris* L. in Tanzania. II. Response to N and P fertilisers and inoculation with *Rhizobium*. Afr. Crop Sci. J. 6:171–178.

Helyar, K.R. 1987. Nutrition of plants on acid soils. p. 159–171. *In* J.L. Wheeler et al. (ed.) Temperate pastures, their production, use and management. CSIRO Publ., Melbourne, Victoria, Australia.

Herridge, D.R., G. Gemell, and E. Hartley. 2002. Legume inoculants and quality control. p. 105–115. *In* D.R. Herridge (ed.) Inoculants and nitrogen fixation of legumes in Vietnam, ACIAR Workshop Proc. No. 109e, Canberra, Australia. 17–18 Oct. 2001. ACIAR, Canberra.

Herridge, D.R., and I. Rose. 2000. Breeding for enhanced nitrogen fixation in crop legumes. Field Crops Res. 65:229–248.

Howieson, J.G. 1995. Rhizobial persistence and its role in the development of sustainable agricultural systems in Mediterranean agriculture. Soil Biol. Biochem. 27:603–610.

Howieson, J.G. 1999. The host-rhizobia relationship. p. 96–106. *In* S.J. Bennett and P.C. Cocks (ed.) Genetic resources of Mediterranean pasture and forage legumes. Kluwer Academic Press, Dordrecht, The Netherlands.

Howieson, J.G., and M.A. Ewing. 1986. Acid tolerance in the *Rhizobium meliloti-Medicago* symbiosis. Aust. J. Agric. Res. 37:55–64.

Howieson, J.G., I.R.P. Fillery, A.B. Legocki, M.M. Sikorski, T. Stepkowski, F.R. Minchin, and M.J. Dilworth. 1998. Nodulation, nitrogen fixation and nitrogen balance. p.149–180. *In* J.S. Gladstones et al. (ed.) Lupins as crop plants: Biology, production and utilization. CAB Int., Cambridge, UK.

Howieson, J.G., A.C. Loi, and S.J. Carr. 1995. *Biserrula pelecinus* L.—A legume pasture species with potential for acid duplex soils which is nodulated by unique root-nodule bacteria. Aust. J. Agric. Res. 46:997–1009.

Howieson, J.G., G.W. O'Hara, and S.J. Carr. 2000. Changing roles for legumes in Mediterranean agriculture: Developments from an Australian perspective. Field Crops Res. 65:107–122.

Knights, E.J., and K.H.M. Siddique. 2002. Chickpea status and production constraints in Australia. p. 33–41. *In* M.A. Bakr et al. (ed.) Integrated management of botrytis grey mould of chickpea in Bangladesh and Australia: Summary Proc. of a Project Inception Workshop, Gazipur, Bangladesh. 1–2 June 2002. Bangladesh Agricultural Research Institute (BARI), Joydebpur, Gazipur, and Centre for Legumes in Mediterranean Agriculture (CLIMA), Crawley, Western Australia, Australia.

Loss, S.P., and K.H.M. Siddique. 1997. Adaptation of faba bean (*Vicia faba* L.) to dryland Mediterranean-type environments. I. Seed yield and yield components. Field Crops Res. 52:17–28.

Materne, M.A., J.B. Brouwer., and J. Hamblin. 1999. The flowering response of Australian lentil cultivars. p 53–54. *In* P. Langridge et al. (ed.) Proc., 11th Australian Plant Breeding Conf., Vol. 2, Adelaide, South Australia. 19–23 Apr. 1999. CRC for Molecular Plant Breed., Univ. of Adelaide, Glen Osmond, South Australia.

Martinez, J., D. Vidal, and E. Simon. 1996. Nitrogenase activity, nodulation and seed production in *Vicia faba* as affected by methabenzthiazuron. J. Agric. Sci. (Cambridge) 127:319–324.

O'Hara, G.W. 2001. Nutritional constraints on root nodule bacteria affecting symbiotic nitrogen fixation: A review. Austr. J. Exp. Agric. 41:417–433.

O'Hara, G.W., N. Boonkerd, and M. Dilworth. 2001. Mineral constraints to nitrogen fixation. Plant Soil 108:93–110.

O'Hara, G.W., T.J. Goss, M. Dilworth, and A. Glenn. 1989, Maintenance of intracellular pH and acid tolerance in *Rhizobium meliloti*. Appl. Environ. Microbiol. 55:1870–1876.

Peltzer, S.C., L.K Abbott, and C.A. Atkins. 2002. Effect of low root-zone temperature on nodule initiation in narrow-leafed lupin (*Lupinus angustifolius* L.). Aust. J. Agric. Res. 53:355–365.

Peoples M.B., R.R. Gault, G.J. Scammell, B.S. Dear, J. Virgona, G.A. Sandral, J. Paul, E.C. Wolfe, and J.F. Angus. 1998. The effect of pasture management on the contributions of fixed N to the N-economy of ley-farming systems. Aust. J. Agric. Res. 49:459–474.

Peoples, M.B., J.K Ladha, and D.F. Herridge. 1995. Enhancing legume N_2 fixation through plant and soil management. Plant Soil 174:83–101.

Rai, R. 1991. Effects of soil acidity factors on interaction of chickpea (*Cicer arietinum* L.) genotypes and *Rhizobium* strains: Symbiotic N-fixation, grain quality and grain yield in acid soils. p.597–601. *In* R.J. Wright et al. (ed.) Proc., 2nd Int. Symp. on Plant-Soil Interactions at Low pH. Kluwer Academic Publ., Dordrecht, The Netherlands.

Reeves, T.G., A. Ellington, and H.D. Brooks. 1984. Effects of lupin-wheat rotations on soil fertility, crop disease and crop yields. Aust. J. Exp. Agric. Anim. Husb. 24:595–600.

Richardson, A.E., and R.J. Simpson. 1988. Enumeration and distribution of *Rhizobium trifolii* under a subterranean clover-based pasture growing in an acid soil. Soil Biol. Biochem. 20:431–438.

Ridley, A.M., K.R. Helyar, and W.J. Slattery. 1990. Soil acidification under subterranean clover (*Trifolium subterraneum* L.) pastures in north-eastern Victoria. Aust. J. Exp. Agric. 32:1061–1068.

Riffkin, P.A., P.E. Quigley, G.A. Kearney, F.J. Cameron, R.R. Gault, M.B. Peoples, and J.E. Thies. 1999. Factors associated with biological nitrogen fixation in dairy pastures in south-western Victoria. Aust. J. Agric. Res. 50:261–272.

Sarmah, A.K., R.S. Kookana, and A.M. Alston. 1998. Fate and behaviour of trisulfuron, metsulfuron-methyl, and chlorsulfuron in the Australian soil environment: A review. Aust. J. Agric. Res. 49:775–790.

Sessitsch, A., J.G. Howieson, X. Perret, H. Antoun, and E. Martinez-Romero. 2002. Advances in *Rhizobium* research. Crit. Rev. Plant Sci. 21:323–378.

Siddique, K.H.M. 2000. Understanding growth stresses of cool season pulses. p.12–16. *In* L. O'Connell (ed.) Shared Solution manual. Australian Grain. Berekua Pty. Ltd., Brisbane.

Siddique, K.H.M., R.B. Brinsmead, R. Knight, E.J. Knights, J.G. Paull, and I.A. Rose. 2000. Adaptation of chickpea (*Cicer arietinum* L.) and faba bean (*Vicia faba* L.) to Australia. p.289–303. *In* R. Knight (ed.) Linking research and marketing opportunities for pulses in the 21st century. Kluwer Academic Publ., Dordrecht, The Netherlands.

Siddique, K.H.M., and S.P Loss. 1999. Studies on sowing depth for chickpea (*Cicer arietinum* L.), faba bean (*Vicia faba* L.) and lentil (*Lens culinaris* Medik) in a Mediterranean-type environment of south-western Aust. J. Agron. Crop Sc. 182:105–112.

Siddique, K.H.M., S.P Loss, D.L. Pritchard, K.L. Regan, D. Tennant, R.L. Jettner, and D. Wilkinson. 1998. Adaptation of lentil (*Lens culinaris* Medik.) to Mediterranean environments: Effect of time of sowing on growth, yield and water use. Aust. J. Agric. Res. 49:613–626.

Siddique, K.H.M., S.P Loss, K.L. Regan, and R.L. Jettner. 1999. Adaptation and seed yield of cool season grain legumes in Mediterranean environments of south-western Australia. Aust. J. Agric. Res. 50:375–387.

Siddique, K.H.M., and J. Sykes 1997. Pulse production in Australia past, present and future. Aust. J. Exp. Agric. 37:103–111.

Slattery, J.F., and D.R. Coventry. 1995. Acid-tolerance and symbiotic effectiveness of *Rhizobium leguminosarum* bv *trifolii* isolated from subterranean clover growing in permanent pastures. Soil Biol. Biochem. 25:1725–1730.

Slattery, J.F., and D.R. Coventry. 1999. Persistence of introduced strains of *Rhizobium leguminosarum* bv *trifolii* in acidic soils of north-eastern Victoria. Aust. J. Exp. Agric. 39:829–837.

Slattery, J.F., D.R. Coventry, and W.J. Slattery. 2001. Rhizobial ecology as affected by the soil environment. Aust. J. Exp. Agric. 41:289–298.

Slattery, J.F., and D.J. Pearce. 2002. Development of elite inoculant strains of *Rhizobium* in southeastern Australia. p. 86–94. *In* D.R. Herridge (ed.) Inoculants and nitrogen fixation of legumes in Vietnam. ACIAR Workshop Proc.109e, Canberra. 17–18 Oct. 2001. ACIAR, Canberra.

Slattery, J.F., and D.J. Pearce. 2003. Selection of *Rhizobium* strains for improved lentil production in Nepal. *In* A. Sarker (Ed.) Lentil improvement in south Asia. Workshop Proc. 24–25 Feb. 2003. ICARDA, Aleppo, Syria.

Slattery, J.F., D.J. Pearce, M. Raynes, D. Carpenter, G. Dean, and M. Materne. 2003. Variation in yield of faba bean across southern Australia following rhizobial inoculation (CD). *In* Solutions of a better environment. Proc., 11th Australian Agron. Conf., Geelong, Victoria, Australia, Australian Soc. Agron.

Slattery, W.J., D.G. Edwards, L.C. Bell, D.R. Coventry, and K.R Helyar. 1998. Soil acidification and the carbon cycle in a cropping soil of north-eastern Victoria. Aust. J. Soil Res. 36:273–290.

Thies, J.E., P.W. Singleton, and B. Bohlool. 1991. Influence of the size of indigenous rhizobial populations on establishment and symbiotic performance of introduced rhizobia on field grown legumes. Appl. Environ. Microbiol. 57:19–28.

Thompson, B.D., R.W. Bell, and M.D.A Bolland. 1997. Low seed phosphorus concentration depresses early growth and nodulation of narrow-leafed lupin (*Lupinus angustifolius* cv. Gungurru). J. Plant Nutr. 15:1193–1214.

Unkovich, M.J., J.S. Pate, and P. Sandford. 1997. Nitrogen fixation by annual legumes in Australian Mediterranean agriculture. Aust. J. Agric. Res. 48:267–293.

Williams. C.H. 1980. Soil acidification under clover pasture. Aust. J. Exp. Agric. Anim. Husb. 20:561–567.

16 Forage Legumes for Dryland Agriculture in Central and West Asia and North Africa

Ali M. Abd El Moneim and John Ryan
ICARDA
Aleppo Syria

ABSTRACT

The agriculture of the Mediterranean zone is dominated by rainfed cereal cultivation in conjunction with livestock rising. As legumes are indigenous to the region, they are alternated with cereals and provide an essential source of biological nitrogen (N) for the cereals. Food legumes such as Kabuli-type chickpea (*Cicer arietinum* L.) and lentil (*Lens culinaris* Medik.) are basic foods for humans, while annual and perennial forage legumes provide valuable animal feed as grazing and hay. Since its inception, research at the International Center for Agricultural Research in the Dry Areas (ICARDA) has focused on development of forage legumes in association with the national agricultural systems of the West Asia–North Africa region. While much effort has been expended on perennial self-regenerating medic (*Medicago* spp.), adoption was limited because of technical difficulties. The greatest potential was with annual vetch (*Vicia* spp.) as a viable animal feed source and a rotation crop with cereals. Emphasis was also placed on forage legumes that survive harsh conditions by their unique underground growth habit, for example, *V. amphicarpa* and *Lathyrus ciliolatus*. Efforts to improve forage legumes were based on both management/cultural factors and breeding. Development of forage legumes is essential to agricultural sustainability in the Mediterranean region and in other areas of the world where grazing livestock area a dominant feature.

INTRODUCTION

The Mediterranean region is the center of origin of settled agriculture and has been cultivated intensively for millennia. Most of the world's major crops, for example, cereals and pulses, involved there. The major constraint to agricultural production in the region is limited rainfall, and therefore drought is invariably an ever-present threat (Kassam, 1981). Terminal drought is common at the end of the cropping season, which occurs during the cool rainy season (200–600 mm yr^{-1}) from November to May/June.

 Challenges and Strategies for Dryland Agriculture. CSSA Special Publication no. 32.

The farming system is based on dryland wheat, either bread wheat (*Triticum aestivum* L.) or durum wheat (*T. turgidum* Desf. var. *durum*), and barley (*Hordeum vulgare* L.). These crops are grown in rotation with food and forage legumes, or fallow to conserve moisture (Cooper et al., 1987). Increasingly due to population pressure fallow is giving way to continuous cropping. Animal production, mainly sheep (*Ovis aries*) and goats (*Capris hircus*) in drier areas, is an integral part of the system. Therefore, not only is food production for humans a pressing concern, but so also is the provision of feed for animals

Shortage of animal feed and the degradation of natural grazing lands are serious and interrelated problems in West Asia and North Africa. These regions are characterized by a rapidly growing livestock population but inadequate sources of feed (FAO, 1987). This has imposed a heavy burden on the rangelands, which have become the main source of feed for livestock, which has accelerated their degradation. Also, in times of high livestock prices, feed legumes, which may be fed to small ruminants in the form of straw and grains, are increasingly attractive to farmers in West Asia and North Africa (Abd El Moneim et al., 1988). In recent decades, the rapidly growing livestock population and degradation of rangelands are the two major reasons for the severe feed deficits (Joubert, 1987). At the same time, the demand for increased crop production in West Asia and North Africa is expected to lead to increased cropping on marginal lands (Oram, 1988). A possible solution to animal feed deficits is the development of indigenous forage legumes, for grazing and stored feed, and in some cases the introduction of new species or cultivars (Thompson et al., 1992). This brief review highlights the role of annual sown forage legumes with brief reference to self-generating medic in grazing pastures. While major attention is given to these forage legumes as feed sources, their value in rotations for economic and sustainable cropping will also be mentioned.

VETCHES AND CHICKLINGS

The ICARDA pays particular attention to annual feed legume species such as *Vicia* spp. (vetches) and *Lathyrus* spp. (chicklings) for dry areas where rainfall is between 250 to 350 mm. These areas are between steppe and high potential cereal-growing regions in West Asia and North Africa. These low rainfall areas have a very fragile agro-ecosystem and are currently threatened by further degradation and erosion because of the increasing annual cropping of barley in response to increasing population pressure (Ghassali et al., 1999). However, they provide opportunities for introduction of drought- and/or cold-tolerant annual feed legume species such as *Vicia* spp. and *Lathyrus* spp. for augmenting feed resources and preventing soil degradation.

Both *Vicia* spp. and *Lathyrus* spp. are recognized for their potential to produce extra feed from fallow lands (Abd El Moneim et al., 1988, 1990b), and are one of the major options being considered, either sown to interrupt barley monoculture or replace fallow in the fallow-barley rotations. These species are sown and harvested in a single year and can be used for grazing during winter, harvested for hay in spring, or harvested for grain and straw at maturity. In view of the huge diversity of Mediterranean legume species, few have been used as feed crops, and these have received virtually little attention by agronomists and plant breeders. Kernick (1978) reported that three species of *Lathyrus* and nine of *Vicia* are potentially important, but few of these have been tested and used. In areas where annual rainfall is <300 mm, *Lathyrus* spp. is common, whereas in higher rainfall areas there are large areas of common vetch (*Vicia sativa* L.) and bitter vetch [*V. ervilia* (L.) Willd.] (Table 16–1).

Although precise estimates of the area of production of these forage legumes are not available, some reports suggest that total areas under *Vicia* spp. might be more than 600 000 ha in six countries in West Asia and North Africa (Algeria, Turkey, Cyprus, Iraq, Tunisia, and Jordan). *Lathyrus sativus* L. (grasspea) is common in India and covers 1.6 to 2.0 million ha, and is also common in Pakistan, Nepal, Afghanistan, China, Ethiopia, Greece, Portugal, and France, *L. cicera* L. is found in Cyprus, Greece, Libya, Iran, Iraq, Spain, and Syria, while *L. ochrus* (L.) DG is the main species in Cyprus and Greece.

An important research activity at ICARDA has been to study the use of feed legumes for feed production, and their potential role in developing sustainable agriculture systems in West Asia and North Africa. The general objective of our breeding program is to develop and disseminate a range of improved feed legume crops adapted to various agro-ecological zones and target these crops to feed livestock in areas receiving average annual rainfall ranging downwards from 400 to 250 mm. It is also desirable to have widely adapted cultivars that can be recommended for different locations with similar agro-ecological conditions.

In feed legumes improvement, we deal with two major genera, that is, *Vicia* and *Lathyrus*. Within each genus we deal with several species to assess a wide range of feed legume crops in low rainfall areas (<350 mm rainfall) for different utilizations and niches. Of the vetches, we are selecting or hybridizing genotypes of *V. sativa* L. (common vetch), *V. narbonensis* L. (narbon vetch), *V. villosa* spp. *dasycarpa* Tan (wooly-pod vetch), *V. ervilia* L. (bitter vetch), *V. palaestina* (Palestinian vetch), *V. panonica* GR (Hungarian vetch), and chicklings such as *L. sativus* L. (common chickling or grasspea), *L. cicera* L. (dwarf chickling) and *L. ochrus* (L.) DG (ochrus vetch).

Underground Growth Habit

Underground vetch is a Mediterranean species found in Western Asia and Southern Africa. During the evaluation of new *Vicia* spp. germplasm a few genotypes of *V. sativa* ssp. *amphicarpa* were discovered. The words "amphicarpa" means producing two kinds of fruits. The plants not only flower aboveground but also flower underground by producing cleistogamous flowers that are never exposed to the light. Unlike subclover, the underground pods are protected from overgrazing at all time. Underground vetch is native to Mediterranean grasslands and confined to poor soils and rocky slopes, and it does best in microhabitats that are relatively too dry for other vetches. Its natural distribution includes most grassland areas receiving <300 mm rainfall in southeast Turkey, central Anatolian region, and northern Syria. The possibility of using underground vetch as a reseeding annual feed legume in certain pastures and rangelands is intriguing. Its ability to produce

Table 16–1. Adaptation and research of *Lathyrus* and *Vicia* spp. in the WANA region.

Species	Use†	Adaptation (rainfall, mm)	Research
Lathyrus sativus (grasspea)	GZ, G, S	<300, moderate cold	Resistance to *Orobanche* and foliar diseases; high HI‡ (%); low β-ODAP content.
L. cicera (dwarf chickling)	G, S	<300, moderate cold	Resistance to *Orobanche* and foliar diseases; high HI (%); low β-ODAP content.
L. ochrus (ochrus chickling)	G, S	<300, mild winters	Improve cold tolerance.
V. sativa ssp. *amphicarpa* (underground chickling)	GZ	250, cold, marginal	Improve biomass yield, hard seededness.
L. ciliolatus (underground chickling)	GZ	250, cold, marginal	Improve biomass yield, hard seededness.
V. sativa (common vetch)	GZ, G, S, H	>300, moderate cold	Leafiness, nonshattering pods; resistance to foliar diseases and nematode; low β -cyano-alanin (BCA).
Vicia narbonensis (narbon vetch)	G, S	<300, moderate cold	Earliness, improve HI (%), botrytis, downy mildew resistance, low tannin and GEC content.
V. villosa ssp. *dasycarpa* (wooly-pod vetch)	GZ, H	>300, high elevation	Earliness, increase leaf retention, cold tolerance and improve HI (%).
V. ervilia (bitter vetch)	G, S	>300, cold	Reduce pod shattering.
V. panonica (*Hungarian vetch)*	GZ, G, S	>300, high elevation, severe cold	Improve HI (%), ascochyta blight resistance.
V. hybrida (hybrida vetch)	GZ	<300, high elevation, severe cold	Improve HI (%).
V. palaestina (Palestinian vetch)	GZ, H	>300, mild winter	Improve pod-shattering and cold tolerance.

† G = grain; GZ = grazing; H = hay; S = straw.
‡ HI = harvest index.

both aerial and underground seeds should enhance drought resistance and persistence under heavy grazing.

Chicklings (*Lathyrus* spp.) are feed legumes with high yield potential in Mediterranean region, especially in areas where the annual rainfall is <300 mm (Abd El Moneim and Cocks, 1992). There are three species of *Lathyrus* that are potentially important: *L. sativus* (common chickling or grasspea), *L. cicera* (dwarf chickling), and *L. ochrus* (ochrus vetch). *Lathyrus sativus* is a drought-tolerant feed and food legume. Because of its ability to tolerate drought and produce good yields under adverse conditions, it is one of the major components of human diets in times of drought-induced famine. It is also being tested in Canadian prairies, where it is potentially valuable for feeding livestock (Briggs et al., 1983).

There are two species, *Vicia sativa* ssp. *amphicarpa* Dorth (underground vetch) and *L. ciliolatus* L. (underground chickling), that are characterized by producing both underground and aboveground pods. We have two approaches to developing feed legume crops (i) selection from the wild types to develop cultivated types, and (ii) genetic improvement by hybridization. The process of developing cultivar types from wild species (Robertson et al., 1996) involves preliminary evaluation of germplasm for desired characters and progeny tests for selected genotypes (selections), evaluation of selections in preliminary microplot field trials, evaluation of promising selections in advanced yield trials at two contrasting sites, that is, Tel Hadya (330 mm yr^{-1}) and Breda (280 mm yr^{-1}), and multi-location testing of promising lines in different agro-ecological zones. Recently, more emphasis is being given to genetic improvement with the support of other disciplines, for example, pathology, physiology, and entomology.

Breeding for improved yield is being supplemented by improving the quality. Therefore, palatability; nutritive value of the herbage, hay, grain, and straw; and feeding trials are also considered. We aim to serve national breeding programs through assembling, classifying, maintaining, and distributing germplasm; developing and supplying breeding populations with sufficient diversity to be used in different environments; and coordinating international trials to facilitate multi-location testing and identification of widely adapted cultivars. A total of 4267 vicia accessions are dealt with, as well as 1315 lathyrus accessions.

In our regional testing trials, many breeding lines of *Vicia* spp. and *Lathyrus* spp. are evaluated in different locations and years (environments) before the final selection of suitable lines (Abd El Moneim, 1987, 1992; Abd El Moneim et al., 1988; Abd El Moneim and Cocks, 1992). In testing 25 promising lines in each of *V. sativa*, *V. ervilia*, and *V. villosa* ssp. *dasycarpa* with contrasting rainfall, which varied from 233 to 504 mm, and absolute minimum temperature from 5.8 to 9.9°C, we found considerable variation between species and lines within the same species. *Vicia sativa* was the species most affected by frost, while *V. villosa* ssp. *dasycarpa* proved to be cold tolerant (Table 16–2). Although *V. villosa* ssp. *dasycarpa* produces high herbage yield in spring its grain yield is low, resulting in a low harvest index. Both *V. ervilia* and *V. sativa* produce high grain yield with a high harvest index.

There were also differences in phenology (Table 16–3), with *V. ervilia* being the earliest in flowering and maturity, reaching 50% flowering when the other two species had just commenced. *Vicia villosa* ssp. *dasycarpa* flowered 2 to 3 wk later than common vetch and bitter vetch and reached maturity after 180 d. The average

Table 16–2. Variability in quantitatively scored traits in three (*Vicia* spp.) species.

Species/traits†	Range	Means ± SE
Vicia sativa		
Seedling vigor	2.5–4.5	3.79 ± 0.30
Winter growth	3.0–5.0	4.01 ± 0.40
Cold effect	2.1–4.5	3.60 ± 0.31
Spring growth	4.3–5.0	4.46 ± 0.37
Leafiness	3.5–5.0	4.25 ± 0.40
Vicia ervilia		
Seedling vigor	1.9–3.5	3.01 ± 0.30
Winter growth	2.5–5.0	3.75 ± 0.46
Cold effect	0.5–2.0	1.01 ± 0.10
Spring growth	2.1–5.0	3.55 ± 0.30
Leafiness	2.0–5.0	3.70 ± 0.28
Vicia villosa ssp. *dasycarpa*		
Seedling vigor	1.1–3.2	2.72 ± 0.20
Winter growth	0.9–1.0	1.57 ± 0.13
Cold effect	0.5–1.5	1.01 ± 0.09
Spring growth	0.6–5.0	3.42 ± 0.26
Leafiness	1.0–4.0	3.13 ± 0.30

† Visual scale basis where 0 = poor, and 5 = very good, and for cold effect 0 = no damage, and 5 = nearly all plants killed by frost.

flowering period was 37, 23, and 45 d for common vetch, bitter vetch, and wooly-pod vetch, respectively.

These findings demonstrate the potential of *Vicia* spp. to replace fallow or interrupt monoculture barley rotations in the dry areas of West Asia and North Africa. The great variation in their qualitative and quantitative characters gives an opportunity to select suitable genotypes for cultivation in different niches in the prevailing farming systems. For example, the rapid winter and spring growth of *V. sativa* suggest that it could be grown for early grazing when feed shortages are acute in areas with mild winters. It could also be recommended for haymaking or straw and grain production (Jones, 2000). It is widely used in mixture with oat (*Vicia*/oat mixture) in Tunisia (Halila et al., 1990). This forage mixture is as profitable as any other cash crop, especially during dry years' cropping, when there is a severe shortage in animal feed. It can be more expensive than the subsidized bread prices.

In areas where winters are severe, cold and frost occurrence is frequent, and cold tolerances is considered a major requirement and *V. villosa* ssp. *dasycarpa* and

Table 16–3. Range of variability in phenological traits in three *Vicia* spp.

	V. sativa		*V. ervillia*		*V. villosa* ssp. Dasycarpa	
Trait†	Range	Mean ± SE	Range	Mean ± SE	Range	Mean ± SE
Days to:						
start flowering	105–115	110 ± 1.03	95–109	102 ± 1.02	114–136	120 ± 1.10
50% flowering	122–133	127 ± 1.09	105–114	112 ± 1.04	120–160	140 ± 1.17
100% flowering	134–160	147 ± 1.10	115–130	125 ± 0.95	140–168	165 ± 1.20
full maturity	170–189	160 ± 1.18	122–140	131 ± 1.09	162–196	180 ± 1.28

† Measured as number of days from germination.

V. ervilia are the most suitable because of their high degree of cold tolerance (Ratinam et al., 1994). Keatinge et al. (1991) indicated that wooly-pod vetch Selection 683, which was developed at ICARDA, could survive the extreme conditions of frost and could be a highly productive forage crop in the highlands of Baluchistan in West Asia. Variation in phenology can also affect the utilization of any species by farmers (Abd El Moneim, 1992). Bearing in mind that feed legume species can be used for grazing, hay, straw, and grain production, consideration of the phenology of the various species can easily identify the utilization of a particular species and hence its suitability for the farming system. For example, the early flowering and maturing cultivars of *V. ervilia* with high grain and straw yields suggest that it could be used by farmers who require straw and grain for summer feeding. Because of its high degree of leafiness and relatively high harvest index, *V. sativa* could be recommended for haymaking or straw and grain production. *Vicia villosa* ssp. *dasycarpa* would be most suitable for grazing because of its long flowering period, high herbage production, and low harvest index. *Vicia hybrida* L. and *V. palaestina* are also suitable for grazing because of their long flowering period and prostrate growth habit.

Harvest index and grain yield are likely to determine the way in which the farmers use feed legumes. Crops such as narbon vetch (*V. narbonensis*) with high grain yield and harvest index (Table 16–4) could be used as a dual-purpose crop for grain and straw production. Its upright growth habit enables easier mechanical harvesting (Castleman, 1987; Abd El Moneim, 1993a). The lines with high harvest index also had a potential to give high biological yield.

Narbon vetch possesses high seedling vigor with rapid winter growth and negligible cold damage. In our studies, grain yield varied from 0.5 to 1.9 t ha^{-1} with harvest index varying from 30 to 40%; below 300 mm rainfall the grain yield varied from 0.5 t ha^{-1} when rainfall was 195 mm to 1.4 t ha^{-1} when rainfall was 245 mm. Most of our lines had wide adaptation to dry areas in terms of both grain yield and its stability. Climate, except early spring rains had little effect on biological and grain yields. This combination makes narbon vetch a valuable species for dry areas. Because of its susceptibility to the parasitic broomrape (*Orobanche crenata* Frosk), the late maturing lines of *V. sativa* produced low seed yield and had low har-

Table 16–4. Mean grain yield and harvest index of narbon vetch for eight environments and their ranks.

Environments						
Sites	Years	Rainfall	Grain yield	Rank	Harvest index	Rank
		mm	t ha^{-1}		%	
Tel Hadya	1985–1986	316	1.55	3	38	2
Tel Hadya	1986–1987	358	1.90	1	40	1
Tel Hadya	1987–1988	504	1.10	6	34	6
Tel Hadya	1988–1989	233	0.90	7	32	7
Breda	1985–1986	218	1.29	5	36	4
Breda	1986–1987	245	1.40	4	36	4
Breda	1987–1988	415	1.61	2	38	3
Breda	1988–1989	195	0.47	8	30	8
Mean			1.28		35	
LSD(0.05)			0.54		4	

vest index. The early lines were also susceptible but were able to set seeds before the worst effects of broomrape occur. In contrast, *V. villosa* ssp. *dasycarpa* is resistant to broomrape (Linke et al., 1993), but still produces low seed yields. This crop is recommended for areas heavily infested with broomrape.

Losses of seeds from matured pods (pod shattering) is common in *V. sativa*. This pod shattering restricts its use as leguminous forage crop to replace fallow in fallow cereal rotations (Abd El Moneim, 1993b). Incorporation of nonshattering genes into agronomically promising lines was achieved by ICARDA. Lines having 95 to 97 nonshattering pods were obtained as opposed to 40 to 45% in the original cultivated lines. The practical potential of developing nonshattering lines will be the increase of grain yield which results in reducing the price of seed and allows the farmers to increase the area cultivated with common vetch, and in turn increase livestock production. Operationally, it facilitates mechanical harvesting and allows farmers to defer vetch harvesting until they finish lentil harvest, which otherwise coincides with common vetch harvest and is almost invariably delayed in preference to feed legume crop. In contrast to annual medics soft seededness is one of the most important attributes we are searching for to facilitate the use of herbicide to kill vetch weeds germinated in the subsequent cereal crop.

Our observations on underground vetch (*V. amphicarpa* Dorth) revealed that drier conditions favored earlier subterranean flowering; consequently, the proportion of subterranean pods was higher. Clipping of aerial shoots stimulates the basal branching both above-and belowground level, decreases the total number of fruits, and increases the percentage of subterranean ones. Under heavy clipping, the populations were maintained by underground pods. During the 1989/1990 growing season, underground vetch was grown in large plots (100 m^2) replicated three times and allowed to be grazed by sheep at the end of each month: February, March, and April. Plots were also left without grazing along with barley plots. The productivity of this vetch was determined and the amount of underground seeds under each grazing treatments were estimated. During the 1990/1991 season, barley variety Atlas 46 was planted after underground vetch on the same plots and also after the barley plots of 1989/1990. During the barley phase the seed bank of underground vetch was monitored.

Barley (*Hordeum vulgare* L.) after barley produced significantly less yield both in terms of grain as well as straw than barley grown after underground vetch (Table 16–5). Grazing of underground vetch had no effect on the productivity of barley. Early grazing of vetch greatly affected the yield of underground seeds. The difference in vetch seeds found buried in the soil at the beginning and at the end of the barley phase gives an indication of germination and hard-seededness of buried seed during the barley phase. The yield of self-regenerated vetch was determined in 1991/1992, which varied from 3258 to 3900 kg ha^{-1}. This indicates that underground vetch could play a great role in drier areas as a self-reseeding pasture plants.

One of the most important advantages is that it is naturally adapted to a moderate level of disturbance which makes it an ideal candidate for ley-farming systems, but differs from most other legumes such as annual medics (*Medicago* spp.), which are naturally found in undisturbed habitats and may be better adapted by virtue of its large underground seeds to deep ploughing—a common feature in the West

Table 16–5. Herbage yield of *Vicia sativa* ssp. *amphicarpa* in the establishment year 1988/1989, under three grazing treatments†.

Yields	Underground vetch grazed				Barley	±SE	LSD
	Feb.	Mar.	April	Zero grazing			($P < 0.05$)
	kg ha^{-1}						
Herbage yield (establishment year)	830	730	860	2020	--	57	140
Grain yield of barley	1966	2035	1925	1909	1599	98	227
Total barley biomass	4346	4193	3947	3877	3143	215	497
Seed bank barley phase, 1990,1991							
Start	50	130	160	240	--	27	75
End	32	95	141	218	--	34	85
Herbage, vetch 1991/1992	3258	3879	3708	3900	--	320	590

† Barley yield grown after underground vetch in comparison with barley after barley, seed banks at the beginning and end of the barley phase on 1990/1991 and herbage production of self-regenerated underground vetch in 1991/1992.

Asia-North Africa (WANA) region. While much effort was expanded in developing medic technology (adoption was poor due to on-farm constraints such as grazing management and poor harvesting (Christiansen et al., 2000b).

More emphasis is being given by ICARDA to these legume species. A breeding program was initiated at ICARDA in 1985 and promising lines were identified and tested under different environments (Table 16–6). *Lathryus ochrus* is susceptible to frost and when rainfall is below 300 mm, its grain yield decreases. It is resistant to broomrape. For these reasons it is recommended only for regions of mild winters such as Western Australia and Southern Australia. *Lathrus sativus* and *L. cicera* produce high grain and straw yields when rainfall is below 300 mm, they are recommended for producing grain and straw in drier areas. They are also moderately cold tolerant.

Table 16–6. Mean herbage and seed yields for the eight environments.

No.	Location	Rainfall	Year	Herbage Yield	Herbage Rank	Seed Yield	Seed Rank
		mm		kg ha^{-1}			
1.	Tel Hadya	361	1985/1986	2676	3	846	4
2.	Tel Hadya	358	1986/1987	2809	2	1358	1
3.	Tel Hadya	504	1987/1988	2579	4	912	3
4.	Tel Hadya	233	1988/1989	1980	5	497	6
5.	Breda	218	1985/1986	1222	8	373	8
6.	Breda	245	1986/1987	1332	7	392	7
7.	Breda	415	1987/1988	3282	1	1250	2
8.	Breda	195	1988/1989	1465	6	547	5
	Mean			2418		772	
	LSD (0.05)			623		211	

GRAIN AND FORAGE QUALITY

One of the main objectives of ICARDA's grain legume breeding program is identification of nutritious crops that are capable of giving good yields under adverse conditions. One of the drawbacks of *L. sativus* is that its excessive use in the diet or animal feed causes "lathyrism", a nervous disorder resulting in incurable paralysis of the lower limbs (Briggs et al., 1983; Roy, 1981; Roy and Kirby, 1989). The occurrence of lathyrism in human beings or domestic animals is caused by the presence of a neurotoxin compound, a free amino acid known as 3-N-Oxalyl-L-2, 3 diaminopropionic acid (β-ODAP). *Lathyrus* is of special interest because of its ability to tolerate drought, and screening germplasm for low β-ODAP is considered worthwhile (Aletor et al., 1994a, 1994b). The chemical estimation of β-ODAP is quite laborious and expensive; this limits its use in the screening and identification of lines that are low in neurotoxin. In 1988/1989, promising breeding lines of *L. sativus*, *L. cicera,* and *L. ochrus* were screened for low β-ODAP content in seeds using near-infrared method. Some lines contained very low amounts of β-ODAP in comparison with lines that had high concentrations. The identification of lines having very low or nearly free from β-ODAP with high yield potential in dry areas would help prevent the development of lathyrism and provide a safe food for humans and feed for animals, and hence alleviate protein-calorie malnutrition in such regions of the world that are subjected to frequent drought (Abd El Moneim et al., 2000). This will also make the crops attractive to marginal farmers and provide an important reserve staple in times of drought and impending famine.

Ideally, high yield should be supplemented by high quality. Although wooly-pod vetch is mainly recommended for grazing, its dry herbage quality is relatively low in terms of crude protein percentage and in vitro dry matter disappearance (Table 16–7). This is due to high proportion of leaf drop, leaf:stem radio, and rapid accumulation of nondigestible fibers contributing to its high dry matter (Abd El Moneim et al., 1990b). The crude protein (CP) content of the grains and straw of narbon vetch varied from 26 to 32% and from 6.4 to 12%, respectively. The CP% in straw of narbon vetch is considerably greater than that of wheat, barley, lentil, and wooly-pod vetch as estimated by Rees et al. (1991). Genotypic differences in quality aspects of other feed legume species were reported earlier by Abd El Moneim et al. (1990b). The protein content in grain and straw of narbon vetch emphasizes the potential of this crop as a source of supplementary animal feed in dry areas. This attribute and its adaptability to drier areas (<300 mm rainfall) illustrate its huge potential in drier areas, especially areas receiving rainfall between 250 to 300 mm

Table 16–7. Quality characteristics in three *Vicia* spp.

Character	V. *sativa* Means ± SE	*V. ervillia* Mean ± SE	V. *villosa ssp. dasycarpa* Mean ± SE
		%	
Crude protein	19.5 ± 2.1	21.5 ± 2.3	16.5 ± 1.8
In vitro dry matter digestibility	69.0 ± 5.9	72.0 ± 6.3	46.0 ± 5.8
Neutral detergent fiber	31.0 ± 2.9	28.0 ± 2.6	40.0 ± 3.1
Acid detergent fiber	22.0 ± 1.9	20.0 ± 1.9	31.0 ± 2.3

where other legumes do not perform well. More research is needed on the nutritional value of the grain and straw as they are known to carry anti-nutritional or anti-palatable factors.

In developing highly productive feed legumes, it would be desirable if some attributes could be identified at the morphological level that could prove to be useful selection criteria. Our investigations revealed that a high degree of leafiness and leaf retention could improve the palatability and nutritive value of a feed legume. Development of a high potential legume would normally require a multidisciplinary approach involving quantity as well as quality aspects, the latter being of special importance, as it relates to animal performance. Feed legume crops have little relevance until these are evaluated in terms of factors such as digestibility and utilization.

Soil Properties and Fertility Buildup

Replacement of the fallow phase or interruption of barley monoculture in rotations by well-adapted legumes reduces the hazard of soil erosion by providing protective ground cover during critical periods and improving soil organic matter and therefore soil aggregates (Ryan, 1998). As compared to cereal-cereal rotation, a cereal crop in a cereal-legume rotation develops a deeper root system because of better soil physical conditions and is thus better able to exploit soil water and nutrients within the soil profile.

The key soil fertility factor constraining productivity and water-use efficiency of the cropping systems in the dry areas of WANA is the limited N inputs to the production system. The average use of fertilizer N in major agricultural countries of Africa (Algeria, Ethiopia, Libya, Sudan, and Tunisia) and Asia (Afghanistan, Syria, and Yemen) in the WANA region for the arable lands and permanent crops does not exceed 20 kg ha^{-1}. The major share of this N use should in fact be applied towards the permanent crops and crops grown under assured moisture supply. Hence, the N cycle of the drier areas has its major input from symbiotic N_2 fixation. It is here that the food and feed legumes have their major role in ensuring sustained productivity of rainfed farming systems.

Studies using N^{15} technique at ICARDA and elsewhere in the region have demonstrated that under optimum management, well-adapted legume species yield 60 to 120 kg N ha^{-1} by symbiotic N_2 fixation with a seasonal precipitation of about 300 mm or more (Ryan, 1997). Rotational trials at ICARDA and elsewhere in the region have confirmed that the total N yield is higher in ‘cereal-legume’ rotation that in ‘cereal-cereal’ or ‘cereal-fallow’ rotations (Ryan et al., 2002). Rough N balance studies at ICARDA with food legumes have shown that if harvested only for the grain, most leave a positive soil N balance, but when harvested for both grain and straw, the balance was invariably negative or near zero. This latter fact notwithstanding, the total influx of N in the system is higher with legumes incorporated in the rotations compared to cereal-fallow rotation, and considerable soil N savings are achieved compared to continuous cereal (Harris, 1995).

Research on appropriate combinations of Rhizobium strain x legume genotype, use of P fertilizer either in the cereal phase or applied directly to the legume on P-deficient soils (Materon and Ryan, 1995), appropriate adjustment of sowing dates, control of cyst and root-knot nematodes, Orobanche and Sitona larvae damage to root nodules, has shown promise of improving the symbiotic N_2 fixation as well as the fraction of total plant N derived from fixation (Abd El Moneim and Bellar, 1993). Studies have now been initiated at ICARDA and in the region that would permit quantification of N conservation because of incorporation of legumes in the cereal-based cropping systems. This may permit assessment of the 'N' component of the legume effect in rotations.

Forage Legumes as Break Crops

It is now recognized that forage legumes also play a major role in improving subsequent cereal yields by acting as a break crop to cereal-to-cereal root diseases, while maintaining soil fertility. Forage legume and cereal crops are complementary when it comes to cropping rotations. Legumes play a vital role in controlling major cereal root diseases, particularly cereal eelworm or cereal cyst nematodes, *Heterodera avenae*. The combination of high soil N and reduced nematodes population is cumulative and can result in a big increase in subsequent cereal yield. *Vicia* spp. and *Lathyrus* spp. can be used as clearing crops. The nematodes will not reproduce on these plants. Such legumes are profitable for nematode control because grassy weed control is obligatory for good yields. Medics are less reliable as cleaning phases, not because the plants themselves but because of the lack of reliability of getting a dense self-sown stand as a result of such problems as lack of rain, poor seed reserves, and *Sitona* weevil (Alan, 1979); this allows grass establishment and consequently some nematode reproduction.

Also, intensification of rainfed agriculture by replacement of 'fallow-cereal' rotation by 'continuous cereal' has several undesirable consequences in terms of the buildup of noxious weeds, pests and pathogens, besides accumulation of allelopathic compounds. The cereal cyst nematode (*Heterodera avenae*), soil-inhabiting fungi such as *Cochlibolus sativus* syn. *Helminthosporium sativum* and "take-all" disease pathogen (*Gaeumannomyces gramines* var. *tritici*), and wheat ground beetle (*Zabrus tenebroides*) can all cause considerable yield decline in continuous cereal cropping systems. As a break crop, legumes can reduce the cereal yield decline by suppressing the build-up of these pests and pathogens and by preventing the build-up of allelopathic compounds.

FUTURE PERSPECTIVE

While the role of forage legumes in relation to animal feed supplies and ancillary benefits in terms of soil quality is now well established, the future challenge is to promote forage-based systems at farmers level. Despite the focus on ley farming with medics and its benefit for soil N, the technology did not meet farmers' acceptance and is now de-emphasized, though such legumes have a function in native dryland pastures.

The work on vetches ranged from its role in cropping systems (Jones and Singh, 2000a, 2000b), hard seededness (Christiansen et al., 1996) disease (Ahmed et al., 2000), and nematode (Abd El Moneim and Bellar, 1993) resistance, effects on soil quality (Ryan, 1998; Ryan et al., 2002), and on-farm testing (Thomson et al., 1992; Christiansen et al., 2000a). With common vetch now increasing in adoption at farmers' level, the prospects for the crop contributing to animal feed supplies in the Mediterranean zone are bright. However, Jones (2000) expressed caution about vetch adoption with conservation tillage in relatively drier areas.

The work with underground vetch provides a basis for it being used in degraded lands in harsh dry environments, providing a forage source for grazing and at the same two protecting the soil from erosion forces (Abd El Moneim, 1992). While lathyrus is a durable forage crop, its seeds are eaten when other foods are scarce. Breeding for reducing the B-ODAP toxin has made considerable strides and will continue to be a research focus.

In summary, forage legumes are a vital source of animal feed in the Mediterranean region, and can contribute to sustainable use of the land resources. Future efforts will focus on technology adoption with farmers.

REFERENCES

Abd El Moneim, A.M. 1987. Improving genetic potential of pastures and forage plants. p.165–196. *In* Pasture, Forage and Livestock Program Ann. Rep. ICARDA, Aleppo, Syria.

Abd El Moneim, A.M.1992. Narbon vetch (*Vicia narbonensis* L.): A potential feed legume crop for dry areas in West Asia. J. Agron. Crop Sci. 169:347–352.

Abd El Moneim, A.M. 1993a. Agronomic potential of three vetches (*Vicia* spp.) under rainfed conditions. J. Agron. Crop Sci. 170:113–120.

Abd El Moneim, A. 1993b. Selection of non-shattering common vetch, *Vicia sativa L.* Plant Breed. 110:168–171.

Abd El Moneim, A., and M. Bellar. 1993. Response of forage vetches and forage peas to root-knot nematode (*Meloidogyne artiellia*) and cyst nematode (*Heterodera ciceri*), Nematol. Medit. 21: 67–70.

Abd El Moneim, A.M., and P.S. Cocks 1992. Adaptation and yield stability of selected lines of *Lathyrus* spp. under rainfed conditions. Euphytica. 66:89–97.

Abd El Moneim, A.M., P.S. Cocks, and B. Mawlawi. 1990a. Genotype–environment interactions and stability analysis for herbage and seed yields of forage under rainfed conditions. Plant Breed. 104:231–240.

Abd El Moneim, A.M., P.S. Cocks, and Y. Swedan 1988. Yields stability of selected forage vetches (*Vicia* spp.) under rainfed conditions in West Asia. J. Agric. Sci. (Cambridge) 111:295–301.

Abd El Moneim, A.M., M.A. Khair, and S. Rihawi 1990b. Effect of genotypes and plant maturity on forage quality of certain forage legume species under rainfed conditions J. Agron. Crop Sci. 164:85–92.

Ahmed, S., C. Akem, and A.M. Abd El Moneim. 2000. Sources of resistance to downy mildew in narbon (*Vicia narbonensis*) and common (*Vicia sativa*) vetches. Genet. Res. Crop Evol. 47:153–156.

Alan, D. 1979. Cereal eelworm. South Australia Dep. Agric. Sheet 9/79. Dep. of Agric., Adelaide, South Australia.

Aletor, V.A., A. Abd El Moneim, and A.V. Goodchild. 1994a. Evaluation of the seeds of selected lines of three Lathyrus spp for B-N-Oxalylam-ino-l-alanine (BOAA) tannins, trypsin inhibitor activity and certain *in-vitro* characteristics. J. Sci. Food Agric. 65:143–151.

Aletor, V.A., A.V. Goodchild, and A.M. Abd El Moneim. 1994b. Nutritional and anti-nutritional characteristics of selected Vicia genotypes. Anim. Feed Technol. 47:125–139.

Briggs, C.J., N. Parenno, and C.G. Campbell. 1983. Phytochemical assessment of lathyrus species for the neurotoxin agent B-N-Oxalyl-1-α-B-diamino-propionic acid. J. Med. Plant Res. 47:188–190.

Christiansen, S., A.M. Abd El Moneim, P. Cocks, and M. Singh. 1996. Seed yield and hardseededness of two amphicarpic pasture legumes (*Vicia sativa* ssp. *amphicarpa* and *Lathyrus citiolatus*) and two annual medics (*Medicago* rigidula and M. *noena*). J. Agric. Sci. (Cambridge) 126:421–427.

Christiansen, S., M. Bounejmate, F. Bahhady, E. Thompson, B. Mawlawi, and M. Singh. 2000a.On-farm trials with forage legume-barley compared with fallow-barley rotations and continuous barley in north-west Syria. Exp. Agric. 36:195–204.

Christiansen, S., M. Bounejmate, H. Sawny-Edo, B. Mawlawi, F. Shomo, P.S. Cocks, and T.N. Nordbeom. 2000b.TAH Village project in Syria: Another unsuccessful attempt to introduce ley-farming in the Mediterranean basin. Exp. Agric. 36:181–193.

Food and Agriculture Organization. 1987. Agriculture towards 2000. Food and Agriculture Organization, UN C87/27, Rome.

Ghassali, F., P.S. Cocks, A.E. Osman, G. Gintzburger, S. Christiansen, A. Semaan, and M. Leybourne. 1999. Rehabitation of degraded grasslands in north Syria: Use of farmer participatory research to encourage the sowing of annual pasture legumes. Exp. Agric. 35:489–506.

Halila, M.H., A.B.K. Dahmane, and H. Seklani. 1990. The role of legumes in the farming systems of Tunisia. p. 115–129. *In* A.E. Osman et al. (ed.) The role of legumes in the farming systems of the Mediterranean areas. Kluwer Academic Publ., The Netherlands.

Harris, H.C. 1995. Long-term trials on soil and crop management at ICARDA. Adv. Soil Sci. 19:447–469.

Jones, M.J. 2000. Comparison of conservation tillage systems in barley-cropping systems in northern Syria. Exp. Agric. 36:195–204.

Jones, M.J., and M. Singh. 2000a. Long-term yield patterns in barley-based cropping systems in northern Syria. 1. Comparison of rotations. J. Agric. Sci. (Cambridge) 135:223–236.

Jones, M.J., and M. Singh. 2000b. Long-term yield patterns in barley-based cropping systems in northern Syria. 2. The role of feed legumes. J. Agric. Sci. (Cambridge) 135:237–249.

Joubert, R. 1987. The semi-arid areas of Syria: Farming systems in decline and issues in research design. *In* C.B. Flora and M. Tomecek (ed.) Proc. 1984 Symp. on Farming System Res. Kansas State Univ., Manhattan.

Keatinge, J.D.H., A. Asghar, K.B. Roidar, A.M. Abd El Moneim, and S. Ahmed. 1991. Germplasm evaluation of annual sown forage legumes and environmental conditions marginal for crop growth in highland of West Asia. J. Agron. Crop Sci. 166:48–57.

Kernick, M.D. 1978. Indigenous and semi-arid forage plants of north Africa, the Near and Middle East. p. 519–689. *In* Ecological management of arid and semi-arid rangelands in Africa and Near and Middle East. Vol. 4. FAO, Rome, Italy.

Linke, K.H., A.M. Abd El Moneim, and M.C. Saxena 1993. Variation in resistance of some forage legume species to *Orobanche crenata* Frosk. Field Crop Res. 32:277–285.

Materon, L., and J. Ryan. 1995. Rhizobial inoculation and phosphorus and zinc nutrition of annual medics (*Medicago* spp.) adapted to Mediterranean-type environments. Agron. J. 87:692–698.

Oram, P. 1988. Agricultural production and food deficits in West Asia and North Africa: Future prospects and the role of high elevation areas.p. 99–131. *In* J.P.S. Srivastava et al. (ed.) Winter cereals and food legumes in mountainous areas. ICARDA, Aleppo, Syria.

Ratinam, M., A.M. Abd El Moneim, and M.C. Saxena. 1994. Variation in sugar content and dry matter distribution in roots and their associations with frost tolerance in certain forage legume species. Agron. Crop Sci. 173:345–353.

Rees, D.J., M. Islam, A. Samiullah, R. Fahema, S.H. Raza, Z. Quresh, and S. Mehmood 1991. Rainfed crop production systems of upland Baluchistan: Wheat (*Triticum aestivum*), barley (*Hordeum vulgare*) and forage legumes (*Vicia* spp.). Exp. Agric. 27:53–69.

Robertson, L.D., K.B. Singh, W. Erskine, and A.M. Abd El Moneim. 1996. Useful genetic diversity in germplasm collections of food and forage legumes from West Asia and North Africa. Genet. Res. Crop Evol. 43:447–460.

Roy, S. 1981. Toxic amino acids and proteins from *Lathyrus* plants and other leguminous species: A literature. Nat. Abst. Rev. Ser. A-51:691–707.

Roy, D., and G.E. Kirby 1989. Toxicology of *Lathyrus sativus* and neurotoxin BOAA. p. 76–85. *In* The grasspea: Threat and promise. Proc. Int. network of the improvement of *Lathyrus sativus* and eradication of lathyrism. Third World Med. Res. Foundation, New York.

Ryan, J. (ed.) 1997. Accomplishments and future challenges in dryland soil fertility research in the Mediterranean area. Proc., Int. Soil Fertility Workshop. 19–23 Nov. 1995. ICARDA, Aleppo, Syria.

Ryan, J. 1998. Changes in organic carbon in long-term rotation and tillage trials in northern Syria. p. 285–295. *In* R. Lal et al. (ed.) Management of carbon sequestration in soil. Adv. Soil Sci. CRC Press, Boco Raton, FL.

Ryan, J., S. Masri, M. Pala, and M. Bounejmate. 2002. Barley-based rotations in a typical Mediterranean ecosystem: Crop production trends and soil quality. Options Mediterraneennes 50:287–296.

Thompson, E.F., R. Jaubert, and M. Oglah. 1992. Using on-farm trials to study the benefits of feed legumes in barley based rotations in north-west Syria. Exp. Agric. 28:143–154.

17 Drought Tolerance in Chickpea and Lentil—Present Status and Future Strategies

Rajinder S. Malhotra, Ashutosh Sarker, and Mohan C. Saxena

ICARDA
Aleppo, Syria

ABSTRACT

Chickpea (*Cicer arietinum* L.) and lentil (*Lens culinaris* Medik. subsp. *culinaris*) are important cool-season food legumes globally. These crops encounter numerous biotic and abiotic stresses among which drought is the most widespread. Substantial yield losses have been reported due to drought stress. Various techniques for evaluation of drought tolerance and various mechanisms of drought tolerance are reviewed, and their merits and demerits are discussed in this chapter. Most laboratory screening techniques have limitations for use in breeding programs and evaluation of a large number of materials. The field screening techniques are, however, more satisfactory and are in use in crop improvement programs in different countries. To combat drought in chickpea and lentil in West Asia and North Africa region, where rainfall takes place in winter months, the key to success of crop improvement is to maximize the water-use efficiency in water-limited environments, thus making these crops remunerative to the farmers and able to retain their place in the cropping systems. Our studies indicate that winter sowing of chickpea in low- to medium-altitude areas and lentil in high altitude areas increase productivity due to increased water-use efficiency and escape of terminal drought. Using various techniques at ICARDA, it has been possible to select drought-tolerant chickpea and lentil genotypes, which are being shared with national programs. As drought is a very complex phenomenon, use of biotechnological tools for increasing efficiency of selection for drought tolerance is advocated.

INTRODUCTION

Drought is a challenge to agricultural scientists, particularly in the context of global warming and climate changes. Drought is the major abiotic stress in many parts of the world (Johansen et al., 1994). In cereal-based farming systems in South and West Asia, and North and East Africa, grain legumes make a valuable contribution to the human diet, animal feed, crop diversification, and to a sustainable production system through breaking cereal monoculture. Traditionally, food legumes are cultivated as rainfed crops depending on either winter rainfall or grown on conserved soil moisture. These crops mostly suffer due to scarcity of water; due to global warming and climate change, the scarcity may further increase. Therefore, the key to success is to develop drought-tolerant cultivars with high water-use efficiency.

Chickpea and lentil are among important cool-season food legumes in the cropping systems of the dry areas (Saxena, 1984, 1985). These crops are grown as rainfed in <500 mm rainfall areas, and are generally subjected to drought. Drought depends on many factors including the amount and distribution of rainfall during cropping season, evaporative demand of the atmosphere, and the capacity of the soil to conserve moisture, and thus is most unpredictable (Ceccarelli and Grando, 1996). In fact, severity, timing, and duration of drought vary from year to year, and cultivars successful in one year may fail in another. The worst aspect is that drought seldom occurs in isolation; it often interacts with other stresses, particularly high temperature. Thus, breeding for drought resistance becomes complex because of the interactions of drought with other stresses (Arnon, 1980).

In South Asia, the major chickpea and lentil-producing region, these crops are grown in winter season on conserved soil moisture after monsoon rains. In the Mediterranean region, the second largest producing region for these crops, lentil is traditionally grown in winter and chickpea in spring (Saxena, 1981, 1984). Although, chickpea and lentil are most important food legumes grown and well adapted in dry areas, the crops face considerable yield loss due to drought stress. More than 60% yield loss in chickpea and 54% in lentil are reported (Johnson et al., 1994). However, severe and prolonged drought spells can result in complete crop failure.

DROUGHT AND DROUGHT TOLERANCE

There are several definitions of drought based on precipitation, potential evapo-transpiration and temperature and wind velocity for the whole year or a season (WMO, 1975). Drought may also be the resultant effect of many factors including, precipitation, potential evapo-transpiration, temperature, and wind velocity at a particular location in a particular year or a season in which a particular crop is grown (Khanna-Chopra and Sinha, 1998). For the crop breeder or agronomist, tolerance of the crop and cultivar in question are of great concern. Thus, there is a need to define drought tolerance and then find ways and means to overcome drought. Although drought resistance has been defined in many ways in the context of field crops, the definition given by Quisenberry (1982) appears to be most appropriate: "the ability of a genotype within a species to be relatively more productive than others under water-deficit conditions".

Drought is a natural recurring phenomenon more frequent in arid and semi-arid regions and it may be aggravated by mismanagement of resources. Drought may lead to poor growth and development, low productivity, poor quality, and in extreme cases, plant death and no yield. On a national basis, severe drought may result in economic losses and chaos in people's lives, and desertification. Recently, most of the areas in West Asia and North Africa region suffered severe droughts during 1999,

2000, and 2001 seasons, and caused heavy losses (Sakr and Malhotra, 2003). This incidence resulted in reduction in the area of these crops in North Africa, which previously exported chickpea and lentil to other countries.

Drought Tolerance Mechanisms

Resistance to drought can be achieved by various mechanisms including escape, dehydration avoidance, and dehydration tolerance (Blum, 1988; Ludlow and Muchow, 1990; Wery et al., 1994). However, traits coping with drought and dehydration avoidance and tolerance are considered as components of drought resistance (Levitt, 1980). *Drought escape* is particularly an important strategy for matching phenological development with the period of soil moisture availability to minimize the impact of drought stress. Early flowering and maturity with high yield potential are the components of drought escape in chickpea (Saxena et al., 1993; Silim and Saxena, 1993b) and lentil (Silim et al., 1993a). Progress has been made at ICARDA to develop early maturing genotypes without penalizing yield. Such chickpea and lentil genotypes are put in the International Drought-Tolerant Nursery and are being shared with the national programs. Breeding for short-duration cultivars is not only to match phenology to season length, but also for other reasons such as to fit crops/genotypes into more intensive cropping systems.

Crop species have evolved several mechanisms to maintain plant water status within the limits required for normal metabolic functioning when exposed to limited water supply or an atmosphere of excessive evaporative demand. One of the components of dehydration avoidance is related to root attributes (root size, depth, length, density, hydraulic conductance). Genotypic variation for root characters has been reported in chickpea (Nagarajarao et al., 1980; Brown et al., 1989) and lentil (Sarker and Erskine, 2000). A chickpea genotype, ICC 4958, has been identified as drought tolerant with 30% higher root dry weight than the cv. Annegeri (Saxena et al., 1994). Wide variation in taproot length and lateral root number in lentil is directly related to yield performance (Table 17–1). Shoot traits such as canopy structure, leaf movements, leaf surface characteristics, stomatal and cuticular characteristics, osmotic adjustment, also have significant roles in dehydration tolerance. But, in lentil, no significant correlation is observed between yield performance or harvest index and osmotic adjustment (Clements, 1997).

Table 17–1. Variability in shoots and root characteristics in lentil.

Characters	Mean	SD†	Min.	Max.	H%
Stem length, cm §	7.50 ± 0.13	1.05	4.70	13.9	66
Stem weight, g ‡	0.38 ± 0.04	0.18	0.15	0.82	39
Tap root length, cm §	28.50 ± 0.74	5.20	11.6	47.0	76
Lateral root number‡	31.50 ± 0.73	7.78	16.0	66.0	68
Total root length, m ‡	3.01 ± 0.18	0.72	0.68	7.05	53
Total root weight, g‡	0.35 ± 0.03	0.09	0.12	0.92	41
Seed yield/plant, g§	0.46 ± 0.05	0.12	0.16	0. 95	58

† SD = Standard Deviation, H% = Heritability estimate.
‡,§ Significance at the $P < 0.05$ and 0.01, respectively.

Dehydration tolerance is another mechanism of drought resistance (Wery et al., 1994; Malhotra, 1995). This relates to the ability of cells to continue metabolism at low water supply. Most crop plants belong to the dehydration-intolerant category. In environments where water deficit can occur at any stage of growth, dehydration tolerance may have some role in survival of the crop until soil moisture levels improve with succeeding rains (Turner, 1979). Simple measurements of electrolyte leakage can be used to determine dehydration tolerance in chickpea and lentil.

DROUGHT MITIGATION

Mitigation of drought in the dry areas is one of the most important challenges faced by crop breeders and agronomists in the context of changing climatic conditions resulting from global warming. Keeping in view the importance of drought, an attempt has been in this presentation to give an overview of some of the strategies, which can help in mitigation of drought.

Agronomic Considerations

To combat drought, various agronomic practices leading to reduction of evaporative losses through management of plant density and row spacing, shifting of planting time to better manage the water-use efficiency; and supplemental irrigation where irrigation facilities are available are found successful.

Supplemental Irrigation

Where available, supplemental irrigation is the major means for combating drought. Our research on spring sown chickpea done at an average rainfall site (average annual long-term rainfall of 338 mm), Tel Hadya, the main research site of ICARDA, has shown that the yield of spring chickpea can be increased by 40% with the application of only 50 mm of supplemental irrigation (Malhotra et al., 1997). Supplemental irrigation can be applied to save the crop in case of unexpected drought or as a planned practice to supplement the expected total seasonal rainfall in low rainfall areas. Use of supplemental irrigation to relieve soil moisture stress during the reproductive phase in chickpea was found promising in improving the crop productivity (Saxena et al., 1990).

Reducing Soil Evaporative Loss

Reducing evaporative loss from the soil surface to avoid depletion of soil moisture before the crop successfully completes its life cycle is also an important agronomic management strategy for obtaining satisfactory yields in drought conditions. Recommended management practices for reducing evaporation from bare soil and thus improving water use in the Mediterranean climates include mulching and planting the crops in narrow-row spacing to attain fast ground cover at early stage of crop growth.

Varietal Deployment

Different cultivars differ in their response to a given environment and this difference is used to reduce the risk associated with growing a particular variety. In Syria, Jordan, Iraq, and Central Asia and the Caucasus countries, a transient heat wave that normally occurs around flowering time of the chickpea can cause heavy yield losses due to poor seed-set. The reduction in damage can be achieved when the variety with different flowering and maturity date is grown, thus escaping that stress period, or has a buffering capacity to adjust the phenology. We have two scenarios to deal with such situations, change of planting season (winter sowing) and change of variety (a variety having high buffering capacity) to adjust with changing environmental conditions.

Winter Sowing

In the recent years, drought with high temperatures is becoming more frequent phenomenon than in the past. Recently in many countries in the West Asia and North Africa region where these crops are grown on conserved soil moisture in spring, which causes heavy yield loss. This uncertainty resulted in heavy reduction in area, especially in Algeria, Jordan, Morocco, and Tunisia. Shifting planting time from spring to winter can successfully mitigate drought stress (Sakr and Malhotra, 2003). As the rainfall generally takes place in the winter months in this region, the spring-sown crops cannot get full benefit from this water. Earlier sowing can permit the use of this water by the crop and can thus sustain productivity in drought-prone environments.

Research at ICARDA and national programs has clearly demonstrated the superiority of early spring or winter sowing of chickpea (Saxena, 1984; Singh and Saxena, 1996; Singh et al., 1997). This applies to other food legume crops grown in the Mediterranean region (Saxena, 1981), but cold tolerance and other biotic stress resistance factors should be considered. The average seed yields obtained from winter and spring trials conducted across numerous years in Syria and Lebanon revealed that winter sowing can produce almost double than the traditional spring-sown chickpea (Fig.17–1), provided the cultivars for winter sowing possess resistance to ascochyta blight and tolerance to cold. Most of the traditional spring-sown cultivars in West Asia and North Africa, however, lack tolerance to cold as well as ascochyta blight, the two pre-requisites for winter planting.

ICARDA scientists have successfully developed such winter-sown cultivars with resistance to these stresses and have shared them with various national programs. We observed in the last few years (1998–2000), when drought effects were more pronounced in different parts in the region, farmers who planted chickpea early in spring or winter harvested much higher seed yield (>1000 kg ha^{-1}) as opposed to very low yields (<300 kg ha^{-1}) from spring-sown chickpea. Most of the countries in the region have identified and released varieties from ICARDA-supplied chickpea materials that are being cultivated by farmers. Winter chickpea as a new technology to mitigate drought is further moving to cover still new dry environmental niches, and also in traditional chickpea-growing areas in many West Asian, North African, over southern European counties. So far, 40 varieties have been identified and released by national programs in 20 countries for winter sowing. The Syrian

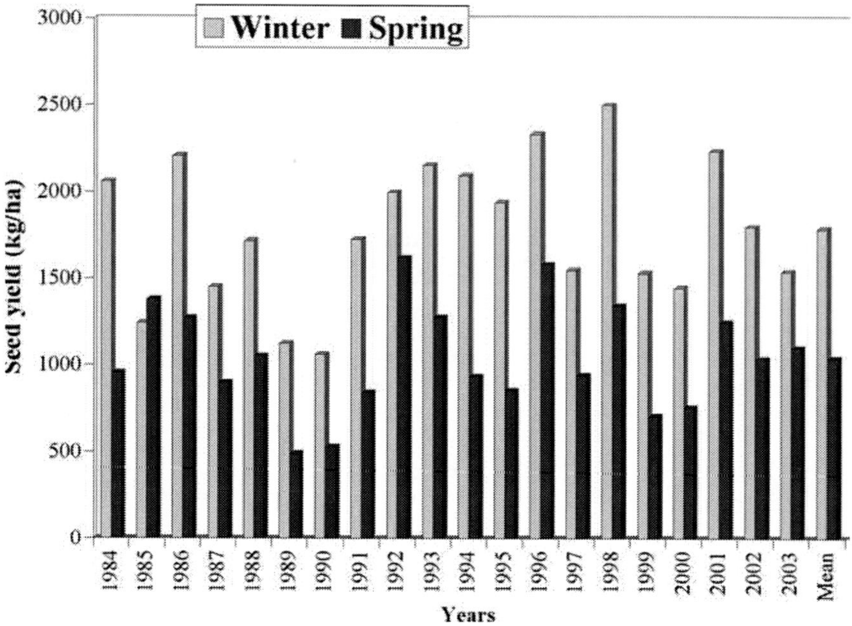

Fig. 17–1. Mean seed yield (kg ha^{-1}) of chickpea grown during winter and spring in Syria and Lebanon, for the period of 1984 to 2003.

Ministry of Agriculture has recently released two new varieties, "Ghab 4" and "Ghab 5", for winter sowing in different zones in Syria.

Lentil production can also be increased significantly by shifting planting from spring to early spring or fall sowing in highlands of Central Anatolia of Turkey, Iran, Afghanistan, and Beluchistan province of Pakistan. This environment allows optimum vegetative growth and development of higher yield potential and provides higher water-use efficiency. Spring lentil crops in the highlands frequently suffer from terminal drought, which can be avoided by growing them in winter. This also allows taller canopy development suitable for mechanical harvest. With significantly higher biomass development from the winter crop, the additional residue is highly priced for feeding of small ruminants. Appropriate varieties with a high level of winter-hardiness have been released in Turkey (e.g., Kafcas, Cifci, and Uzbek) and are being cultivated by farmers. About 400 000 ha of spring crop can be replaced by winter lentil in the highlands of West Asia. Lines with sufficient winter-hardiness, early growth vigor, and faster ground coverage have been identified from ICARDA nurseries in Iran. These lines (ILL 662, -857, -975, -1878) are under on-farm evaluation for future release in Iran.

BREEDING FOR DROUGHT RESISTANCE

As explained earlier, drought resistance can be obtained with various types of mechanisms, including *escape*, *dehydration avoidance*, or *dehydration tolerance*.

For environments exhibiting terminal drought, shortening the crop duration can help to escape drought. Dehydration avoidance can be achieved by selecting suitable cultivars with desirable root and shoot attributes. Selection for dehydration tolerance is one of the important keys to success in breeding for drought tolerance.

Drought tolerance/resistance refers to the ability of a genotype within a species to be relatively more productive than others under limited/deficit water conditions (Quisenberry, 1982). Although the genetic and physiological bases of different mechanisms have not been established precisely, they have been exploited by plant breeders to select for drought-resistant genotypes. For the success of a breeding program in dry areas, selection for drought tolerance is a key consideration. As drought does not occur every year at a particular location or in a particular region, selection of cultivars using a particular site or location under natural conditions is extremely difficult (Sarker et al., 2002). Thus, efforts have been made at various places to develop suitable and reliable screening techniques and scoring indices for evaluation for drought to select suitable genotypes or genetic materials, which can resist drought and also give reasonable yield.

Drought tolerance research in chickpea at ICARDA has addressed various strategic options, and indicated that productivity under drought conditions can be improved using two steps: (i) through rejection of less productive lines in drought-prone, low-input marginal conditions, and (ii) evaluation of remaining lines in high-input, drought-prone marginal environments for selection of the high productivity from the high-input environments. Following this procedure, we have achieved good success. One such variety, Gokce (FLIP 87-8C), has very high buffering ability to adjust to the changing environmental conditions.

As there is no single trait, morphological, physiological, or biochemical, which alone is responsible for drought tolerance, we have been following a gene pyramiding approach for development of breeding materials with high yield and drought tolerance. Relatively stable resistance is achieved by pyramiding different genes for early seedling establishment, early growth vigor, early flowering and maturity etc., which contribute to escape, dehydration avoidance, and tolerance aspects (Wery et al., 1994). Empirical breeding has exhibited limited success in improvement of drought tolerance in drought-prone areas. As a result, yield improvement and stability of rainfed crop varieties has been negligible. Therefore, we propose that breeding and selection for drought resistance should address the following areas given below.

Selection in Target Environment

The first step in designing strategies to alleviate drought stress is characterization of the drought pattern of the target environment. The complexity of this task has been reduced in recent years with development of characterization tools such as soil-water balance model and geographic information systems. Variability in soil moisture deficit must be considered over years for the entire cropping season. This knowledge allows estimation of long-term crop losses due to drought stress and the potential gains from alleviating drought stress through genetic and management options. Therefore, probability analysis may be an aid in decision making in drought research. In the Mediterranean region, decision-making can be improved through knowledge of the relationship between the onset of rains, the length of growing season, and appropriate phenology of the genotype.

We at ICARDA have addressed the question of the optimum environment for selection for higher productivity on chickpea in drought-prone and low-input, marginal conditions. Our results demonstrate that alleles conferring high yields in low-yielding environments are partially different from those conferring high yields under favorable environments (observations of R.S. Malhotra in chickpea nursery planted in spring with and without supplemental irigation in the Years 2001, 2002, and 2003; and nursery planted under drought stress and supplemental irrigation conditions at Tel Hadya, Syria in 2003). We found that some chickpea genotypes yielded high in low input and also responded well when sown under supplemental irrigation. This finding has an important bearing or implications for improvement of drought tolerance and high yield in crops grown in drought prone environments. Therefore, we recommend that selection for drought tolerance be done in low-yielding environments, followed by testing their response in high-yielding environments to discard the ones showing poor response. Thus, selection must be done in the target marginal environments under both conditions (low input and high input).

Selection of Drought Tolerant Genotypes

Drought tolerance research at ICARDA has indicated that lentil is superior to faba bean and field pea in drought tolerance (Saxena et al., 1993), but with genotypic differences (Silim et al., 1993a). Further studies on morphological and physiological indicators of drought resistance have elucidated the importance of early seedling vigor and high early biomass (Saxena et al., 1993; Silim et al., 1993a). Early vigor, early flowering, and early maturity with high biomass have been taken into consideration during selection for drought tolerance. A genotype with a small improvement in growth rate in the early phase of growth can attain a considerable increase in dry matter in later stages due to the exponential nature of growth. Any additional biomass produced by a genotype with rapid vigorous growth is an advantage in dry environments (Turner and Nicholas, 1987). Traits associated with lentil yield under rainfed conditions indicate that early vigor is strongly associated with biomass and seed yield (Silim et al., 1993a). However, developmental and phenological stages of crop growth must be matched with agroecological conditions of the target environment.

Drought-resistant lentil lines, ILL 590 and ILL 7200, have been selected in Australia (which has a Mediterranean climate) from Drought-Tolerant Nursery sent by ICARDA. These are characterized by early and rapid biomass and leaf area development and high photosynthetically active radiation interception (Clements, 1997). Precoz (ILL 4605), an early-maturing lentil genotype with early vigor and rapid biomass development has been selected by various national programs.

Adopting a delayed-spring sowing technique in chickpea, about 2150 cultivated accessions of chickpea were screened for drought tolerance using a 1 (= resistant) to 9 (= susceptible) scale. The materials screened originated from a wide range of environments. Repeated screenings continued over the years have resulted in a total of 22 drought-tolerant lines (Table 17–2). Such lines with drought tolerance are shared with national partners through the International Testing Pro-

Table 17–2. Drought tolerant sources (with rating 3, on 1 to 9 scale, where 1 = resistant and 9 = no pod setting) in kabuli chickpea, Tel Hadya, Syria.

Entry name	Origin	Entry name	Origin
ILC 19+	Jordan	ILC 3843	Morocco
ILC 588	India	ILC 4291	Mexico
ILC 1306	Turkey	ILC 4945	ICARDA/ICRISAT
ILC 1799	Syria	ILC 5766	Pakistan
ILC 3101	Turkey	ILC 6023	Mexico
ILC 3105	Turkey	ILC 6056	USA
ILC 3182	Turkey	ILC 7067	ICARDA/ICRISAT
ILC 3210	Turkey	FLIP 87-51C	ICARDA/ICRISAT
ILC 3216	Turkey	FLIP 87-58C	ICARDA/ICRISAT
ILC 3321	Syria	FLIP 87-85C	ICARDA/ICRISAT
ILC 3832	Morocco	FLIP 88-42C	ICARDA/ICRISAT

† 1 to 9 scale, where 1 = resistant and 9 = no pod setting, + all scored 3.

gram of ICARDA for the evaluation of their drought tolerance in different countries.

Pyramiding for Drought Tolerance

There is no single morphological, physiological, or biochemical trait that can serve as a criterion for selection for drought resistance (Khanna-Chopra and Sinha, 1998). An analogy can be drawn from disease resistance breeding; horizontal resistance can be achieved by pyramiding different genes carrying resistance to individual physiological races into one cultivar. We emphasize the need to integrate the traits like early seedling establishment; early growth vigor and canopy development; leaf area maintenance; early flowering and maturity, which contribute to escape; dehydration avoidance; and tolerance aspects. These traits are easier to quantify in a breeding program where many lines must be evaluated.

Breeding which considers pyramiding of these traits would provide stable and higher yield in drought-prone environments of the Mediterranean region (Sarker et al., 2003). Rapid root development and growth would facilitate successful establishment of seedlings. There are genotypic differences in the ability to germinate and establish seedlings of chickpea (Saxena, 1987), dry bean (Seong et al, 1988), and faba bean (Soja et al., 1988) under suboptimal moisture levels. Early vigor and rapid canopy development lead to improvement in water-use efficiency because water use early in the season, when vapor pressure deficits are smaller, would improve transpiration efficiency. Genetic variation in several grain legumes, indicate the feasibility of manipulating this trait as required for specific environments (Onim, 1983; Silim et al., 1993a). Rapid leaf area expansion and prolonged leaf retention up to maturity are important attributes for drought resistance. The leaf retention trait is easy to assess visually among large numbers of test lines common in a breeding program.

Developmental and phenological plasticity to match with the growing environment is key to develop drought-tolerant genotype for a particular environment. Wide range of variation for these traits has been recorded in various grain legumes.

Plant breeders have been quite successful in incorporating various levels of phenological plasticity in many food legumes (ICRISAT, 1991).

Wild Relatives

Accessions of wild relatives of lentil were studied for drought resistance at ICARDA's drier research site, Breda (long-term average rainfall 267 mm), using supplemental irrigation (Hamdi and Erskine, 1996). The accessions of four wild relatives were assessed for drought tolerance on the basis of drought susceptibility index. *Lens culinaris* ssp. *orientalis* is more drought resistant than other wild relatives and the cultigen. However, the cultivated lentils contrasted with their wild relatives by producing markedly higher biomass. It is clear that direct selection of wild lentil germplasm for biomass yield under dry conditions is of little value, but they can, however, be used in a hybridization program for enhancement of drought tolerance in lentil.

Our preliminary screening for drought tolerance among eight annual wild chickpea species showed that *Cicer reticulatum*, *C. bijugum,* and *C. judaicum* are more drought-tolerant than the other species. Two accessions each of *C. bijugum* (ILWC 34 and ILWC 65) and *C. reticulatum* (ILWC 36 and ILWC 116) gave highest seed yield under drought conditions (ICARDA, 1996). As *C. reticulatum* can be readily crossed with the cultigen, we used high-yielding accessions of this species in hybridization program to improve the response of the cultivated species to water stress.

SCREENING FOR DROUGHT TOLERANCE

For the last few decades, scientists around the world have developed and used various screening techniques to screen for drought tolerance (Wery et al., 1994; Malhotra, 1995). The techniques differ from crop to crop and extent of drought to be combated. Some of the key methodologies that are followed at ICARDA and elsewhere are described below.

Drought Escape

The ability of a crop to complete its life cycle before serious soil water deficit develops is termed as drought escape. Early maturity is an important trait to avoid drought stress. Yield potential and early flowering are the two major components of drought escape in lentil (Silim et al., 1993a) and chickpea (Silim and Saxena, 1993b). Other plant responses associated with drought escape include photoperiod sensitivity, developmental plasticity, and re-mobilization of assimilates. Compared to medium- to late-maturing cultivars, short-duration cultivars, in general, give low yields in seasons with higher rainfall. This is due to the fact that these cultivars cannot make full use of available water and leave a large proportion of the potentially transpirable water unused in the soil. Thus, emphasis should be laid on selection of earlier flowering lines with plasticity for maturity that gives large yield under high moisture supply (Silim and Saxena, 1993a).

Screening of lentil genotypes with early seedling vigor, faster ground cover, early flowering and maturity, with high biomass development is being carried out at ICARDA; Lentil International Drought-Tolerant Nursery is formulated using these materials. Genetic differences in developmental plasticity for chickpea under rainfed conditions (with terminal drought) and assured moisture supply have been reported (ICRISAT, 1977; Saxena and Sheldrake, 1980), but there is a need to evaluate a large number of genotypes and establish the advantages of such plastic genotypes in drier environments.

A line-source sprinkler irrigation system developed at ICARDA has been very effective for screening lentil and chickpea (Silim and Saxena, 1993a; Silim et al., 1993b) genotypes for yield under different moisture gradients to observe the drought stress and response to increased moisture supply. In this system a gradient in soil moisture is created, and the genotypes are evaluated for seed yield under different moisture gradients to observe drought stress and response to increased moisture supply. Different genotypes are scored and the ones with high Drought Response Index are selected. Using this technique, we observed that early flowering is the main component of drought escape in chickpea, and it is associated with harvest index, the number of pods per unit area, and the seed yield. As a result of the evaluation of chickpea using this technique, good progress has been made at ICARDA to develop early-maturing genotypes without penalizing yield.

Dehydration Avoidance

Dehydration avoidance is the mechanism that helps to decrease the plant water content during drought stress to maintain the essential plant productive functions, and is mainly influenced by different root attributes including root size, morphology, depth, length, density, and hydraulic conductance (Subbarao et al., 1995).

Screening for dehydration avoidance involved a sand culture technique whereby the genotypes are grown in sand culture, measurements of different morphological and root traits are recorded, and the genotypes that possess more branches and greater root growth (root length, root density, and root weight), are considered as drought-tolerant, has been used for screening for dehydration avoidance (Saxena et al., 1993); genotypes with a deep root system had a high leaf water potential (Silim and Saxena, 1993b). Since screening for differences in root length and density is laborious and time-consuming, the indirect selection through related traits seems more practical, and so, being cumbersome, the sand culture method has not become popular for common use.

Thus, on the basis of studies on relationships of different root, shoot, and some morpho-physiological parameters, we found that a combination of various traits, including earliness-to-flower, high harvest index and deep rooting, which are responsible for large average yields, should be used as criteria for improvement of dehydration avoidance. Some traits such as stomata number and size, leaf rolling, leaf movement, and high level of reflectance, which have been reported (Subbarao et al., 1995) as good indicators of drought avoidance in other crops, can also be explored in lentil and chickpea.

Stem length, taproot length, and lateral root number are key traits for drought tolerance in lentil (Sarker et al., 2003). These traits are also highly correlated with

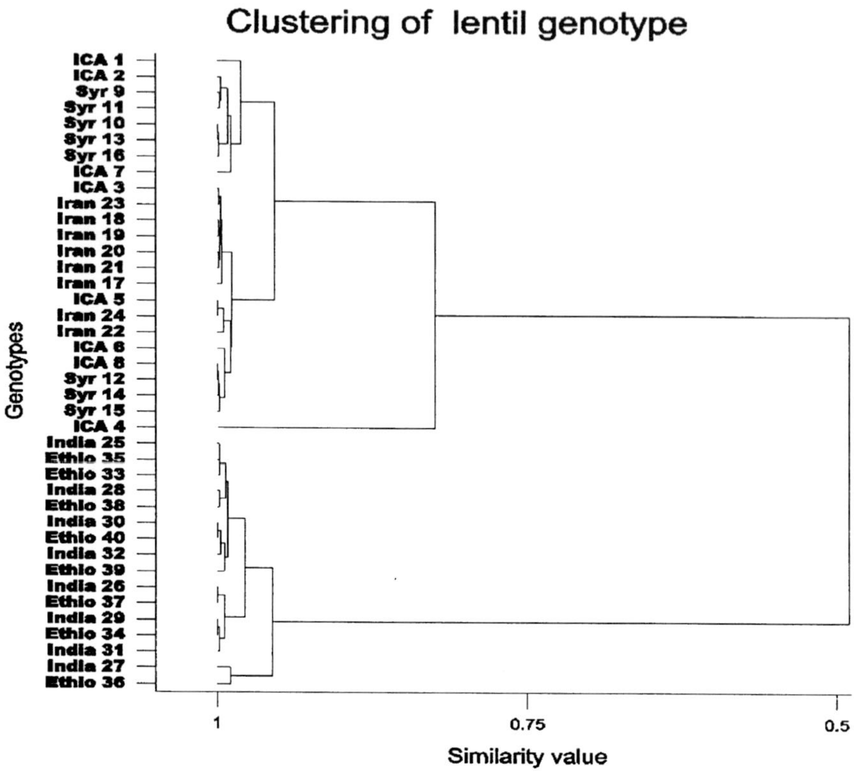

Fig.17-2. Dendrogram showing inter-relationships of the lentil genotypes originating from different geographical regions

yield. High heritability estimates provide reliability in screening based on these traits. Regression analysis using these traits showed that stem length alone accounted for 85% of the variance in seed yield per plant. The importance of these traits, for their association with yield in drought condition, is in the order of stem length > taproot length > lateral root number. Thus, to screen a large number of genotypes, one can rely on stem length alone. Cluster analysis showed that the landraces originated in Iran and Syria, and the breeding lines developed at ICARDA are distinctly different from the lentil accessions originated in southerly latitude countries, India and Ethiopia (Fig. 17–2). Among a total of 40 genotypes, ILL 6002 was strikingly different from all other test genotypes, and exhibited significantly superior root and shoot traits and yield and, therefore, is a valuable germplasm for breeding drought-tolerant cultivars.

Characters such as stomata number and size, leaf rolling, leaf movement, and high level of reflectance have been reported as good indicators for drought avoidance in some of the crop plants (Ludlow and Muchow, 1990), which can also be studied in these crops and their importance in drought tolerance, if any, can be looked into. Morphological traits such as reduced leaf area or small leaf size, which can produce high grain yield under low rainfall and shallow soils with reduced transpiration, especially in the vegetative phase, can be used as criteria for drought screening. These plant types are, in general, poor in productivity under increased moisture supply; thus, their use is limited to consistently low rainfall areas.

Dehydration Tolerance

The mechanism that allows the plant to withstand dehydration stress, especially at the end of the reproductive phase is dehydration tolerance; it relates to the ability of cells to continue metabolism at low water supply. Simple screening tests like electrolyte leakage measurements after stress can be used for dehydration tolerance in cool-season food legumes. The box screening technique provides a more reliable test for dehydration tolerance in field crops. Wooden boxes of convenient size with at least 20-cm depth are filled with 1:1 mixture of sterilized soil and sand. Seeds of different genotypes are planted in equally spaced rows and in equal depth and the soil-sand mixture in the boxes is saturated with moisture. No more water is applied after planting seeds. The evaluations for drought tolerance are done when the repeated susceptible check shows wilting. The leaf samples from different genotypes are used for cell leakage studies using an electro-conductivity test. Based on days to wilting and electro-conductivity values, the lines are scored as drought tolerant or susceptible. Many drought-tolerant lentil genotypes (ILL 759, ILL 6465, ILL 6467, ILL 7005, ILL 7940, ILL 7955, ILL 7972, ILL8072, ILL7980, ILL8075, and ILL 8090), selected at ICARDA following this technique, stayed alive 12 to16 d more compared to the most susceptible line.

Most of the techniques mentioned above are useful and reliable but at the same time either these are time-consuming, expensive, cumbersome, or only good for evaluation of a limited number of genotypes under controlled conditions. To overcome this problem, we developed a simple field screening technique involving two steps:

1. The preliminary screening of materials planted late in the spring by about 3 wk, and evaluation of the lines on a 1 to 9 scale (1 = free, very good early plant vigor, 100% pod setting; 9 = highly susceptible, 100% plants killed, lack of early plant vigor, no flowering, no pod setting, no yield) to discard susceptible lines.
2. In final screening, promising lines (with ratings 1–5) are sown with and without supplemental irrigation in replicated trials in late spring (late March). Lines that produce high yields (more than the mean of the trial) under drought (late sown conditions) as well as irrigated conditions, and respond to supplemental irrigation by producing substantially higher yield, are selected.

Based on field screening a good number of lines have been identified and shared with the national programs through ICARDA's Legume International Testing Program in the form of special nursery known as Chickpea International Drought-Tolerance Nursery. Many lines were identified as drought tolerant from different countries (Table 17–2).

Several physiological and phenological characters, which appear to be associated with genetic resistance to drought, have been identified. These traits include osmotic adjustment, stomata resistance, transpiration efficiency, rapid seed filling, early maturity, root growth, and high leaf water potential. Although, the genetic bases for these traits have not yet been established, some of them have been used to identify drought tolerant. Among these lines ICCV-2, ILC2293, ILC2537 were identified in Morocco (Dahan 1997).

Other techniques using isozymes, RFLP markers-assisted screening, polyethylene glycol, C^{13} discrimination, and effects of abscisic acid in regulating the response to drought are worth exploring. As drought is a complex trait and its occurrence is unpredictable and screening not very reliable, there is a need to exploit molecular techniques including identification of Quantitative Traits Loci (QTLs), genes, and molecular markers for marker-assisted selection. Osmoregulation is determined from measurements of osmotic potentials and relative water content in leaves. Genotypic differences in osmoregulation have been reported in different legume crops (Subbarao et al., 1995). As osmoregulation measurements are time and resource consuming, only the advanced materials or parental lines should be subjected to such tests.

Using multi-location testing and delayed sowing of chickpea by 3 wk during spring at a relatively dry site, we evaluated a large number of materials for drought tolerance on 1 to 9 scale (1 = resistant, 9 = susceptible), to discard susceptible lines. Based on this technique, 2144 germplasm lines were evaluated in 1997, 1998, and 1999. A total of 22 lines with a 3 rating were found resistant during the last 3 yr (Table 17–2). Using this technique in Morocco (Sakr and Malhotra, 2003), the following lines were identified as sources of drought tolerance: PCH 80, PCH 102, ILC588, ILC3101, ILC3105, ILC1799, ILC3182, ILC 1929, FLIP 87-59C, ILC 6104 and ILC 6118.

FUTURE DIRECTIONS

Research at ICARDA and elsewhere has successfully identified drought-tolerant genotypes of chickpea and lentil, and made available to the national programs. However, until today it has not been possible to quantify the effect of drought on yield losses. More effective genetic improvement may be possible by incorporating in the breeding program specific agronomic and physiological traits, which enhance drought resistance. Additionally, a better understanding of the target production systems, and our ability to define and characterize more precisely the target environment could facilitate genetic improvement for resistance to drought stress. The last two decades of research on drought resistance in legumes have led to a vastly improved understanding of the various physiological mechanisms and strategies that crop plants have evolved during the course of their adaptation to water-limited environments. Despite these developments, more needs to be done in the field of drought tolerance research in grain legumes. Particularly, the following aspects need to be addressed:

- Identification of specific physiological traits having functional significance in productivity under drought conditions and their characterization.
- Study of genetic basis and heritability of different physiological traits related to drought tolerance to determine the feasibility of these traits in a breeding program.

- Development of a conceptual "physiological ideotype" for breeding for drought tolerance.
- Development of genetic markers for various physiological and morphological traits which confer drought resistance, for use in Marker Assisted selection in order to increase the efficiency of selection in a breeding program.
- Tagging drought tolerance with quantitative trait locis (QTLs).
- Assembling "physiological trait collections" and development of genetic stocks using gene pyramiding concept. This would be advantageous for long-term crop improvement in the context of continuous climatic change.
- Development of crop models for chickpea and lentil which enable the clustering of various genotypes and environments which help in prediction of performance of specific genotypes under specific environmental conditions in the Mediterranean environments.

We believe that this work, when built on past drought-resistance research with proper integration of physiological, genetic, breeding, molecular biology, and management strategies could form a strong basis for tailoring new food legume genotypes which will adapt well and be more productive under drought-prone Mediterranean environments.

REFERENCES

Arnon, I. 1980. Breeding for higher yields. p. 77–81. *In* Physiological aspects of crop productivity. Proc. 15th colloquium of the International Potash Institute. IPI, Wageningen, The Netherlands.

Blum, A. 1988. Plant breeding for stress environments. CRC Press, Boca Raton, FL.

Brown, S.C., P.J. Gregory, P.J.M. Cooper, and J.D.H. Keatinge. 1989. Root and shoot growth and water use of chickpea (*Cicer arietinum*) grown in dryland conditions: Effect of sowing date and genotype. J. Agric. Sci. (Cambridge) 113:41–49.

Ceccarelli, S., and S. Grando. 1996. Drought as a challenge for the plant breeder. Plant Growth Regul. 20:149–155.

Clements, J. 1997. Studies on drought tolerance in lentil. p. 1–10. Research report for Centre for Legumes in Mediterranean Agriculture (CLIMA), Perth Australia.

Dahan, R. 1997. Identification of sources of resistance to drought. Activity report. Centre Regional de la Researche Agronomique, Settat, Morocco.

Hamdi, A., and W. Erskine. 1996. Reaction of wild species of the genus Lens to drought. Euphytica 91:173–179.

International Center for Agricultural Research in the Dry Areas. 1996. Legume Program Annual Report for 1995. ICARDA, Aleppo, Syria.

International Crops Research Institute for the Semi-arid Tropics. 1991. Annual report 1990. ICRISAT, Patancheru, India.

International Crops Research Institute for the Semi-arid Tropics. 1997. Annual report 1996. ICRISAT, Patancheru, India.

Johansen, C., B. Baldev, J.B. Brouwer, W. Erskine, W.A. Jermyn, L. Li-Juan, B.A. Malik, A.A. Miah, and S.N. Silim. 1994. Biotic and abiotic stresses constraining productivity of cool-season food legumes in Asia, Africa and Oceania. p.175–194 *In* F.J. Muehlbauer and W.J. Kaiser (ed.) Expanding the production and use of cool-season food legumes. Kluwer Academic Publ., Dordrecht, The Netherlands.

Khanna-Chopra, R., and S.K. Sinha. 1998. Progress in drought resistance research in India: Integration of physiological, gentic and molecular approaches. p. 526–539. *In* V.L. Chopra et al. (ed.) Proc. of 2nd Int. Crop Science Congress. Crop productivity and sustainability–Shaping the future. Oxford and IBH Publ. Co. Pvt. Ltd., New Delhi, India.

Levitt, J. 1980. Responses of plants to environmental stresses. Vol. II. Water, radiation, salt and other stresses. 2nd ed. Academic Press, New York.

Ludlow, M.M., and R.C. Muchow. 1990. A critical evaluation of traits for improving crop yields in water-limited environments. Adv. Agron. 43:107–153.

Malhotra, R.S. 1995. Evaluation techniques for abiotic stresses in cool season food legumes. p. 459–474. *In* A.N. Asthana and Masood Ali (ed.) Recent advances in pulses research. Indian Soc. of Pulses Res. and Develop., Kanpur, India.

Malhotra, R.S., K.B. Singh, and M.C. Saxena. 1997. Effect of irrigation on winter-sown chickpea in a Mediterranean environment. J. Agron. Crop Sci. 178:237–243.

Nagarajarao, Y., S. Mallick, and G.C. Singh. 1980. Moisture depletion and root growth of different varieties of chickpea under rainfed conditions. Indian. J. Agron. 25:289–293.

Onim, J.F.M. 1983. Association between grain yield and drought resistance in pigeonpea in marginal rainfall areas of Kenya. p. 864–872. *In* J.C. Holmes and W.M. Tahor (ed.) More food for better technology. FAO, Rome, Italy.

Quisenberry, J.E. 1982. Breeding for drought resistance and plant water use efficiency. p. 193–212 *In* M.N. Christiansen and C.P. Lewis (ed.) Breeding plants for less favorable environments. John Wiley & Sons, New York.

Sakr, B., and R.S. Malhotra. 2003. Managing drought stress in chickpea in Morocco. p. 210–216. *In* R.N. Sharma et al. (ed.) Chickpea research for the millennium: Proc. of the Int. Chickpea Conf., Raipur, India. 20–22 Jan. 2003. Indira Gandhi Agric. Univ., Raipur, Chhattisgarh, India.

Sarker, A., and W. Erskine. 2000. Drought tolerance in lentil: Root parameters. p. 180. *In* O. Christen and F. Ordon (ed.) Book of abstracts: 3rd Int. Crop Sci. Congr. 2000. ICSC-III, Hamburg, Germany. 17 22 Aug. 2000., Hamburg, 17.8.2001.S. CCH-Congress Centrum Hamburg, Germany. European Soc. of Agron., Hamburg, Germany.

Sarker, A., W. Erskine, and M. Singh. 2003. Variation in root and shoot traits and their relationship in drought tolerance in lentil. Genetic Res. Crop Evolution. (In press.)

Sarker, A., R.S. Malhotra, W. Erskine, and M.C. Saxena. 2002. Drought resistance in chickpea and lentil in Mediterranean environments. p. 119–126. *In* Symp. Proc., LEGUME–Grain Legumes in the Mediterranean Agriculture. 25–27 Oct. 2001. IAV Hassan II, Rabat Morocco. Eur. Assoc. For Grain Legumes (AEP), Paris, France.

Saxena, M.C. 1981. Agronomy of lentils. p. 111–130. *In* C. Web and G. Hawtin (ed.) Lentils. CAB Int., Wallingford, UK and ICARDA Aleppo, Syria.

Saxena, M.C. 1984. Agronomy of chickpea. p. 207–232. *In* M.C. Saxena and K.B. Singh (ed.) The chickpea. CAB Int., Wallingford, UK.

Saxena, M.C. 1985. Food Legume Improvement Program at ICARDA—An overview. *In* M.C. Saxena and S. Varma (ed.) Proc. Faba beans, Kabuli chickpeas and Lentils in the 1980. An International Workshop. 16–20 May 1983. ICARDA, Aleppo, Syria.

Saxena, N.P. 1987. Screening for adaptation to drought: Case studies with chickpea and pigeon pea. p. 63–76. *In* N.P. Saxena and C. Johansen (ed.) Adaptation of chickpea and pigeonpea to abiotic stress. ICRISAT, Patancheru, India.

Saxena, N.P., and A.R. Sheldrake. 1980. Physiology of growth, development and yield of chickpea in India. p. 89–96. *In* J.M. Green et al. (ed.) Proc. of the Int. Workshop on Chickpea Improvement. ICRISAT, Patancheru, India.

Saxena, M.C., S.N. Silim, and K.B. Singh. 1990. Effect of supplementary irrigation during reproductive growth on winter and spring chickpea (*Cicer arietinum*) in a Mediterranean environment. J. Agric. Sci. (Cambridge) 114:285–293.

Saxena, N.P., C. Johansen, M.C. Saxena, and S.N. Silim. 1993. Selection for drought and salinity: A case study with chickpea. p. 11. *In* K.B. Singh and M.C. Saxena (ed.) Breeding for stress resistance in cool-season food legumes. John Wiley & Sons, New York.

Saxena, N.P., L. Krishnamurthy, and C. Johansen 1994. Registration of drought resistant chickpea germplasm. Crop Sci. 33:1424.

Seong, R.C., H.J. Chung, and E.H. Honh. 1988. Varietal responses of soybean germination and seedling elongation to temperature and polyethylene glycol solution. Korean J. Crop Sci. 33:31–37.

Silim, S.N., and M.C. Saxena. 1993a. Adaptation of spring-sown chickpea to the Mediterranean basin. I. Response to moisture supply. Field Crops Res. 34:121–136.

Silim, S.N., and M.C. saxena. 1993b. Adaptation of spring-sown chickpea to the Mediterranean basin. II. Factors influencing yield under drought. Field Crops Res. 34:137–146.

Silim, S.N., M.C. Saxena, and W. Erskine. 1993a. Adaptation of lentils to the Mediterranean environments. I. Factors influencing yield under drought conditions. Exp. Agric. 29:9–19.

Silim, S.N., M.C. Saxena, and W. Erskine. 1993b. Adaptation of lentils to the Mediterranean environments. II. Response to moisture supply. Exp. Agric. 29:112–118.

Singh, K.B., R.S. Malhotra, M.C. Saxena, and G. Bejiga. 1997. Superiority of winter sowing over traditional spring sowing of chickpea. In Mediterranean region. Agron. J. 89:112–118.

Singh, K.B., and M.C. Saxena. 1996. Winter chickpea in Mediterranean-type environments. Tech. Bull. ICARDA, Aleppo, Syria.

Soja, G., A. Soja, and R. Zarghami. 1988. Early screening of faba bean (*Vicia faba* L.) for drought resistance. FABIS Newsl. 22:20–24.

Subbarao, G.V., C. Johansen, A.E. Slinkard, R.C. Nageswara Rao, N.P. Saxena, and Y.S. Chauhan. 1995. Strategies for improving drought resistance in grain legumes. Crit. Rev. Plant Sci. 14(6):469–523.

Turner, N.C. 1979. Drought resistance and adaptation to water deficits in crop plants. p. 343–372. *In* H. Mussell and R.C. Staples (ed.) John Wiley & Sons, New York.

Turner, N.C., and M.E. Nicholas. 1987. Drought resistance in wheat for light textured soils in a Mediterranean climate. p. 203–216. *In* J.P. Srivastava (ed.) Drought tolerance in winter cereals. John Wiley & Sons, Manchester, UK.

Wery, J., S.N. Silim, E.J. Knights, R.S. Malhotra, and R. Cousin. 1994. Screening techniques and sources of tolerance to extremes of moisture and air temperature in cool-season food legumes. Biotic and abiotic stresses constraining productivity of cool-season food legumes in Asia, Africa and Oceania. p. 339–456. *In* F.J. Muehlbauer and W.J. Kaiser (ed.) Expanding the production and use of cool-season food legumes. Kluwer Academic Publ. Dordrecht, The Netherlands.

World Meteorological Organization. 1975. Drought and agriculture. Tech. note 138. prepared by C.E. Hounam (Chairman), J.J. Burgos, M.S. Kulik, W.C. Palmer and J. Rodda.

18 Soil Fertility Enhancement in Mediterranean-type Dryland Agriculture: A Prerequisite for Development

John Ryan
ICARDA
Aleppo, Syria

ABSTRACT

Historically, the major factors dictating the growth of the various civilizations that flourished in the Middle East were the availability of water and fertile soils. Sustained cropping was only possible when the soils were rich in nutrients or where fertility was regenerated through flood-borne sediments. While the dominant constraint today to grow crops is still inadequate moisture, due to low and erratic rainfall and limited water supplies, economic crop production is not possible without an adequate supply of the essential nutrients, either from the soil or added as fertilizers or manures. As with soils elsewhere in the world, the native fertility of the Mediterranean region is insufficient to continuously support economic yields of modern crops. Research during the past three to four decades, a time when chemical fertilizers began to be used extensively in the region, has established the essential need for fertilizer nitrogen (N) in all but the most drought—stressed areas. Similarly, the calcareous nature of most soils in the region is such that native phosphorus (P) fertility is low, and thus without added P fertilizer only meager crop yields are possible. Fortunately, other important crop growth elements such as potassium (K), calcium (Ca), magnesium (Mg), and sulfur (S) are well supplied in the region's weakly weathered and poorly leached soils. In recent years, awareness has developed of the importance of micronutrients for crop production, and deficiencies of iron (Fe) and zinc (Zn) are common, while both boron (B) deficiency can occur, as well as toxicity of the element. This brief overview of various aspects of soil fertility and plant nutrition in the primarily dryland Mediterranean region gives a chronology of developments in soil fertility and related plant nutrition and a "birds-eye" view of the main research accomplishments. While much of the work reported comes from countries of the region where the author has worked (Syria, Morocco, and Lebanon), the findings are applicable to the region as a whole, given the similarity of the region's soils and climate; indeed much of the work reported was from collaboration with scientists in other countries of the Middle East region (e.g., Libya, Tunisia, Algeria, Iraq, Iran, Jordan, Turkey, Cyprus, and Spain) through various research networks. Future soil fertility and fertilizer research will involve greater use of soil analysis to identify nutrient constraints; adaptation to the nutrient needs of improved crop cultivars and indeed new crop interaction of nutrients, especially N with soil moisture; more efficient use of fertilizers in a systems context and with conservation tillage; and an awareness of environmental and human health issues.

 Challenges and Strategies for Dryland Agriculture. CSSA Special Publication no. 32.

INTRODUCTION

Maintaining soil fertilizer in agriculture is analogous to adequate nutrition and health care in humans; only in its absence do we appreciate its significance in either case. Soil fertility and food security are inextricably linked, though not always in a dramatic attention-attracting fashion. Nowhere is the issue of soil fertility so crucial as in Africa today, where poverty, hunger, and malnutrition are common. Though the causes of poverty in Africa are many and complex (Buresh et al., 1997), declining soil fertility through cropping without using adequate quantities of animal manure or fertilizers to replenish the lost nutrients is a key factor in low-output agriculture (Sanchez, 2002). Without adequate nutrient supply, the impact of other technologies, such as improved varieties, is minimal under current African conditions; *fertilizer costs are the main obstacle.* Notwithstanding the potential for nutrient cycling in Sub-Saharan Africa (Powell et al., 1995), the fundamental problem of widespread nutrient depletion (Stoorvogel and Smaling, 1990) cannot be ignored. The awareness of soil fertility in cropping production systems in Africa was taken further by Tian et al. (2001), who indicated ways in which soil fertility could be maintained within existing systems of production. The lessons learned from Africa are applicable elsewhere.

While soil fertility is also a major issue in another feed-deficit area of the world, the moisture-starved area of the Middle East, its significance is masked by the dominance of drought as a constraint—and indeed by the erroneous perception that the region is rich in economic terms and can import its own food supplies. The irony is that this historical region of the world—the center of origin of settled agriculture where many of the world's major crops (cereals, pulses, nuts) evolved, and where civilizations flourished through its bountiful agriculture (White, 1970)—is now a major food-crisis area. Nevertheless, soil fertility had a major influence on the evolution of the various people's of the region (Carter, 1974); where soils were fertile and where there was an adequate supply of water to irrigate them, societies prospered, as did people in areas where rainfall was favorable. Today, soil fertility and adequate crop nutrition are no less important than in former times, and yet are often taken for granted in terms of research support, often taking a "backseat" to more glamorous disciplines such as biotechnology and research that involves "integrated" and "participatory" approaches. As a background to considering the role of soil fertility in the agriculture of the Mediterranean region, it is pertinent to outline the context in which it is considered, that is, the region's farming systems, and the climatic and soil resources.

Farming Systems: Tradition and Current

The manner in which people in the Middle East tilled the soil and grew crops to support themselves is an age-old method going back to Biblical times (Grove, 1996). Cereals and animals were part of the scene then as they are now. The denuded landscape of the region today is stark testimony of man's efforts to eke a living from the soil of this harsh region, and has been worn threadbare in the process. Much has been written about the soil and the agriculture of the Middle East (Clawson et al., 1971; Carter, 1974; Dregne, 1976). Reviews of the cropping systems from a historical context and a relatively current perspective are provided by Gibbon (1981) and Cooper et al (1987).

In essence, the system that evolved to grow crops in a moisture-stressed region was to alternate cereals, wheat (*Triticum aestivum* L.) and barley (*Hordeum vulgare* L.), with fallow; this uncropped year allows for some moisture carryover, thus ensuring a harvestable crop once every 2 yr. Food legumes were also grown in a rotation. The system has not changed much over the centuries, except for increasing pressure currently on land use due to burgeoning populations. The range of food legumes currently used includes lentil (*Lens culinaris* Medik.), chickpea (*Cicer arietineum* L.), pea (*Pisum sativum* L.), and faba bean (*Vicia faba* L.), while forage legumes include vetch (*Vicia* spp.) and medics (*Medicago* spp.). Much research has focused on integrating such alternative crops to avoid continuous cropping and at the same time increase total crop and annual output from the system (Harris, 1995; Ryan and Abdel Monem, 1998; Jones, 1993), as well as maximizing water-use efficiency (Harris, 1994). Soil and climatic factors are important in this process.

Soils

The soil types in the Mediterranean, while having some common features, differ widely depending on parent material, landscape position, and previous climatic conditions. Thus, one finds most of the major soil orders in the world (Kassam, 1981), with Inceptisols, Lithosols, Aridisols, and Entisols being dominant, as well as large areas of Vertisols. Soil depth varies widely from shallow on hills and hillslopes to deep alluvial soils in the valleys. Soil depth is of particular importance in semi-arid moisture stressed environments as it dictates the capacity of the soil to store residual moisture from the winter season's rainfall and controls the extent to which roots can exploit the soil volume for moisture and nutrients. Thus, under similar rainfall conditions, Abdel Monem et al. (1990a) showed that cereal yields were directly related to depth in a range of Moroccan soils: Rendoll (20–30 cm), Calcixeroll (40–60 cm), and Vertisol (>100 cm). In a controlled greenhouse study (Ryan and Masri, 2003), soil depth was shown to influence the extent of response to P at any level of moisture. In a current publication (Ryan et al., 2004), updated descriptions are given of the distribution of soil properties in the Mediterranean region that impact crop production.

Climate

The general pattern for climatic variables in the Mediterranean is cold, wet conditions from November to April/May and hot dry summers (Cooper et al., 1987; Harris, 1995). However, there are, in fact, several variants of the typical Mediterranean climate due to variation in latitude and altitude (Kassam, 1981). Thus, in lowland coastal regions close to the Mediterranean Sea, the climate in warm and humid, while in highland areas (e.g., Anatolian Plateau) winters are longer and extremely cold; inland areas tend to have characteristics of a continental climate. The typical Mediterranean climate allows for cropping from November to May/June, due to increasing rainfall with a maximum in December-January, and decreasing temperature and evapotranspiration. Thus, growth occurs in the cooler wetter period, with limited evaporative loss of the moisture from rainfall (Harris, 1994). Across the Mediterranean zone, dryland cropping is possible between 200 and 600 mm rainfall per year (Cooper et al., 1987). As rainfall decreases, so too does it variability; in the lower rainfall areas, drought is frequent, often causing complete crop failure. Even in more favorable rainfall zones (<350 mm), terminal drought is common at the end of the growing season (April-May). Efforts are needed to expand drought research to include forecasting, characterization and spatialization, and vulnerability from an agroecological and livelihood perspective (De Pauw, 2004, this publication). Rainfall and its distribution are the single most important determinants of cropping in the Mediterranean region, and also dictate the efficiency of use of fertilizers, without which economic crop yields are not possible.

SOIL FERTILITY AND FERTILIZER RESEARCH

Agricultural research in the countries of the Mediterranean region reflect the state of development of any particular country's research and education institutions, as well as the intensity of the agricultural sector and the overall importance of agriculture in the country's economy. Dryland research, including soil fertility and plant nutrition research, has always received lower priority than irrigated agriculture, and understandably so. Nevertheless, fertilizer use is essential for increasing agricultural output, indeed for stimulating overall national economic development (Pala et al., 2004). Fertility-related research tended to be dictated by the type of organization involved. National Ministry of Agriculture organizations tend to conduct applied on-station, agronomic-type trials demonstrating the need for nutrients such as N and P, the appropriate fertilizer rate to obtain optimum yields of various crops, and methods of fertilizer application and timing of such applications (Ryan and Matar, 1990, 1992). Such trials later evolved to consider fertilizer-use efficiency, long-term fertilizer effects, fertilizers within cropping systems and, to a minor extent, implications for the environment (Ryan, 1997). Later, and depending on resources, there was a gradual shift to off-station, more representative conditions in farmers' fields. While more control could be exercised with on-station trials, such sites are generally high in native soil fertility or added as fertilizers where nutrients are not an experimental variable (Ryan et al., 1997a, 1980, 1990) and are often unrepresentative of conditions in farmers' fields (Abdel Monem et al., 1990b). In contrast to Ministry of Agriculture research, soil-related research in universities tended to be less problem-oriented and more theoretical.

As countries of the region entered the era of chemical fertilizer use, the question whether a particular fertilizer nutrient was needed was not the issue, but how much to apply and how effectively should it be applied. In that regard, the Soil Test Calibration Program, operated by ICARDA in conjunction with various national programs and funded by the United Nations Development Program (UNDP), and partly by the Institut Mondial du Phosphate (IMPHOS), established valid soil tests calibrated for giving recommendations for various field crops. Several issues directly and peripherally related to this program were reported in a number of regional workshops (Ryan and Matar, 1990, 1992; Ryan 1997). The program set the scene for common protocols for dryland field experimentation on nutrient and fertilizer needs and helped form a communications network among countries of the region (Ryan et al., 1995a).

The following brief glimpse of soil fertility and fertilizer reflects both an evolution and a cross-section—indeed a flavor—of research issues common to many countries of the region with substantial areas of dryland agriculture. Much of what has been learned has been communicated to researchers in the various national research and educational programs. The challenge of transferring fertility and fertilizer-related research to the region's farmers remain a daunting task and is as much dependent on socioeconomic factors as technical ones.

Nitrogen

While drought is the major and ever-present constraint in dryland agriculture of the Mediterranean region, the evidence that has accumulated since the 1970s or so of fertilizer research and fertilizer use by farmers suggests that N is virtually always needed for nonlegume crops and that economic yields are impossible without fertilization (Shroyer et al., 1990; Ryan et al., 1992a). This general observation has been confirmed by the review of Harmsen (1984) and by workshop proceedings that involved scientists from various countries of the Mediterranean who reported the findings of N-related research primarily from on-station and some from on-farm trials (Ryan and Matar, 1990, 1992; Ryan, 1997). In essence, the almost ubiquitous response to fertilizer N in all but the driest of conditions is related to the low level of organic matter (Ryan, 1998) and, consequently, the limited reserves of potentially mineralizable N (Matar et al., 1991). The early stages of adoption of mineral fertilizers in the mid-1970s (Ryan, 2003) especially N, coincided with the initiation of dryland N research at ICARDA in conjunction with national scientists in the region.

With the establishment of ICARDA in Syria in 1977, with its main station at Tel Hadya (330 mm yr^{-1}) and substations in wetter and drier areas in Syria as well as colder ones in Lebanon (Ryan et al., 1997a), the importance of seasonal rainfall for crop yields was well documented (Keatinge et al., 1985, 1986; Pala et al., 1996). From the initial research priority-setting (Monteith and Webb, 1981), the importance for using fertilizer N to obtain acceptable economic yields was acknowledged and validated by both on-station and on-farm trials (Pala et al., 1986; Jones and Wahbi, 1992; Wahbi et al., 1993). The review of N in dryland systems by Harmsen (1984) was a major milestone in the institution's N research program; it highlighted the various agronomic and physiological parameters that reflect efficient N fertilizer use.

Considerable attention was given to cool-season food and feed legumes in the cereal-production system common throughout the Mediterranean region (Abd El Moneim and Ryan, 2004, this publication). A basic concern was how much N any legume crop added to the succeeding cereal crop as a result of biological N_2 fixation. Beck et al. (1991) estimated the proportion of N derived from the atmosphere (Ndfa) to be about 70% in both chickpea and lentil; however, these values varied with cultural practices. With winter sowing of chickpea, Ndfa increased to 72% compared to 26% for spring sowing, and from 52 to 72% with rhizobial inoculation. Good soil and crop management, such as P fertilization, increased Ndfa from 55 to 69% (Keatinge et al., 1988). Such food legumes contribute at least 10 kg N ha^{-1} to the cereal crop.

Numerous studies addressed soil N behavior. While early work of Ryan et al. (1981a) had demonstrated under laboratory conditions that volatile loss of urea could be substantial and was influenced by soil type and properties (calcium carbonate, iron oxides), the field study of Abdel Monem (1986) under cool growing season conditions showed actual losses of only11 to 18% of the N applied. In related work, Matar et al. (1991) showed that soils such as Mollisols had a relatively high potential to release N through mineralization, but the actual rate of mineralization varied little between the soils studied. Similarly, based on soil test calibration work in the field, Matar et al. (1990) showed that mineral N in the top 0 to 60 cm soil was best correlated with yield and crop response, and thus could be used as an indicator of N sufficiency; following a legume or a low-yielding crop, about 8 mg kg^{-1} of nitrate (NO_3^-)-N was considered the critical value, while a value of about 15 mg kg^{-1} was suitable after a summer crop or a high-yielding cereal.

Given the importance of rotational cropping, many long-term trials were designed to assess various legumes grown alternatively with wheat and barley (Jones, 1998); similar trials were conducted throughout the region from Morocco to Jordan (Ryan and Abdel Monem, 1998). Virtually all trials had a N component to evaluate its response in terms of crop yield and quality, its residual effects, and indirect influence on soil properties (Harris, 1995, Harris et al., 1995; Ryan, 1998; Jones, 1998). A few brief observations on N in such trials are pertinent.

In the main "Cropping Systems Productivity" rotation trial, there were consistent cereal grain and biomass responses to applied N fertilizer (Harris, 1995, Harris et al., 1995); yields were maximized at 30 kg N ha^{-1} in dry years and up to 90 kg N ha^{-1} in favorable rainfall years. Because of the addition of N to the soil through N_2 fixation in the legume phase, relative responses to applied N were lower compared to continuous cereals or cereal after fallow. In addition, some legumes, notably vetch and medic, increased both total soil N and organic matter (Ryan, 1998), including the total, labile, and biomass fractions (Ryan et al., 2002a), and the N-mineralization potential (Ryan et al., 2003). A side benefit of the increased organic matter was an improvement in aggregate stability and related physical parameters (Masri et al., 1998).

Other studies in Syria (Garabet et al., 1998a, 1998b) and Morocco (Abdel Monem and Ryan, 1990a; Mergoum et al., 1994; Ryan et al., 1989, 1991a, 1992a, 1993a, 1994a) have shown a similar N response with increasing rainfall. Indeed, various studies have shown that N can reduce disease incidence (Jones et al., 1990), improve grain quality in durum wheat (Mahdi et al., 1996; Ryan et al., 1997b), and reduce damage from Hessian fly (*Mayetiola destructor*), a major pest of cereals in North Africa (Ryan et al., 1991b, 1998a). A cross-section of country reports from countries as diverse as Tunisia, Turkey, Yemen, and Jordan (Ryan et al., 1990, 1992; Ryan, 1997) clearly showed the influence of seasonal rainfall in dryland crop yields in the Mediterranean region. While N fertilizers can increase total grain in protein, the extent to which it can influence essential amino acids is limited (Barg et al., 1982).

As dryland cropping in Syria is being transformed through supplemental irrigation during the normal rainfed year, the concept of deficit irrigation is to apply a limited amount of water during relatively dry periods to stabilize yields at an economically optimum level (Perrier and Salkini, 1991). Thus, the distinction between purely rainfed agriculture and irrigated agriculture has become blurred. While recent research (Oweis et al., 1998, 1999; Garabet et al., 1998a, 1998b) has clearly shown the judicious use of a few irrigations in the growing season, along with adequate N fertilization and early sowing, could greatly increase water-use efficiency and produce acceptable and economic yields, the practice will remain a theoretical one until water has a cost; without water charges, and having invested in pumping equipment, there is no incentive for farmers to limit the amount of water they use.

Soil and Fertilizer Phosphorus

Because of its complex chemical nature and soil reactions that restrict availability in soils, as well as its essential need and importance, soil and fertilizer P has been the subject of innumerable studies at both applied and basic levels. In their natural unfertilized condition, most soils are invariably low or deficient in plant-available P. In soils of the Mediterranean region, the factor adversely constraining P availability is solid-phase calcium carbonate ($CaCO_3$), which invariably occurs in such soils, often up to 50% or more of the soil volume (Ryan, 1983; Afif et al., 1993; Matar et al., 1992). Despite the predominance of $CaCO_3$, iron oxides, which occur in smaller amounts, usually 1 to 5%, can have a disproportionate role in adsorption/precipitation of soluble P (Ryan et al., 1985a; Torrent, 1987), as well as influencing the rate of reversion of soluble P to relatively insoluble forms (Ryan et al., 1985c, 1986). Notwithstanding such basic-type studies of P in Mediterranean soils as reviewed by Matar et al. (1992), applied studies have focused on field-crop responses and assessment of plant-available P, while a recent review of Ryan (2003) focused exclusively on P research in Syria. In this brief overview of P research, studies from various countries around the Mediterranean area are cited.

Logically, the more obvious studies conducted in most countries were field ones, and most have conclusively demonstrated the need for P fertilizer in countries such as Lebanon (Ryan and Hamzé, 1987), Syria (Matar, 1977; Matar et al., 1988; Gregory et al., 1986; Matar and Brown, 1989a, 1989b), Cyprus (Krentos and Orphanos, 1979), and Morocco (Shroyer et al., 1990, Abdel Monem and Ryan, 1990a, 1990b; Ryan et al., 1992b, 1992c, 1993b, 1995a, 1995b, 1995c). While most of these studies involved cereals, forage legumes were also shown to respond well to fertilizer P (Materon and Ryan, 1995, 1996). Many of these studies also showed that P improved root growth and water-use efficiency, and that banding was more efficient than broadcasting (Ryan, 2002). Numerous field trials conducted in countries of the Mediterranean region and reported in various workshops (Ryan and Matar, 1990, 1992; Ryan, 1997) indicate that fertilizer application rates for dryland crops generally range from 0 to 40 kg P ha^{-1} normally as superphosphate, triple-superphosphate, or NPK compounds, depending on the extent of soil P availability. Notwithstanding some potential advantages of using other sources of P in Middle Eastern soils, for example, urea phosphate (Ryan and Tabbara, 1989; Ryan et al., 1988), and rock phosphate (Habib et al., 1999), there is little likelihood of any replacement of the common commercial P fertilizers for field crops.

The concept of assessing P availability, or diagnosing deficiency, was introduced to the Mediterranean region since the 1980s or so. Of the many tests available for soil P, including those commonly used in the region (Ryan, 1983; Matar et al., 1992), the Olsen procedure (0.5 M $NaHCO_3$) has been shown to be the most effective indicator of the available P status of the soil (Matar et al., 1988; Ryan and Ayubi, 1981). The many soil test calibration studies conducted in the region (Ryan and Matar, 1990, 1992) suggest a critical range of 5 to 7 mg kg^{-1} for dryland cereals and pulses, below which P fertilizer is needed and above which it is adequate.

Other studies have shown how spatially variable available soil P is in the field (Ryan et al., 1998c) as well as temporally (Ryan et al., 1993a), and that soil P forms are related to weathering and soil development (Ryan and Zghard, 1980). A more recent study (Ryan and Masri, 2003) indicated the importance of soil depth for interpreting soil test values for dryland crops, primarily because depth dictates the soils's moisture-holding capacity. Other studies on P mineralization in pots (Habib et al., 1994) and under rotational conditions in the field (Kabengi et al., 2003) showed that while the process can increase P availability, depending on total organic matter and environmental conditions, the effect can also be inconsistent.

Numerous developments of practical importance regarding soil and fertilizer P are worthy of note. Despite the notion of "fixation" of soluble P based on laboratory and greenhouse studies, observations in the field indicate that with regular P fertilizer application, residual P builds up fairly quickly, thereby eliminating a current-year response to P (Ryan et al., 1994b). Thus, soil testing is seen as an effective tool to monitor plant-available P and thus more effectively use this costly, but essential, input. However, one *caveat* remains for soil testing, and one that is the need for having reliable tests and good management of soil laboratories, with internal and external standardization and quality control (Ryan and Garabet, 1994; Ryan et al., 1999a; Ryan, 2000a). Clearly, there is a need for collaboration between international research centers and agencies involved in quality assurance to effective use of soil and plant analysis in agricultural development (Ryan et al., 2002b). Though this concern applies to all soil properties and nutrients, it is especially important for P in both soils and plants.

As irrigation, especially supplemental irrigation, is making inroads in previously dryland cropping areas, there are some implications for P fertilizer use. With higher crop yields from irrigation, fertilizer application rates need to be adjusted upwards, as well as critical soil P levels; a figure of 15 mg P kg^{-1} is considered a critical value from applied research elsewhere. Even apart from such irrigation using conventional water sources and irrigations, there is a growing awareness of the possibilities of using wastewater for irrigation (Ryan et al., 1999b) and indeed using P in modern drip irrigation systems (Ryan and Saleh, 2000; Ryan, 2000b).

Potassium

The K status of any soil is largely related to the soil mineralogy and soil texture and the extent of weathering or leaching. By comparison with other regions of the world, the soils of the Mediterranean are generally well supplied with available

K. For example, early research in Lebanon showed that many soils were not only high in available K (Ryan and Hamzé, 1987), but had a large capacity to release K for crop K uptake (Sahyouni and Ryan, 1983). Similarly, surveys of experiment stations (Ryan et al., 1997a) and farmers' fields (Ryan et al., 1996, 1997b) across a range of rainfall zones in northern Syria confirmed the general adequacy of available K in the region's soils.

While the satisfactory status of available K was also reflected in reports from various countries of the Mediterranean region (Mengel and Krauss, 1993; Johnston, 1999), one has to consider that K fertilization is likely to be needed under intensively irrigated conditions and on relatively lighter-textured soils and where high-K demanding crops such as potatoes (*Solanum tuberosum* L.) and sugarbeet (*Beta vulgaris* L.) are grown. While K fertilizer use is likely to remain small by comparison with N and P in the Mediterranean region (Ryan, 2003), especially in dryland or rainfed cropping, one cannot overlook the additional and indirect physiological benefits of K in terms of confirming or enhancing drought, cold and disease resistance in crops, as well as improving crop quality. Balanced fertilization is a current concept that implies having all essential and beneficial nutrients in adequate amounts and proportions.

Micronutrients

While the role of micronutrients in agriculture in the West has been widely recognized and appreciated for the past half century or so (Mortvedt et al., 1972), the dawning of such an awareness in the Middle Eastern region has been relatively new, particularly in dryland agriculture where, in the hierarchy of nutrient constraints for crop production, an influence of micronutrients was obscured by drought and, to a lesser extent, N and P. Yet deficiencies of micronutrients such as zinc (Zn), iron (Fe), boron (B), and possibly manganese (Mn) and copper (Cu) are likely to occur in some circumstances. With the exception of B, the solubility of these metals is lower under the high pH conditions buffered by solid-phase $CaCO_3$ that occurs in most soils of the Mediterranean region.

In anticipation of the significance of micronutrients, several surveys and laboratory studies of elements such as Mn (Khan and Ryan, 1978; Curtin et al., 1980), (Bhatti et al., 1982), Cu (Curtin et al., 1993), and B (Khan et al., 1979) were conducted in Lebanon, on the instigation of the late Professor Kermit Berger, who was one of the earliest micronutrient researchers in the USA, especially involving B. The conclusion from these and related studies (Ryan et al., 1981b) was that B was unlikely to be a significant constraint to crop production in Lebanon's mainly irrigated agriculture, and even more unlikely under dryland cropping conditions. However, Fe deficiency was deemed to be important in ornamental plants (Hamzé et al., 1985) and in citrus, where grafting onto resistant rootstocks was the preferred solution to lime-induced chlorosis (Hamzé et al., 1986), or selection of resistant cultivars, as in the case of dryland chickpea (Hamzé et al., 1987). While various studies addressed the issue of evaluating the effectiveness of micronutrient sources in soils (Ryan and Prasad, 1979; Ryan and Hariq, 1983; Ryan et al., 1985b), only some chelates, for example, Fe EDDHA, are effective when soil-applied, but because of costs they are not feasible in low crop-output conditions.

While the early work in Lebanon did not indicate any major concerns with micronutrients, it did stimulate an increased awareness of their possible importance in the wider geographic area of West Asia and North Africa, where conditions are generally drier than those in Lebanon. As forage legume, production is a major research area at ICARDA, and as such crops are known to be sensitive to Fe and Zn, several studies of medics (*Medicago* spp.) incorporated assessing Zn as a factor in production. Thus significant growth responses were evoked by adding Zn at low levels (e.g., 10 mg kg^{-1}) in greenhouse (Materon and Ryan, 1985) and field studies (Materon and Ryan, 1996); crop growth responses are unlikely where DTPA-extractable Zn is above 0.5 mg kg^{-1}. Thus, soil testing is a good basis for problem diagnosis. Later studies (Abd El Moneim and Ryan, 2000) indicated a possible role of Zn in ameliorating "lathyrism" toxicity in grasspea (*Lathyrus sativa*).

Another major area of endeavor was with B, this time the concern was with B toxicity. Early observations and experience from Australia had suggested that fungal disease-like symptoms on cereals in Syria and Turkey-might be due to high or toxic levels of B, a phenomenon associated only with dry areas (Gupta, 1993). Thus, studies at ICARDA demonstrated the toxic effect of relative moderate levels of soluble B in the soil (Mahalaksmi et al., 1995). In field conditions, soluble B is variable, while toxic concentrations are often found in subsoils and are not detected by normal sampling in the surface horizon, for example, 0 to 20 cm (Ryan et al., 1998c). However, as it is impractical to remove excess B from the root zone, the only feasible solution is to adapt crops to such conditions. Thus, various landraces and cultivars of durum and bread wheat (Yau et al., 1995a, 1995b, 1997) were identified as being tolerant to excess B, based on screening in variable B media. Resistance to toxicity is genetically controlled by a limited number of genes, and breeding for resistance has been incorporated into the cereal development program.

Future efforts will involve identification of probable micronutrient "hotspots' in various agroecological zones, the incorporation of genetic resistance, or tolerance to both deficiency and toxicity, where feasible, into new crop cultivars, and the implications of nutrients such as Fe and Zn in the entire food chain, from soil to plant to human. Efforts are underway to identify "micronutrients-dense" crops to improve human nutrition.

Future Perspective

It is abundantly clear that much has been learned about soil fertility and fertilizer use in the Mediterranean conditions, both for dryland and irrigated conditions. Given the broad commonalities in the region and the limited resources for research, future efforts should concentrate on sharing what is already known with the region's scientists and focusing on transferring that knowledge in a useable package to farmers; soil fertility is one area where there is little or no justification for "re-inventing the wheel" as far as research is concerned. The use of models will be invaluable to synthesize information and predict behavior under any circumstances as done by Daroub et al. (2003) for P in soils of the Mediterranean. While there is now a good basis for efficient and economic application of fertilizers through soil analysis (Ryan, 2000a), and to a lesser extent with plant analysis (Papastylianou, 1986; Papastylianou and Puckridge, 1983), the potential of using such technologies at farmers' level need to be exploited. Other areas of endeavor will include assessing the nutritional needs of new crop cultivars being developed, an emphasis on crop quality from the human and animal nutrition standpoint, management of nutrients so as to avoid any negative impact on the environment, the use of crops from the standpoint of C sequestration, and the use of wastewater as a source of both water and nutrients. While every effort should be made to use any potential sources of nutrients in a resource-poor region as the Middle East, such as municipal wastes and various organic materials (Khuri et al., 1987; Ryan et al., 1984, 1985d), there is no alternative to commercial fertilizers in crop production. The onus is on researchers and extension agents to provide farmers with the means to use P fertilizers in the most efficient and economic manner in the interests of maximizing crop production within the limits of the Mediterranean dryland environment.

REFERENCES

Abd El Moneim, A.M., and J. Ryan. 2000. Nutritional influences on neurotoxin concentrations in grasspea (*Lathyrus sativus*). p. 39. *In* Third Int. Crop Sci. Congr., Hamburg, Germany. 17–22 Aug. 2000. Abstracts. European Soc. of Agron.

Abd El Moneim, A.M., and J. Ryan. 2004. Forage legumes for dryland agriculture in West Asia and North Africa. p. 243–256. *In* S.C. Rao and J. Ryan (ed.) Challenges and strategies for dryland agriculture. CSSA Spec. Publ. 32. CSSA and ASA, Madison, WI.

Abdel Monem, M. 1986. Labeled urea fertilizer experiments on soils of the Mediterranean region. Ph.D. thesis. Colorado State Univ., Fort Collins.

Abdel Monem, M., and J. Ryan. 1990a. Profitability of Hessian fly-resistant wheat in rainfed areas of Morocco: Nitrogen and location effects. Agric. Medit. 120:429– 435.

Abdel Monem, M., and J. Ryan. 1990b. Field crop response to phosphorus. Rachis. 9(2):38–39.

Abdel Monem, M., J. Ryan, and M. El Gharous. 1990b. Preliminary assessment of the soil fertility status of the mapped area of Chaouia. Al-Awamia 72:85–107.

Abdel Monem, M., M.A. Azzaoui, M. El-Gharous, J. Ryan, and P.N. Soltanpour. 1990a. Response of wheat to N and P in some Moroccan soils. p. 52–65. *In* J. Ryan and A. Matar (ed.) Proc., Third Regional Soil Test Calibration Workshop, Amman, Jordan. 3–9 Sept. 1998. ICARDA, Aleppo, Syria.

Afif, E., A. Matar, and J. Torrent. 1993. Availability of phosphate applied to calcareous soils of North Africa–West Asia. Soil Sci. Soc. Am. J. 57:756–760.

Barg, M., G.S., J. Ryan, and K.C. Berger. 1982. Effect of nitrogen fertilization on yield, protein content, and amino acids of irrigated Mexipack wheat. Iran Agric. Res. 1:17–240.

Beck, D. P., J. Werry, M. C. Saxena, and A. Ayadi. 1991. Dinitrogen fixation and nitrogen balance in cool-season legumes. Agron. J. 83:334–341.

Bhatti, A., K. C. Berger, and J. Ryan. 1982. Available zinc status of Lebanese soils. Iran Agric. Res. 1:41–47.

Buresh, R.J., P.A. Sanchez, and F. Calhoun. 1997. Replenishing soil fertility in Africa. SSSA Spec. Publ. 51. SSSA, Madison, WI.

Carter, V.G. 1974. Topsoil and civilization. Univ. of Oklahoma, Norman.

Clawson, M., H.H. Landsberg, and L.S. Alexander. 1971. The agricultural potential of the Middle East. Elsevier Sci., New York.

Cooper, P.J.M., P.J. Gregory, D. Tully, and H.C. Harris. 1987. Improving water use efficiency of annual crops on the rainfed farming systems of West Asia and North Africa. Exp. Agric. 23:113–158.

Curtin, D., J. Ryan, M. Ahmed, and I. Piracha. 1993. Retention and extractability of copper and zinc in calcareous Lebanese soils. Lebanese Sci. Bull. 6(1):7–16.

Curtin, D., J. Ryan, and R.A. Chaudhary. 1980. Manganese adsorption and desorption in calcareous Lebanese soils. Soil Sci. Soc. Am. J. 44:947–950.

Daroub, S.H., A. Gerakis, J. Ritchie, D. Friesen, and J. Ryan. 2003. Development and testing of a soil-plant phosphorus simulation model. Agric. Syst. 76:1157–1181.

De Pauw, E. 2004. Drought early warming systems for the Near East. p. 93–112. *In* S.C. Rao and J. Ryan (ed.) Challenges and strategies for dryland agriculture. CSSA Spec. Publ. 32. CSSA and ASA, Madison, WI.

Dregne, H.E. 1976. Soils of arid regions. Elsevier Sci., New York.

Garabet, S., J. Ryan, and M. Wood. 1998a. Nitrogen and water effects on wheat yield in a Mediterranean-type climate. II. Nitrogen-use efficiency by the direct and difference methods. Field Crops Res. 58:213–221.

Garabet, S., M. Wood, and J. Ryan. 1998b. Nitrogen and water effects on wheat yield in a Mediterranean-type climate. I. Dry matter yield and nitrogen accumulation. Field Crops Res. 57:309–318.

Gibbon, D. 1981. Rainfed farming systems in the Mediterranean region. Plant Soil 58:59–80.

Gregory, P.J., K.D. Shepherd, and P.J. Cooper. 1986. Effects of fertilizer on root growth and water use of barley in northern Syria. J. Agric. Sci. 103:429–438.

Grove, A.T. 1996. The historical context: Before 1850. p. 13–17. *In* C.J. Brandt and J.B. Thornes (ed.) Mediterranean desertification and land use. Wiley and Sons Publ., Chichester, UK.

Gupta, U. 1993. Boron and its role in crop production. CRC Press, Ann Arbor, MI.

Habib, L., S.H. Chien, G. Carmona, and J. Henao. 1999. Rape response to a Syrian phosphate rock and its mixture with triplesuperphosphate in a lined alkaline soil. Commun. Soil Sci. Plant Anal. 30(3&4):449–456.

Habib, L., S. Hayfa, and J. Ryan. 1994. Temporal change in organically amended soil: Implications for phosphorus solubility and adsorption-desorption. Commun. Soil Sci. Plant Anal. (19&20):3281–3290.

Hamzé, M., J. Ryan, R. Mikdashi, and M. Solh. 1987. Evaluating of chickpea (*Cicer arietinum* L.) genotypes for lime-induced chlorosis. J. Plant Nutr. 10:1031–1039.

Hamzé, M., J. Ryan, R. Shwayri, and M. Zaabout. 1985. Iron treatment of lime-induced chlorosis: Implications for chlorophyll, Fe^{2+}, Fe^{3+}, and K content in leaves. J. Plant Nutr. 8(5):437 – 448.

Hamzé, M., J. Ryan, and M. Zaabout. 1986. Screening of citrus rootstocks for lime-induced chlorosis tolerance. J. Plant Nutr. 9(2):459– 489.

Harmsen, K. 1984. Nitrogen fertilizer use in rainfed agriculture. Fert. Res. 5(4):371–382.

Harris, H.C. 1994. Water use efficiency of crop rotations in a Mediterranean environment. Aspects Appl. Biol. 38:165–172.

Harris, H. 1995. Long-term trials on soil and crop management at ICARDA. R. Lal and B.A. Stewart (ed.) Adv. Soil Sci. 19:447–469. CRC Lewis Publ., Boca Raton, FL.

Harris, H., J. Ryan, A. Matar, and T. Teacher. 1995. Nitrogen in dryland farming systems common in northwestern Syria. p. 323–335. *In* J.M. Powell et al. (ed.) Livestock and Sustainable Nutrient Cycling in Mixed Farming Systems of Sub-Saharan Africa. Vol. II. Technical Papers.Proc. of an Int. Conf., Addis Ababa, Ethiopia. 22–26 Nov. 1995. ILCA, Addis Ababa, Ethiopia.

Johnston, A.E. 1999. Food security in the WANA region, the essential need for balanced fertilization. Int. Potash Inst., Basel, Switzerland.

Jones, J., M. Abdel Monem, and J. Ryan. 1990. Nitrogen effects on tan spot of durum wheat lines. Med. J. Plant Prot. 8:10–13.

Jones, M. 1993. Sustainable agriculture: An explanation of a concept: Crop protection and sustainable agriculture. p. 30–47. Tropical Agric. Res. Ser. 24, John Wiley, Chichester, UK.

Jones, M.J. (ed.) 1998. The challenge of production sustainability: Long-term studies in agronomic research in dry areas. Abstracts and conclusions of a workshop held at ICARDA, Aleppo, Syria. 8–11 Dec. 1997. ICARDA, Aleppo, Syria.

Jones, M.J., and A. Wahbi. 1992. Site-factor influences on barley response to fertilizer in on-farm trials in northern Syria: descriptive and predictive models. Exp. Agric. 28:63–87.

Kabengi, N., R. Zurayk, R. Baalbaki, and J. Ryan, 2003. Phosphorus availability and characterization a under long-term rotation trial. Commun. Soil Sci. Plant Anal. 34(3&4):375–392.

Kassam, A.H. 1981. Climate, soil and land resources in the West Asia and North Africa. Plant Soil 58:1–28.

Keatinge, J.D.H., N. Chapanian, and M.C. Saxena. 1988. Effect of improved management of legumes in a cereal-legume rotation on field estimates of crop nitrogen uptake and symbiotic nitrogen fixation in northern Syria. J. Agric. Sci. (Cambridge) 110:651–659.

Keatinge, J.D.H., M.D. Dennett, and J. Rogers. 1985. The influence of precipitation regime on the management of three-course crop rotations in northern Syria. J. Agric Sci. (Cambridge) 104:281–287.

Keatinge, J.D.H., M.D. Dennett, and J. Rogers. 1986. The influence of precipitation regime on the crop management in dry areas in northern Syria. Field Crops Res. 12:239–249.

Khan, M.A., and J. Ryan. 1978. Manganese availability of calcareous soils of Lebanon. Agron. J. 70:79–82.

Khan, Z.D., J. Ryan, and K.C. Berger. 1979. Available boron in calcareous soils of Lebanon. Agron. J. 71:688–690.

Khouri, N., A.T. Shamas, and J. Ryan. 1987. Greenhouse evaluation of Beirut municipal compost. Lebanese Sci. Bull. 3(2):53–63.

Krentos, V.D., and P.I. Orphanos. 1979. Nitrogen and phosphorus fertilizers for wheat and barley in a semi-arid region. J. Agric. Sci. 93:711–717.

Mahalaksmi, V., S.K. Yau, J. Ryan, and J.M. Peacock. 1995. Boron toxicity in barley (*Hordeum vulgare* L.) seedings in relation to soil surface temperature. Plant Soil 177:151–156.

Mahdi, L., C. J. Bell, and J. Ryan. 1996. Non-vitreousness ("Yellow berry") in durum wheat as affected by both depth and date of planting. Cereal Res. Commun. 24(3):347–352.

Masri, Z., J. Ryan, and S. Masri. 1998. Changes in soil physical properties with continuous cereal cropping in northern Syria. p. 643–648. *In* N. Munsuz et al. (ed.) The "M. Sefik Yesilsoy" Int. Symp. on Arid-Region Soils (held in honor of Prof. Yesilsoy deceased 1993), Izmir, Turkey. 21–24 Sept. 1998. Soil Sci. Soc. of Turkey, Ankara.

Matar, A.E. 1977. Yields and response of cereal crops to phosphorus fertilization under changing rainfall conditions. Agron. J. 69:879–881.

Matar, A.E., D. Beck, M. Pala, and S. Garabet. 1991. Nitrogen mineralization potentials of selected Mediterranean soils. Commun. Soil Sci. Plant Anal. 21:33–36.

Matar, A.E., and S.C. Brown. 1989a. Effect of rate and method of phosphate placement on productivity of durum wheat in Mediterranean environments. I. Crop yields and P uptake. Fert. Res. 20:75–82.

Matar, A.E., and S.C. Brown. 1989b. Effect of rate and method of phosphate placement on productivity of durum wheat in a Mediterranean climate. II. Root distribution and P dynamics. Fert. Res. 20:83–88.

Matar, A.E., S. Garabet, S. Riahi, and A. Mazid. 1988. A comparison of four soil test procedures for determination of available phosphorus in calcareous soils of the Mediterranean region. Commun. Soil Sci. Plant Anal. 19:127–140.

Matar, A.E., M. Pala, D. Beck, and S. Garabet. 1990. Nitrate-N test as a guide to N fertilization of wheat in the Mediterranean region. Commun. Soil Sci. Plant Anal. 21:1171–1130.

Matar, A.E., J. Torrent, and J. Ryan. 1992. Soil and fertilizer phosphorus and crop responses in the dryland Mediterranean zone. Adv. Soil Sci. 18:81–146.

Materon, L.A., and J. Ryan. 1995. Rhizobial inoculation, and phosphorus and zinc nutrition for annual medics (*Medicago* spp.) adapted to Mediterranean-type environments. Agron. J. 87:692–698.

Materon, L., and J. Ryan. 1996. Effects of rhizobial inoculation and mineral nutrition of common Mediterranean pasture and forage legumes. Agric. Medit. 126(1):64–74.

Mengel, K., and A. Krauss. 1993. Potassium availability in soils of West Asia and North Africa: Status and perspectives. Proc., Regional Symp., Tehran, Iran. 19–22 June 1993. Int. Potash Inst., Basel, Switzerland.

Mergoum, M., J. Ryan, M. El-Gharous, and M. Amrani. 1994. Dryland triticale: Varying seeding rates and nitrogen fertilization. Al-Awamia 90:89–96.

Monteith, J., and C. Webb. 1981. Soil water and nitrogen in Mediterranean-type environments. Developments in plant and soil sciences. Vol. 1. Martinus Nijhoff/Dr. W. Junk, Publ., The Hague, Holland.

Mortvedt, J., P.M. Giordano, and W. Lindsay. 1972. Micronutrients in agriculture. SSSA, Madison, WI.

Oweis, T., M. Pala, and J. Ryan. 1998. Stabilizing rainfed wheat yields with supplemental irrigation in the Mediterranean region. Agron. J. 90:672–681.

Oweis, T., M. Pala, and J. Ryan. 1999. Impact of supplemental irrigation, nitrogen and planting date on yield and quality of durum wheat. Eur. J. Agron. 11:255–266.

Pala, M., A. Matar, and A. Mazid. 1996. Assessment of the effects of environmental factors on the response of wheat to fertilizer in on-farm trials in a Mediterranean-type environment. Exp. Agric. 32:339–349.

Pala, M., J. Ryan, A. Mazid, O. Abdallah, and M. Nachit. 2004. Wheat farming in Syria: An approach to economic transformation and sustainability. Renewable Agric. Food Syst. 18(3):(in press).

Papastylianou, I. 1986. Diagnosis of nitrogen deficiency in barley in different rotation systems by plant analysis. Fert. Res. 9:241–250.

Papastylianou, I., and D.W. Puckridge. 1983. Stem nitrate nitrogen and yield of wheat in a permanent rotation experiment. Aust. J. Agric. Res. 34:599–606.

Perrier, E.R., and A.B. Salkini. 1991. Supplemental irrigation in the Near East and North Africa. Kluwer Academic Press, Dordrecht, The Netherlands.

Ryan, J. 1983. Phosphorus in soils of arid regions. Geoderma 19:341–354.

Ryan, J. (ed.) 1997. Accomplishments and future challenges in dryland soil fertility research in the Mediterranean area. Proc., Int. Soil Fertility Workshop, Aleppo, Syria. 19–23 Nov.1995. ICARDA, Aleppo, Syria.

Ryan, J. 1998. Changes in organic carbon in long-term rotation and tillage trials in northern Syria. p. 285–295. *In* R. Lal et al. (ed.) Management of carbon sequestration in soil. Adv. Soil Sci. CRC, Boca Raton, FL.

Ryan, J. 2000a. Soil and plant analysis in the Mediterranean region: Limitations and potential. Commun. Soil Sci. Plant Anal. 31(11 – 14):2147–2154.

Ryan, J. (ed.) 2000b. Plant nutrient management under pressurized irrigation systems in the Mediterranean region. Proc., Int. Workshop, Amman, Jordan. 25–27 Apr. 1999. ICARDA, Aleppo, Syria and Inst. Mondial du Phosphate, Casablanca, Morocco.

Ryan, J. 2002. Methods of fertilizer application. p. 553–556. *In* R. Lal (ed.) Encylopedia of soil science. Marcel Dekker, New York.

Ryan, J., 2003. Phosphorus fertilizer use in dryland agriculture: the perspective from Syria. p. 500–515. *In* J.M. Lynch et al. (ed.) Innovative soil-plant systems for sustainable agricultural practices. OECD Publ., Paris, France.

Ryan, J., and M. Abdel Monem. 1998. Soil fertility for sustained production in West Asia-North Africa region: Need for long-term research. p. 155–174. *In* R. Lal. (ed.) Soil quality and agricultural sustainability. Ann Arbor Press, Chelsea, MI.

Ryan, J., M. Abdel Monem, and A. Amri. 1994a. Barley and nitrogen fertilization in Morocco's semi-arid zone. Al-Awamia 85:3–14.

Ryan, J., M. Abdel Monem, A. Azzaoui, and M. El-Gharous. 1995a. Impact of phosphorus fertilizer on barley, wheat and triticale in a phosphorus-deficient dryland zone soil. Al-Awamia 90:81–88.

Ryan, J., M. Abdel Monem, M. Dafir, M. Mergoum, and B. Sali. 1995b. Response of local and improved corn varieties in Morocco to phosphorus and zinc. Al-Awamia. 90:69–80.

Ryan, J., M. Abdel Monem, and M. El-Gharous. 1990. Soil fertility assessment at agricultural experiment stations in Chaouia, Adba, and Doukkala. Al-Awamia 12:1–47.

Ryan, J., M. Abdel Monem, and M. El-Gharous. 1993a. Seasonal variation in nitrogen and phosphorus in a vertic Calcixeroll: Implications for soil testing. Al-Awamia, 80:91–100.

Ryan, J., M. Abdel Monem, and K. El Mejahed. 1989. Nitrogen fertilization of Hessian fly-resistant Saada wheat in a shallow semi-arid soil in Morocco. Rachis 8(2):23–26.

Ryan, J., M. Abdel Monem, and M. Mergoum. 1995c. Nitrogen and phosphorus fertilization of triticale varieties in the Settat area of Chaouia. Al-Awamia 88:93–101.

Ryan, J., M. Abdel Monem, M. Mergoum, and M. El Gharous. 1991a. Comparative triticale and barley responses to nitrogen under varying rainfall locations in Morocco's dryland zone. Rachis 10(2):3–7.

Ryan, J., M. Abdel Monem, and J.P. Shroyer. 1992a. Using visual assessment of nitrogen deficiency in dryland cereals as a basis of action in Morocco. J. Nat. Res. Life Sci. Educ. 21:31–33.

Ryan, J., M. Abdel Monem, J.P. Shroyer, M. El Bouhssini, and M.M. Nachit. 1998a. Potential for nitrogen fertilization and Hessian fly resistance to improve Morocco's dryland wheat yields. Eur. J. Agron. 8(3&4):153–159.

Ryan, J., and A.G. Ayubi. 1981. Phosphorus availability indices in calcareous Lebanese soils. Plant Soil. 62:141–145.

Ryan, J., D. Curtin, and M.A. Cheema. 1985a. Significance of iron oxides and calcium carbonate particle size in phosphorus sorption and desorption in calcareous soils. Soil Sci. Soc. Am. J. 49(1):74–76.

Ryan, J., D. Curtin, and I. Safi. 1981a. Ammonia volatilization as influenced by calcium carbonate particle size and iron oxides. Soil Sci. Soc. Am. J. 45:338–341.

Ryan, J., E. De Pauw, H. Gomez, and R. Mrabet. 2004. Drylands of the Mediterranean zone: Biophysical resources and cropping systems. Dryland agriculture. ASA monograph (in review).

Ryan, J., M. Derkaoui, A.W. Chriyaa, and M. Mergoum. 1992b. Phosphorus fertilization of vetch and medic cultivars in Chaouia. "Actes", Inst. Agron. Vet. 12(3):17–21.

Ryan, J., M.A. Eisa, A. Tabbara, and M. Baasiri. 1988. Phosphorus movement following water-applied urea phosphate and phosphoric acid in a calcareous clay soil. J. Fert. Issues 5(3):89–96.

Ryan, J., L. El-Fattal, and N. Harik. 1981b. Micronutrient studies in Lebanese soils. Bull. No. 65. Fac. Agric. Food Sci., Am. Univ. Beirut, Lebanon.

Ryan, J., and S. Garabet. 1994. Soil test standardization in West Asia-North Africa. Commun. Soil Sci. Plant Anal. 25(9&10):1641–1655.

Ryan, J., S. Garabet, A. Rashid, and M. El Gharous. 1999a. Soil laboratory standardization in the Mediterranean region. Commun. Soil Sci. Plant Anal. 30(5&6):885– 894.

Ryan, J., and M. Hamzé. 1987. Soil fertility studies in Lebanon: A review. Lebanese Sci. Bull. 3(2):93–104.

Ryan, J., M. Hamzé, S. Daroub, and S.N. Harik. 1986. Nutrient availability in incubated urea phosphate-treated calcareous soil. J. Fert. Issues 3(4):146–150.

Ryan, J., M. Hamzé, R. Shwayri, and S.N. Hariq. 1985b. Behavior of iron-supplying materials in incubated calcareous soils. J. Plant Nutr. 8(5):425–436.

Ryan, J., and S.N. Hariq. 1983. Transformation of micronutrient chelates in calcareous soil as influenced by selected soil properties, temperature, and wetting-drying during incubation. Soil Sci. Soc. Am. J. 47:806–810.

Ryan, J., H. Hasan, M. Baasiri, and H.S. Tabbara. 1985c. Availability and transformation of applied phosphorus with time in calcareous Lebanese soils. Soil Sci. Soc. Am. J. 49(5):1215–1220.

Ryan, J., R. Hasbany, and T. Atallah. 2003. Factors affecting nitrogen mineralization under laboratory conditions with soils from a wheat-based rotation trial. Lebanese Sci. J. 4(2):3–13.

Ryan, J., and S. Masri. 2003. Crop response to available phosphorus: Soil depth and moisture interactions. p. 71. *In* Int. Soil and Plant Analysis Meet., Cape Town, South Africa. 13–17 Jan. 2003. Abstracts. Soil and Plant Analysis Council, Lincoln, NE and Agri Lab. Assoc. of South Africa, Cape Town, South Africa.

Ryan, J., S. Masri, and S. Garabet. 1996. Geographical distribution of soil test values in Syria and their relationship with crop response. Commun. Soil Sci. Plant Anal. 27(5–8):1579–1593.

Ryan, J., S. Masri, S. Garabet, J. Diekmann, and H. Habib. 1997a. Soils of ICARDA's agricultural experimental stations and sites: Climate, classification, physical-chemical properties and land use. Tech. Bull. ICARDA, Aleppo, Syria.

Ryan, J., S. Masri, L. Habib, and M. Pala. 1994b. Long-term phosphorus fertilization over a rainfall gradient in dryland farming systems in northwestern Syria. p. 369–370. *In* Trans., Int. Soil Sci. Congr., Acapulco, Mexico. 10–16 July 1994. Int. Soil Sci. Soc. and Mexican Soc. of Soil Sci.

Ryan, J., S. Masri, and F. Karajeh. 1999b. Use of untreated sewage water in Syria: A bane or blessing? p. 332 – 333. *In* 6th Int. Meet., Soils with a Mediterranean-Type of Climate, Barcelona, Spain. 4–9 July. Extended Abstracts. Univ. De Barcelona, Barcelona, Spain.

Ryan, J., S. Masri, and Z. Masri. 1997b. Potassium in Syrian soils and crops. p. 134–145. *In* A.E. Johnston (ed.) Food security in the WANA region, the essential need for balanced fertilization. Int. Potash Inst., Basel, Switzerland.

Ryan, J., S. Masri, M. Pala, and M. Bounejmate. 2002a. Barley–Based rotations in a typical Mediterranean agroecosystem: Crop production trends and soil quality. Options Medeterrenees, Series A No. 50:287–296.

Ryan, J., and A. Matar (ed.) 1990. Proc., Third Regional Soil Test Calibration Workshop, Amman, Jordan. 2–9 Sept.1990. ICARDA, Aleppo, Syria.

Ryan, J., and A.E. Matar (ed.) 1992. Fertilizer use efficiency under rainfed agriculture. Proc., Fourth Regional Soil Test Calibration Workshop. 5–11 May 1991. Agadir, Morocco. ICARDA, Aleppo, Syria.

Ryan, J., L. Materon, and S. Christiansen. 1995c. The networks for research collaboration in the dryland West Asia-North Africa region. J. Nat. Resour. Life Sci. Educ. 24:155–160.

Ryan, J., M. Mergoum, M. El Gharous, and A. Azzaoui. 1993b. Importance of combined phosphorus and nitrogen fertilization of barley in semi-arid deficient soils. Al-Awamia 80:101–111.

Ryan, J., M. Mergoum., and N. Nsarellah. 1992b. Responses of rainfed triticale cultivars to nitrogen and phosphorus in Morocco. Rachis 11(1&2):77–79.

Ryan, J., G. Mushrafieh, and A. Barsumian. 1980. Soil fertility characterization of the Agricultural Research and Education Center of the American University of Beirut. Fac. Agric. Tech. Bull No. 64. Am. Univ. Beirut, Lebanon.

Ryan, J., N. Nsarellah, and M. Mergoum. 1997c. Nitrogen fertilization of durum wheat cultivars in the rainfed area of Morocco; biomass, yield and quality components. Cereal Res. Commun. 25(1):85–90.

Ryan, J., and J.D. Prasad. 1979. Factors affecting release and plant availability of sulfur-coated micronutrients. Soil Sci. Soc. Am. J. 43:1039–1043.

Ryan, J., and M. Saleh. 2000. Use of phosphorus fertilizers in pressurized irrigation systems: Problems and possible solutions. p. 249–259. *In* J. Ryan (ed.) Plant nutrient management under pressurized irrigation systems in the Mediterranean region: Proc., Int. Fertigation Workshop, Amman, Jordan. 25–27 Apr. 1999. Inst. Mondial du Phosphate, Casablanca, Morocco and ICARDA, Aleppo, Syria.

Ryan, J., J.P. Shroyer, and M. Abdel Monem. 1991b. Interaction between nitrogen level and Hessian fly protection in Morocco. J. Farm. Sys. Res. 2:47–55.

Ryan, J., R. Shwayri, and S.N. Hariq. 1984. Preliminary assessment of microbial–Based soil additives. Iran Agric. Res. 3(1):57–64.

Ryan, J., R. Shwayri, and S.N. Hariq. 1985d. Short-term evaluation of non-conventional organic wastes. Agric. Wastes 12:241–249.

Ryan, J., M. Singh, J. Diekmann, and S. Masri. 1998b. Spatial variability of soil chemical properties: Implications for field trials under semi-arid conditions. (20–26 Aug.). Int. Soil Sci. Congr., Montpellier, France.

Ryan, J., M. Singh, S. K. Yau, and S. Masri. 1998c. Spatial variability of soluble boron in Syrian Soils. Soil Tillage Res. 45(3&4):407– 417.

Ryan, J., P. Smithson, B. Mandac, and J. Uponi. 2002b. Role of international research centers' soil laboratories in agricultural development. Commun. Soil Sci. Plant Anal. 33(15–18):3213–3225.

Ryan, J., and H. Tabbara. 1989. Influence of urea phosphate on infiltration and sodium parameters of a calcareous sodic soil. Soil Sci. Soc. Am. J. 53:1531–1536.

Ryan, J., and M.A. Zghard. 1980. Phosphorus transformations with age in a calcareous soil chronosequence. Soil Sci. Soc. Am. J. 44:168–169.

Powell, J.M., S. Fernandez-Rivera, T.O. Williams, and C. Renard (ed.) 1995. Livestock and sustainable nutrient cycling in mixed farming systems of Sub-Saharan Africa. Vol. II. Technical papers. Proc., Int. Conf., 22–26 Nov. 1993. Addis Ababa, Ethiopia. Int. Livestock Center for Africa, Addis Ababa, Ethiopia.

Sahyouni, M., and J. Ryan. 1983. Potassium in Lebanese soils as reflected by electro-ultrafiltration (EUF), chemical extraction, and cropping. Iran Agric. Res. 2(1):11–22.

Sanchez, P. 2002. Soil fertility and hunger in Africa. Science (Washington, DC) 295:2019–2020.

Shroyer, J.P., J. Ryan, M. Abdel Monem, and M. El Mourid. 1990. Production of fall-planted cereals in Morocco and technology for its improvement. J. Agron. Educ. 19:32–40.

Stoorvogel, J.J., and E.M.A. Smaling. 1990 Assessment of nutrient depletion in Sub-Saharan Africa: 1983–2000. Wageningen, The Netherlands.

Tian, G., F. Ishida, and D. Keatinge (ed.) 2001. Sustaining soil fertility in West Africa. SSSA Spec. Publ. 58, SSSA, Madison, WI.

Torrent, J. 1987. Rapid and slow phosphate sorption by Mediterranean soils effect of iron oxides Soil Sci. Soc. Am. J. 51:78–82.

Yau, S.K., M.M. Nachit, J. Ryan, and J. Hamblin. 1995a. Phenotypic variation of boron toxicity tolerance in durum wheat at seedling stage. Euphytica 83:185–191.

Yau, S.K., M. Nachit, G. Ortiz-Ferrara, J. Ryan, and M.C. Saxena. 1995b. Differential responses of durum and bread wheat to excess soil boron. Wheat Newsl. 41:204–207.

Yau, S.K., M. Nachit, and J. Ryan. 1997. Variation in growth, development and yield of durum wheat in response to high soil boron. II. Differences between genotypes. Aust. J. Agric. Res. 48:951–957.

Wahbi, A., A.E. Matar, and M.J. Jones. 1993. Responses of a forage hay crop to the residual effects of nitrogen and phosphorus fertilizers in on-farm trials in northern Syria. Exp. Agric. 29:429–435.

White, H. 1970. Following crop rotation and crop yields in Roman times. Agric. Hist. 44:281–290.

19 Optimizing Soil Water Balance Components for Sustainable Crop Production in Dry Areas of South Africa

Danie J. Beukes

Agriculture Research Council (ARC)-Institute for Soil, Climate, and Water
Pretoria, South Africa

Alan T. P. Bennie and Malcolm Hensley

University of the Orange Free State
Bloemfontein, South Africa

ABSTRACT

The wide variation in its natural agricultural resources makes South Africa a country of great diversity. It is exemplified in the rainfall and production potential maps presented, and also by the wide variety of crops grown and production techniques employed. The latter range from advanced technology to traditional subsistence procedures. Of the total area, <14% is arable and <4% of high potential, with rainfall the main limiting factor. Maize (*Zea mays* L.) and wheat (*Triticum aestivum* L.) are the main cereal crops. Results presented describe extensive on-station and on-farm research to quantify water losses by evaporation, runoff, and deep drainage, and on measures to minimize these losses such as mulching, soil surface modification, and planting patterns. Success depends on correct technique/soil type matching. Increased soil water storage through fallowing, improved infiltration, and water harvesting has received much attention. Further research priorities include studies on: (i) simultaneous optimization of soil water and nutrient use, water harvesting, and mulching to reduce evaporation losses; (ii) refinement of seasonal outlook forecasting to allow for the prediction of seasonal onset and distribution of rainfall; and (iii) determination of the effects of different crop rotation systems on long-term water-use efficiency under dryland conditions under different soil and climatic conditions.

DRYLAND AND RAINFED CROP PRODUCTION SYSTEMS

Introduction

South Africa occupies the southern most part of the African continent and lies between latitude 22°S and 35°S and longitudes 17°E and 33°E. It comprises nine

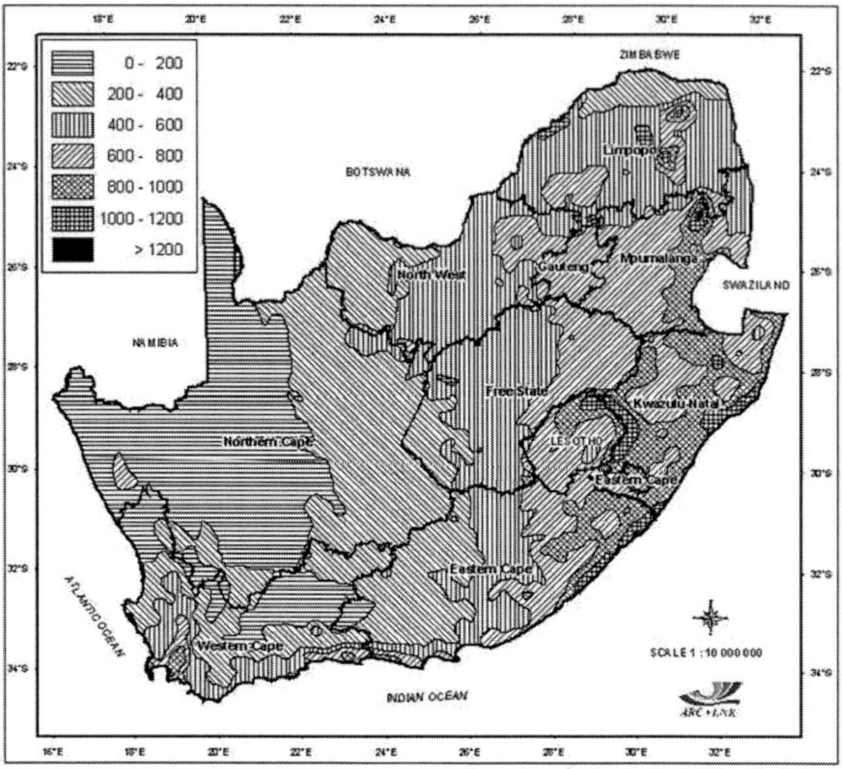

Fig. 19–1. Mean annual rainfall (mm) in South Africa.

provinces (Fig. 19–1) and shares boundaries with Lesotho, Swaziland, Mozambique, Zimbabwe, Botswana, and Namibia. The wide variation and abundance in its natural agricultural resources, especially climate, soils, and topography, contribute to making South Africa a country of amazing diversity. Conditions range from semi-desert to subtropical rain forests, floods to severe droughts, snow in winter to heat waves in summer, winter to summer rainfall, and barren sand dunes to soils of high productivity. Such variety enables the country to produce a wide range of agricultural commodities, but at the same time demands managerial skills of a high order to prevent over-exploitation of the natural resources. South Africa is a world in miniature, a fragile and unique system of nature with resources of great value to man. It poses many challenges for the future. One such challenge is how to optimize the use of one of the country's scarce natural resources, namely water (Scotney et al., 1990).

The Natural Resource Base

South Africa covers an area of 122.34 million ha, with <14% suitable for dryland and rainfed cropping (as defined by Stewart and Burnett, 1987), of which only about a quarter is land of high productive potential. Rainfall, both total and seasonal distribution, plays a dominant role in determining the resource situation, crop selection, yield horizons, and risk of agricultural production. There are two main rainfall regions, namely, a winter rainfall region in the southern part of the country, which includes a small all-year-round rainfall area, and a summer rainfall region which covers the remainder of the country. There is an increase in annual rainfall from <125 mm along the arid west coast to more than 1000 mm on the eastern seaboard (Fig. 19–1). Mean annual rainfall is 511 mm, but more than 60% of the country receives <500 mm per annum. Rainfall is extremely variable, with wide deviations from the mean annual values, especially in low rainfall areas. The country is characterized by the occurrence of regular droughts of varying intensity, some of which may have devastating consequences. In contrast, severe floods are also not uncommon (Scotney et al., 1990).The composition of the soil mantle of South Africa is well understood. An efficient soil classification system has been developed (SCWG, 1991), and the task of mapping the soil patterns in the form of a 'land type' survey at a scale of 1:250 000 for the whole country was completed in 2002. Around 7100 land types and 3000 climatic zones have been defined, and more than 2300 modal soil profiles have been described and analyzed (Paterson and van der Walt, 2003). In the areas of high potential for cropping, the two main soil types are: (i) fairly deep red and yellow well-drained soils with a medium to high clay content, and (ii) medium textured soils with a plinthic horizon at depths of between 600 and 1200 mm. The main soils in the medium potential cropping areas are: (i) soils with a plinthic catena as described above, but with a lower rainfall, and (ii) soils with a high clay content, many of them Vertisols. Figure 19–2 reflects the distribution of several crop potential classes over the country, although the extent of land suitable for cropping may vary considerably within each class. The latter was derived from soil, terrain, and climatic data, with the aid of crop simulation models, assuming good management practices. The scarcity of high potential agricultural land in South Africa is quite evident from the map, making it a critical resource that should be preserved at all costs.

CROPS AND MANAGEMENT PRACTICES

Winter Rainfall Region

Under the Mediterranean climate of the southern coastal (Fig. 19–1) area the annual rainfall can vary from 200 mm to more than 3000 mm in the mountainous areas, with the main (ca. 80–85%) incidence in the months April to September. This gives rise to hot and dry summer months. The eastern part is primarily a sowing and grazing region where small grains, that is, wheat and barley (*Hordeum vulgare* L.) and stock farming on established pastures are practiced complementary to each other under an annual rainfall of 350 to 500 mm. In the western part, wheat is the major crop grown primarily in a monoculture system under an annual rainfall of 250 to 500 mm. In both areas, dryland viticulture is historically and culturally a very important farming activity. The fallowing of lands is a practice on 6 to 20% of about 830 000 ha under winter cereal production. Although dryland agricultural production (on ca. 1.8 million ha) is the major enterprise, intensive vegetable, viticulture, and fruit farming under irrigation make a valuable contribution to agricultural production. In both parts, winter cereals are grown mainly on shallow duplex or stony soils derived from shale and schists. These soils have low water and nutrient storage capacities, low organic carbon (C) and pH values, and are prone to crusting and water erosion. Heavy downpours often lead to waterlogging and runoff, thus aggravating the situation. Cultivation is by conventional mouldboard and disc ploughing, while there is a trend towards shallow tine and no-till to cut down production costs.

SUMMER RAINFALL REGION

This region is situated on the central high plateau of South Africa, encompassing parts of Free State, KwaZulu-Natal, Northern and North West Provinces, as well as the provinces of Gauteng and Mpumalanga (Fig. 19–2). Annual rainfall varies from 400 mm in the west to more than 900 mm in the east. More than 75% of the rainfall occurs between November and March. Midsummer drought is a general phenomenon, coinciding with the flowering period of the summer crops, often

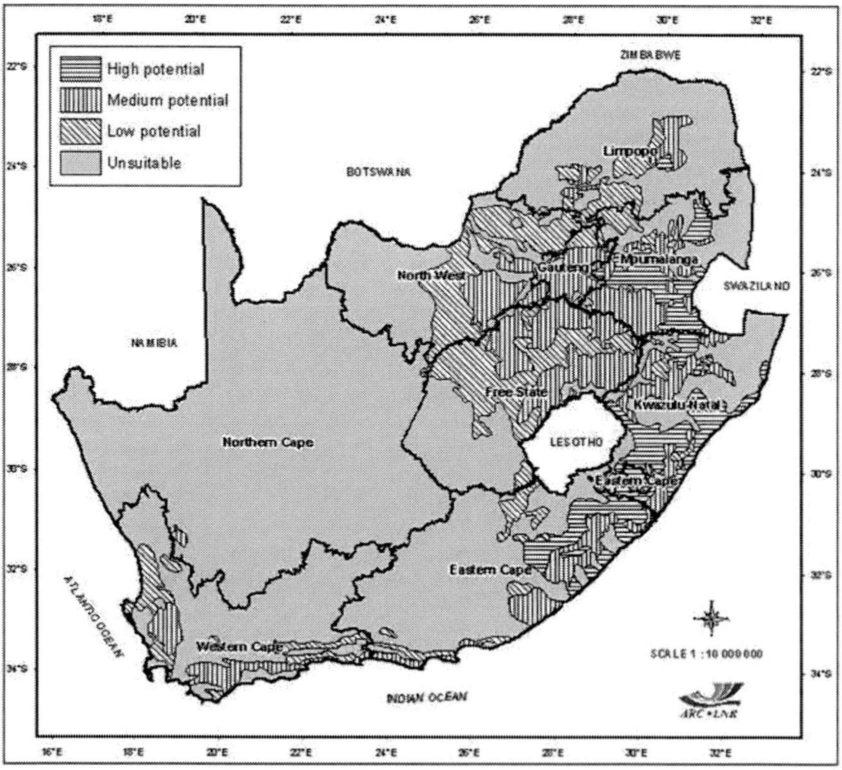

Fig. 19–2. Generalized crop production potential of South Africa.

causing poor flowering and consequent low yields. In this region maize, wheat, grain sorghum [*Sorhum bicolor* (*L.*) Moench], groundnuts (*Arachis hypogea* L.), dry bean (*Phaseolus vulgaris* L.) and soya bean [*Glycine max* (L.) Merr.] are being produced on about 5.9 million ha of land. Maize and grain sorghum are grown in monoculture systems, while the other crops are produced in rotation because of problems with soil-borne diseases.

Cultivation is done by moldboard and disc plowing. Reduced tillage with residue mulching, under tine or chisel plough cultivation, is also practiced but to a lesser degree. Although the beneficial effects of mulching due to organic C build-up, protection against raindrop impact and improved soil water storage are well understood by farmers, residue-borne diseases of maize and wheat have led to a decline of this practice. On the vast areas of sandy soils tine implements are extensively used in a controlled traffic system to conserve soil water, combat wind erosion, and to alleviate subsoil compaction. Other management practices employed to optimize rainfall in maize production are to plant area specific cultivars and make use of wide (2.2 m) rows.

In the central and northern parts there is an estimated 3.6 million ha of Vertisols, a large portion of which is arable. Mostly grain sorghum, sunflower (*Helianthus annus* L.), wheat, cotton (*Gossypium hirsutum* L.), and tobacco (*Nicotiana tobaccum* L.) are grown under dryland conditions on these soils. Both self-mulching and crusting have been dramatically increased by long-term tillage, causing a decrease in soil organic matter, resulting in structural degradation with consequent impeded infiltration, enhanced run-off, lower soil water storage, and increased wind and water erosion. The climatic constraints such as high summer temperatures, low erratic rainfall, and high evaporation rates often lead to low crop yields and sometimes total crop failures on these soils. It is therefore essential that tillage practices should aim at maximum soil water storage and conservation (Jacobs and Beukes, 1997). Of the different crops grown by commercial farmers on about 6.7 million ha dryland and rainfed land, maize (50% of area), and wheat (19%) are by far the most important, followed by crops such as sunflower, oat (*Avena sativa* L.), sugarcane (*Saccharum officinarum* L.), grain sorghum, and some six other crops. Thus, this review of water use will focus on these important crops.

SOIL WATER BALANCE RESEARCH

Principles of Efficient Soil Water Use

The basic principle of efficient water use for plant production lies in optimizing each of the components of the soil water balance. There are two distinct management periods. The first is the period of rain storage lasting from harvesting of the previous crop till planting of the next crop. Under semi-arid climatic conditions, the soil and water management strategies during this period should be to maximize the gains and minimize the losses in Eq. [1]. The soil water balance during period of the rain storage can be written as follows (adapted from Hillel, 1982):

$$\Delta S = P + I \pm D \pm R - E - T \qquad [1]$$

where, ΔS = change in water content in the potential root zone; P = precipitation; I = irrigation; D = downward drainage out of the root zone (–) or upward capillary flow into the root zone (+); R = runoff (-) or run-on (+); E = evaporation from the soil surface; T = transpiration.

The growing season is the second management period, lasting from planting till harvesting of the crop. The soil water balance can then be re-arranged in the following form:

$$T = P + I \pm DS \pm D \pm R - E \qquad [2]$$

To allow for the maximum amount of water to be available for transpiration (T), and therefore maximum plant production, the parameters on the right hand side of Eq. [2] should be optimized. A wide range of soil and water management practices are currently being applied or tested in South Africa to achieve these goals.

Optimizing Soil Water Balance Components

Rainfall

In South Africa, an area of about 3.4 million ha can be considered as high potential agricultural land with rainfall more than 800 mm per annum (Fig. 19–1). Rainfed commodities such as natural and commercial forests, perennial agricultural crops, horticultural crops, and vegetables are mainly grown. Only small areas are used for the production of cereal crops. The main cereal producing areas have a semi-arid climate with a mean annual rainfall varying between 25 and 50% of the mean annual evaporation. However, rainfall distribution is erratic with short seasonal droughts being a common phenomenon and longer droughts occurring in 6 to 12-yr cycles (Tyson and Dyer, 1978). Long-term weather forecasting is still virtually impossible but progress is being made in relating regional weather behavior to the El Nino and Southern Oscillation Index (SOI) phenomena.

Irrigation

Although it is the aim of this review to concentrate on efficient utilization of rainfall, the collection of runoff by large dams constructed in rivers, and using this water for irrigation, also contributes towards achieving the original aim. The South African Government had since the beginning of the 20th century, embarked on a program of building large dams in rivers, and in water transfer schemes between catchments. This ensured the retention of sufficient runoff to support the irrigation of 1.3 million ha, or almost 10% of the cultivated area. Of the total irrigated area, 22% is under supplemental irrigation. Full irrigation requires between 750 and 1000 mm irrigation additional to the rainfall, whereas supplemental irrigation, on average, only needs about 350 mm. Irrigated agriculture is responsible for 15% of the total cereal production in South Africa.

Pre-Plant Soil Water Storage

The majority of the soils cultivated for annual crop production, especially cereal crops in the semi-arid regions, have sandy topsoils with clay contents lower than

25%. Approximately 2 million ha, or 15% of the cultivated soils, contain pedogenic horizons below the root zone that restrict deep percolation. The plant-available water storage capacity of these soils varies between 120 and 200 mm. In approximately 1 million ha of these soils, shallow, perched water-tables can be observed in wet seasons. In an extensive research program on the efficiency of the storage and utilization of rainfall for dryland crop production, Bennie et al. (1994) reported pre-plant rain storage efficiencies varying between 2 and 37%. They defined the latter term as the change in soil water content over the potential rooting depth (ca. 1.8 m) of cereal crops from harvesting of the previous crop till planting of the present crop, expressed as a percentage of the rainfall over the same period. The advantage of increasing the length of fallowing for pre-plant rain storage from 5 to 10 mo was illustrated by Bennie et al. (1995), who reported yield increases of maize varying between 26 and 50%, and of wheat varying between 0 and 68%. A problem associated with fallowing is the large evaporation losses from the bare soil during this period. As a result of this, the mean PUE, measured over three seasons of maize production decreased from 5.98 kg seed ha^{-1} mm^{-1} for the short 5-mo fallowing, to 5.05 for the long 10-mo fallowing. For wheat production over the same period, the decrease was even greater, viz. from 6.19 to 3.14 kg seed ha^{-1} mm^{-1} for the short and long fallowing, respectively. Under the Mediterranean climate of the southern coastal area, it was found that a long (12 month) fallow practice was unsuccessful in increasing winter wheat yields because water storage was insufficient due to the low water holding capacities of the soils and the very low summer rainfall (Agenbag, 1987).

Deep Percolation

The quantification of deep percolation from field measurements is difficult. When a drainage curve is available, good estimates can be made. Bennie et al. (1994) reported values ranging from 0 to 20% of the seasonal rainfall under semi-arid conditions measured on well-drained sandy aeolian soils. The magnitude of deep percolation depends on antecedent soil wetness and the amount of seasonal rainfall. Less deep percolation occurred in soils with clayey horizons in or below the root zone. The same authors reported upward fluxes of soil water into the root zone equivalent to between 0 and 8% of the mean seasonal rainfall.

Runoff and Infiltration

A reduction in runoff will result from practices that successfully increase the infiltration capacity of the soil, increase the contact time, and/or reduce surface sealing. It is commonly accepted that covering the soil with a mulch, for example, crop residue, will achieve these goals (Unger, 1990). The most promising results have been obtained in the high rainfall subhumid climatic regions (Fig. 19–1) where Lang and Mallett (1984) reported a reduction in runoff and soil loss when the residue cover exceeded 30% on a highly weathered and weakly structured clay loam soil in the KwaZulu Natal Province. In contrast, Bennie et al. (1994) found with studies on sandy soils with slopes $<2\%$ in a semi-arid climate in the Free State Province, higher runoff from shallow-tilled residue mulching and no-till than from the deeper tilled conventional moldboard plowing. Compared with the runoff values measured from

Table 19–1. Long-term runoff values on similar soils at Glen (after Du Plessis and Mostert, 1965) and at Pretoria (after Haylett, 1960).

Site	MAP†	Period‡	Slope	Soil	Treatment	Mean annual	
						Runoff§	Soil loss
	mm	yr	%			%	t ha^{-1}
Glen	508	18	5.0	Red, well-drained sandy loam	Natural veld	4.4	0.26
					Continuous maize	8.5	8.56
					Bare-tilled plots	10.3	13.22
Pretoria	721	27	3.75	Red, well-drained sandy loam	Natural veld	4.2	1.12
					Continuous maize	26.7	22.62
					Bare-tilled plots	24.4	22.40
Pretoria	737	21	7.0	Red, well-drained sandy loam	Natural veld	6.8	0.90
					Continuous maize	23.6	65.41
					Bare-tilled plots	24.4	86.02

† MAP = Mean annual precipitation during the duration of the experiment.
‡ Period = The duration of the experiment.
§ Expressed as a percentage of the rainfall.

undisturbed natural grass cover, tillage of the soil increased the runoff. The amount of crop residue retained over all 4 yr was only 4 t ha^{-1}. Botha et al. (1999) have shown that employing an in-field water harvesting technique whereby in-field run-off is captured in micro basins, runoff can be reduced to zero by converting it to stored soil water, and consequently to increased yields, compared to conventional tillage.

Long-term runoff measurements have also been made elsewhere (Table 19–1). These results are particularly valuable because of the long duration of the experiments, and the similar soils and slopes at the two sites. As expected the runoff and soil loss at both sites are much lower from natural veld than for continuous maize. The much higher rainfall at the Pretoria Experimental Station resulted in a three times higher runoff than at the Glen Experimental Station under continuous maize cropping. Another significant observation is that runoff on the Pretoria plots was not significantly affected by doubling the slope, whereas the soil loss was tripled where the soil had been cultivated. The influence of gravity on raindrop splash action evidently dominated the erosion process on the latter soil with its rapid infiltration rate. In order to address the nonsustainability of conventional tillage practices on Vertic soils evidenced by the deterioration of physical properties, Beukes (1987, 1992) evaluated the effects of stubble and conventional (moldboard + disc) tillage (CT) on these properties and on growth and yield of grain sorghum in a long-term tillage trial. Nutrient management was based on soil analyses, crop withdrawal, and commercially accepted norms for fertilization that ranged from 10 to 15 kg N ha^{-1} and from 40 to 50 kg P_2O_5 ha^{-1} in general. The no-till and tine-tillage practices, both with stubble retained, gave the highest soil water contents (Fig. 19–3a). The same cultivations, but with stubble removed, gave the lowest water contents.

The favorable effect of stubble mulch on soil water contents was particularly noticeable during the early part of the growing season. With the exception of CT+stubble (CTS), the retention of stubble significantly increased infiltration (Table 19–2). This can best be explained in terms of increased soil aggregation and aggregate stability due to stubble mulching (Fig. 19–3b). Soil C and nitrogen (N)

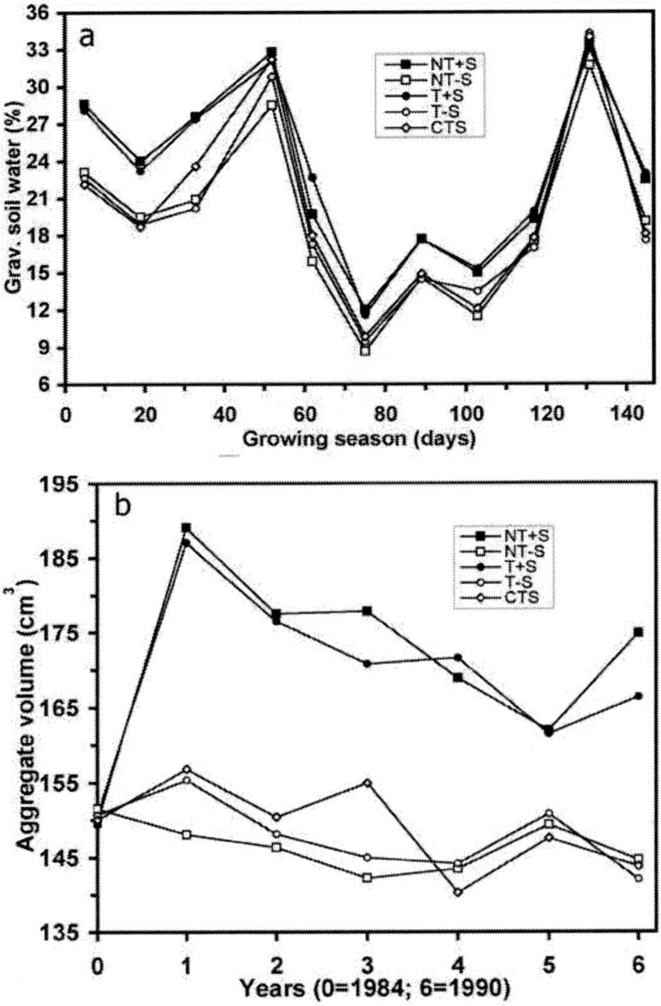

Fig. 19–3. Tillage and soil water contents and changes in aggregate stability: (a) tillage and soil water contents and (b) changes in aggregate stability (After Beukes, 1987).

contents were maintained and even increased under stubble mulching. The conventional CTS and tillage practices, where stubble was removed, led to a major reduction in soil C over years. The retention of stubble (CTS excluded) in the tillage practices led to a dramatic increase in grain yield, especially in dry years (Table 19–2). The sequence in plant height and grain yields in terms of the treatments was similar to that of soil water contents.

This indicates that available soil water is a major factor that determines growth and yield of grain sorghum on a Vertic soil. Higher N-contents might also have contributed to higher yields. Under the same semi-arid climate, Hattingh

Table 19–2. Sorghum grain yield and cumulative infiltration values after two experimental seasons (after Beukes, 1987).

Treatment	Yield	Cumulative infiltration†
	kg ha^{-1}	mm
Tine-tillage, stubble removed	833a‡	99a
No tillage, stubble removed	567a	118a
Conventional tillage, stubble retained	855a	139a
No-tillage, stubble retained	1880b	188b
Tine-tillage, stubble retained	1838b	205b
Mean	1195	150
LSD ($P < 0.05$)	345	44
LSD ($P < 0.01$)	466	59
F-value	26.8§	8.0§

† Cumulative infiltration after a 40-min measurement period.
‡ Column values not followed by the same letter are significantly different at $P < 0.05$.
§ $P < 0.001$.

(1995) evaluated in a long-term trial the effects of various levels of stubble mulch and plant population densities on the soil water regime and yield of maize on a Hutton soil form (SCWG, 1991) with 12% clay. She found an increase in soil water storage under stubble, which led to an increase in maize yield. Most encouraging was the increase in organic C in the topsoil layer from the third season onwards, due to stubble mulching.

In a comprehensive study of Vertisols in the northern summer rainfall region, Grubb (1993) found that tillage and cropping markedly affected soil structure and other properties. Soil degradation manifested itself primarily in the form of reduced dry-aggregate stability, leading ultimately to increase self-mulching. He found that the water storage, especially of the gilgai mounds (micro-relief produced by swelling clays), was very poor. Given the low and erratic rainfall, this had a direct effect on the observed crop failures. On these heavy clay soils, Birch et al. (1986) found that basin tillage was the most effective way to retain rainfall, thereby improving soil water storage and, consequently, sunflower yields.

Soil Surface Evaporation

Soil surface evaporation (E_S) is by far the most important water loss contributing to inefficient water use for dryland crop production. Under semi-arid climatic conditions in South Africa, evaporation from bare soils during the fallowing period can amount to 60 to 75% of the rainfall in the driest summer cropping areas (Bennie et al., 1994). Percentages of the rainfall lost by evaporation during the period of water storage for different soil types, tillage, and cropping practices, are presented in Table 19–3. These are mean values for 4 yr, comprising two dry, one average, and one above-average, rainfall years. Evaporation was highest for the CT treatment on the Westleigh soil with a more clayey topsoil. The small amounts of crop residue (1024–3880 kg ha^{-1}) retained on both soils shaded <30% of the soil surface.

Hoffman (1997) used microlysimeters to measure the evaporation from a wide range of soils with silt+clay contents ranging from 4 to 66% under similar evapo-

Table 19–3. Percentage of rainfall lost by evaporation during water storage periods under various tillage systems (Bennie et al., 1994).

	Bainsvlei†			Westleigh†		
Cropping practices	CT‡	SM	NT	CT	SM	NT
Continuous wheat (5 mo fallow)	69.0	68.9	72.0	70.0	62.7	70.5
Continuous summer grain (5 mo fallow)	68.4	74.1	76.5	74.6	62.2	60.7
Summer grain/wheat rotation (10 mo fallow)	72.5	69.3	67.6	78.9	75.5	72.2
Mean	70.0	70.8	72.0	74.5	66.8	67.8

† Soil form (SCWG, 1991).
‡ CT = Conventional moldboard plowing and shallow tine harrowing; SM = Shallow sweep tillage retaining crop residue on the surface; NT = No tillage, chemical weed control.

rative demand conditions. He found that the cumulative evaporation increased with increasing silt+clay contents or water-holding capacities. He also determined that a minimum of 80% shading is required to ensure significant decreases in the cumulative evaporation within the first 10 d after wetting under dry climatic conditions. When the drying-out period exceeded 15 to 20 d, the beneficial effect disappeared. Berry and Mallett (1988) found that with maize residue covers greater than 70%, evaporation losses decreased for wetting frequencies shorter than 14 d, under subhumid climatic conditions. In order to obtain a 70% ground cover, a minimum of 6 t crop residue ha^{-1} was required.

Van Averbeke and Mkile (1996) compared the potential of straw and stones for use as a mulch to reduce evaporation during the growing season in a lysimeter experiment involving maize. Four soil surface treatments, consisting of a control, straw, or stones on the soil surface, and stones placed in trays for easy removal before tillage, were maintained throughout the growing season. Relative to the control, stone mulching reduced evapotranspiration by 18%. The water-use efficiency (WUE) of maize was 9.8 kg ha^{-1} mm^{-1} in the control, compared to 13.6 kg grain ha^{-1} mm^{-1} (straw and stone mulching) and 16.6 kg grain ha^{-1} mm^{-1} (tray mulch). When separating the evaporation and transpiration components, using microlysimeters, it was shown that the soil evaporation component comprises between 50 and 67% (Haarhoff, 1989; Hattingh, 1993) and 42 and 50% (Hoffman, 1990; Hattingh, 1993) of the total evapotranspiration of maize and wheat, respectively. Bennie et al. (1994) have shown that, of the total rainfall during the rain storage plus growing period, 73% is lost through evaporation from the soil when winter wheat is grown on stored soil water under semi-arid summer rainfall conditions. For maize the equivalent evaporative loss is 60%.

Botha et al. (2001) quantified the influence of different mulches on evaporation from the soil surface of a sandy loam and clay for both a summer and winter observation period. Four treatments were imposed comprising of a bare soil, a stone mulch covering 50%, and organic mulches covering 50 and 100% of the surface, respectively. Installations were made to measure soil water content and soil temperature. Figures 19–4(a) and (b) show that the 100% organic mulch is superior to reduce E_s on the clay (summer) and sandy loam (winter) compared to the bare soil. Smaller reductions in E_s were recorded for the 50% (organic and stone) mulches. The latter two mulches performed very similarly, a phenomenon that has

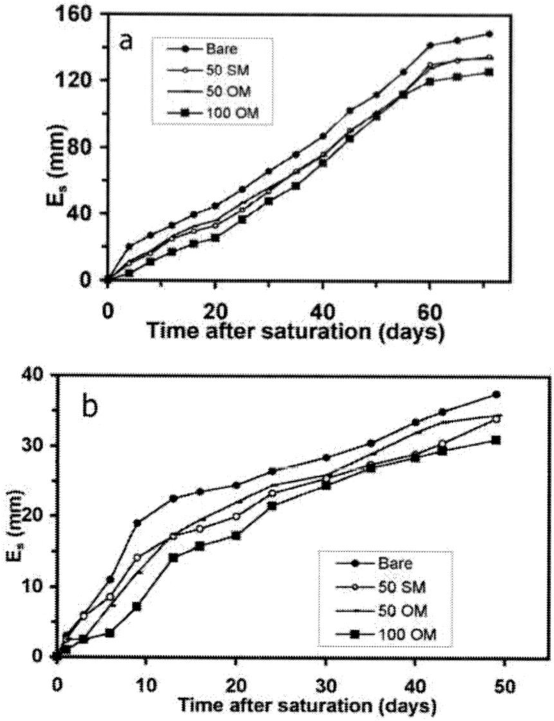

Fig. 19–4. Effect of mulches on evaporation from the soil surface: (a) clay soil during summer and (b) sandy loam during winter (SM = stone mulch; OM = organic mulch, with 50 and 100% soil cover)(After Botha et al., 2001).

a very practical implementation value. In rural areas where crop residues are a scarce commodity, apart from the urgent need as animal feed during winter, a stone mulch can be used to reduce E_S. The consequent effect on soil water storage with the 100% mulch yielding the highest, and the bare soil the lowest, values over time can be seen in Fig. 19–5(a) and (b) for the clay soil, for the summer and winter observation periods, respectively.

Evapotranspiration and Yield

The amount of water transpired by a crop is a function of the total biomass produced and the climatic conditions controlling the evaporative demand of the atmosphere (De Wit, 1958, after Hanks and Rasmussen, 1982) Therefore any cropping or tillage practice that affects crop growth and development will have a similar effect on transpiration (and on evaporation from the soil surface) under comparable climatic conditions.

Tillage Effects. Bennie et al. (1994) determined, under optimal soil fertility conditions, the effects of tillage and cropping practices on the yields and ET for wheat, maize, and grain sorghum. The average of 4 yr, yield and ET results are presented, respectively, in Tables 19–4 and 19–5. Yields and evapotranspiration (ET)

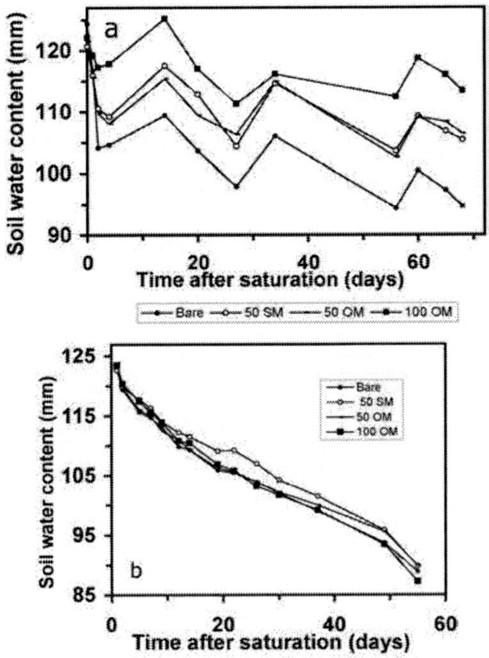

Fig. 19–5. Effect of mulches on soil water storage: (a) clay soil during summer and (b) clay soil during winter (SM = stone mulch; OM = organic mulch, with 50 and 100% soil cover) (After Botha et al., 2001).

of both wheat and grain sorghum were in the order of CT > SM > NT for the more sandy Bainsvlei soil. The reduction in growth in the shallow sweep and no tillage treatments, compared with the conventional tillage treatment, was attributed to

Table 19–4. Total biomass and grain yields of wheat (W), grain sorghum (GS), and maize (M) grown on two soils (Bennie et al., 1994).

	Bainsvlei						Westleigh					
	CT†		SM†		NT†		CT		SM		NT	
Cropping practice	W	GS	W	GS	W	GS	W	M	W	M	W	M
	t ha^{-1}											
Continuous wheat	3.6	--	1.4	--	1.4	--	2.7	--	2.4	--	3.0	--
(5 mo fallow)	1.0	--	0.4	--	0.4	--	0.8	--	0.7	--	1.0	--
Continuous summer	--	7.7	--	6.1	--	3.2	--	5.9	--	6.1	--	5.1
grain (5 mo fallow)	--	1.9	--	1.6	--	0.9	--	2.6	--	2.7	--	2.2
Summer grain/wheat	4.4	1.1	3.2	9.4	2.5	6.4	3.3	8.0	2.8	6.8	2.4	5.5
rotation (10 mo fallow)	1.2	3.3	1.0	2.5	0.8	1.8	1.0	3.5	0.8	3.0	0.7	2.3
Mean	4.0	9.7	2.3	7.8	2.0	4.8	3.0	7.0	2.6	6.5	2.7	5.3
	1.1	2.6	7.1	2.0	6.3	1.4	0.9	3.0	0.7	2.8	0.9	2.3

† CT = Conventional moldboard and shallow–tine harrowing; SM = Shallow sweep tillage retaining crop residue on the surface; and NT = No tillage, chemical weed control.

Table 19–5. Evapotranspiration values for the growing seasons of wheat, grain sorghum, and maize grown on two soils (Bennie et al., 1994).

	Bainsvlei						Westleigh					
	CT†		SM‡		NT‡		CT†		SM‡		NT§	
Cropping practice	W	GS	W	GS	W	GS	W	M	W	M	W	M
	mm											
Rainfall (mm)	143	254	143	254	143	254	159	342	159	342	159	342
Continuous wheat (5 mo fallow)	233	--	194	--	183	--	212	--	203	--	214	--
Continuous summer grain (5 mo fallow)	--	261	--	267	--	217	--	345	--	358	--	348
Summer grain/wheat (10 mo fallow)	239	382	218	322	208	351	214	394	227	392	210	360
Mean	236	322	206	295	196	284	213	370	215	375	212	354

† CT = Conventional moldboard and shallow-tine harrowing.
‡ SM = Shallow sweep tillage retaining crop residue on the surface.
§ NT = No tillage, chemical weed control. Crops: Wheat = W; Grain sorghum = GS; Maize = M.

poorer root development due to the shallower tillage depth. This was to a lesser extent also the case in the Westleigh soil with a 6% higher silt+clay content. Agenbag and Stander (1988) and Agenbag and Maree (1991) also ascribed lower wheat yields, obtained with stubble mulching and no-till during dry seasons, to poorer root development due to higher penetration resistance values in the topsoil. During wet seasons no-till equalled, and in some years out-yielded, conventional-tilled wheat on a shallow stony soil with 30% silt+clay in the topsoil. Working under the Mediterranean climate of the southern coastal areas, they concluded that in the long-term, tillage-induced differences in soil water content had no significant effect on wheat yield. This is because of the low water storage capacity of the predominantly shallow stony soils of the region and the well-distributed rainfall during the growing season. Bennie et al. (1995) reported lower wheat and maize yields with stubble mulching compared with conventional tillage on a sandy soil, although deep ripping and controlled wheel traffic were used in both practices to alleviate the effect of soil compaction.

Berry et al. (1988) and Mallett et al. (1987) obtained better, or the same, maize yields with no-till compared with conventional tillage. These results were obtained on a well-drained and fertilized soil with 60% silt+clay under subhumid climatic conditions. From the limited research on conservation tillage practices, where the crop residues are retained on the surface, it can be concluded that lower yields can be expected on soils with less than approximately 20% silt+clay in the topsoil. Soils with higher silt+clay contents seem to have a better structure, allowing for better root development. Engelbrecht et al. (1986) showed that, for wheat, a long fallow (planting every second year) resulted in significantly higher yields. Testing five different conservation tillage practices for wheat, Snyman et al. (1992) found that no-till treatments were not beneficial and that higher grain yields were obtained as the intensity of tillage increased, probably mainly due to improved infiltration and decreased evaporation losses.

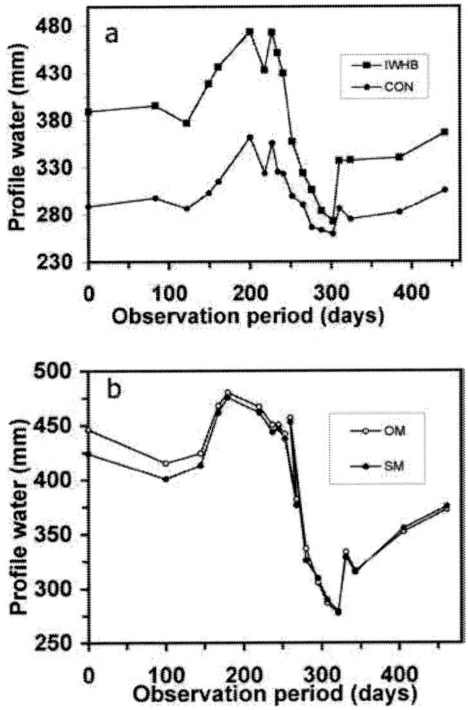

Fig. 19–6. Soil water storage: (a) conventional vs. in-field water harvesting (IWHB) tillage practices and (b) organic vs. stone mulches (SM) in basin (After Botha et al., 2003).

In-field Water Harvesting. Botha et al. (1999) employed an in-field, water-harvesting (IWH) technique making use of a 'tram line' row spacing (1 m by 2 m), whereby in-field run-off from the 2-m strips between the plant rows is captured in microbasins. Planting was done at 1-m row spacing in the basins. They found, for example, that during a wet year IWH increased seed yields by 39, 27, and 65% for maize, sunflower, and sorghum, respectively, above the 3160, 1870, and 1880 kg ha^{-1} with conventional tillage. In a severe dry year, IWH induced 76, 18, and 51% higher biomass yields for maize, sunflower, and sorghum, respectively, compared to conventional tillage (1220, 1420, and 1140 kg ha^{-1}).

In a recent study, Botha et al. (2003) compared (a) IWH + basin tillage (IWHB) with conventional tillage (CON), and (b) organic (OM) with stone (SM) mulches in basins, in terms of evapotranspiration, water conservation, and maize yield, respectively. The observation period included a fallow (Days 0–214) and a summer growing season (Days 215–461). Figure 19–6(a) shows that soil water storage was much higher for IWHB, resulting in statistically significantly higher maize seed and biomass yields, as well as WUE terms (Table 19–6) compared to conventional tillage. Figure 19–6(b) shows that soil water storage in the basins under organic and stone mulches was very similar. Table 19–7 indicates that there are no

Table 19–6. Conventional vs. in-field water harvesting + basin tillage in terms of yield and water balance parameters for maize (After Botha et al., 2003).

	Parameters Yield		Water balance components							Water-use efficiencies			
Treatment	Seed	Biomass	P	ΔS	D	R	E_s	E_v	$E_s + E_v$	WUE_{ET}‡	WUE_{EV}‡	PUE_{gf}§	RSE†
	kg ha^{-1}		mm							kg ha^{-1} mm^{-1}			%
CON¶	1837	4034	193.5	87	0	16.2	202	63	264	6.95	29.3	3.24	24.18
IWHB	3369	6878	193.5	136	0	0	222	107	329	10.23	31.4	5.57	37.9
t Value	5.0**	4.73**	ND#	ND	ND	ND	1.7NS	ND	4.3*	4.0*	1.1NS	5.2**	

*,**,*** t(0.05) = 2.78, t(0.01) = 4.60, t(0.001) = 8.61, respectively.
† RSE = rainfall storage efficiency.
‡ WUE_{ET} = water-use efficiency in terms of water used for evapotranspiration; WUE_{Ev} = water-use efficiency in terms of water used for transpiration.
§ PUE_{gf} = precipitation use efficiency during the fallow and growing season.
¶ CON = conventional tillage; IWHB = in-field water harvesting + basin tillage.
# ND = not determined; NS = statistically not significant.

Table 19–7. Organic vs. stone mulches in microbasins in terms of yield and water balance parameters for maize (after Botha et al., 2003).

	Parameters Yield		Water balance components							Water-use efficiencies			
Treatment	Seed	Biomass	P	ΔS	D	R	E_s	E_v	$E_s + E_v$	WUE_{ET}‡	WUE_{EV}‡	PUE_{gf}§	RSE†
	kg ha^{-1}		mm							kg ha^{-1} mm^{-1}			%
CON¶	3360	7853	193	151	0	0	221	123	343	9.79	27.4	4.92	5.75
IWHB	3496	7817	193	147	0	0	218	122	340	10.30	28.7	5.29	10.56
t Value	−1.55NS#	0.12NS	ND	ND	ND	ND	0.25NS	ND	ND	−1.10NS	−0.90NS	−2.11NS	−1.15NS

*,**,***t(0.05) (5 d.f.) = 2.78; t(0.01) (5 d.f.) = 4.60; t(0.001) (5 d.f.) = 8.61, respectively.
† RSE = rainfall storage efficiency; IWHB = in-field water harvesting with basins.
‡ WUE_{ET} = water-use efficiency in terms of water used for evapotranspiration; WUE_{Ev} = water-use efficiency in terms of water used for transpiration.
§ PUE_{gf} = precipitation use efficiency during the fallow and growing season.
¶ CON = conventional tillage; IWHB = in-field water harvesting + basin tillage.
# NS = statistically not significant; ND = not determined; (t>0: OM>SM; t<0: OM<SM).

statistical differences in terms of yield and WUE terms between the two mulches. This finding has an important implementation value in practice, especially for rural farmers. As pointed out earlier, crop residues are a scarce commodity in rural areas, apart from the urgent need as animal feed during winter. Hence, stone, instead of organic, mulches can be used in the basins to reduce evaporation from the soil surface, thereby increasing water storage to be used for crop production.

Sowing Density. Efficient water use also requires that the plant population density should be matched with the water supply, which is determined by the stored soil water at planting plus the expected rainfall. An experiment involving maize planted in well-fertilized soil at seven population densities ranging from 4000 to 111 000 plants ha^{-1}, and different levels of water supply, showed a positive interaction between planting density and water supply on the yield of maize (Van Averbeke and Marais, 1992). Plant population for optimum yields decreased from 60 000 plants ha^{-1} with 650 mm water supply to 10 000 plants ha^{-1} when 238 mm water is available (Fig. 19–7). The authors concluded that matching planting density to the prevailing water supply was indeed a management practice that could result in an increase in yield and reduce risk in local maize production. De Bruyn (1974) found over 10 seasons with annual rainfall of 540 mm on a deep soil that the optimum row width and plant population for maize were 1.83 m and 17 000 plants per ha, respectively, and that long fallow increased the average yield from 1665 to 2250 kg ha^{-1}.

Well-established guidelines for fertilization and plant population densities for all the major crops and production regions are being followed by the commercial farming community of South Africa. In the low potential production regions the recommended maize plant populations will vary between 10 000 and 18 000 plants ha^{-1} planted in 1.5 or 2.3 m row spacing. Under these conditions yields of between 2000 and 6000 kg ha^{-1} can be obtained, depending on the rainfall.

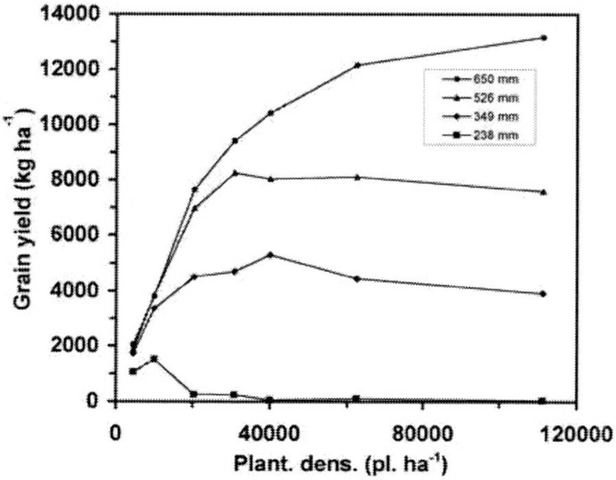

Fig. 19–7. Relationship between grain yield of maize at different planting densities and levels of water supply (Adapted from Van Averbeke and Marais, 1992).

SIMULATION MODELS IN EFFICIENT SOIL WATER USE

Locally and internationally developed crop growth and soil water balance computer models are currently being used in South Africa by several institutions. Long-term results are necessary for reliable production technique recommendations and production risk quantification under semi-arid conditions. These are obtainable using tested crop models and long-term climate data. Attempts to improve the reliability of the PUTU (locally developed) and Decision Support System of Agrotechnology Transfer (DSSAT) wheat and maize models by comparing measured and simulated yields on a range of ecotopes (i.e., a land-use unit with homogenous climate, soil, and aspect characteristics) have also been made (Hattingh, 1993; Botha et al., 1997, 1998; Hensley et al., 1997).

The resultant cumulative distribution functions (CDFs) of yield for each of these scenarios provide valuable information about the productivity and management of an ecotope. For example, Anderson et al. (2003) used PUTU to compare expected maize yields in a marginal ecotope under conventional (CON) and in-field water harvesting + basin tillage (IWHB) tillage practices when planting on full to empty root zones. At a 50% probability [Fig. 19–8(a) and (b)] CON and IWHB give yields of 1700 and 2300 kg ha^{-1}, respectively, when PUTU was run with a full profile, and 950 and 1800 kg ha^{-1}, respectively, when starting with an empty profile. The superiority of IWHM over CON is clearly demonstrated, as well as the significance of the root zone water content at planting.

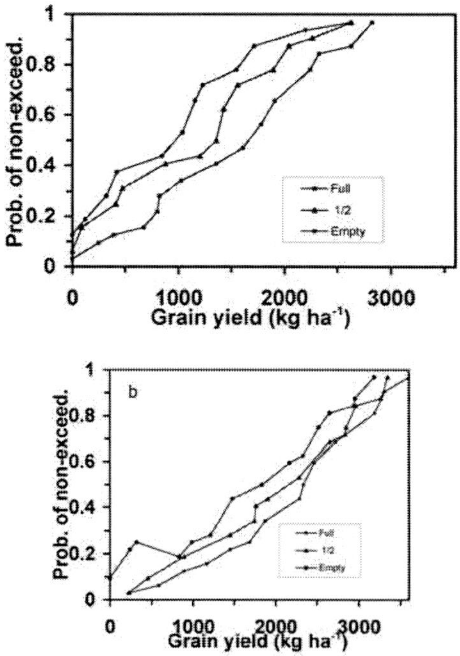

Fig. 19–8. Cumulative distribution functions of long-term maize yields on a clay soil: (a) conventional tillage and (b) in-field water harvesting with basins (After Anderson et al., 2003).

DISCUSSION AND CONCLUSIONS

The basic principle of efficient soil water use for plant production lies in maximizing the gains and minimizing the losses of water. It was shown that crop production risks are lower with practices that assure good soil water storage at planting. The largest soil water losses are due to evaporation from the soil surface. These losses can be in the order of 60 to 75% of the annual rainfall under semi-arid climatic conditions. Soil water evaporation losses can be decreased over periods shorter than 14 d, provided that a 70%, or higher, shading is maintained through mulching practices. Yield reductions of several crops have been reported with crop residue mulching and reduced tillage on sandy to loamy sand textured soils, while opposite results were obtained on soils with higher clay contents in the topsoil. Runoff from cultivated fields can amount to 10% of the annual precipitation.

The practice of using crop residue as a mulch to control runoff or reduce evaporation from the soil surface is seldom economically feasible, when compared with using it as animal feed under low rainfall situations. However, it has been shown that stone mulching is very effective in reducing evaporation. Poor control of weeds during the period of preplant water storage, as well as during the growing season, will almost always result in reduced yields or even total failure during below-average rainfall years. Plant populations should always be adapted to the available supply of stored soil water plus the rainfall that can be expected. Drainage losses can be as high as 20% on deep well-drained sandy soils. Cultivation of soil types, with clay horizons below the potential rooting depth that will impede deep percolation, could be advantageous because of increased water supply to crops grown in low rainfall areas.

On clay soils in areas with marginal and erratic rainfall, especially where a large proportion falls as thunderstorms, the major forms of water losses are runoff and evaporation. Realizing this, researchers have successfully employed water-harvesting techniques to increase yields in various parts of Africa (Kronen, 1994). Similar work is currently being undertaken at the Glen Experimental Station in the Free State Province on clay soils, focusing on the needs of small farmers, and combining the benefits of water harvesting, basin tillage with mulch in the basins to reduce evaporation, and long fallow. Results indicate increases in soil water storage and, hence, in improving yields of various crops compared to conventional tillage.

There is limited scope to optimize stored soil water use for wheat production on the shallow, stony soils of the southern coastal areas. Because of the low water-storage capacities of these soils, the contribution of efficient stored soil water use to production is inferior to the amount and distribution of rainfall. However, reduced tillage and residue mulching are gaining momentum because of their beneficial effects on biological and physical properties of the soils. The beneficial effects of crop rotation including wheat and leguminous species are also well known to farmers.

In addressing the adverse effects of long-term conventional tillage on Vertic soils, Beukes (1987, 1992) found that available soil water is a major factor that determines growth and yield of grain sorghum. Stubble mulch practices improved aggregation on these soils, thereby increasing infiltration and soil water contents, with consequent increases in yield. Grubb (1993) was of the opinion that the nonsustainable cultivation of Vertisols can, inter alia, be arrested by retention of stubble mulch, restricting tillage methods to tine and disc plow only, and crop rotation practices.

The South African experience has shown that for small-scale crop production under subhumid climatic conditions with an annual rainfall exceeding 800 mm, the goals of economic viability and natural resource sustainability are relatively easy to achieve. In the lower rainfall areas (<600 mm per annum), and based on a PUE for maize of 3.5 kg ha^{-1} mm^{-1} (Bennie et al., 1994), the conclusion can be made that the large area of the farms needed to sustain a living standard comparable to the rest of the population, will require large-scale extensive practices when using conventional tillage. However, it is well known that by reducing runoff and evaporation and increasing soil water storage, for example, with various mulch practices, PUE can be improved, thereby reducing the area needed for a sustainable living. The various water-harvesting techniques to improve PUE, especially on marginal soils and under erratic and low rainfall, have not yet been sufficiently investigated in South Africa.

PRIORITY ISSUES FOR FURTHER RESEARCH

Emerging from his brief overview and analysis of soil water balance in relation to efficient crop production in South Africa, a number of key concerns need further investigation:

1. There is an urgent need for an integrated approach to simultaneously optimize soil water and nutrient use by crops in dry areas for greater efficiency and sustainability. For too long these two aspects have been researched independently. Furthermore, an integrated catchment management approach, where soil water and nutrient balances are determined per land use pattern in a catchment, has never been investigated in South Africa. This is a broader natural resource management perspective that can also supply information to off-site, downstream soil and water users.
2. Under the Mediterranean climate of the southern coastal areas, the storing of light rainfall prior to planting should be investigated. Not only will the accumulated soil water prompt earlier planting of wheat, but N mineralization can also be enhanced. The beneficial effects of reduced tillage and the use of crop residues on inter alia soil organic C, soil structure and soil water properties, such as water retention, need further clarification. The positive effects of crop rotation systems to improve soil conditions and combat soil erosion should also be studied in long-term trials.
3. Due to their unique features, Vertic soils are inherently fertile and productive when properly managed. However, their high clay contents, and structural degradation due to long-term conventional tillage, cause low infiltration capacities and, consequently, low water storage. The beneficial effects of reduced tillage and stubble mulching should further be investigated, especially on Vertic soils with strong crusting and self-mulching properties.

4. Various water harvesting techniques, such as planting in microbasins and varying ratios of catchment area (runoff) to cropped area (run-on), should also be evaluated. The latter techniques, when adapted for local conditions, could be of great benefit to sustain food production by small farmers who have to make a living on high clay soils and under marginal rainfall.
5. The critical levels of different types of mulching materials to effectively reduce evaporative losses under different climatic conditions on different soil types must be determined. The water-use efficiency in large-scale crop production can be improved if the problem of the lower yields with stubble mulching on sandy soils can be solved.
6. With regard to modeling there is a need to develop and refine models in collaboration with international expertise in terms of: (1) a resource-based model that determines the minimum farm size for sustainable production, (2) refinement of seasonal outlook forecasting to allow for the prediction of seasonal onset and distribution of rainfall, and (3) improvement of the WUE modeling capability of crop and systems models.
7. Develop a system to evaluate production systems at various levels and scales (in terms of environmental, institutional, economic, and sociocultural) with regard to their sustainability.
8. Determine the effects of different crop rotation systems on long-term water-use efficiency under dryland conditions with field trials under different soil and climatic conditions in collaboration with various stakeholders.

ACKNOWLEDGMENTS

The following persons and their institutions are gratefully acknowledged for their contributions: Prof. G.A. Agenbag, University of Stellenbosch; Prof. P. Fouche, University of the North; Dr. M. Marake, University of Lesotho; Dr. W. van Averbeke, ARDRI, University of Fort Hare; Dr. N. Miles from KwaZulu, Department of Agriculture; Mr. P.J. Snyman from Free State Department of Agriculture; Messrs. J. Blomerus and F. Knight from Western Cape Department of Agriculture; Messrs. W. du Toit and A. Nel from the ARC-Grain Crops Institute; Dr. T. Fyfield and Mr. R. Kuschke from ARC-Institute for Soil, Climate and Water; Dr. J. van Heerden from ARC-Range and Forage Institute; Mr. W. Kilian from ARC-Small Grain Institute; Dr. M. Dippenaar from ARC-Tobacco and Cotton Research Institute.

REFERENCES

Agenbag, G.A. 1987. The effects of tillage on soil factors and the consequent development and yield of wheat in the Swartland (African). Ph.D. thesis. Univ. of Stellenbosch, Stellenbosch, South Africa.

Agenbag, G.A., and P.C.J. Maree. 1991. Effect of tillage on some soil properties, plant development and yield of spring wheat (*Triticum aestivum* L.) in stony soil. Soil Tillage Res. 21:97–112.

Agenbag, G.A., and J.J. Stander. 1988. The effect of tillage on some soil properties and the consequent development and yield of wheat in the Southern Cape. (African). S. Afr. J. Plant Soil 5:142–146.

Anderson, J.J., J.J. Botha, L.D. van Rensburg, M. Hensley, and P.P. van Staden. 2003. The application of crop modelling technology to quantify risk for marginal crop production areas. *In* Final Programme and Abstracts of Golden Jubilee Combined Congress of SSSSA, SASCP, SASHS, Stellenbosch, South Africa. 20–23 Jan. 2003. Univ. Documentation Manage. Serv., Observatory, Cape Town, South Africa.

Bennie, A.T.P., J.E. Hoffman, and M.J.Coetzee. 1995. Sustainable crop production on aeolian semi-arid soils in South Africa. Afr. Crop Sci J. 3:67–72.

Bennie, A.T.P., J.E. Hoffman, M.J. Coetzee, and H.S. Vrey. 1994. Effects of different soil management practices on components of the soil water balance. (African.) p. 31–110. *In* Storage and utilization of rain water in soils for stabilizing crop production in semi-arid areas. (African.) Report no. 227/1/94. Water Res. Comm., Pretoria. WRC, Pretoria, South Africa.

Berry, W.A.J., and J.B. Mallett. 1988. The effect of the tillage: Maize residue interactions upon soil water storage. S. Afr. J. Plant Soil. 5:57–64.

Beukes, D.J. 1987. The effect of different stubble mulch tillage practices on certain soil properties, as well as on growth and yield of grain sorghum on a vertisol in the Highveld Region (African). p 47. *In* Abstracts of 14th National Congress of the Soil Sci. Soc. of South Africa, Nelspruit, 13–17 July 1987. SSSSA, Pretoria, South Africa.

Beukes, D.J. 1992. Long-term effects of stubble and conventional tillage on aggregate stability and other properties of a vertic soil. p. 1. *In* Abstr. of 17th Natl. Congr. of the SSSSA, Stellenbosch. 28–30 Jan. 1992. Soil Sci. Soc. of South Africa, Pretoria, South Africa.

Birch, E.B., J.C. Van der Sandt, and F.M. Strauss. 1986. The potential of basin tillage in sunflower production. p. 32–38. *In* Proc. Int. Symp. on Agric. Eng., Pretoria. 20–24 Jan. 1986. South African Inst. for Agric. Engineers, Pretoria, South Africa.

Botha, J.J., J.J. Anderson, M Hensley, and W.H.O. Du Toit. 1997. Secondary wheat roots in sandy soils. p. 681–683. *In* African Crop Science Congress Proc. Vol. 3, Pretoria. 13–17 Jan. 1997. South African Soc. for Crop Production, Pretoria, South Africa.

Botha, J.J., J.J. Anderson, M. Hensley, and P.P. van Staden. 1999. Crop production on high drought risk clay soils using an in-field water harvesting production technique. p.17–20. *In* Programme and abstracts of 22nd National Congress of the SSSSA, Pretoria. 28 June–1 July 1999.

Botha, J.J., J.J. Anderson, L. D. van Rensburg and D. J. Beukes. 2003. Assessment and modelling of water harvesting techniques to optimize water use in a semi-arid crop production area in South Africa. *In* Research report GW/A/2002/131. ARC-Inst. for Soil, Climate and Water, Pretoria, South Africa.

Botha, J.J., J.J. Anderson, P. P. van Staden, L. D. van Rensburg, D. J. Beukes, and M. Hensley-2001. Quantifying and modelling the influence of different mulches on evaporation from the soil surface. *In* Research report GW/A/2001/5. ARC-Inst. for Soil, Climate and Water, Pretoria, South Africa. Agric. Res. Council-Inst. For Soil, Climate, and Water, Pretoria, South Africa.

Botha, J.J., and M. Hensley. 1998. Quantification of the stress curve for wheat on a Hebron/Bainsvlei ecotope (African). p. 27. *In* Abstracts of combined Congress of the SSSSA and the S.A. Soc. for Crop Prod., Alpine Heath, KwaZulu-Natal. 20–22 Jan. 1998. Soil Sci. Soc. of South Africa/South African Soc. for Crop Production, Pretoria, South Africa.

Du Plessis, M.C.F., and J.W.C. Mostert. 1965. Runoff and soil loss at the Agricultural Research Center Glen (African). S. Afr. Tydskr. Land bouvetenskap. 8:1051–1060.

Engelbrecht, C., A.J. Van der Westhuizen, and G.D. Joubert. 1986. Evaluation of fallow systems in dryland wheat production in the Orange Free State (African). S. Afr. J. Plant Soil. 3:198–200.

Grubb, P.L.C. 1993. Vertisols in the Springbok Flats, Northern Transvaal: Some aspects of their nature, cultivation and structural degradation. p. 204–221. *In* Research report GW/A/94/32. ARC-Inst. for Soil, Climate and Water, Pretoria, South Africa.

Haarhoff, D. 1989. The water regime of a Clansthal series soil under dryland maize production in the Western Transvaal (African). p. 58–61. *In* M.Sc. Agric. Thesis. Univ. of the Orange Free State, Bloemfontein, South Africa.

Hanks, R.J., and V.P. Rasmussen. 1982. Predicting crop production as related to plant water stress. Adv. Agron. 35:193–215.

Hattingh, A.M. 1995. Soil: The effects of stubble mulching and planting density on the soil water regime and yield of maize (African). p. 4–12. *In* Research progress report 7. Highveld Region, Dep. of Agric., Potchefstroom.

Hattingh, H.W. 1993. Estimating the evaporation from the soil surface under dryland wheat and maize production (African). p. 108–113. *In* M.Sc. Agric. thesis. Univ. of the Orange Free State, Bloemfontein, South Africa.

Haylett, D.G. 1960. Run-off and soil erosion studies at Pretoria. S. Afr. J. Agric. Sci. 3:379–394.

Hensley, M., J.J. Anderson, J.J. Botha, P.P. Van Staden, A. Singels, M. Prinsloo, and A. Du Toit 1997. Modelling the water balance on benchmark ecotopes. Water Res. Comm. Rep. 508/1/97. WRC, Pretoria, South Africa.

Hillel, D. 1982. Infiltration and surface runoff. p. 211–234. *In* D. Hillel (ed.) Introduction to soil physics. Academic Press, New York.

Hoffman, J.E. 1997. Quantification and prediction of soil water evaporation under dryland crop production (African). p. 91–97. *In* Ph.D. Diss., Univ. of the Orange Free State, Bloemfontein, South Africa.

Hoffman, J.E. 1990. The influence of soil tillage practices on the soil water balance of an Avalon Series soil under wheat production at Bethlehem (African). p. 48–94. *In* M.Sc. Agric. thesis. Univ. of the Orange Free State, Bloemfontein, South Africa.

Jacobs, E.O., and D.J Beukes 1997. The nature and sustainable management of Vertisols in South Africa. p. 1–5. *In* Proc. 23th Meet. South African Regional Commission for the Conserv. and Utilization of the Soil Standing Committee for Soil Science, Annex. B, Pretoria, South Africa. 10–13 Nov. 1997. SARCCUS, Pretoria, South Africa.

Kronen, M. 1994. Water harvesting and conservation techniques for smallholder crop production systems. Soil Tillage Res. 32:71–86.

Lang, P.M., and J.B. Mallett. 1984. The effect of increasing the amount of surface residue on infiltration and soil water loss. S. Afr. J. Plant. Soil 1:97–98.

Mallett, J.B., P.M. Lang, and A.J. Arathoon. 1987. Changes in a Doveton clay loam after 12 years of direct drill maize production. S. Afr. J. Plant Soil 4:188–192.

Paterson, G.P., and M. van der Walt. 2003. Soil patterns of South Africa from the Land Type Survey. *In* Final Programme and Abstracts of Golden Jubilee Combined Congress of SSSSA, SASCP, SASHS, Stellenbosch. 20–23 Jan. 2003. UDMS, Observatory, South Africa.

Scotney, D.M., J.E. Volschenk, and P.S. Van Heerden. 1990. The potential and utilization of the natural agricultural resources of South Africa. Dep. of Agricultural Development, Pretoria, South Africa.

Snyman, P.J., C. Engelbrecht, and S.W.J. Van der Merwe. 1992. Evaluation of conservation cultivation practices for a fallow-wheat system in the Central Free State (Africa). Appl. Plant Sci. 10:65–68.

Soil Classification Working Group. 1991. Soil classification. A taxonomic system for South Africa. Memoirs on the Agric. Natural Resource of SA no. 15. Dep. of Agric., Pretoria, South Africa.

Stewart, B.A., and E. Burnett. 1987. Water conservation technology in rainfed and dryland agriculture. p. 355–359. *In* W.R. Jordan (ed.) Water and water policy in world food supplies. Proc. of the Conf., Texas A&M Univ., College Station. 26–30 May 1995. Texas A&M Univ., College Station.

Tyson, P.D., and T.G.J. Dyer. 1978. The predicted above-normal rainfall of the seventies and the likelihood of drought in the eighties in South Africa. S. Afr. J. Sci. 74:372–377.

Unger, P.W. 1990. Conservation tillage systems. Adv. Soil Sci. 13:27–68.

Van Averbeke, W., and J.N. Marais 1992. Maize response to plant population and soil water supply: 1. Yield of grain and total above-ground biomass. S. Afr. J. Plant Soil 9:186–192.

Van Averbeke, W., and Z. Mkile 1996. Effect of stone mulching on the consumptive water use by a sparse stand of maize grown in lysimeters. p. 52. *In* Abstracts of S.A. Soc. for Crop Prod. Conf., Bloemfontein, South Africa. 23–25 Jan. 1996. SASCP, Pretoria, South Africa.

20 Carbon Sequestration in Dryland Agriculture

Rattan Lal

Carbon Management and Sequestration Center
Columbus, Ohio

ABSTRACT

Drylands cover about 4886 million hectares (Mha) in ice-free land areas of the world. Of this, 2124 Mha is in semi-arid regions, 2180 Mha in arid regions, and 581 Mha in extremely arid regions. Predominant soils of dryland areas include 307 Mha of Alfisols, 1657 Mha of Aridisols, 1915 Mha of Entisols, 547 Mha of Mollisols, 188 Mha of Vertisols, and 290 Mha of miscellaneous soils. Most soils in dryland regions are young, characterized by less weathering, and dominated by physical weathering processes. In general, soils have low soil organic carbon (SOC) concentration ranging from 0.05 to 0.5% in the surface horizon. However, dryland soils have high concentrations of soil inorganic carbon (SIC) especially in the calcic/petrocalcic horizon which also has high concentrations of secondary/pedogenic carbonates. The strategy to improving SOC and SIC pools in these soils involves improving water and nutrient-use efficiencies by decreasing losses and improving biomass production. Land use/farming practices to achieve these goals include conservation tillage, mulch farming, including cover crops in the rotation cycle, mixed farming/cropping, agroforestry, ley farming, and adoption of integrated nutrient and pest management practices. There are also options for management of grazing lands and forest lands such as controlled grazing, planting improved species, fire management etc. Conversion to recommended land use and management practices can bring about a modest increase in SOC pool, and in SIC pool through formation of secondary carbonates. Enhancing SOC pool is more challenging in dryland soils than in humid temperate climates. Principal limitations to SOC sequestration are low precipitation, high temperatures, low clay content in soil, and low biomass productivity due to harsh climate, drought prone soils of low inherent fertility, and often resource-poor farmers.

INTRODUCTION

Dryland agriculture implies production of crops and animals under rainfed conditions without any supplemental irrigation in areas characterized by prolonged dry season, and negative water balance. Therefore, agriculture is dependent on the vagaries of weather, especially the rainfall amount and its distribution. Because of the high risks involved, dryland agriculture has received relatively less attention of policymakers, researchers, and land managers. Yet, dryland agriculture is the last frontier in the quest to achieve global food security.

Dryland agriculture is practiced in developed and developing countries in all continents. The drylands of Africa, the Middle East, Latin America, and South Asia are inhabited by more than one billion people (Noin and Clarke, 1998). Most of the farmers in developing countries are small landholders, resource-poor and practice subsistence agriculture involving mixed farming systems based on some combination of crops, trees, and livestock. Crop yields are low and highly variable depending primarily on the rainfall amount and its distribution.

The green revolution of the 1960s and 1970s brought about a quantum jump in food production in irrigated agriculture in South Asia and elsewhere in the world (Hazel and Ramasamy, 1991). Yet, the green revolution technology bypassed the regions where dryland agriculture is practiced. Production gains have not been realized in sub-Saharan Africa (SSA), South Asia, Central Asia, West Asia, and North Africa, and elsewhere in regions with predominantly rainfed agriculture. On a global scale, there exists a vast potential to increase crop yields in dryland agriculture. The gap between attainable and actual yield in several regions of dryland agriculture show that application of the recommended management practices (RMPs) can bring about a quantum jump in food production. An important by-product of agricultural intensification in these regions will be enhancement of the terrestrial carbon (C) pool through C sequestration in biota and soil. The objective of this chapter is to review the available information on the potential of C sequestration through conversion of agriculturally marginal soils to restorative land use, and adoption of RMPs in regions with favorable soils and moisture regime. Soil, climate, and other constraints to realizing the potential of terrestrial C sequestration are also discussed.

CLIMATE AND SOILS OF DRYLAND REGIONS

Low rainfall and the annual water deficit are the predominant features of drylands (Dutton et al., 1998). The water deficit is attributed to excess of evapotranspiration over the precipitation due to a wide array of interacting factors including low and erratic precipitation, high temperature, low relative humidity, high winds etc. On the basis of the aridity index (AI), defined as the ratio of precipitation (P) to potential evapotranspiration (PET), dry lands are classified into semi-arid region with AI of 0.20 to 0.50, arid regions with AI of 0.05 to 0.20 and hyper-arid regions with AI of <0.05 (UNEP, 1992; Meigs, 1953). Excluding cold deserts, land area under these three climates is as follows (Shantz, 1956):

Semi-arid	2124.3 Mha
Arid	2180.3 Mha
Extremely arid	581.2 Mha
Total	4885.8 Mha

In addition, the glacier-free terrestrial area of polar region deserts include the following (Péwé, 1974):

Arctic region	430.0 Mha
Antarctic region	60.0 Mha

Soils of dryland regions are generally derived from wind or water-transported material. Most soils are young, less weathered, and poorly developed. Weathering is characterized by predominantly slower physical than rapid chemical processes. Most physical weathering processes include abrasion by wind and water, and fracturing of rock/parent material by heating and cooling. The sparse vegetation cover leads to a very small return of biosolids to the soil. Thus, biotic factors may play a relatively smaller role in dry land soils than in formation of humid region soils.

Predominant soils of the drylands include Alfisols, Aridisols, Entisols, Mollisols, and Vertisols (Table 20–1). Agriculturally important soils are alluvial soils or Entisols (Fluvents) formed along flood plains of rivers and streams. Important distinguishing features of dryland soils include the following: (i) sand dunes are a common feature of the dryland soils, (ii) desert pavement is a regular feature and consists of gravels, pebbles, and stones lying on the top of the soil, (iii) calcic, petrocalcic, or indurated horizons rich in calcium carbonate and gypsum occur in most dryland soils, (iv) high base saturation leads to high soil pH, and (v) the low soil organic matter content is due to sparse vegetation cover.

Soil Organic Carbon

Soils of the dryland regions contain lower soil organic matter (SOM) concentration than their counterparts in humid climates. The SOC pool in soils under tropical savannas and forests are low, with a wide range of 20 to 150 Mg C ha^{-1}, with an average value of 25 Mg C ha^{-1} in the top 20-cm layer (Tiessen et al., 1998). The SOC pool decreases logarithmically with increase in mean annual temperature (Jenny and Raychaudhuri, 1960). The large variation is also attributed to a difference in soil properties and site characteristics (Feng et al., 2002). The data on SOC concentration and pool for some soils of Arizona are shown in Table 20–2. The SOC concentration of surface horizon (0–30 cm) is generally <0.5%, but there can be some exceptions as in Mollisols containing as high as 4% SOC concentration. Some arid zone soils contain a high amount of SOC pool. Consequently, the SOC pool to 1-m depth in warm dryland soils can be as low as 20 Mg ha^{-1} (Table 20–2), compared with 80 to 100 Mg ha^{-1} for soils of the cool temperate climates. All other factors remaining the same, the SOC concentration in surface horizon increases with increase in clay content. The data in Table 20–3 show that SOC concentrations of clay and clay loam soils is much greater than those of sandy loam and sandy texture. In addition to low concentration, the SOC profile of dryland soils differ markedly from those of humid climates. Charley and Cowling (1968) observed that dryland soils of Australia contain lower SOC concentration than in equivalent soils in other dryland regions (Fig. 20–1). They also observed that SOC, total nitrogen (N) and organic phosphorus (P) are generally concentrated in the surface layer with sharp decline in the subsoil, compared with relatively higher concentrations with depth in soils of the subhumid and humid climates. The SOC pool of world dryland soils is 522 petagram (Pg = 10^{15} g) which is about one third of the global SOC pool (Table 20–4) for regions that occupy about 37.3% of the ice-free land. Aridisols account for almost 25% of the area of all soils but contain only 7% of the SOC pool (Eswaran et al., 2000).

Table 20–1. Arid soils of the world (adapted from Dregne, 1976; Buringh, 1979; Beaumont, 1989; Eswaran et al., 1999; Van Wambeke, 1990).

Soil order	Africa	Asia	Australia	Europe	North America	South America	Total area in arid regions	Arid region area	World: Total area of the order	World: World total in arid regions
	Mha							%	Mha	%
Alfisol	209	--	44	--	16	38	307	6.7	1262	24.3
Aridisol	489	592	277	26	194	79	1657	36.1	1657	100
Entisol	1032	486	228	16	35	118	1915	41.6	2114	90.5
Mollisol	12	285	--	22	179	49	547	11.5	901	60.7
Vertisol	24	78	76	--	10	--	188	4.1	316	59.5
Total	1766	1441	625	64	434	284	4614			
% of the continent	59.2	33.0	82.1	6.6	18.0	16.2				

Table 20–2. Organic carbon (C) pool in soils of Arizona (recalculated from Hendricks, 1985).

Series	Horizon	Depth	SOC† concentration	Soil bulk density	Clay	SOC pool
		cm	%	Mg m^{-3}	%	Mg ha^{-1}
Ajo	A	0–5	0.27	1.26	18	1.7
	Bt	5–33	0.32	1.26	35	11.3
Antho	Ap	0–18	0.14	1.55	9	3.9
	C	18–127	0.08	1.70	8	14.8
Anthony	Ap	0–30	0.10	1.65	10	4.9
	C	30–107	0.09	1.52	8	10.5
Arp	A	0–5	0.81	1.60	27	6.5
	Bt	5–40	0.81	1.68	53	47.6
	C	40–75	0.20	1.73	26	12.1
Balon	A	0–8	0.63	1.54	21	7.8
	Bt	8–58	0.31	1.64	35	25.4
	BCt	58–90	0.09	1.60	32	4.6
Tubac	A	0–28	0.35	1.52	11	14.9
	Bt	28–53	0.26	1.45	52	9.4
	Btk	53–112	0.04	1.50	24	3.5
	C	112–178	0.02	1.40	41	1.8
Soldier	A	0–10	4.79	1.16	12	55.6
	E	10–38	1.86	1.30	17	67.7
	Bt	38–70	0.49	1.57	53	24.6
	Bc	70–110	0.31	1.63	34	20.2
Pinamt	A	0–7	0.20	1.60	8	2.2
	BA	7–22	0.12	1.64	9	2.9
	Bt	22–35	0.15	1.60	13	3.1
	2Bt	35–81	0.25	1.60	21	18.4
	3Bct	81–117	0.07	1.62	17	4.1
	3CK	117–170	0.05	1.50	8	4.0

† SOC = soil organic carbon.

Soil Inorganic Carbon

Dryland soils are characterized by high concentration of soil inorganic carbon (SIC). These soils contain high concentrations of soluble salts in the form of carbonates and bicarbonates of Ca^{+2}, Mg^{+2}, K^{+}, and Na^{+}. The SIC pool in dryland soils is estimated at 916 Pg or 97% of the global pool (Table 20–4). There are two types of inorganic C in soil: primary or lithogenic and secondary or pedogenic carbonates. Formation of pedogenic carbonates in soil is a process of carbon sequestration, and depends on land use and soil/crop management systems. Addition of

Table 20–3. Effect of texture on soil organic carbon (C) concentration in irrigated soils of New Mexico (adapted from Dregne and Maker, 1955).

Soil texture	Soil organic C concentration
	g kg^{-1}
Clay, silty clay	10.4
Clay loam, silty clay loam, sandy clay	11.6
Loam, sandy clay loam, silt loam	9.3
Sandy loam	5.8
Loamy sand, sand	1.7

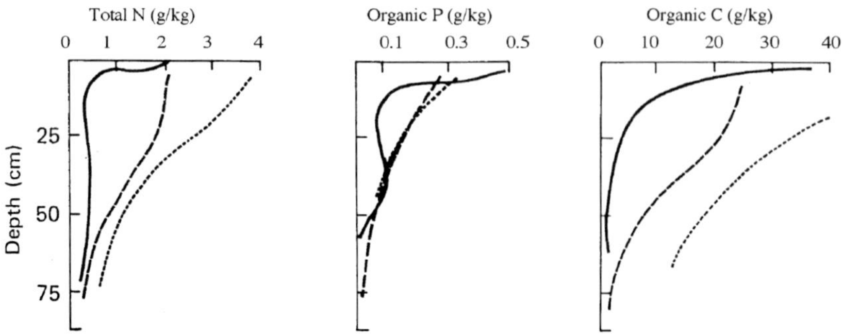

Fig. 20–1. Climatic effects on soil organic C, N, and P profile (adapted from Charley and Cowling, 1968). —— Arid ---- Subhumid ------- Humid.

biosolids and their decomposition in soil increases the partial pressure of CO_2 in the soil air. Depending on soil wetness, dissolution of CO_2 leads to formation of carbonic acid and eventual precipitation of calcium carbonate ($CaCO_3$), magnesium carbonate ($MgCO_3$) as secondary carbonates.

In some dryland soils, carbonates occur as a solid layer called "caliche" or "calcic" horizon. These horizons have accumulations of secondary $CaCO_3$ or $MgCO_3$, and are formed by leaching of carbonates from surface soils or by upward movement of carbonates with the carbonate-rich capillary water. The term "caliche" refers to a crust of Ca and sometimes also of Mg carbonate, usually cemented and hard. This crust is widely observed in arid soils. Gile (1966) described four distinct but sequential stages of development of calcic horizon:

1. Thin pebble coatings in gravelly layers and few filaments or faint coatings in nongravelly soils.
2. Thick pebble coatings or some inter-pebble fillings in gravelly soils, and few to common nodule formation in nongravelly soils.
3. Many inter-pebble fillings in gravelly soils, and many nodules and inter-nodular fillings in non-gravelly soils, and
4. Laminar horizon overlying plugged horizon in gravelly soils, and increasing carbonate impregnation in nongravelly soils.

Table 20–4. The soil organic and inorganic carbon (C) pools of dryland soils of the world to 1-m depth (adapted from Eswaran et al., 1999).

Climate	Soil organic C	Soil inorganic C
	Pg	
Semi-arid	337	134
Mediterranean	40	50
Arid	145	732
Total	522	916
Percent of world total	34.2	96.8

Land Use

Extensive agriculture, based on mixed farming and agroforestry, are widely practiced throughout the dryland regions comprising developing countries. Heathcote (1983) identified five principal land uses: (i) uninhabited wasteland, (ii) nomadic pastoralism, (iii) ranching, (iv) rainfed agriculture, and (v) irrigated agriculture. Livestock raising is practiced on two-thirds of the arid world (Noin and Clarke, 1998). Rainfed agriculture mostly involves cultivation of drought tolerant cereals. Irrigated agriculture is practiced in South Asia, West Asia, North Africa, China, Australia, and North America. The irrigation potential of sub-Saharan Africa (SSA) remains to be developed. Extensive land use and subsistence agriculture based on no or low external input may lead to mining of soil fertility, depletion of SOC pool, and emission of CO_2 into the atmosphere.

Soil Degradation and Desertification

There is a widespread problem of degradation of vegetal cover (Mainguet, 1991; Thomas and Middleton, 1994). Desertification is land degradation in arid, semi-arid and dry subhumid areas resulting from various factors, including climatic variations and human activities (UNEP, 1992, 1997; UNCCD, 2001), is a serious problem (Dregne and Chou, 1992). However, the extent of desertification is a controversial issue (Thomas, 1993). Soil degradation by wind and water erosion, and salinization (Oldeman, 1994) is also a serious issue in these regions. Soil degradation and desertification deplete the SOC pool (Fig. 20–2 and 20–3), expose caliche layers to the climatic elements, and emit CO_2 and other greenhouse gases into the atmosphere.

SOIL CARBON DYNAMICS AND THE GLOBAL CARBON CYCLE

The atmospheric concentration of carbon dioxide (CO_2), which has increased from 280 parts per million by volume (ppmv) during the pre-industrial era to 370 ppmv in the Year 2000, depends on a complex system of interactions among atmosphere, biosphere, hydrosphere, lithsphere, cryosphere, and the pedosphere (Fig. 20–4). Among all the interactions, processes within biosphere and the pedosphere have a drastic impact on the atmospheric concentration of greenhouse gases (GHGs). Important biospheric processes are photosynthesis, respiration, and decomposition of dead/detritus plant material, which govern the net primary productivity (NPP) and net ecosystem productivity (NEP). Important pedospheric processes are humification, aggregation, erosion, leaching, and formation of secondary carbonates. Anthropogenic intervention through adoption of RMPs alter biospheric and pedospheric processes and may enhance photosynthesis, NPP, humification, and aggregation. Simultaneously, losses due to erosion and decomposition/mineralization can be curtailed leading to increase in SOC and SIC pools.

There is a close relationship between the atmospheric concentration of CO_2 and the biospheric and pedospheric processes. Land use changes and drastic disturbance of soil lead to a net flux of CO_2 from the terrestrial biosphere into the atmosphere. In contrast, soil restoration and conversion of degraded ecosystems through afforestation can reverse the trend and lead to a net flux of C into the terrestrial biosphere notably in biota and soils. Indeed, many researchers working on the global C cycle believe that the so-called "missing carbon" estimated at 1 to 2 Pg C yr^{-1} (Tans et al., 1990) is sequestered in terrestrial biosphere and the pedosphere. The net positive C flux into the terrestrial biosphere and pedosphere is attributed to numerous processes including CO_2 fertilization effect, N fertilization probably due to industrial pollution, and better plant growth due to global warming over the 20th century. The positive impact of N fertilization is more in N-deficient than fertile soils (Innes, 1991) and that of global warming in soils of the cold

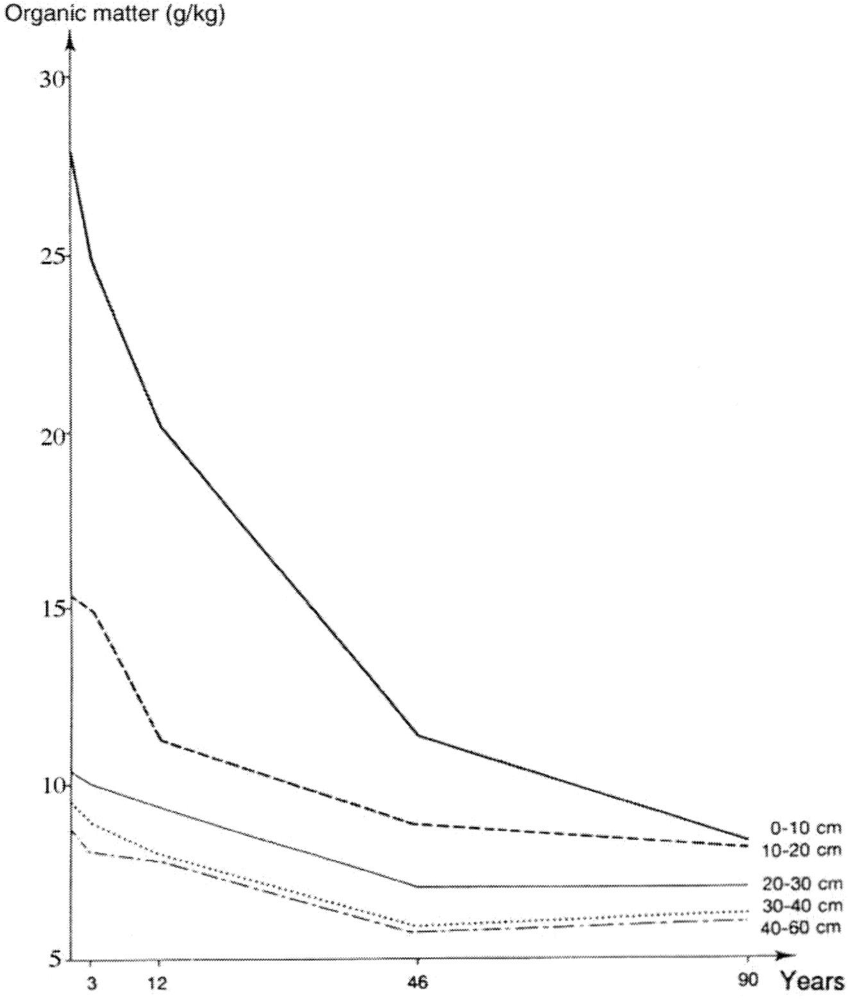

Fig. 20–2. Technological options leading to formation of secondary carbonates in soils of dryland ecosystems. (The organic matter can be converted to soil organic concentration by multiplying with a factor 0.724.) (Redrawn from Siband, 1974.)

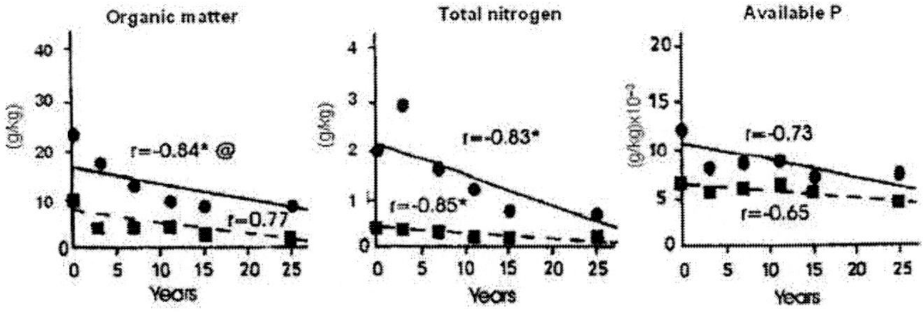

Fig. 20–3. Variation of mean values of soil properties with length of cultivation (adapted from Jaiyeoba, 2003).

than warm regions. The impact of a slight increase in global mean temperature (0.6 + 0.2°C over the 20th century) may have only a minor impact on photosynthesis in the cold regions (Wigley, 1994). Nonetheless, these observations indicate the importance of terrestrial and biospheric processes on moderating the global C cycle.

It is this interaction between the biospheric/pedospheric processes and the global C cycle that highlights the importance of dryland agriculture in moderating the atmospheric concentration of CO_2. Increasing biomass production through in-

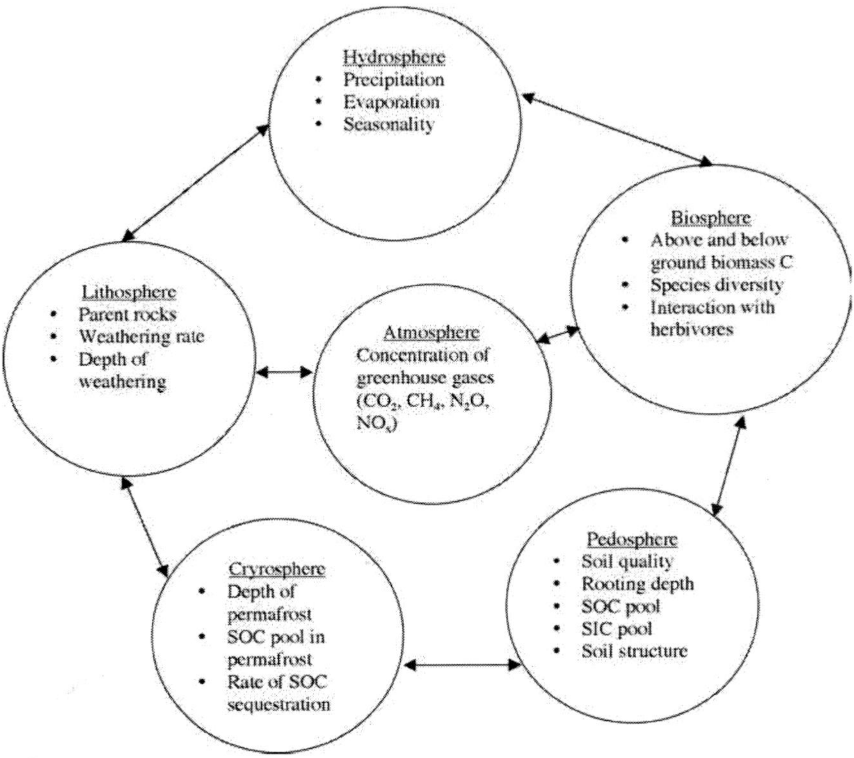

Fig. 20–4. Atmospheric concentration of greenhouse gases depend on a complex system of interactions.

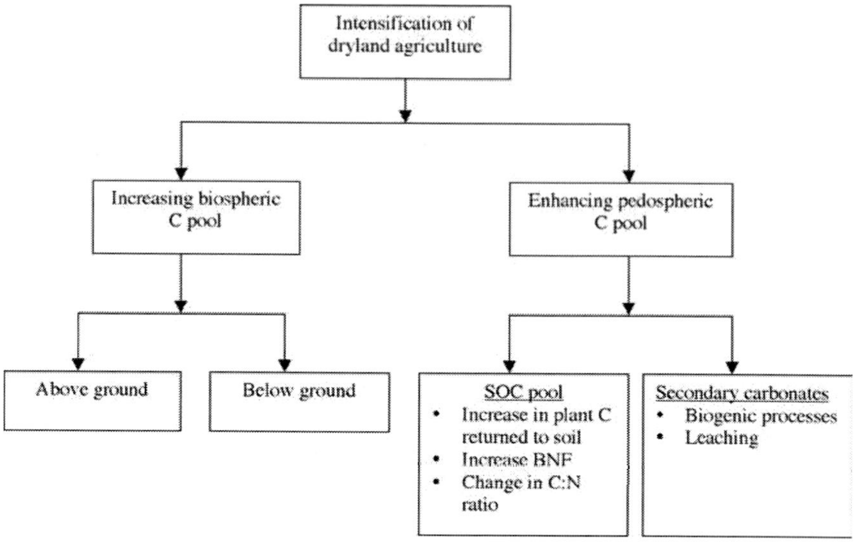

Fig. 20–5. Enhancing soil carbon pool through agricultural intensification.

tensification of dryland agriculture would enhance biospheric and pedospheric C pool, thereby reducing the rate of enrichment of atmospheric concentration of CO_2 (Fig. 20–5). In addition to adopting RMPs, dryland agriculture would also benefit from the CO_2 fertilization effect, especially the C_3 plants would respond positively to increased concentration of CO_2. These plants include rice (*Oryza sativa* L.), wheat (*Triticum aestivum* L.), barley (*Hordeum vulgare* L.), soybean [*Glycine max* (L.) Merr.], sunflower (*Helianthus annuus* L.), potato (*Solanum tuberosum* L.), cotton (*Gossipium hirsutum* L.), most leguminous and woody plants, and most horticultural crops. In addition, there is also a positive CO_2 anti-transpirant effect due to improved water use efficiency (WUE). The gains in NPP due to the fertilization effect leads to reduced transpiration because of the partial closure of stomata (Wolfe and Erickson, 1993). Increase in biological nitrogen fixation (BNF) would alter the C:N ratio and increase humification efficiency of the crop residue returned to the soil.

WATER AND NUTRIENT-USE EFFICIENCIES AND SOIL CARBON DYNAMICS

The SOC pool depends on the quantity of biosolids returned to the soil, which in turn depends on biomass production. Water scarcity is the most limiting factor to enhancing crop/biomass production in dryland farming, and strongly interacting with water scarcity is the nutrient deficiency especially that of N and P. Therefore, the strategy to enhancing biomass production and increasing SOC pool is enhancing water and nutrient-use efficiencies. Hillel (1997) defined agronomic efficiency of water use as follows:

$$B_w = P/u_w \qquad [1]$$

where B_w is the overall biomass efficiency for water use, P is total biomass production, and u_w is the volume of water applied (irrigation plus precipitation). Under field conditions, there are several components of u (Eq. [2]).

$$u_w = (R + D + E_p + E_s + T_w + T_c) \qquad [2]$$

where R is runoff, D is deep drainage, E is evaporation from the conveyance channel, E_s is soil evaporation, T_w is water uptake by weeds and T_c is water uptake by crops. Substituting Eq. [2] in Eq. [1] leads to Eq. [3] which details the factors affecting water-use efficiency of the biomass produced.

$$B_w = \frac{P}{(R + D + E_p + E_s + T_w + T_c)} \qquad [3]$$

Therefore, enhancing B_w implies decreasing losses due to runoff, deep drainage, evaporation from channels and soil and water uptake by weeds. Similar analyses can be made for nutrients, such as N (Eq. [4]).

$$B_N = \frac{P}{U_N} \qquad [4]$$

where B_N is the overall biomass efficiency for N use, P is total biomass production, and U_N is the N applied (Eq. [5]).

$$U_N = (F_N + M_N + R_N + S_N + V_N + L_N + W_N + C_N) \qquad [5]$$

where F_N is the chemical fertilizer applied, M_N is the manure/compost used, R_N is the losses in runoff, S_N is the losses in soil erosion, V_N is the losses due to volatilization, L_N is the losses due to leaching, W_N is the uptake by weeds, and C_N is the uptake by crop. Substituting Eq. [5] in Eq. [4] leads to Eq. [6] that details the factor affecting nutrient-use efficiency. Thus, enhancing nutrient-use efficiency implies decreasing losses of nutrients by water runoff, soil erosion, volatilization, deep leaching, and weeds.

$$B_N = \frac{P}{(F_N + M_N + R_N + S_N + V_N + L_N + W_N = C_N)} \qquad [6]$$

Further, there is a strong relationship between U_W and U_N, because losses by surface runoff, deep percolation, evaporation/volatilization, and weed uptake are strongly interconnected. In the context of enhancing biomass production and SOC pool in soil, the strategy is to enhance B_w and B_n through agricultural intensification and adoption of RMPs.

SOIL/CROP MANAGEMENT FOR AGRICULTURAL INTENSIFICATION

Continuous cultivation, especially with low external input of biosolids and inorganic fertilizers, usually leads to drastic reduction in SOC pool (Nye and Greenland, 1960). In Senegal, Siband (1974) observed a drastic decline in SOC pool upon conversion from natural to agricultural ecosystems (Fig. 20–2). In northern Nigeria, Jaiyeoba (2003) reported a progressive decline in SOC over a 25-yr period from 1.3 to 0.4% in the 0 to 10 cm layer and 0.6 to 0.06% in 20 to 30 cm layer (Fig. 20–3). Similar observations were reported by Jones (1971).

Improving efficiencies of water and nutrients involve two-pronged strategies: (i) decreasing losses, and (ii) improving production (Fig. 20–6). Losses of water and nutrients can be minimized through adoption of conservation-effective measures of controlling runoff, decreasing erosion and evaporation, and reducing losses by weeds. These losses can be reduced through adoption of conservation tillage, use of crop residue mulch, and elimination of bare/uncropped summer fallow through incorporation of an appropriate cover crop in the rotation cycle. Enhancing biomass production in dryland agriculture involves using adaptable species of plants and animals and growing improved cultivars with high production potential in the harsh environments, following recommended cropping/farming systems, and enhancing soil fertility. These objectives can be achieved through adoption of mixed/ley farming practices, integrated nutrient management (INM) and integrated pest management (IPM). While enhancing biomass/agronomic production, adoption of practices outlined in Fig. 20–6 also increase the biomass returned to the soil and thus the SOC pool. Improving soil biodiversity, especially the activity of termites, may be important to minimize crusting and enhancing turnover of biosolids returned to the soil (Lee, 1981).

Conservation Tillage

Conversion from plow tillage to conservation tillage can enhance biomass production, and increase SOC pool (Havlin et al., 1990; Shroyer et al., 1990; So et al., 2001; Mrabet et al., 2001a, 2001b). Mrabet et al. (2001a) reported a 13.6% increase in SOC concentration in 0 to 20 cm layer over an 11-yr period in Morocco. They also observed the stratification of SOC and total N in no-till in the surface 2.5-cm layer without their depletion at the deeper horizon, and depletion of SOC under fallow-wheat compared with wheat-wheat rotation (Table 20–5). Adoption of no-till system increased SOC pool and improved the overall soil quality (Mrabet et al., 2001b). Conservation tillage generally leads to stratification of SOC pool in the surface layer (Mrabet, 2002). In the western part of Argentine Pampas characterized by subhumid and semi-arid climate, Diaz-Zorita et al. (2002) observed that SOC concentration increased in 0 to 20 cm layer by conversion to a no-till system. The SOC concentration also increased with more years of maize (*Zea mays* L.) and wheat than sunflower or soybean. Further, crop productivity was related to SOC concentration in the top 0 to 20 cm layer. In a rice-barley rotation under dryland farming in Varanasi, India, Kushwaha et al. (2000) reported an increase in microbial biomass C in minimum tillage residue retained treatment. Reduced tillage systems have

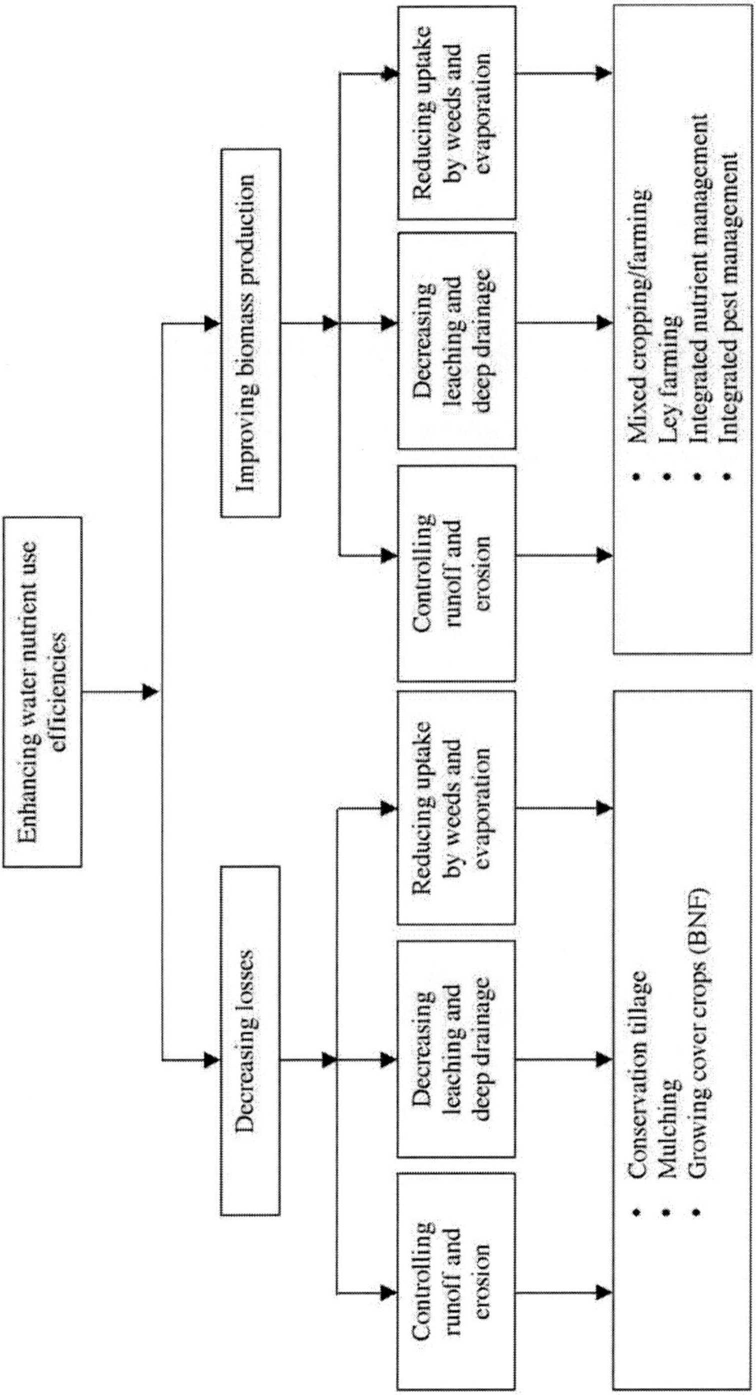

Fig. 20–6. Strategies for enhancing water and nutrient-use efficiencies.

Table 20–5. Tillage effects on soil organic carbon (C) profile in a soil in semi-arid Morocco (adapted from Mrabet et al., 2001b).

Depth	Soil organic C concentration	
	No till	Plow till
cm	g kg^{-1}	
0–2.5	23.1	14.5
2.5–7.0	14.2	14.5
7.0–20.0	12.3	12.2

also been shown to increase SOC pools in semi-arid soils of Australia (Dalal and Carter, 2000; Chan et al., 2001), USA (Havlin et al., 1990; Unger, 1991; Potter et al., 1997; Rasmussen et al., 1998), India (Swarup et al., 2000), and Syria (Jenkinson et al., 1999). The effectiveness of conservation tillage in SOC sequestration is enhanced when used in conjunction with appropriate crop rotations, especially with incorporation of leguminous cover crops in the rotation cycle (Jenkinson et al., 1999).

Water Conservation

Water conservation in the root zone is important to improving biomass productivity. Conservation tillage, mulch farming, and other engineering practices of runoff management are important to enhancing water infiltration rate and increase plant available water in the root zone. Storey (2002) describes several options of soil-water conservation and water harvesting. Techniques for water conservation include using mulches, installing windbreaks, adopting ridge-furrow and tied-ridge systems, using planting pit method etc. There is also a strong interaction between water conservation and soil fertility enhancement.

Soil Fertility Management

Enhancing soil fertility is important to improving biomass productivity, increasing agronomic yield, and returning root biomass and crop residue to the soil. Most dryland soils are deficient in N and P, and improving the nutrient bank is essential to enhancing the SOC pool. Several long-term experiments have documented the positive impact of using chemical fertilizers on drastic increase in crop yields (Pieri, 1992; Vlek, 1993; Bationo et al., 2000; Swarup et al., 2000). Regular use of farmyard manures and biosolids also improves SOC pool (Fig. 20–7), enhances soil fertility and crop yields (Swarup et al., 2000; Pieri, 1992). If feasible, application of organic manures, in combination with a judicious use of chemical fertilizers, can improve soil fertility, enhance SOC pool, and sequester C in soil and biota.

Grazing Management

Grazing is an important land use in dryland areas. Sustaining food production from grasslands involves maintenance of SOC pool at a high level. Conversion

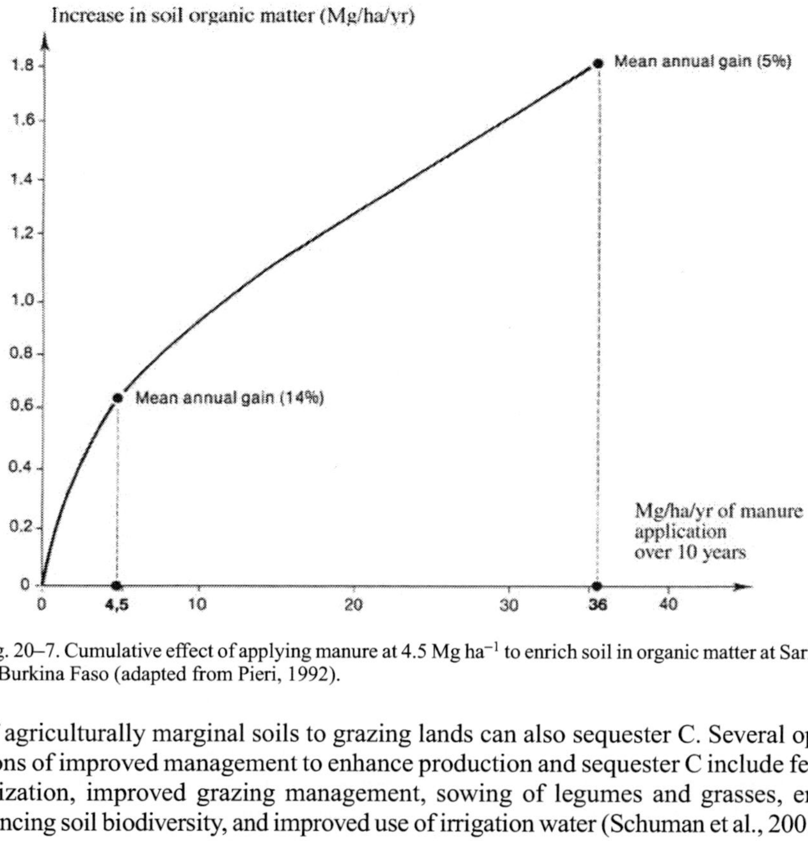

Fig. 20–7. Cumulative effect of applying manure at 4.5 Mg ha^{-1} to enrich soil in organic matter at Saria, Burkina Faso (adapted from Pieri, 1992).

of agriculturally marginal soils to grazing lands can also sequester C. Several options of improved management to enhance production and sequester C include fertilization, improved grazing management, sowing of legumes and grasses, enhancing soil biodiversity, and improved use of irrigation water (Schuman et al., 2001; Reeder and Schuman, 2002). However, excessive grazing can shift the climax vegetation leading to predominance of unpalatable weeds and disappearance of perennial grasses. Such a shift may decrease SOC pool and reduce soil quality (Hiernaux et al., 1999; Abril and Bucher, 2001). Conant et al. (2001) observed that rates of SOC sequestration with improved management of grazing lands range from 0.11 to 3.04 Mg C ha^{-1} yr^{-1} with an average of 0.54 Mg C ha^{-1} yr^{-1}.

Forestry and Agroforestry

There are also some agroforestry systems which can enhance the SOC pool. Schroeder (1994) estimated that median C storage potential by agroforestry practices is 9 Mg C ha^{-1} in semi-arid, 21 Mg C ha^{-1} in subhumid, 50 Mg C ha^{-1} in humid and 63 Mg C ha^{-1} in temperate ecozones. In addition to biomass, there must also be potential of SOC sequestration. In Niger, Guillaume et al. (1999) reported that soils under tiger bush had a capacity to store SOC despite a moderate primary production. The SOC content was higher within the vegetation area than within the bare area. The benefits of agroforestry systems have been documented in a wide range of soils and climatic conditions in the Sahel (Breman and Kessler, 1997). The importance of mesquite (*Prosopis glandulosa* Torr.) on SOC enhancement has been documented in southwestern USA (Geesing et al., 2000) and in Argentina (Felker, 2000).

ROLE OF SECONDARY CARBONATES

Secondary carbonates in calcic/petrocalcic horizons are formed through dissolution of CO_2 generated by decomposition of biosolids returned to the soil. While some consider that the rate of C sequestration through formation of secondary carbonates is too small to be important (Schlesinger, 1995), others feel that secondary carbonates play an important role in soil C sequestration and the global C cycle (Fitzpatrick and Merry, 1999; Nordt et al., 1999; Pal et al., 1999; Pan and Guo, 1999). Furthermore, there is a strong interaction between enhancement of SOC pool and formation of secondary carbonates, because of the increase in concentration of CO_2 in soil, caused by application of biosolids (Monger and Gallegos, 1999). While the rates of formation of secondary carbonates may range widely, the importance of SOC sequestration cannot be ignored. Management practices which accentuate formation of secondary carbonates include irrigation, manuring, multiple cropping, integrated nutrient management (INM) to enhance soil fertility, and measures to conserve soil and water resources (Lal and Kimble, 1999).

SOIL AND CLIMATIC CONSTRAINTS TO CARBON SEQUESTRATION IN DRYLAND FARMING

The SOC pool in dryland soils declines rapidly upon conversion from natural to agricultural ecosystems, and with duration of cultivation (Jones, 1971; Siband, 1974; Westerman, 1992; Albaldejo et al., 1998; Rasmussen et al., 1998; Solomon et al., 2000; Jaiyebo, 1995, 2003). Cultivation year after year, without growing cover crops or seeded/planted fallows, can drastically lower the SOC pool even with the adoption of RMPs (Allison, 1973). Surface application of biosolids (Rostagno and Sosebee, 2001) and adoption of conservation tillage can improve soil quality and reduce the rate of depletion of SOC pool. The absolute SOC capacity in these soils is strongly influenced by the climate (Jobbagy and Jackson, 2000). The total SOC pool increases with precipitation and clay content and decreases with temperature. However, SOC pool is more sensitive to temperature changes than precipitation (West et al., 1994; Schimel et al., 1994). In addition to the effect of clay content, availability of N and P to convert biomass C into humus is an important soil control in determining the SOC sink capacity. Because management plays an important role in soil C sequestration (Paustian et al., 1996), resource-poor farmers in sub-Saharan Africa and elsewhere in the dryland tropics may not be able to afford the input required for agricultural intensification. Conversion of degraded and marginal soils to natural/native vegetation can recover some of the historically depleted SOC pool, as is the case with the conversion to Conservation Reserve Program (CRP) (Robles and Burke, 1997; Lal, 2002). However, the rate of recovery of SOC pool is more in fertile than nutrient-depleted fields (Robles and Burke, 1997). In general, realization of the SOC sequestration potential is a more challenging task in dryland tropics than in humid temperate climates (Stewart et al., 1991; Stewart and Robinson, 2000).

CONCLUSIONS

Conversion of natural to agricultural ecosystems depletes the SOC pool in dryland soils. Adoption of restorative land use and RMPs can enhance SOC pool and also lead to formation of secondary carbonates. The rates of SOC sequestration are promising with adoption of conservation tillage, elimination of bare fallows, incorporation of cover crops, integrated nutrient management and manuring, controlled grazing, and use of improved species of pastures and trees. Further, the rates of SOC sequestration are higher in fertile than nutrient depleted soils. In addition, temperature regime and clay contents are strong controlling factors in SOC sequestration. Therefore, attainable SOC sequestration rates in dryland agriculture are much lower than in humid temperate climate or irrigated agriculture. The SOC sequestration rates are also constrained by low input of nutrients (N, P, S) and biosolids (crop residues and manure) by the resource-poor farmers of developing countries.

REFERENCES

Abril, A., and E.H. Bucher. 2001. Overgrazing and soil carbon dynamics in the western Chaco of Argentina. Appl. Soil Ecol. 16:243–248.

Albaldejo, J., M. Martinez-Mena, A. Roldan, and V. Castillo. 1998. Soil degradation and desertification induced by vegetation removal in a semi-arid environment. Soil Use Manage. 14:1–5.

Allison, F.E. 1973. Soil organic matter and its role in crop production. Elsevier Scientific Publ. Co., Amsterdam, The Netherlands.

Bationo, A., S.P Wani, C.L. Bielders, P.L.G. Vlek, and A.U. Mokwunye. 2000. Crop residue and fertilizer management to improve soil organic carbon content, soil quality and productivity in the desert margins of West Africa. p. 117–146. *In* R. Lal et al. (ed.) Global Climate change and tropical ecosystems. CRC/Lewis Publ., Boca Raton, FL.

Breman, H., and J.J. Kessler. 1997. The potential benefits of agroforestry in the Sahel and other semi-arid regions. Eur. J. Agron. 7:25–33.

Beaumont, P. 1989. Drylands: Environmental management and development. Routledge, London.

Buringh, P. 1979. Introduction to the study of soils in tropical and sub-tropical regions. PUDOC, Wageningen, Holland.

Chan, K.Y., A.M. Bowman, W. Smith, and R. Ashley. 2001. Restoring soil fertility of degraded hard-setting soils in semi-arid areas with different pastures. Aust. J. Exp. Agric. 41:507–514.

Charley, J.L., and S.W. Cowling. 1968. Changes in soil nutrient status resulting from over-grazing and their consequences in plant communities of semi-arid areas. Proc. Ecol. Soc. Australia 3:28–38.

Conant, R.T., K. Paustian, and E.T. Elliott. 2001. Grassland management and conversion into grassland: Effects on soil carbon. Ecol. Appl. 11:343–355.

Dalal, R.C., and J.O. Carter. 2000. Soil organic matter dynamics and carbon sequestration in Australian tropical soils. p. 283–316. *In* R. Lal et al. (ed.) Global climate change and tropical ecosystems. CRC/Lewis Publ., Boca Raton, FL.

Diaz-Zorita, M., G.A. Duarte, and J.H. Grove. 2002. A review of no-till systems and soil management for sustainable crop production in the sub-humid and semi-arid Pampas of Argentina. Soil Tillage Res. 65:1–18.

Dregne, H.E. 1976. Soils of arid regions. Elsevier Scientific Publ. Co., Amsterdam, The Netherlands.

Dregne, H.E., and H.J. Maker. 1955. Fertility levels of New Mexico soils. Bull. 396. New Mexico Agric. Exp. Stn., Los Crucas.

Dregne, H.E., and N.-T. Chou. 1992. Global desertification: Dimensions and costs. p. 249–282. *In* H.E. Dregne (ed.) Degradation and restoration of arid lands. Texas Tech Univ., Lubbock.

Dutton, R.W., J.I. Clarke, and A. Battikhi (ed.) 1998. Arid land resources and their managmeent. Kegan Paul Int., London.

Eswaran, H., P.F. Reich, J.M. Kimble, F.H. Beinroth, E. Padamnabhan, and P. Moncharoen. 1999. Global carbon stocks. p. 15–25. *In* R. Lal et al. (ed.) Global climate change and pedogenic carbonates. CRC/Lewis Publ., Boca Raton, FL.

Felker, P. 2000. An investment based approach to *Prosopis* agroforestry in arid lands. Ann. Arid Zone 38:383–395.

Feng, Q., K.N. Endo, and G.D. Cheng. 2002. Soil carbon in desertified land in relation to site characteristics. Geoderma 106:21–43.

Fitzpatrick, R.W., and R.H. Merry. 1999. Pedogenic carbonate pools and climate change in Australia. p. 105–120. *In* R. Lal et al. (ed.) Global climate change and pedogenic carbonates. CRC/Lewis Publ., Boca Raton, FL.

Geesing, D., P. Felker, and R.L. Bingham. 2000. Influence of mesquite (*Prosopis glandulosa*) on soil nitrogen and carbon development: implications for global carbon sequestration. J. Arid Environ. 46:157–180.

Gile, L.H. 1966. Cambic and certain non-cambic horizons in desert soils of southern New Mexico. Soil Sci. Soc. Am. Proc. 30:773–781.

Guillaume, K., L. Abbadie, A. Marriott, and H. Nario. 1999. Soil organic matter dynamics in tiger bush (Niamey, Niger). Preliminary results. Acta Oecologica-Int. J. Ecol. 20:185–195.

Havlin, J.L., D.E. Kissel, L.D. Maddux, M.M. Classen, and J.H. Long. 1990. Crop rotations and tillage effects on soil organic carbon and nitrogen. Soil Sci. Soc. Am. J. 53:1515–1519.

Hazel, P.B.R., and C. Ramasamy. 1991. The Green Revolution reconsidered: The impact of high yielding varieties in South India. John Hopkins Univ. Press, Baltimore.

Heathcote, R. 1983. The arid lands: Their use and abuse. Longman, London.

Hendricks, D.M. 1985. Arizona soils. The Univ. of Arizona, Tucson.

Hiernaux, P., C.L. Bielders, C. Valentin, A. Batino, and S. Fernandez-Rivera. 1999. Effects of livestock grazing on physical and chemical properties of sandy soils in Sahelian rangelands. J. Arid Environ. 41:231–245.

Hillel, D. 1997. Small scale irrigation for arid zones: Principles and options. FAO Development Ser. 2. FAO, Rome, Italy.

Innes, J.L. 1991. High-altitude and high-latitude tree growth in relation to past, present and future global climate change. The Holocene 1:168–173.

Jaiyeoba, I.A. 1995. Changes in soil properties related to different land uses in part of the Nigerian semi-arid Savannah. Soil Use Manage. 11:84–89.

Jaiyeoba, I.A. 2003. Changes in soil properties due to continuous cultivation in Nigerian semi-arid savannah. Soil Tillae Res. 70:91–98.

Jenkinson, D.S., H.C. Harris, J. Ryan, A.M. McNeil, C.J. Pilbeam, and K. Coleman. 1999. Organic matter turnover in calcareous soil from Syria under a two-course cereal rotation. Soil Biol. Biochem. 31:643–649.

Jenny, H., and S.P. Raychaudhuri. 1960. Effect of climate and cultivation on nitrogen and organic mater reserves in Indian soils. Indian Council of Agric. Res., New Delhi, India.

Jones, M.J. 1971. The maintenance of soil organic matter under continuous cultivation at Samar Nigeria. J. Agric. Sci. 77:473–482.

Kushwaha, C.P., S.K. Tripathi, and K.P. Singh. 2000. Variations in soil microbial biomass and N availability due to residue and tillage management in a dryland rice agroecosystem. Soil Tillage Res. 56:153–166.

Lal, R. 2002. Carbon sequestration in dryland ecosystems of West Asia and North Africa. Land Degrad. Dev. 13:45–59.

Lal, R., and J.M. Kimble. 1999. Pedogenic carbonates and the global carbon cycle. p. 1–14. *In* R. Lal et al. (ed.) Global climate change and pedogenic carbonates. CRC/Lewis Publ., Boca Raton, FL.

Lee, K.E. 1981. Effects of biotic components on abiotic components. p.105–123. *In* D.W. Doodall et al. (ed.) Arid land ecosystems: Structure, functioning and management. Cambridge Univ. Press, Cambridge.

Mainguet, M. 1991. Desertification: Natural background and human mismanagement. Springer Verlag, Berlin.

Meigs, P. 1953. World distribution of arid and semiarid homoclimates. P. 203–210. *In* Arid zone hydrology. Ser. 1. UNESCO-Paris.

Monger, H.C., and R.A. Gallegos. 1999. Biotic and abiotic processes and rates of pedogenic carbonate accumulation in the southwestern United States—Relationship to atmospheric CO_2 sequestration. p. 273–290. *In* R. Lal et al. (ed.) Global climate change and pedogenic carbonates. CRC/Lewis Publ., Boca Raton, FL.

Mrabet, R. 2002. Stratification of soil aggregation and organic matter under conservation tillage systems in Africa. Soil Tillage Res. 66:119–128.

Mrabet, R., N. Saber, A. El-Brahli, S. Lahlou, and F. Bessam. 2001a. Total, particulate organic matter and structural stability of a calcixeroll soil under different wheat rotations and tillage systems in a semi-arid area of Morocco. Soil Tillage Res. 57:225–236.

Mrabet, R., K. Ibno-Namr, F. Bessam, and N. Saber. 2001b. Soil chemical quality changes and implications for fertilizer management after 11 years of no-tillage wheat production in semi-arid Morocco. Land Degrad. Dev. 12:505–517.

Noin, D., and J.I. Clarke. 1998. Population and environment in arid regions of the world. p. 1–18. *In* J. Clarke and D. Noin (ed.) Population and environment in arid regions. Man and the Biosphere Series of UNESCO. The Parthenon Publ. Group, Carnforth, UK.

Nordt, L.C., L.P. Wilding, and L.R. Drees. 1999. Pedogenic carbonate transformations in leaching soil systems: Implications for the global carbon cycle. p. 43–64. *In* R. Lal et al. (ed.) Global climate change and pedogenic carbonates. CRC/Lewis Publ., Boca Raton, FL.

Nye, P.H., and D.J. Greenland. 1960. Soils under shifting cultivation. Tech. Comm. Commonwealth Bureau of Soils, Harpenden, England.

Oldeman, L.R. 1994. The global extent of soil degradation. p. 99–118. *In* D.J. Greenland and I. Szabolcs (ed.) Soil resilience and sustainable land use. CAB Int., Wallingford, UK.

Pal, D.K., G.S. Dasong, S. Vadivelu, R.L. Ahuja, and T. Bhattacharyya. 1999. Secondary calcium carbonate in soils of arid and semi-arid regions. p. 149–186. *In* R. Lal et al. (ed.) Global climate change and pedogenic carbonates. CRC/Lewis Publ., Boca Raton, FL.

Pan, G., and T. Guo. 1999. Pedogenic carbonate of aridic soils in China and its significance in carbon sequestration in terrestrial ecosystems. p. 135–148. *In* R. Lal et al. (ed.) Global climate change and pedogenic carbonates. CRC/Lewis Publ., Boca Raton, FL.

Paustian, K., E.T. Elliott, G.A. Peterson, and K. Killian. 1996. Modelling climate, CO_2 and management impacts on soil carbon in semi-arid ecosystems. Plant Soil 187:351–365.

Péwé, T.L. 1974. Geologic and geomorphic processes of polar deserts. p. 33–52. *In* T.L. Smiley and J.H. Zumberge (ed.) Polar deserts. Univ. of Arizona Press, Tucson.

Pieri, C.M.M.G. 1992. Fertility of soils: A future for farming in West African Savannah. Springer Verlag, Berlin.

Potter, K.N., O.R. Jones, H.A. Torbert, and P.W. Unger. 1997. Crop rotation and tillage effects on organic carbon sequestration in the semi-arid southern Great Plains. Soil Sci. 162:140–147.

Rasmussen, P.E., S.L. Albrecht, and R.W. Smiley. 1998. Soil carbon and nitrogen changes under tillage and cropping systems in semi-arid Pacific Northwest agriculture. Soil Tillage Res. 47:197–205.

Reeder, J.D., and G.E. Schuman. 2002. Influence of livestock grazing on carbon sequestration in semi-arid mixed-grass and short-grass rangeland. Environ. Pollut. 116:457–463.

Robles, M.D., and I.C. Burke. 1997. Legume, grass and conservation reserve program effects on soil organic matter recovery. Ecol. Appl. 7:345–357.

Rostagno, C.M., and R.B. Sosebee. 2001. Surface application of biosolids in the Chihuahuan desert: Effects on soil physical properties. Arid Land Res. Manage. 15:233–244.

Schimel, D.S., B.H. Braswell, E.A. Holland, R. McKeown, D.S. Ojima, P.H. Painter, W.J. Parton, and A.R. Townsend. 1994. Climatic, edaphic and biotic controls over storage and turnover of carbon in soils. Global Biogeochem. Cycles 8:279–293.

Schlesinger, W.H. 1995. Biogeochemistry: An analysis of global change. Academic Press, San Diego, CA.

Schroeder, P. 1994. Carbon storage benefits of agroforestry systems. Agrofor. Syst. 27:89–97.

Schuman, G.E., H.H. Janzen, and J.E. Herrick. 2001. Soil carbon dynamics and potential carbon sequestration by rangelands. Environ. Pollut. 116:391–396.

Shantz, H.L. 1956. History and problems of arid lands development. *In* G.F. White (ed.) The future of arid lands. Am. Assoc. Adv. Sci. Publ. 43:3–25.

Shroyer, J.P., J. Ryan, M. Abdel Monem, and M. El Mourid. 1990. Production of fall-planted cereals in Morocco and technology for its improvement. J. Agron. Ed. 19:32–40.

Siband, P. 1974. Evolution des caractères et de la fertilité d' un sol rouge de Casamanca. L' Agron. Trop. 29:1228–1248.

So, H.B., G. Kirchhof, R. Baker, and G.D. Smith. 2001. Low-input tillage cropping systems for limited resource areas. Soil Tillage Res. 61:109–123.

Solomon, D., J. Lehmann, and W. Zech. 2000. Land use effects on soil organic matter properties of chronic luvisols in semi-arid northern Tanzania: carbon, nitrogen, lignin and carbohydrates. Agric. Ecosystems Environ. 78:203–213.

Stewart, B.A., R. Lal, and S.A. El-Swaify. 1991. Sustaining the resource base of an expanding world agriculture. p. 125–144. *In* R. Lal and F.J. Pierce (ed.) Soil management for sustainability. Spec. Publ. Soil Water Conserv. Soc., Ankeny, IA.

Stewart, B.A., and C.A. Robinson. 2000. Land use impacts on carbon dynamics in soils of the arid and semi-arid tropics. p. 251–260. *In* R. Lal et al. (ed.) Global climate change and tropical ecosystems. CRC/Lewis Publ., Boca Raton, FL.

Storey, P.J. 2002. The conservation and improvement of sloping land. Science Publ., Enfield, NH.

Swarup, A.C., M.C. Manna, and G.B. Singh. 2000. Impact of land use and management practices on organic carbon dynamics in soils of India. p. 261–282. *In* R. Lal et al. (ed.) Global climate change and tropical ecosystems. CRC/Lewis Publ., Boca Raton, FL.

Tans, P.P., I.Y. Fung, and T. Takahashi. 1990. Observational constraints in the global atmospheric CO_2 budget. Science (Washington, DC) 247:1431–1438.

Thomas, D.S.G. 1993. Sandstorm in a teacup? Understanding desertification. Geogr. J. 159:318–331.

Thomas, D., and N. Middleton. 1994. Desertification, exploding the myth. John Wiley & Sons, Chichester, UK.

Tiessen, H., C. Feller, E.V.S.B. Sampaio, and P. Garin. 1998. Carbon sequestration and turnover in semi-arid savannas and dry forest. Climatic Change 40:105–117.

United Nations Convention to Combat Desertification. 2001. Assessment of the status of land degradation in arid, semi-arid and dry sub-humid areas. Land degradation assessment in drylands and millennium ecosystem assessment. UNCCD. ICCD/COP (5), Geneva, Switzerland, 12 Sept. 2001.

United Nations Environment Program. 1992. World atlas of desertification. N. Middleton and D. Thomas (ed.) Arnold, London.

United Nations Environment Program. 1997. World atlas of desertification. N. Middleton and D. Thomas (ed.) Second ed. Oxford Univ. Press, Oxford.

Unger, P.W. 1991. Organic matter, nutrient and pH distribution in no- and conventional-tillage semi-arid soils. Agron. J. 83:186–189.

Van Wambeke, A. 1990. Soils of the tropics: Properties and appraisal. McGraw Hill, New York.

Vlek, P.L.G. 1993. Strategies for sustaining agriculture in sub-Saharan Africa: The fertilizer technology issue. p. 265–277. *In* J. Ragland and R. Lal (ed.) Technologies for sustainable agriculture in the tropics. ASA Spec. Publ. 56. ASA, CSSA, and SSSA, Madison, WI.

West, N.E., J.M. Stark, D.W. Johnson, M.M. Abrams, J.R. Wight, D. Heggem, and S. Peck. 1994. Impacts of climatic change on the edaphic features of arid and semi-arid lands of western North America. Arid Soil Res. Rehab. 8:307–351.

Westerman, R.L. 1992. Efficient use of fertilizers. Agronomy 92-1, Oklahoma State Univ., Norman, OK.

Wigley, T.M.L. 1994. How important are the carbon cycle uncertainties? p. 169–190. *In* T. Hanisch (ed.) Climate change and the agenda for research. Westview Press, Boulder, CO.

Wolfe, D.W., and J.D. Erickson. 1993. Carbon dioxide effects on plants: uncertainties and implications for modeling crop response to climate change. p. 153–178. *In* H.M. Kaiser and T.E. Drennen (ed.) Agricultural dimensions of global climate change.

21 Acidification and Its Evolution under Australian Dryland Cropping Systems

William J. Slattery

Rutherglen Research Institute
Rutherglen, Victoria, Australia

Keith R. Helyar

Wagga Wagga Agricultural Institute
NSW Agriculture, Wagga Wagga NSW

ABSTRACT

In this chapter, we review the development of acid soils in dryland cropping environments in Australia. These soils have become strongly acid in the surface soil layers, affecting some 90 million hectares of agriculturally important land. The main causes of soil acidification in farmed soils are through the leaching of nitrate nitrogen (NO_3^- N), removal of excess alkali in farm produce, accumulation and leaching of organic acid anions, and the use of ammonium-based N fertilizers. These processes have accelerated the rate of pH decline and in some cases have reduced surface soil pH by more than 1.5 units. The impact of this acidification being greatest in high rainfall environments. We review the impact of declining soil pH on farm production and discuss amelioration strategies such as liming, lime with gypsum, organic matter additions and the frequency of application of these materials. We also explore economic and management decisions that will provide solutions for the continued use of acid soils in the future.

INTRODUCTION

The Australian continent is comprised of a land mass that is generally of low relief with alpine regions confined to the coastal margins. This topography, together with a global location that is in the mid to low latitudes, has restricted the development of young soils over the past 500 million years due to the lack of glaciation (Beckmann, 1983). In addition, the landscape has been weathered dramatically over this time period. During the more recent 30 000 year time frame much of the continent has received <300 mm of annual rainfall, often in episodic events lasting for up to 6 mo, with wind and water erosion contributing to the major changes in land-

---

scape and soil formation observed today (Beckmann, 1983). As a consequence, much of the interior of the continent is composed of arid sands with the majority of fertile heavier textured soils confined to the southern and eastern margins of the continent (Hubble et al., 1983). Many of the soils in the higher rainfall regions (>400 mm) were poor in nutrient content and required the addition of fertilizer (particularly P, N, S, and Mo) to sustain highly productive agricultural crops. They were also often moderately acid (pH 5.0–6.0) in the surface profile (Williams and Andrew, 1970).

ACID ADDITION, ITS CAUSES, AND EFFECTS ON THE SOIL

Evidence of Soil pH Decline

Soil acidification has been recorded across a range of duplex, gradational and uniform textural soil types and crop rotations in Australia over the past 50 yr. Some of the more significant findings are as follows:

- The introduction of legumes in the rotation can accelerate the rate of acidification almost threefold (Slattery et al., 1998).
- The use of ammonium fertilizers will double the maximum potential rate of acidification compared with ammonium nitrate (NH_4NO_3), urea, ammonia (NH_3) or aqueous ammonia based fertilizers (Helyar, 1976).
- Increasing the concentration of soil organic matter increases the soil pH buffering capacity and provides some protection against aluminium (Al) toxicity (Kwong and Huang, 1979; Ritchie and Dolling, 1985; Hue et al., 1986; Slattery et al., 1998) as well as causing a lowering of soil pH (Helyar and Porter, 1989).
- Soils have acidified deeper in the soil profile under permanent pastures than under crop production, largely due to nitrate (NO_3^-) leaching (Helyar et al., 1997).
- The rate of pH decline decreases as the soil pH decreases, reflecting the increase in buffering capacity at pH_{ca} below 4.5 as Al minerals become more soluble (Aitken, 1992; Ridley et al., 1990c; Conyers et al., 2000). Some of the more widely reported data showing pH change over time on long-term sites is given in Table 21–1. It can be seen from these results that in general the higher acidification rates and largest declines in soil pH are on soils that contain a legume in the rotation.

Over the many years of agronomic studies involving the field evaluation of crops on acid soils there have been several consistent field observations that have been made and appear regularly in newsletters and special farmer advice publications. Although these observations are not necessarily substantiated with rigorous scientific investigation, they are worth noting. These observations include the following: (i) increase in the incidence of fungal disease, (ii) increase in the breakdown of soil structure due to loss of fertility, and (iii) increase in weedy species such as sorrel that are more tolerant of low pH. More specific and scientifically substantiated impacts of soil acidification are the less visibly obvious soil chemical changes

Table 21–1. Soil pH decline (0–10cm layer) under different farming practices in Australia (Data taken from Slattery et al., 1999).

Location	Rotation	Initial pH_{Ca}	pH decline	Years	Depth	Acidification	Reference
		0–10 cm			cm	kmol H^+ ha^{-1} yr^{-1}	
North East Victoria	Lupins	5.96	1.5	15	0–60	11.4	Coventry and Slattery (1991)
North East Victoria	Wheat-Lupin	5.96	0.9	12	0–20	9.0	Coventry and Slattery (1991)
North East Victoria	Wheat	5.96	0.6	15	0–60	4.1	Coventry and Slattery (1991)
South Eastern Queensland	Pasture grass-legume	5.47			0–90	11.0	Moody and Aitken (1997)
Southern tablelands of NSW	Pasture Subclover	5.30	1.4	55	0–60	3.5	Helyar and Porter (1989)
Southern tablelands of NSW	Pasture Subclover	4.85	0.8	33	0–60	4.2	Helyar and Porter (1989)
South Eastern Queensland	Wheat-pasture	5.08			0–50	1.7	Moody and Aitken (1997)
South West WA	Pasture-wheat	5.00				0.4	Dolling (1995)
South West WA	Lupin-wheat	5.00				1.6	Dolling (1995)
Central NSW	Wheat	4.93			0–20	1.4	Conyers et al. (1996)
Southern tablelands NSW	Wheat-clover			18	0–30	2.5	Helyar et al. (1997)
Upper Murray, Vic	Annual pasture	4.60	0.5	50	0–60	4.7	Ridley et al. (1990a)
Upper Murray, Vic	Phalaris pasture	4.60	0.5	50	0–60	3.7	Ridley et al. (1990b)
South West WA	Wheat-lupin	4.50			0–60	0.6	Loss et al. (1993)
South West WA	Pasture	4.50			0–60	0.2	Loss et al. (1993)
Central West WA	Wheat					3.1	Poss et al. (1995)

can be monitored using soil testing, such as: (i) measured decline in soil pH, (ii) an increase in Al and Mn levels in the soil, (iii) lowering of the exchangeable concentrations of Ca, Mg, and even K, (iv) reduced availability of Mo, and (v) decrease in the number of nodules appearing on legumes.

Soil Acidification

Agriculture's Impact on Acidification Rate

Degradation of agricultural soils in many parts of the world is related to several processes, including water and wind erosion, waterlogging, salinization, and acidification. Acidification of arable lands under intensive agricultural production in medium (400–600 mm) rainfall environments has been accelerated beyond that of the normal slow rates found in natural ecosystems. Some or most of this pH decline in agricultural soils can be attributed to the use of intensive farming practices (Slattery et al., 1998). Farming practices such as continuous cropping, long-term cultivation (Dalal and Mayer, 1986; Coventry and Slattery, 1991; Chan et al., 1992; Loss et al., 1993; Dolling, 1996) and the introduction of permanent annual pastures and legumes into the crop rotation (Bromfield et al., 1987; Scott et al., 2000) have all had an enormous impact on the rate of soil acidification. There is a considerable volume of information in the Australian literature and elsewhere linking poor plant growth to low soil pH and the associated high concentrations of Al and manganese (Mn) ions that are toxic to plant roots under these conditions. Likewise, it is not a new concept that more and more soils are becoming more strongly acid and need to be managed appropriately to continue farming them. However, economic solutions for the amelioration of soils supporting grazed pasture systems are less obvious than for broadacre dryland agricultural crops (Scott et al., 2000).

Impact of Soil Acidification in Australia

In Australia, approximately 77 to 90 million hectares of agriculturally productive land is currently affected by soil acidity (Von Uexkll and Mutert, 1995; AACM, 1995), with some 29 to 35 million hectares of agriculturally productive land possessing a surface soil pH_{Ca}<5.0 (pH indicates pH measured in 1:5 soil:0.01 *M* $CaCl_2$) and exhibiting an adverse effect on plant growth (Chartres et al., 1990a; AACM, 1995). The statewide distribution of acid soils (million hectares), estimated to be affected by losses in productivity due to acidity is approximately: NSW, 9.5 million hectares; Queensland, 7 million hectares; Victoria, 4.8 million hectares; WA, 4 million hectares; SA, 2.8 million hectares; and Tasmania, 1 million hectares (Chartres et al., 1990a). The distribution of strongly acidic soils and other soils at risk from acidification, with a pH in water (pH_W) <6.0 in Australia, are shown in Fig. 21–1.

In 1980, farmers in the Holbrook area of southern NSW were experiencing difficulties in growing productive wheat (*Triticum aestivum* L.) crops. The problem of soil acidity and the associated effects of increasing Al and Mn toxicity were becoming more widespread across the tablelands, lower slopes, and coastal regions of southeastern Australia (Lee, 1980). It has been estimated that the decline in soil pH by one unit in the surface 10 to 20 cm soil layer of these podzolic soils under

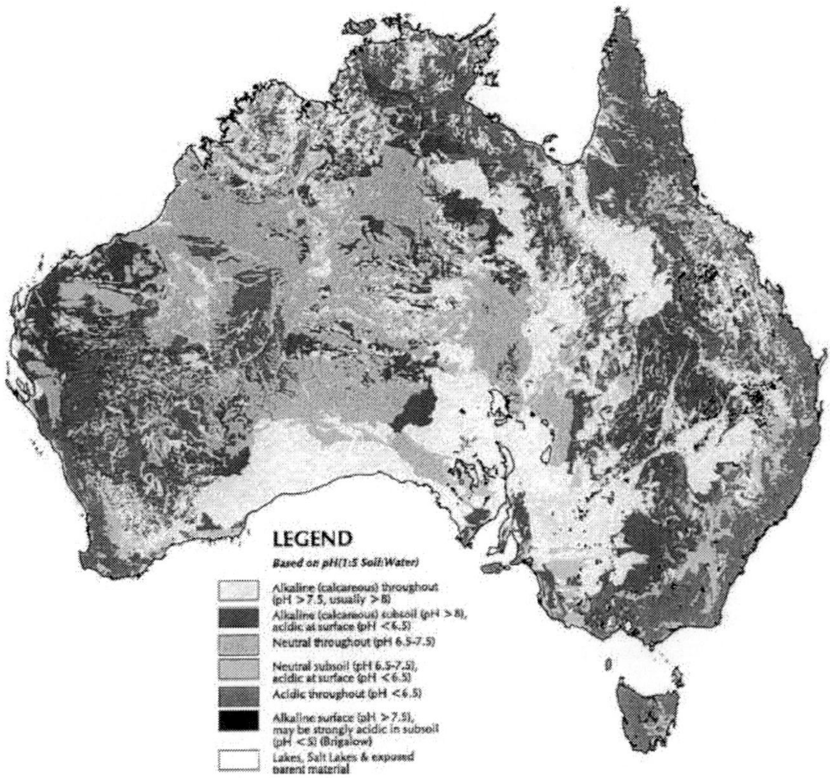

Fig. 21–1. Extent of soil acidity in Australia. Source: Spouncer et al. (1996).

continuous improved pasture would take between 30 and 60 yr (Williams, 1980; Lee, 1980). Although not considered to be a major component of acidification, the application of superphosphate at rates of 126 kg ha^{-1} was associated with a decrease in soil pH of between 0.03 and 0.06 units each year (Lee, 1980; Slattery et al., 1991). The pH decline was probably caused by the elimination of P and S deficiencies, and sometimes Mo deficiency leading to increased biological fixation of N_2, increased plant yields, product removal and soil organic matter concentrations, and increased rates of NO_3^- leaching (Helyar and Porter, 1989). At this rate of pH decline it would take approximately 2 t ha^{-1} of superphosphate applied over a period of 16 yr to reduce the soil pH by one unit.

Acidification Processes

It is generally recognized that acid inputs into the soil arising from the biological C and N cycles are the most significant in pastoral and crop systems (Helyar and Porter, 1989). Unlike some of the more obvious forms of land degradation, such as soil erosion and salinization, soil acidification is largely unseen. Although identified as a land degradation problem nearly a half century ago by Donald and

Williams (1954), acceptance has been slow due to the ability to maintain production on acidifying soils. Soil acidification is a naturally occurring process that has been accelerated by changed landuse such as the removal of native perennials for broadacre agriculture, and forest plantations on many different soil types throughout the high rainfall ecosystems of Australia.

Declining soil pH is caused by several processes that involve the production of acids, the reaction of H^+ with soil materials, transport of reaction products through the soil, and the uptake of nutrients by plants. The rate at which a soil acidifies is dependent on initial pH and upon the rainfall and the soil type. The pH is buffered against H^+ additions by reactions of H^+ with pH-dependent cation exchange sites on clay and organic matter surfaces (Aitken, 1992) and by mineral dissolution reactions with H^+. Declining soil pH in agricultural soils is functionally correlated with the following processes:

- Leaching of NO_3^- N, assuming the source of N in the system is through biological fixation of N_2 or addition of organic N or NH_4^+ (Helyar, 1976; Ball and Ryder, 1984; Robson and Abbott, 1989; Ridley et al., 1990c; Whitehead and Bristow, 1990; Black, 1992),
- Removal of excess alkali in farm produce (Pierre and Banwart, 1973; Slattery et al., 1991; Dolling, 1995),
- Leaching of organic residues containing organic anions (Herbauts, 1980; Slattery et al., 1998),
- Rhizosphere hydrogen ion (H^+) production (McLay et al., 1997),
- Accumulation of organic anions in soil organic matter or living biomass (Helyar and Porter, 1989; Russell, 1960),
- Use of ammonium-based N fertilizers (Helyar, 1976),
- For soil layers with pH above 6.0 the net absorption of CO_2 by water leaching through the layer leads to addition of carbonic acid to the soil (Zabowski and Sletten, 1991).

Unlike the Northern Hemisphere, soil acidification due to the pollution from acid rain is not considered a major problem in Australia (Porter et al., 1995). Most acid soils in the high rainfall regions of Australia are naturally acid due to leaching of NO_3^- and organic C accumulation, and have high organic matter content (3–5%), relative to the more arid regions (Baldock and Skjemstad, 1999). Their low pH buffering capacity and their coarse textural contrast within the surface soil horizons (Beckmann, 1983) facilitates subsurface water movement and nutrient leaching. In contrast to these duplex soils, there are also soils of a more uniform textured profile that lead to deep drainage of water movement and nutrient leaching. Both soil types can develop acidified surface soil layers due to NO_3^- leaching. Additional rapid rates of pH decline have been recorded since the commencement of land clearing however, associated with the introduction of grazed pastures dependant upon the fixation of N by annual clovers and with production of annual cereal, pulse, and canola crops (Williams and Donald, 1957; Williams, 1980; Bromfield et al., 1983; Loss et al., 1993; Dolling, 1995). Where these changed land-use systems have accelerated the rate of both NO_3^- leaching and product removal and thus the rate of soil acidification (Scott et al., 2000).

Impacts of Dryland Farming

Soil acidification processes are accelerating under farming practices, and it is only in the last 25 to 30 yr that soil pH has declined to such an extent that plant productivity has decreased. Varieties within species can vary in their tolerance to soil acidity, such that fewer varieties are able to survive as the soil becomes more acid. When soil pH_{Ca} falls below a critical value of 4.8, concentrations of Al and Mn begin to rise and express their toxic effects on plant roots (Slattery et al.,1995). Evidence of a slow and gradual decline in soil pH has been masked over time by the use of more acid-tolerant plant species in both pastures and broadacre crops. Donald and Williams (1954) first pointed out that pasture systems containing subterranean clover fertilized with P, S, and Mo to overcome nutrient deficiencies gradually became more acid. However, it was not until 1980 that some soils had reached acidity levels low enough to depress plant growth and in some cases cause serious losses in crop or pasture production (Cregan et al., 1989; Williams, 1980).

The differences between cereal crops and pasture species, and varieties in their tolerances to soil acidity has been demonstrated in numerous field experiments on duplex soils in southeastern Australia (Helyar and Anderson, 1971; Slattery and Coventry, 1993). Data from these field trials provided the opportunity to collate critical pH and Al levels that would inhibit plant growth and cause a reduction in crop yields. These data are presented in Table 21–2 and have been used as a guide for farmers when making crop choices as they begin to manage a liming program on their properties. Ideally, however, plant tolerance to low soil pH and high Al levels should only be used as a short-to-medium term management option to enable profitable pasture and crop production during the period required for application of lime to the surface layer to lead to amelioration of subsurface acid layers.

The maintenance of production on acidifying soils has not been obtained without cost. Soil below the surface 10-cm soil layer has become very acid and is much more difficult to ameliorate directly with lime. The differences between the pasture and crop systems largely arise from the ease of lime application and incorporation through cultivation with crops and the ease with which positive cash flow benefits of lime can be demonstrated with crops. Measurement of small lime responses

Table 21–2. Tolerance ratings for different plant species and their critical pH_{Ca} (pH measured in 0.01*M* $CaCl_2$) and exchangeable Al (Al_{Ex} measured in 1*M* KCl) ranges below, which yield declines (Data taken from Slattery et al., 1999).

Sensitivity	Plant species	Critical pH_{Ca}	Critical Al_{Ex}
			mg kg^{-1}
Very sensitive	Canola, lucerne barley, medic	4.5–4.8	0.8–4.0
Sensitive	Wheat (sensitive cvs.) Phalaris	4.1–4.5	4.0–10.0
Tolerant	Wheat (tolerant cvs.) Perennial ryegrass Subterranean clover	4.0–4.3	10.0–18.0
Very tolerant	Cocksfoot, lupins Oat, Triticale	4.0–4.2	18.0–25.0

of about 20% by pasture production systems is difficult, and factors such as changes in botanical composition with changes in soil pH (Li et al., 2003) complicate biological and economic interpretation of the results of field experiments. In general, permanent pasture soils contain a higher organic matter content than cropping soils (Slattery, 2001); this provides a higher buffering capacity and reduces the plant response to a given rate of lime applied to the surface.

Factors Affecting Acidification Rate

The rate of acidification varies with rainfall and with (i) the amount of acid added to the system (agricultural system) and (ii) the rate of pH change with acid addition or the pH buffering capacity (pHBC) of the soil. The addition of acid to the farming system is controlled by the total amount and type of fertilizer applied, the frequency of legume, and thus biological N_2 fixation, used in the rotation, the quantity of alkali exported in the form of produce, and the degree of perenniality used in the crop system. The buffering capacity of the soil is controlled by the content of exchangeable cations and by the amount of organic matter with functional groups that are capable of participating in soil solution-cation exchange reactions (Aitken and Moody, 1994). The addition of soil conditioners such as lime provides both a neutralizing effect on acid groups as well as increasing the soil Ca levels on the exchange sites of clay surfaces.

The role of increased soil organic matter in stabilizing soil pH by increasing the pH buffering capacity has been clearly demonstrated (Aitken, 1992; Slattery et al., 1998). The effect of changes in soil organic matter on net addition of acid to the soil however is more complex. If an increase in soil organic matter has been derived from plant residues grown on the site, this process is associated with net addition of H^+ from organic acids to the soil. If the increased organic matter is derived from organic residues added to the site, net alkali addition to the soil occurs (Helyar and Porter, 1989). It is proposed that farming systems that increase soil organic matter or at least protect it from declining, such as direct drilling, no-till systems with stubble retention or no burning of crop residues should be adopted, despite the addition of H^+ because of the many other benefits of soil organic matter on structure, pH buffering capacity and the biological fertility of the soil.

It has been reported previously (Conyers et al., 1996) that the rates of soil acidification for a wheat-lupin (*Lupinus angustifolius* L.) rotation have been shown to vary significantly from the Eastern States compared with Western Australia. The soils from Western Australia are much more prone to rapid acidification and are likely to acidify more rapidly to depth than soils in the eastern states due to their lower buffering capacities. Table 21–3 provides an indication of the pH changes observed over time for different agricultural crops in Australia. It was shown by Hochmann et al. (1989) and Ridley et al. (1990c) that the initial soil pH is a significant determinant in predicting the rate of soil acidification for a given enterprise, in this case, permanent pastures. Subsequent research (Conyers et al., 1996; Slattery et al., 1998) has shown that agricultural practice, soil buffering and crop species have a much larger influence on soil acidification than the initial soil pH. It is important to note that calculations of acidification based on the measurement

Table 21–3. Acidification rates and annual lime requirements for different farming systems in high (annual rainfall >550 mm) and low (annual rainfall <300 mm) rainfall environments (Slattery et al., 1999).

	Plant species	Acidification rate	Lime equivalent
		kmol(+) ha^{-1} yr^{-1}	t ha^{-1}
High annual rainfall (>550 mm)	Lucerne hay	12.0	0.60
	Crop/pasture	7.0	0.35
	Grass/legume (hay)	6.0	0.30
	Annual pasture	5.0	0.25
	Wheat/lupin	4.0	0.20
	Perennial pasture	4.0	0.20
	Grapes	2.0	0.10
	Eucalyptus	1.4	0.07
	Eucalyptus/acacia	0.8	0.04
	Tobacco	–0.4	–0.02
Low annual rainfall (<300 mm)	Wheat/legume	3.0	0.15
	Annual pasture	1.0	0.05
	Wheat/pasture	0.4	0.02

of a final change in soil pH, relative to the initial pH before treatments were imposed, are not always possible due to inadequate information.

The pioneering work of Helyar and Porter (1989) made it possible to estimate acidification rates based on relative measures to that of a control soil. In all calculations of acidification, the bulk density and buffering capacity of the soil is needed and requires some assumptions to be made, especially regarding the influence of organic matter. Data in Table 21–3 indicates the acidifying effect of different crop rotations in high-and-low rainfall environments. These data clearly show that production systems in high rainfall environments that produce annual produce that is exported from the paddock produce the highest rates of soil acidification.

Management of Net Acid Additions

There are several strategies that can be used by growers to control the rate of soil acidification on their land. These include:

- The strategic application of lime based upon rates of acidification resulting from product removal and NO_3^- leaching.
- The use of deep rooted perennials where possible in the rotation to access NO_3^- that would otherwise be leached below the rooting zone.
- Consider changing land use where current cropping practices are likely to continue acidifying the soil, especially where lime is not applied.

The calculated rates of acidification for a range of cropping and crop/pasture enterprises determined for the high and low rainfall regions of Australia (Table 21–3) provide a basis for the estimation of liming rates based on enterprise. These rates are determined from calculations of product removal, change in soil pH over a given time period and estimates of NO_3^- leaching.

Influence of pH on Soil Weathering

In pre-agricultural times, weathering of soil minerals was probably quite slow, reflecting low rates of acid addition from the C and N cycle sources (about 1.0 kmol H^+ $ha^{-1}yr^{-1}$ (Prosser et al., 1993)) and from carbonic acid at pH values above 6.0 (Zabowski and Sletten, 1991). However, under intensive agricultural systems, higher rates of addition of acids from the C and N cycles (2–5 kmol H^+ ha^{-1} yr^{-1}; see section on acidification rates) result in high rates of weathering of soil minerals and reduced soil pH. The hard-setting nature of surface soils has been shown to be influenced by acid soil conditions (Chartres et al., 1990b). Under these conditions, the clay mineral constituents of the soil are altered to the extent that there is an increased amount of amorphous silicate and imogolite-like aluminosilicate that play the major cementing role (Chartres et al., 1990b). It has also been shown that some clay minerals are degraded by treatment with solutions at pH < 4.5 (Churchman and Jackson, 1976). The implications of these clay mineral changes for strongly acid soils is that clay loss is likely and that these effects may be irreversible and result in long-term production losses if soils are not ameliorated.

Exudates from ectomycorrhizal roots have also been shown to enhance the dissolution rates of soil clay minerals compared with humic substances and non-mycorrhizal root exudates (Ochs, 1996) and these rates are accelerated below pH 4 to 5 (Drever and Stillings, 1997). Management of soil biomass and organic matter turnover may play an important role in the dissolution of clay minerals in strongly acid soils.

It has been shown in southeastern Australia that soil type, classified according to their great soils groups classification system, indicated that podzolic soils are the most acid and the grey and brown clays and red brown earths are the least acid (Chartres and Geeves, 1992). The latter group are the most susceptible to acidification (Helyar et al., 1987). This knowledge allows us to consider appropriate land-use options for different soil types regarding their productive capacity and how we can manage them appropriately to prevent soil acidification.

Influence of Soil Organic Matter and Its Source

The C cycle is also an important component of soil acidification and has been shown to contribute from as little as 15% to as high as 100% of the acid addition calculated from changes in pH and buffering capacity (Ridley et al., 1990c; Coventry and Slattery, 1991; Dolling, 1996; Moody and Aitken, 1997). In addition, the depletion of soil organic C associated with the oxidation of organic anions in the organic matter is an alkaline process that can be equivalent to almost 40% of the total net acid addition in an Alfisol in southeastern Australia under a continuous wheat crop (Slattery et al., 1998).

Organic matter in the soil is dominated by carboxyl and carbonyl functional groups, which associate with H^+ as the soil pH declines below the pKa for the weak acid groups. It has been shown that losses of organic C from soil result in a loss of negative charge and thereby reduce the buffering capacity of the soil (Chan et al., 1992). Measured values for the acidification rate based on soil C assume that the slope of the titration curve for soil organic matter averages 32 $cmol_c$ kg^{-1} per unit pH, but values could vary from as low as 8 (straw) to as high as 83 (average soil humic acid). The average for soil organic matter depends upon the amount of humic acid present relative to less degraded plant residues (Helyar and Porter, 1989). Acidification calculations based on the total inputs and exports from the system often fail to account for the entire NO_3^- leached below the root zone (Helyar and Porter, 1989; Ridley et al., 1990a; Slattery et al., 1998). In addition, the determination of soil buffering capacity varies according to the method used to estimate its value (Aitken and Moody, 1994). For example, an estimate of the loss of buffering capacity from organic matter loss has been shown to vary from 1.23 to 3.94 $cmol_c\ kg^{-1}$ for a loss of 10 g kg^{-1} organic carbon (OC) in the soil profile (Kapland and Estes, 1985; Chan et al., 1992; Slattery et al., 1998). It would be expected from such data that a soil with a lower OC content would show the impacts of soil acidification on pH earlier than for a soil that had a high soil OC content. In a study of soils in southeastern Australia, it was shown that the pH of soils with high soil OC levels were less affected by acid addition and in some cases the surface soil pH increased relative to control soils (Parnell et al., 1992; Crawford et al., 1994).

In some Australian agricultural systems, it is estimated that soil organic C levels have declined by 20 to 30 g kg^{-1} following clearing of forest and grassland soils in high rainfall areas with concentrations of at least 50 g kg^{-1} originally. For soils in the low-lying plains with at least 20 g kg^{-1} OC originally in the surface 10-cm soil layer OC has declined by 10 g kg^{-1} OC (Grace et al., 1995; Klimowitz and Uziak, 2001; Slattery and Surapaneni, 2002). Figure 21–2 shows a decline in OC of 7 g kg^{-1} over a 40-yr period for a cropping soil in northeastern Victoria and of 5 g kg^{-1} over a 14-yr period in southeastern NSW. These sites demonstrate the impact of intensive cropping practices on soil C depletion over a relatively short time span, where both sites originally had OC concentrations at or near 20 g kg^{-1} when the studies began.

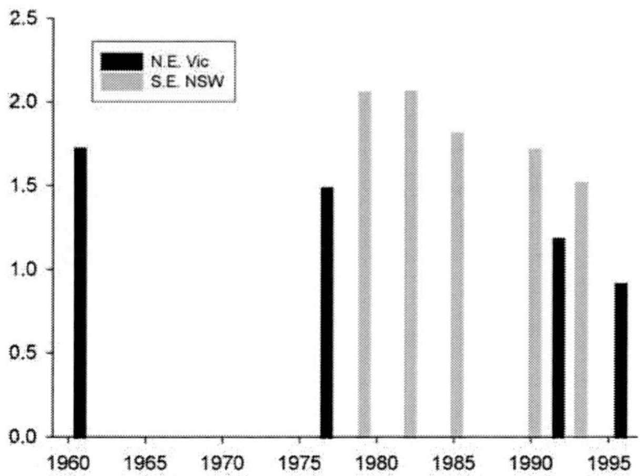

Fig. 21–2. Soil organic C changes with time for cropping soils in northeastern Victoria (Slattery, 2001) and southeastern New South Wales (Heenan et al., 1995).

Practical Implications of Carbon Cycle Losses for Acid Soils

The loss in soil C has led to a decline in soil structure, water-holding capacity, microbial activity, and the ability of the soil to resist external changes, such as acidification. To correct the balances of declining soil C in agricultural soil there are two possibilities: (i) increase soil C levels with long-term pasture rotations and/or (ii) apply new organic materials that will increase soil C levels. Both strategies have the potential to acidify the soil unless managed effectively. The management of cropping soils to provide small increases in C through the use of rotational pastures or the use of imported organic amendments (manures, composts, etc) or incorporated stubbles also have the potential to increase water infiltration and thus leaching losses of mineral N (Carter et al., 1994; Heenan and Taylor, 1995).

Traditional broadacre cropping systems in temperate regions of Australia involve the farming practices of cultivation and stubble burning. These farming practices have been shown to reduce the level of organic matter (Powlson et al., 1987; Haines and Uren, 1990; Chan et al., 1992) and biological activity (Carter, 1986; Wood, 1991; Mele, 1993) in the soil and to increase the potential for soil erosion (Reganold et al., 1990; Carter and Steed, 1992). For these soils, there will be an increase in acidification rate based upon the loss of buffering capacity due to declining soil C. Cultivation and burning practices also increase the potential for soil structural degradation and lead to surface crusting and reduced water infiltration (Carter and Steed, 1992; Carter et al., 1994, Bissett and O'Leary, 1996) and weaker soil aggregation (Boyle et al., 1989; Tisdall and Oades, 1982). The result is increased water runoff and losses of surface soil by either water or wind erosion. These losses of surface soil result in a total loss of soil buffering and a reduced capacity to prevent acidification, while the reduced water infiltration may prevent NO_3^- leaching and acidification.

It has been shown that farming practices that involve the use of stubble burning and conventional cultivation reduce the total amount of OC compared with methods that retain stubble and direct drill crops into uncultivated seedbeds (Dalal, 1989; Chan et al., 1992). In this respect the effect of cultivation is far greater than of stubble burning (Chan et al., 1992). Therefore, minimum tillage agricultural management practices are likely to retain a higher buffering capacity and thus reduce the overall rate of acid soil additions. Conservation of the soil biological diversity within farming systems that employ cultivation and broadacre monocultures, requires alternative production systems that prevent a decline in organic matter but at the same time maintain agricultural productivity. The use of agricultural practices such as stubble burning and cultivation is likely to upset the delicate balance between processes that sustain or degrade one or more aspects of soil physical, chemical, and biological fertility.

AMELIORATION AND MANAGEMENT OF ACID SOILS

Use of Limestone in Agriculture

The traditional remedy for the treatment of acid soils is to apply lime. The use of lime in agriculture has been practiced for many decades in Britain and the

European continent, however, farmers in Australia did not see the need to use lime until the 1980s. Even as more and more research into declining soil pH identified the need to apply lime farmers were reluctant to adopt this change in soil management. The relatively high cost/return ratio for lime use in the cereal cropping and grazing industries (Hochmann et al., 1989) has led to caution by farmers in accepting a new cost in the production system. Even today, many farmers on permanent pastures with strongly acid soil (pH < 4.5) are reluctant to apply lime when the benefits are not clearly demonstrated and obtained over a short period. In the grazing industry, responses in pasture and animal production can take several years to fully develop after lime is applied to the surface (Scott et al., 2000). In contrast, crop producers incorporate lime to a depth of 10 cm and usually obtain a yield response in the year of application. In more recent times, and with the use of legume crops such as lupin and field pea (*Pisum sativum* L.) in broadacre crop rotations, pH declines have become more rapid (Dolling, 1996; Slattery et al., 1998), and liming is now seen as an essential part of the crop rotation.

Assuming there is about 35 million hectares of agriculturally productive soil in Australia requiring lime at a rate of about 2.5 t ha^{-1} every 10 yr, then this would require some 8.75 Tg applied annually. There is a general trend for increasing adoption of liming as a strategy to combat acid soils, however the lime application quantities fall well short of the desired amount to ameliorate the entire strongly acid surface soils in Australian agriculture. Figure 21–3 shows the total lime applied to Australian soils at around 950 000 t annually, or about 11% of the total required amount to ameliorate acid surface soils.

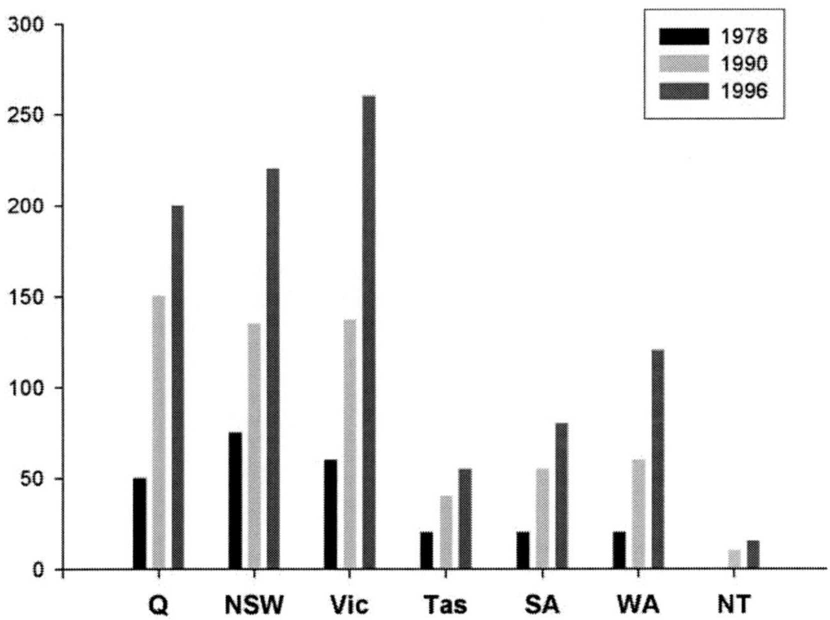

Fig. 21–3. Lime usage for Australian agriculture in each state (Q = Queensland, NSW = New South Wales, Vic = Victoria, Tas = Tasmania, SA = South Australia, WA = Western Australia, NT = Northern Territory) over an 18-yr period from 1978 to 1996 (Lime Assoc. of Australia).

Table 21–4. Grain yield responses to lime for a range of broadacre grain crops in Australia (Data from Conyers et al., 1991; Loss et al., 1993; Poss et al., 1995; McLay et al., 1994; Slattery and Coventry, 1993).

Plant	Locations	Yield increase	Response to lime
		t ha^{-1}	%
Wheat	Vic, WA.	0.5–1.5	31–103
Canola	Vic.	0.1–0.6	9–120
Triticale	Vic.	0.5–1.5	9–100
Barley	Vic, NSW, WA.	0.5	9–30

Soil acidity has been identified as an important factor in the decline of subterranean clover (*Trifolium subterranean* L.) in permanent pastures (Hochmann et al., 1990; Bromfield et al., 1983; Ridley and Coventry, 1992; Burnett et al., 1994), and various studies have demonstrated the benefits of lime on clover yield response (Coventry, 1992; Ridley and Coventry, 1992; Scott and Cullis, 1992). However, differences in soil type, especially the form of parent material and subsequent differences in clay mineralogy and soil buffering capacity (Bromfield et al., 1983; Chartres et al., 1990a; Helyar et al., 1990), can influence the soil acidification rate and subsequently affect plant response to lime.

Results from numerous field experiments have demonstrated that the application and incorporation of lime to a depth of 10 cm with cultivation, results in an increase in yield of cereal crops and pastures on strongly acid soils (Helyar and Anderson, 1971; Horsnell, 1985; Ellington, 1986; Coventry et al., 1987; Brooke et al., 1989; Conyers et al., 1991; Slattery and Coventry, 1993; Moody et al., 1995). These yield increases are principally associated with an increase of soil pH and amelioration of Al toxicity (Kamprath, 1970; Reeve and Sumner, 1970; Webber et al., 1977). The benefits of adding lime to acid surface soils can be as much as 2.0 t of grain ha^{-1} provided that limitations due to Al toxicity are removed (Table 21–4). The opportunity to grow high-value crops that are sensitive to acidity, such as canola and lucerne, can only be realized when soils have a pH_{Ca} above 5.0.

Although crop responses to lime are often variable due to the impact of factors other than surface acidity, there is a need to maintain high soil pH values to optimize crop choice for maximum productivity and steadily ameliorate subsurface acidity. Table 21–5 provides estimates of the amounts of lime required every 10 yr to balance the net rate of acidification for a range of cropping and crop/pasture enterprises.

Applications of lime on acid soils are seen as the only practical means of keeping strongly acid soils productive. For some acid soils in the high rainfall regions that already have subsoil acidity problems, there may be no alternative, but to remove them from productive agriculture because the option to lime is uneconomic. However, for highly productive soils in the medium rainfall region, lime must be seen as a cost of production. Alternative land-use options to that of broadacre farming have been suggested as a means of reducing soil degradation and increasing water use by plants, thus reducing the potential impacts of declining water quality through sedimentation or lowered pH and salinization of discharge areas in the landscape (Scott et al., 2000).

Table 21–5. Total acidification contributions and lime equivalents to balance the acidification for a range of crop and pasture rotations.

Rotation	Average yield	Lime equivalent	Reference
	$t\ ha^{-1}$	$t\ ha^{-1} 10\ yr^{-1}$	
Continuous lupin(Vic)	2.4	5.7	Slattery et al. (1998)
Wheat-Lupin rotations (Vic)	2.8	3.8	Slattery et al. (1998)
Legume based pasture (NSW)	8.8	3.4	Helyar et al. (1990)
Phalaris pasture (Vic)	8.0	1.9	Ridley et al. (1990c)
Continuous wheat (Vic)	2.4	1.6	Coventry and Slattery (1991)
Pasture-Wheat rotations (NSW)	1.3	1.3	Helyar et al. (1997)
Wheat-Lupin rotations (WA)	1.2	0.8	Dolling (1995)
Fertilized pasture(Vic)	5.0	0.7	Ridley et al. (1990c)
Eucalyptus forest	2.0	0.6	Prosser et al. (1993)
Pasture-Wheat rotations (WA)	1.4	0.2	Dolling (1995)

Economics and the Management of Acid Soils

How much, how often, and when should lime be applied to agricultural soils? These are the most commonly asked questions in relation to liming for both pasture and cropping enterprises. Lime should be applied to soils before they reach the critical pH value below which yields decline. Values of soil pH above 5.5 are needed to achieve net leaching of alkaline solutions $[(OH^- + HCO_3^- + CO_3^{2-}) > (H^+ + Al^{3+})]$ to neutralize H^+ added to soil below the layer in which lime is incorporated, thus preventing subsurface acidity and clay loss occurring. On many of the strongly acid soils in the high rainfall regions of southern Australia, the options available to continue to grow crops economically without lime additions are becoming fewer because of acidification of surface and subsurface soil layers. Lime applied to an acid soil in a single application after 12 yr of re-acidification has been shown to maintain higher yields and thus better economic returns than spreading the application in smaller amounts on an annual basis (Coventry et al., 1997). Grain yield responses to the application of lime after 12 to 13 yr of re-acidification have been shown to be up to 103% yield response (Table 21–6), similar to that obtained with the first lime application (Coventry et al., 1997). In high rainfall as well as irrigated cropping systems, the high rate of movement of NO_3^- N through the soil profile can accelerate the rate of NO_3^- loss and hence of acid addition from the N cycle. Such effects need to be taken into account when determining rates and frequencies of lime application.

The danger of overliming should be considered on all soils. Deficiencies of some nutrients (Zn, B, and sometimes Mg and K) can be induced by liming (Castleman et al., 1999). Therefore, careful liming programs need to be employed; in the first instance, to correct a surface soil acidity problem, and then secondly to consider a long-term liming program to prevent creating a subsoil acidity problem. In some cropping soils in southeastern Australia the amount of lime required to maintain a soil pH_{Ca} above 4.5 requires the re-application of about 2.5 $t\ ha^{-1}$ of lime after 8 to 12 yr, depending on the crops used in the rotation (Coventry et al., 1997). In that study, grain yields were highest and Al levels the lowest where lime was re-applied to a depth of 20 cm after 12 yr of continuous cropping (Table 21–6).

Table 21–6. Wheat grain yields, aluminium (Al) levels, and Al percentage of the effective cation exchange capacity (ECEC) after liming in 1980 and re-liming in 1992 on a duplex soil or yellow Sodosol (data from Coventry et al., 1997).

Rate and year(s) of lime application	Yield-1993	Al (0–10 cm)	ECEC†	Al (10–20 cm)	ECEC†
	t ha^{-1}	cmol kg^{-1}	%	cmol kg^{-1}	%
No lime–cropped since 1980	3.12	0.45	31	1.56	62
Lime 2.5 t ha^{-1} in 1980	5.27	0.04	2	0.36	12
Lime 2.5 t ha^{-1} in 1980, 1992	5.92	0	0	0.13	6

†ECEC = equivalent to exchangeable cation exchange capacity.

Gypsum

In some acid soils, such as oxisols, gypsum has been shown to act as an ameliorant of some of the adverse effects related to soil acidity, such as the alleviation of phytotoxic Al and low Ca in the soil solution. In these soils only, it has been shown that gypsum elevate soil solution Ca levels more effectively than lime (Sumner, 1993), which also facilitates the leaching or movement of Ca down the soil profile. In addition, adsorption of sulphate on positively charged sites of Al and Fe oxides causes an increase in soil pH. However, for the majority of Australian soils that have little or no positive charge, gypsum application only decreases the soil pH via the 'salt effect' on pH and is primarily used to improve soil structure, especially on soils that are subject to slaking and dispersion. The primary impact of applying gypsum is the formation of more stable soil aggregates leading to an improved soil structure and allowing better water movement or drainage in heavier soils (Ellington et al., 1997). Where subsoils may be acid or suffer from a deficiency of Ca, there may be a benefit from the addition of gypsum, which is 172 times more soluble than lime. The movement of Ca into the subsoil is enhanced by the movement of water and therefore will be greatest where soils are irrigated. For example, gypsum has been shown to move to a depth of 20 cm in a clay soil in two and a half years when applied at a rate of 10 t ha^{-1} (Valzano et al., 2001). In comparison, lime movement has been shown to move at a rate of around 0.5 to 1.0 cm yr^{-1} in a clay soil of uniform texture profile to a depth of at least 2 m in a 600-mm rainfall environment (Ridley et al., 1990c).

Crop response to gypsum amendment can be variable and influenced by soil type, the surface charge characteristics of those soils and the ability to adsorb both Ca and sulfate (SO_4) (Sumner, 1993). After reducing the effects of sodicity and increased water infiltration, and the alleviation of waterlogging, yield benefits of around 0.9 t ha^{-1} have been shown for gypsum applied to a red brown earth at rates up to 10 t ha^{-1} (Ellington et al., 1997). The impact of gypsum is greatest for soils that have surface or subsurface exchangeable sodium percentages (ESP) higher than 6%, that is, ESP values associated with dispersion of clays (Rengasamy and Churchman, 1999). The application of irrigation water containing high amounts of sodium chloride (NaCl) increases the solubility of gypsum as much as four times (Ostroff and Metler, 1966). This increases the rate of gypsum movement into the profile and is likely to improve subsoil structure much quicker. However, the accelerated movement of gypsum through the soil profile may reduce the length of time that gypsum will be effective.

Gypsum and Lime Together

Combinations of lime plus gypsum have been shown to be beneficial on some soils in enhancing both crop productivity and the Ca nutrition of plants in comparison to both amendments applied separately. The two main soil classes in this category are soils with acid layers in the surface but with sodic subsoils (e.g., some sodic soils or submatric yellow Sodosols (Isbell, 1996)) and deep acid soils dominated by Al and Fe oxides with net cation exchange capacity (CEC) in the surface soil but net anion exchange capacity in the subsoil (e.g., Krasnozems (great soils group), oxisols (U.S. Taxonomy), or Ferrosols (Isbell, 1996). Positive results have been demonstrated in the treatment of Sodosols with acid surface layers such as acid and sodic sandy clay loam duplex soils (Ellington et al., 1997) and acid deep yellow sandplain soils (McLay et al., 1994). In contrast, little response to the combined treatment occurred on an acid, non-sodic sandy loam red duplex soil (Ellington et al., 1997) or on an acid non-sodic grey massive earth (Smith et al., 1994). However, the use of gypsum or lime plus gypsum can also be associated with adverse effects such as loss of exchangeable Mg leading to deficiencies of Mg (Sumner, 1993).

The effectiveness of gypsum can be influenced by the extent of leaching after gypsum application (Smith et al., 1994) and the time required for gypsum to react with soil components. Therefore irrigation will increase the rate of Ca movement into the profile by enhancing the dissolution of gypsum, leading in turn to increased rates of water infiltration and increased potential for lime solubility (Valzano et al., 2001). The benefits of applying lime and gypsum in combination on acid and sodic soils may well exceed the effects of either applied individually. However, there is a risk of inducing nutrient deficiencies as Ca becomes the dominant cation and nutrient imbalances allow the proliferation of root pathogens such as *Septoria* species (Ellington et al., 1997). Careful monitoring of soils is a high priority in any farming system, especially where the alleviation of soil constraints increases crop and pasture yields and may lead to a decline in some other aspect of chemical, physical, or biological fertility.

Organic Amendments

The application of organic amendments in the form of farmyard manure in agriculture has been practiced for thousands of years all over the world as a means of disposing of small quantities of waste and fertilizer for plant growth (Unger and Stewart, 1974). Australian agriculture has been based primarily on the use of pasture and crop legumes to increase the soil N status, and chemical fertilizers to supply other nutrients and some N. Sources of farmyard manure are insufficient to supply crop and pasture requirements, largely because grazing animals are not housed for significant periods during the year. In these production systems, soil organic matter (OM) is managed using crop-pasture rotations; it increases in the pasture phase and decreases in the crop phase (Helyar et al., 1997; Slattery and Surapaneni, 2002).

Different intensities of cropping, cultivation practices and types of crop affect the direction of the trend and the steady-state concentration of soil organic matter approached in the long term (Heenan et al., 1995; Dalal et al., 1995).

Organic ameliorants have been shown to improve soil properties such as organic matter content, aggregate stability, and water infiltration (Cross and Fischbach, 1973; Tiarks et al., 1974; Sommerfeldt and Chang, 1985). Many organic amendments are prepared by composting alone or with a mixture of other organic waste materials such as timber, rice hulls, and straw (Slattery and Surapaneni, 2002). Composting of manures and other waste organic materials is undertaken to stabilize the product, reduce odor and potential pathogens and improve the physical properties of the product that enhance its fertility when applied to soil and its ease of application by machinery.

Soil organic matter plays an important role in providing a pool of nutrients to plants through microbially driven decomposition reactions particularly for P, S, and N. Organic matter also acts as an important source of cation exchange capacity within the soil. Organic matter plays several key roles that when applied to soils can result in immobilization of Al^{3+}, an increase in pH or a decrease in soil pH, depending upon the source and nature of that organic material.

1. The addition of OM results in an increase in CEC that, in turn, increases cation buffering and subsequent immobilization of Al^{3+}. It is this cation exchange potential that enables some sources of organic material to be effective in the detoxification of Al by adsorption or hydrolysis reactions. However, the immobilization of Al^{3+} may only be temporary, as hydrolyzed Al is dissociated according to the reaction $Al(OH)_3 \Rightarrow Al^{3+} + 3OH^-$ under acid soil conditions, but will result in a slight elevation of soil pH. The application of organic matter from a range of sources has been shown to decrease the amount of exchangeable Al and in some cases increase the soil pH. For example, increases in organic matter from 1 to 2% have been shown to decrease the amount of exchangeable Al by 30 to 40% over a range of acid soils (Thomas, 1975; Bloom et al., 1979; Kapland and Estes, 1985).
2. Organic matter imported from off-site sources and containing cation saturation of functional groups will act as a sink for protons in soil solution according to the following reaction; $RCOOM + H^+ \Rightarrow RCOOH + M^+$ and result in a slight elevation of solution pH. Increases in soil pH have been observed after the addition of organic matter to soil (Ritchie and Dolling, 1985; Bessho and Bell, 1992; Pocknee and Sumner, 1997; Slattery and Surapaneni, 2002). It has been postulated that this observed change in soil pH is more directly related to the amount of cations held by functional groups in the organic matter matrix applied to the soil than any other factor (Pocknee and Sumner, 1997). The permanence of this pH change will be determined by subsequent microbial activity that uses this new source of C as an energy supply. It has been shown that some pH effects on soil due to the addition of organic amendments are short-lived (Ritchie and Dolling, 1985; Hue, 1992; Yan et al., 1996), but that a more permanent pH change can be obtained when the composition of the organic material is

dominated by Ca-containing organic molecules (Pocknee and Sumner, 1997).

An important agricultural consequence for the use of organic materials on acid soils is the overall redistribution of alkalinity from one point source to another, as pointed out by Pocknee and Sumner (1997). The disposal of organic waste products from various intensive industries, such as feedlots, onto agricultural land, therefore represents a major transport of nutrients and organic anions from one soil to another. The cost of amelioration of acid soil environments with organic materials that are often considered to be waste products must be balanced by the replacement of these nutrients and bases to the soils from which they originally came. In addition, it has been shown that some organic compost can contain high levels of Na, K, and Cu (Slattery and Surapaneni, 2002) and should be used cautiously to prevent toxicity problems occurring at the expense of increasing a soils buffering capacity.

3. Assimilation of organic matter into the soil from on-site sources will involve decomposition products that result in the formation of organic anions that are likely to associate with cations in solution and result in the production of protons according to the following reaction; $RCOOH + M^+ \Rightarrow RCOOM + H^+$. These reactions will result in a slight lowering of solution pH where dissociation occurs either in the surface or subsoil horizons. Decreases in soil pH have been observed after the decomposition of organic matter in field experiments where residues have been retained (Williams and Donald, 1957; Tyson and Cabrera, 1993; Bevacqua and Mellano, 1994). Subsequent leaching of organic acid anions below the surface horizon to form organo-metallic complexes with metals and displace H^+ from functional groups will also decrease soil pH in the subsoil (McBride, 1994; Slattery et al., 1996). In addition, it has been shown that the leaching of water-soluble organic C fractions below surface horizons is an important mechanism for podzolization under forest soils (Herbauts, 1980).

CONCLUSIONS

A detailed synopsis of the research into acid soils and acidification processes in Australia has revealed the following key outcomes:

- The application of lime required to increase soil pH should be considered well before pH values as low as 4.0 to 4.3 ($CaCl_2$) are reached. In this pH range off-site and intergenerational impacts relating to loss of soil clay through acid weathering, the release of salts to solution and effects on water quality may occur. Optimally, a pH_{Ca} of 4.5 or higher should be maintained to alleviate any Al toxicity problems and retain good ground cover and optimum water use by plants tolerant to moderate Al concentrations. Higher pH values (5.5–6.0) are required for optimum production by plants sensitive to acidity, for maximum microbial activity, to avoid the development of highly acid subsurface layers, and to increase the pH of subsurface layers that are already severely acid.

- Re-liming of continuously cropped soils should be considered at intervals of between 10 and 12 yr, depending upon the farming system. The long-term lime rate depends on the rate acids are being added to the system. The liming rate to achieve a given change in pH is determined by the soil pH buffering capacity that is affected by the clay and organic matter concentrations. The soil organic matter concentration in turn, is affected by the rotation and a range of cultural practices.
- The management of soil acidity and of soil organic matter are ultimately involved with optimizing soil physical, chemical, and biological fertility. The acidity of the soil profile is an indicator of the progress of acid weathering in the soil. Manipulation of soil OC can result in the net addition of H^+ or of OH^- to the soil depending on the cultural practice used, but increased OC has many favorable effects on soil pH buffering and biological fertility in general. Lime or other alkaline ameliorants, including additions of organic amendments from offsite, can be used to maintain soil pH_{Ca} at around 5.0 to 6.0 to stop the addition of acid, produced in the C and N cycles, that are driving acid weathering reactions that decrease the physical, chemical, and biological fertility of the soil.

REFERENCES

AACM-International 1995 Social and economic feasibility of ameliorating soil acidification: A national review. Land and Water Resources Research and Development Corp., Canberra, Australia.

Aitken, R.L. 1992. Relationships between extractable Al, selected soil properties, pH buffer capacity and lime requirement in some acidic Queensland soils. Aust. J. Soil Res. 30:119–130.

Aitken, R.L., and P.W. Moody. 1994. The effect of valence and ionic strength on the measurement of pH buffer capacity. Aust. J. Soil Res. 32:975–984.

Baldock, J.A., and J.O. Skjemstad. 1999. Soil organic carbon/soil organic matter. p. 159–170. *In* K.I. Peverill et al. (ed.) Soil analysis: An interpretation manual. CSIRO, Melbourne, Australia.

Ball, P.R., and J.C. Ryder. 1984. Nitrogen relationships in intensively managed temperate grasslands. Plant Soil 76:23–33.

Beckmann, G.G. 1983. Development of old landscapes and soils. p. 51–71. *In* Soils: An Australian viewpoint. Div. of Soils, CSIRO, Melbourne, Australia.

Bessho, T., and L.C. Bell. 1992. Soil solid and solution phase changes and mung bean response during amelioration of aluminium toxicity with organic matter. Plant Soil 140:183–196.

Bevacqua, R.F., and V.J. Mellano. 1994. Cumulative effects of sludge compost on crop yield and soil properties. Commun. Soil Sci. Plant Anal. 25:395–406.

Bissett, M.J., and G.J. O'Leary. 1996. Effects of conservation tillage on water infiltration in two soils in south-eastern Australia. Aust. J. Soil Res. 34:299–308.

Black, A.S. 1992. Soil acidification in urine- and urea-affected soil. Aust. Soil Res. 30:989–999.

Bloom, P.R., M.B. McBride, and R.M. Weaver. 1979. Aluminum organic matter in acid soils: Buffering in solution aluminum activity. Soil Sci. Soc. Am. J. 43:488–493.

Boyle, M., W.T. Frankenberger, and L.H. Stolzy. 1989. The influence of organic matter on soil aggregation and water infiltration. J. Prod. Agric. 2:290–299.

Bromfield, S.M., R.W.M. Cumming, D.J. David, and C.H. Williams. 1983. Changes in soil pH, manganese and aluminium under subterranean clover pasture. Aust. J. Exp. Agric. Anim. Husb. 23:181–191.

Bromfield, S.M., R.W.M. Cumming, D.J. David, and C.H. Williams. 1987. Long-term effects of incorporated and topdressed lime on the pH in the surface and subsurface of pasture soils. Aust. J. Exp. Agric. 27:533–538.

Brooke, H.D., D.R. Coventry, T.G. Reeves, and D.K. Jarvis. 1989. Liming and deep ripping responses for a range of field crops. Plant Soil. 115:1–6.

Burnett, V.F., D.R. Coventry, J.R. Hirth, and F.C. Greenhalgh. 1994. Subterranean clover decline in permanent pastures in north-eastern Victoria. Plant Soil 164:231–241.

Carter, M.R. 1986. Microbial biomass as an index for tillage-induced in soil biological properties. Soil Tillage Res. 7:29–40.

Carter, M.R., P.M. Mele, and G.R. Steed. 1994. The effects of direct-drilling and stubble retention on water and bromide movement and earthworm species in a duplex soil. Soil Sci.157:224–231.

Carter, M.R., and G.R. Steed. 1992. The effects of direct-drilling and stubble retention on hydraulic properties at the surface of duplex soils in north-eastern Victoria. Aust. J. Soil. Res. 30:505–516.

Castleman, L.J.C., B.J. Scott, W. J. Slattery, and M.K. Conyers. 1999. Nutritional status of wheat on acid and limed soils in southern NSW. p. 661–664. *In* D.L. Michalk and J.E. Pratley (ed.) Proc. 9th Australian Agron. Conf., Wagga Wagga, New South Wales. 20–23 July 1999. Australian Soc. of Agron., Wagga Wagga, NSW.

Chan, K.Y., W. P. Roberts, and D.P. Heenan. 1992. Organic carbon and associated soil properties of a red earth after 10 years of rotation under different stubble and tillage practices. Aust. J. Soil Res. 30:71–83.

Chartres, C. J., W. Cumming, J.A. Beatie, G.M. Bowman, and J.G. Wood. 1990a. Acidification of soils on a transect from plains to slopes, south-western NSW. Aust. J. Soil Res. 28:539–548.

Chartres, C. J., and G. Geeves. 1992. Soil acidification in the higher rainfall wheatbelt of south-east Australia. Aust. J. Soil Water Conserv. 5:39–43.

Chartres, C.J., J.M. Kirby, and M. Raupach. 1990b. Poorly ordered aluminosilicates as temporary agents in hard-setting soils. Soil. Sci. Soc. Am. J. 54:1060–1067.

Churchman, G.J., and M.L. Jackson. 1976. Reaction of montmorillonite with acid aqueos solutions: Solute activity control by a secondary phase. Geochim. Cosmochim. Acta 40:1251–1259.

Conyers, M.K., D.P. Heenan, G.J. Poile, B.R. Cullis, and K.R. Helyar. 1996. Influence of dryland agricultural management practices on the acidification of a soil profile. Soil Tillage Res. 37:127–141.

Conyers, M.K., K.R. Helyar, and G.J. Poile. 2000. pH buffering: The chemical response of acid soils to added alkali. Soil Sci.165:560–566.

Conyers, M.K., G.J. Poile, and B.R. Cullis. 1991. Lime responses to barley as related to available soil aluminium and manganese. Aust. J. Agric. Res. 42:379–390.

Coventry, D.R. 1992. Acidification problems of duplex soils used for crop-pasture rotations. Aust. J. Exp. Agric. 32:901–914.

Coventry, D.R., T.G. Reeves, H.D. Brooke, A. Ellington, and W.J. Slattery. 1987. Increasing wheat yields in North-eastern Victoria by liming and deep ripping. Aust. J. Exp. Agric. 27:679–685.

Coventry, D.R., and W.J. Slattery. 1991. Acidification of soil associated with lupins grown in a crop rotation in north-eastern Victoria. Aust. J. Exp. Agric. 42:391–397.

Coventry, D.R., W.J. Slattery, V.F. Burnett, and G.W. Ganning. 1997. Liming and re-liming needs of acidic soils used for crop-pasture rotations in south-eastern Australia. Aust. J. Exp. Agric. 37:577–581.

Crawford, D.M., T.G. Baker, and J. Maheswaran. 1994. Soil pH changes under Victorian pastures. Aust. J. Soil Res. 32:105–115.

Cross, O.E., and P.E. Fischbach. 1973. Water intake rates on a silt loam soil with various manure applications. Trans. ASAE 16:282–284.

Cregan, P.D., J.R. Hirth, and M.K. Conyers. 1989. Amelioration of soil acidity by liming and other amendments. p. 205–264. *In* A.D. Robson (ed.) Soil acidity and plant growth. Academic Press, Marrickville, Australia.

Dalal, R.C. 1989. Long-term effects of no-tillage, crop residue, and nitrogen application on properties of a vertisol. Soil Sci. Soc. Am. J. 53:1511–1515.

Dalal, R.C., and R.J. Mayer. 1986. Long-term trends in fertility of soils under continuous cultivation and cereal cropping in southern Queensland. V. Rate of loss of total nitrogen from the soil profile and changes in carbon nitrogen ratio. Aust. J. Soil Res. 37:493–504.

Dalal, R.C., W.M. Strong, E.J. Weston, J.E. Cooper, K.J. Lehane, A.J. King, and C.J. Chicken. 1995. Sustaining productivity of a vertisol at Warra, Queensland, with fertilisers, no-tillage, or legumes. 1. Organic matter status. Aust. J. Exp. Agric. 35:903–913.

Dolling, P.J. 1995. Effect of lupins and location on soil acidification rates. Aust. J. Exp. Agric. 35:753–763.

Dolling, P. J. 1996. Effect of legume-cereal rotation on soil acidification in western Australian soils. p. 17–24. *In* Proc. IV Int. Symp. on Plant-Soil Interactions at Low pH' P2. March. Embrapa, Brazilian Agric. Res. Corp., Belo Horizonte, Brazil.

Donald, C.M., and C.H. Williams. 1954. Fertility and productivity of a podzolic soil as influenced by subterranean clover (*Trifolium subterraneum* L.) and superphosphate. Aust. J. Agric. Res. 5:664–687.

Drever, J.I., and L.L. Stillings. 1997. The role of organic acids in mineral weathering. Colloids Surf. A: Physiochem. Eng. Aspects 120:167–181.

Ellington, A. 1986. Effects of deep ripping, direct drilling, gypsum and lime on soils, wheat growth and yield. Soil Tillage. Res. 8:29–49.

Ellington, A., N.S. Badaway, and G.W. Ganning. 1997. Testing gypsum requirements for dryland cropping on a Red-Brown Earth. Aust. J. Soil Res. 35:591–607.

Grace, P.R., J.M. Oades, H. Keith, and T.W. Hancock. 1995. Trends in wheat yields and soil organic carbon in the Permanent Rotation Trial at the Waite Agricultural Research Institute, South Australia. Aust. J. Exp. Agric. 35:857–864.

Haines, P.J., and N.C. Uren. 1990. Effects of conservation tillage farming on soil microbial biomass, organic matter and earthworm populations, in north-eastern Victoria. Aust. J. Exp. Agric. 30:365–371.

Heenan, D.P., W. J. McGhie, F.M. Thomson, and K.Y. Chan. 1995. Decline in soil carbon and total nitrogen in relation to tillage, stubble management, and rotation. Aust. J. Exp. Agric. 35:877–884.

Heenan, D.P., and A.C. Taylor. 1995. Soil pH decline in relation to rotation, tillage, stubble retention and nitrogen fertilizer in south-eastern Australia. Soil Use Manage. 11:4–9.

Helyar, K.R. 1976. Nitrogen cycling and soil acidification. J. Aust. Inst. Agric. Sci. 42:217–221.

Helyar, K.R., and A.J. Anderson. 1971. Effects of lime on growth of five species, on aluminium toxicity, and on phosphorus availability. Aust. J. Agric. Res. 22:707–721.

Helyar, K.R., P.D. Cregan, and D.L. Godyn. 1990. Soil acidity in New South Wales-current pH values and estimates of acidification rates. Aust. J. Soil Res. 28:523–537.

Helyar, K.R., B.R. Cullis, K. Furniss, G.D. Kohn, and A.C. Taylor. 1997. Changes in the acidity and fertility of a red earth soil under wheat-annual pasture rotations. Aust. J. Agric. Res. 48:561–586.

Helyar, K.R., and W.M. Porter. 1989. Soil acidification, its measurement and the processes involved. p. 61–101. *In* A.D. Robson. (ed.) Soil acidity and plant growth. Academic Press, Sydney, Australia.

Helyar, K.R., J.L. Wheeler, C.J. Pearson, and G.E. Robards 1987. Nutrition of plants on acid soils. p. 159–171. *In* J.L. Wheeler and C.J. Pearson. (ed.) Temperate pastures: Their production, use and management. CSIRO Publ., East Melbourne, Australia.

Herbauts, J. 1980. Direct evidence of water-soluble organic matter leaching in brown earths and slightly podzolised soils. Plant Soil 54:317–321.

Hochmann, Z., D.L. Godyn, and B.J. Scott. 1989. The integration of data on lime use by modelling. p. 265–301. *In*. A.D. Robson (ed.) Soil acidity and plant growth. Academic Press, Sydney, Australia.

Hochmann, Z., G.J. Osborne, P.A. Taylor, and B.R. Cullis. 1990. Factors contributing to reduced productivity of subterranean clover (*Trifolium subterranean* L.) pastures on acidic soils. Aust. J. Agric. Res. 41:669–682.

Horsnell, L.J. 1985. The growth of improved pastures on acid soils I. The effect of superphosphate and lime on the establishment and growth of phalaris and lucerne. Aust. J. Exp. Agric. 25:149–156.

Hubble, G.D., R.F. Isbell, and K.H. Northcote. 1983. Features of Australian soils. p. 17–49. *In* Soils: An Australian viewpoint. Div. of Soils, CSIRO, Melbourne, Australia.

Hue, N.V. 1992. Correcting soil acidity of a highly weathered Ultisol with chicken manure and sewage sludge. Commun. Soil Sci. Plant Anal. 23:241–264.

Hue, N.V., G.R. Craddock, and F. Adams. 1986. Effect of organic acids on aluminum toxicity in subsoils. Soil. Sci. Soc. Am. J. 50:28–34.

Isbell, R. 1996. The Australian soil classification. CSIRO Publ., Collingwood, Victoria, Australia.

Kamprath, E.J. 1970. Exchangeable aluminum as a criterion for liming leached mineral soils. Soil Sci. Soc. Am. Proc. 34:252–254.

Kapland, D.I., and G.O. Estes. 1985. Organic matter relationships to soil nutrient status and aluminium toxicity in alfalfa. Agron. J. 77:735–738.

Klimowitz, Z., and S. Uziak. 2001. The influence of long-term cultivation on soil properties and patterns in an undulating terrain in Poland. CATENA 43:177–189.

Kwong Ng Kee, K.F.M and P.M. Huang. 1979. The relative influence of low-molecular-weight complexing organic acids on the hydrolysis and precipitation of aluminum. Soil Sci. 128:337–342.

Lee, B. 1980. Farming brings acid soils. Annual rep. CSIRO Rural Res., 4–9 March. CSIRO, Canberra, Australia.

Li, G.D., K.R. Helyar, C.M. Evans, M.C. Wilson, L.J.C. Castleman, R.P. Fisher, B.R. Cullis, and M.K. Conyers. 2003. Effects of lime on botanical composition of pasture over nine years in a field experiment on the south-west slopes of New South Wales. Aust. J. Exp. Agric. 43:61–69.

Loss, S.P., G.S.P. Ritchie, and A.D. Robson. 1993. Effect of lupins and pasture on soil acidification and fertility in Western Australia. Aust. J. Exp. Agric. 33:457–464.

McBride, M.B. 1994. Soil acidity. p. 169–206. *In* M.B. McBride. (ed.) Environmental chemistry of soils. Oxford Univ. Press, New York.

McLay, C.D.A., L. Barton, and C. Tang. 1997. Acidification potential of ten grain legume species grown in nutrient solution. Aust. J. Exp. Agric. 48:1025–1032.

McLay, C.D.A., G.S. P. Ritchie, and W.M. Porter. 1994. Amelioration of subsurface acidity in sandy soils in low rainfall regions: I. Responses of wheat and lupins to surface-applied gypsum and lime. Aust. J. Soil Res. 32:835–846.

Mele, P.M. 1993. Characterisation of microbial populations in conservation farming practices in north-eastern Victoria. Ph.D thesis. Melbourne Univ., Melbourne, Australia..

Moody, P.W., and R.L. Aitken. 1997. Soil acidification under some tropical agricultural systems: I. Rates of acidification and contributing factors. Aust. J. Soil Res. 35:163–173.

Moody, P.W., R.L. Aitken, and T. Dickson. 1995. Diagnosis of maize yield response to lime in some weathered acidic soils. p. 537–541. *In* R.A. Date., N. J. Grundon., G.E. Rayment, and M.E. Propert (ed.) Plant-soil interactions at low pH: Principles and management. Kluwer Academic Publ., Dordrecht, The Netherlands.

Ochs, M. 1996. Influence of humified and non-humified natural organic compounds on mineral dissolution. Chem. Geol. 132:119–124.

Ostroff, A.G., and A.G. Metler. 1966. Solubility of calcium sulphate dihydrate in the system NaCl-MgCl2-H2O from 28 to 70 C. J. Chem. Eng. Data 11:346–350.

Parnell, C.D., J. Maheswaran, and D.M. Crawford. 1992. Acidity and acidification of subsoil in Victorian pastures. Proc., Australian Soil Acidity Workshop Rutherglen, Victoria, Australia.

Pierre, W.H., and W.L. Banwart. 1973. Excess-base and excess-base/nitrogen ratio of various crop species and part of plants. Agron. J. 65:96–96.

Pocknee, S., and M.E. Sumner. 1997. Cation and nitrogen contents of organic matter determine its soil liming potential. Soil Sci. Soc. Am. J. 61:86–92.

Porter, W.M., C.D.A. McLay, and P.J. Dolling. 1995. Rates and sources of acidification in agricultural systems of southern Australia. p. 75–83. *In* R.A. Date et al. (ed.) Plant-soil interactions at low pH: Principles and management. Kluwer Academic Press, London.

Poss, R., C.J. Smith, F.X. Dunin, and J.F. Angus. 1995. Rate of soil acidification under wheat in a semi-arid environment. Plant Soil 177:85–100.

Powlson, D.S., P.C. Brooke, and B.T. Christensen. 1987. Measurement of soil microbial biomass provides an early indication of changes in total soil organic matter due to straw incorporation. Soil Biol. Biochem. 19:159–164.

Prosser, I.P., K.J. Hailes, M.D. Melville, R.P. Avery, and C.J. Slade. 1993. A comparison of soil acidification and aluminium under Eucalyptus forest and unimproved pasture. Aust. J. Soil Res. 31:245–254.

Reeve, N.G., and M.E. Sumner. 1970. Lime requirements of Natal Oxisols based on exchangeable aluminum. Soil Sci. Soc. Am. Proc. 34:595–598.

Reganold, J.P., R.I. Papendick, and J.F. Parr. 1990. Sustainable agriculture. Sci. Am. 262:112–120.

Rengasamy, P., and G.J. Churchman. 1999. Cation exchange capacity, exchangeable cations and sodicity. p. 147–157. *In* K.I. Peverill et al. (ed.) Soil analysis, An interpretation manual. CSIRO Publ, Collingwood, Melbourne, Australia.

Ridley, A.M., and D.R. Coventry. 1992. Yield responses to lime of phalaris, cocksfoot, and annual pastures in north-eastern Victoria. Aust. J. Exp. Agric. 32:1061–1068.

Ridley, A.M., K.R. Helyar, and W.J. Slattery. 1990a. Soil acidification under subterranean clover pastures in north-eastern Victoria. Aust. J. Exp. Agric. 30:195–201.

Ridley, A.M., W.J. Slattery, K.R. Helyar, and A. Cowling. 1990b. The importance of the carbon cycle to acidification of a grazed annual pasture. Aust. J. Exp. Agric. 30:529–537.

Ridley, A.M, W.J.Slattery and K.R. Helyar 1990c Acidification under grazed annual and perennial grass based pastures. Aust. J. Exp. Agric. 30:539–544.

Ritchie, G.S.P., and P.J. Dolling. 1985. The role of organic matter in soil acidification. Aust. J. Soil Res. 23:569–576.

Robson, A.D., and L.K. Abbott. 1989. The effect of soil acidity on microbial activity in soils. p. 139–65. *In* A.D. Robson (ed.) Soil acidity and plant growth. Academic Press, Sydney, Australia.

Russell, J.S. 1960. Soil fertility changes in the long term experimental plots at Kybybolite, South Australia. I. Changes in pH, total N, organic carbon and bulk density. Aust. J. Agric. Res. 11:902–926.

Scott, B.J., and B.R. Cullis. 1992. Subterranean clover pasture response to lime application on the acid soils of southern New South Wales. Aust. J. Exp. Agric. 32:1051–1059.

Scott, B.J., A.M. Ridley, and M.K. Conyers. 2000. Management of soil acidity in long-term pastures of south-eastern Australia: A review. Aust. J. Exp. Agric. 40:1173–1198.

Slattery, W.J. 2001. The effects of conservation farm management practices on the long-term changes in soil organic matter status and its impact on soil acidification. Ph.D. thesis. Univ. of Queensland, Brisbane, Australia.

Slattery, W.J., M.K. Conyers, and R.L. Aitken. 1999. Soil pH, aluminium, manganese and lime requirement. p. 103–128. *In* K.I. Peverill et al. (ed.) Soil analysis, An interpretation manual. CSIRO Publ., Collingwood, Melbourne, Australia.

Slattery, W.J., and D.R. Coventry. 1993. Response of wheat, triticale, barley and canola to lime on four soil types in north-eastern Victoria. Aust. J. Exp. Agric. 32:609–618.

Slattery, W.J., D.G. Edwards, L.C. Bell, and D.R. Coventry. 1996. Soil acidification due to organic materials leaching from a cropping soil of north-eastern Victoria. p. 253–254. *In* Proc. Australian and New Zealand Natl. Soils Conf., Melbourne. 1–4 July. Australian Soil Sci. Soc. Inc. and the New Zealand Soc. of Soil Sci., Melbourne.

Slattery, W.J., D.G. Edwards, L.C. Bell, D.R. Coventry, and K.R. Helyar. 1998. Soil acidification and the carbon cycle in a cropping soil of north-eastern Victoria. Aust. J. Soil Res. 37:273–290.

Slattery, W.J., G.R. Morrison, and D.R. Coventry. 1995. Liming effects on soil exchangeable and soil solution cations of four soil types in north-eastern Victoria. Aust. J. Soil Res. 33:277–295.

Slattery, W.J., A.M. Ridley, and S.M. Windsor. 1991. Ash alkalinity of plant and animal products. Aust. J. Exp. Agric. 31:321–324.

Slattery, W.J., and A. Surapaneni. 2002. Effect of soil management practices on the sequestration of carbon in duplex soils of southeastern Australia. p. 107–117. *In* J.M. Kimble et al. (ed.) Agricultural practices and policies for carbon sequestration in soil. Lewis Publ., New York.

Smith, G.J., M.B. Peoples, G. Keerthisinghe, T.R. James, D.L. Garden, and S.S. Tuomi. 1994. Effect of surface applications of lime, gypsum and phosphogypsum on the alleviating of surface and subsurface acidity in a soil under pasture. Aust. J. Soil Res. 32:995–1008.

Sommerfeldt, T.G., and C. Chang 1985. Changes in soil properties under annual applications of feedlot manure and different tillage practices. Soil Sci. Soc. Am. J. 49:983–987.

Spouncer, L.R., R.H. Merry, and C.H. Michalski. 1996. Allocation of acidity profile classes to Atlas of Australian soils mapping units. Tech. Rep. 20/1996. Div. of Soils, CSIRO, Canberra, Australia.

Sumner, M.E. 1993. Gypsum and acid soils: The world scene. Adv. Agron. 51:1–32.

Thomas, G.W. 1975. The relationship between organic matter content and exchangeable aluminum in acid soil. Soil Sci. Soc. Am. J. 39:591.

Tiarks, A.E., A.P. Mazurak, and L. Chesnin. 1974. Physical and chemical properties of soil associated with heavy applications of manure from cattle feedlots. Soil Sci. Soc. Am. Proc. 38:826–830.

Tisdall, J.M., and J.M. Oades 1982. Organic matter and water-stable aggregates in soil. J. Soil Sci. 33:141–163.

Tyson, S.C., and M.L. Cabrera. 1993. Nitrogen mineralisation in soils amended with composted and uncomposted poultry litter. Commun. Soil Sci. Plant Anal. 24:2361–2374.

Unger, P.W., and B.A. Stewart. 1974. Feedlot waste effects on soil conditions and water evaporation. Soil Sci. Soc. Am. Proc. 38:954–957.

Valzano, F.P., B.W. Murphy, and R.S.B. Greene. 2001. The long-term effects of lime (CaCO3), gypsum (CaSO4.2H2o) and tillage on the physical and chemical properties of a sodic red-brown earth. Aust. J. Soil Res. 39:1307–1332.

Von Uexküll, H.R., and E. Mutert. 1995. Global extent, development and economic impact of acid soils. p. 5–19. *In* R.A. Date et al. (ed.) Plant soil interactions at low pH: Principles and management. Kluwer Academic Press, The Netherlands..

Webber, M.D., P.B. Hoyt, M. Nyborg, and D. Corneau. 1977. A comparison of lime requirement methods for acid Canadian soils. Can. J. Soil Sci. 57:361–370.

Whitehead, D.C., and A.W. Bristow. 1990. Transformations of nitrogen following the application of 15N-labelled cattle urine to an established grass sward. J. Appl. Ecol. 27:667–678.

Williams, C.H. 1980. Soil acidification under clover pasture. Aust. J. Exp. Agric. Anim. Husb. 20:561–567.

Williams, C.H., and G.S. Andrew. 1970. Mineral nutrition of pastures. p. 321–338. *In* R.M. Moore (ed.) Australian grasslands. ANV Press, Canberra, Australia.

Williams, C.H., and C.M. Donald. 1957. Changes in organic matter and pH in a podsolic soil as influenced by subterranean clover and superphosphate. Aust. J. Agric. Res. 8:179–189.

Wood, M. 1991. Biological aspects of soil protection. Soil Use Manage 7:130–136.

Yan, F., S. Schubert, and K. Mengel. 1996. Soil pH increase due to biological decarboxylation of organic anions. Soil Biol. Biochem. 28:617–624.

Zabowski, A.U., and R.S. Sletten. 1991. Carbon dioxide degassing effects on the pH of Spodosol soil solutions. Soil. Sci. Soc. Am. J. 55:1456–1461.

22 Challenges and Strategies for Dryland Agriculture in Pakistan

Abdul Rashid

Land Resources Research Program
National Agricultural Research Center
Islamabad, Pakistan

John Ryan

ICARDA
Aleppo, Syria

Mushtaq A. Chaudhry

University of Arid Agriculture
Rawalpindi, Pakistan

ABSTRACT

Pakistan is a predominately an arid to semi-arid country of about 140 Mn people, and has 22 Mha of cultivated area, about 5 Mha of which is rainfed and the rest irrigated; irrigated agriculture being predominant. Dryland agriculture is synonymous with rainfed (*barani*) conditions, where land holdings are small and often fragmented. Dryland rainfall (125–1000 mm) is bimodal, mainly (~60%) monsoonal and highly erratic. Rainfed areas are *subhumid* (>500 mm), *semi-arid* (300–500 mm), and arid (<300 mm) with variable constraints and management requirements. As about 70% of the geographical area receives ≤250 mm, and more than 80% receives ≤ 375 mm, the whole country is more or less arid to semi-arid. Dryland crops include wheat (*Triticum* spp.), chickpea (*Cicer arietinum* L.), sorghum [*Sorghum bicolor* (L.) Moench], millet (*Panicum miliaceum* L.), barley (*Hordeum vulgare* L.), maize (*Zea mays* L.), lentil (*Lens culinaris* Medik.), peanut (*Arachis hypogaea* L.), rapeseed-mustard (*Brassica* spp.), and guar seed [*Cyamopsis tetragonoloba* (L.) Taub.]. A dominant fraction of Pakistan's drylands is rangelands, which, despite being overgrazed, sustain a significant livestock population. Dryland soil and crop management practices are mostly primitive, coupled with major problems of moisture stress and/or uncertainty, credit scarcity, soil erosion, and nutrient depletion. Pakistan's dryland agriculture is a high-risk, low-input enterprise for resource-poor farmers, who frequently use poor quality seed, inadequate and imbalanced fertilizers, and poor crop management practices. Consequently, crop yields are much below their demonstrated achievable potentials. Remedial measures for improving crop productivity include effective *rainwater harvesting*, *land consolidation*, *improved credit facilities*, *better soil and water conservation*, *use of good quality seed*, *balanced nutrient management*, and *weed control*.

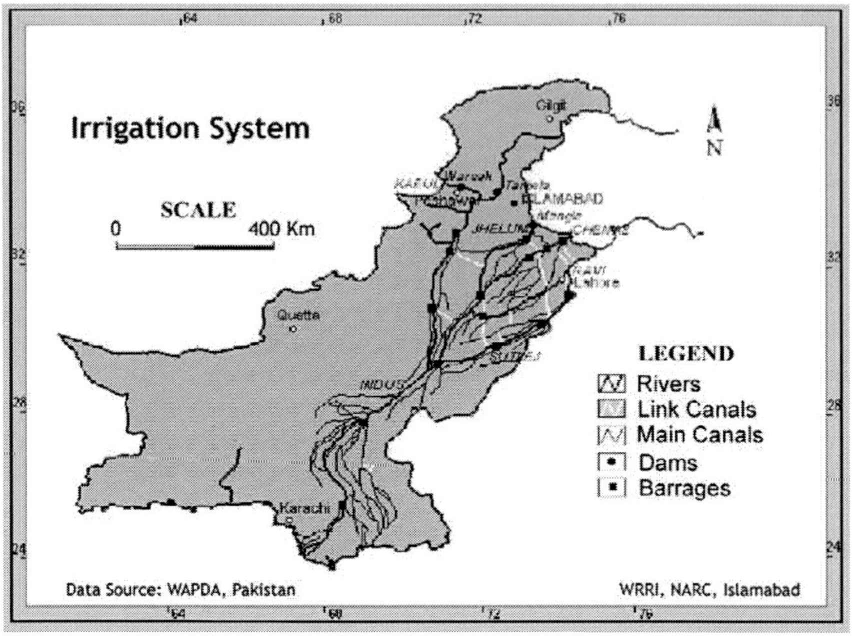

Fig. 22–1. Indus basin irrigation system in Pakistan. Source: Water and Power Development Authority (WAPDA), Pakistan.

INTRODUCTION

Pakistan is a large (80 Mha) and populous (140 Mn people) country of South Asia, lying between 23°5′ and 36°5′ lat and between 61°2′ to 74°5′ long. It is an arid to semi-arid country with 22 Mha cultivated area, of which 5 Mha are rainfed. The country is fortunate to have the Indus basin irrigation system (Fig. 22–1), which is the largest contiguous canal network in the world. Consequently, irrigated agriculture is the mainstay of the country; yet, rainfed agriculture is crucially important because of its significant share in total production of a range of crops except wheat (Table 22–1). This is particularly so because, with a fast growing population,

Table 22–1. Rainfed agriculture's share in total crop production in Pakistan.

Crop	Total production	Rainfed	Total rainfed
	10^3 t		%
Guar-seed	181	181	100
Peanut	96	86	90
Mungbean and blackgram	543	438	81
Chickpea	411	316	77
Sorghum and millet	350	196	56
Barley	164	85	52
Maize	1 284	347	27
Rapeseed-mustard	242	61	25
Wheat	17 550	1 755	10

Source: Government of Pakistan (2002).

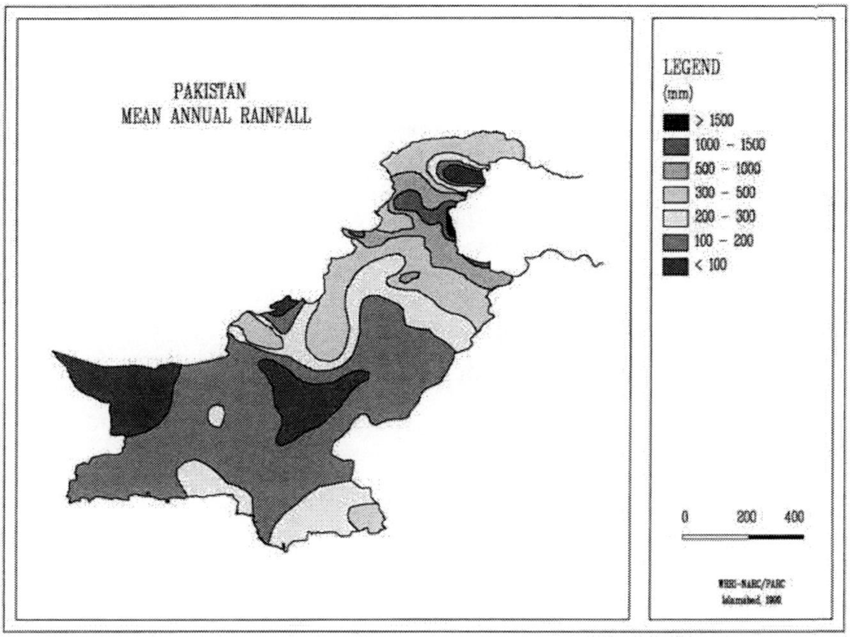

Fig. 22–2. Mean annual rainfall regions in Pakistan. Source: WRRP, NARC, Islamabad, Pakistan.

the country is experiencing a rapid decline in per capita of grain-producing land, that is, 0.58 ha in 1950, 0.25 ha in 1980, and 0.18 ha in 2000. However, better dryland productivity can be achieved with careful management of rainfed resources. This would help ease pressure on the irrigated sector, which can consequently be better utilized for cash crops for improving the overall economic well-being of rural communities.

Rainfall

Pakistan is one of the drier countries of the world as indicated by a rainfall map of the country (Fig. 22–2) that shows limited areas in the northern part with favorable rainfall (>1000 mm yr $^{-1}$), but most of the country with <200 to 300 mm. The gradations in rainfall are more clearly illustrated in a map depicting aridity (Fig. 22–3). Thus, only limited areas are *humid* and *subhumid* (>500 mm), while the vast majority of the land surface is *arid* (<300 mm), especially the vast area of Balochistan, or *semi-arid* (300–500 mm). An analysis of the relative amounts of total rainfall with respect to surface area (Fig. 22–4) area shows that 20% of the country receives <125 mm yr^{-1}, while almost half the country receives between 125 and 250 mm on a yearly basis.

While irrigation water from the Indus counteracts the country's lack of rainfall, the focus of dryland farming is exclusively on areas that are not too dry, that is, semi-arid, where rainfall is sufficient to produce a crop in most seasons. Thus, it is important to examine the seasonal rainfall distribution, which provides for a

Fig. 22–3. Aridity map of Pakistan based on average annual rainfall. Source: WRRP, NARC, Islamabad, Pakistan.

"window of opportunity" in terms of crop planting to harvest. An example for the rainfed Potohar plateau (Fig. 22–5) shows that rainfall has a *bi-modal* pattern, being mainly monsoonal, and highly erratic. These characteristics dictate the agricultural productivity in areas dependent exclusively on rainfall and indeed the viability of dryland communities.

Dryland Characteristics

Dryland soil and crop management practices are mostly primitive, coupled with handicaps of precipitation uncertainty/moisture stress, soil erosion, and nutrient depletion. In this high-risk environment, use of agricultural inputs is minimal: fertilizers and improved seed are scarcely used (Rashid et al., 2002a, 2002b). Crop residues are used primarily as a source of feed for animal, and as a fuel by rural communities (Fig. 22–6). The economic value of such residues, used as feed, can equal or exceed that of the seed, particularly in 'poor' rainfall years (Rees et al., 1991). Uncertain rainfall is the major cause of low productivity and/or of crop failures. For example, in (arid to semi-arid) upland Balochistan, the farmers expect

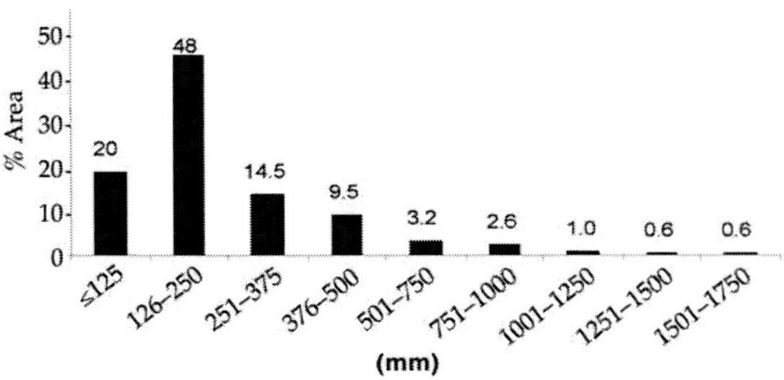

Fig. 22–4. Mean annual rainfalls with corresponding geographical areas in Pakistan. Source: Rashid and Hussain (1994).

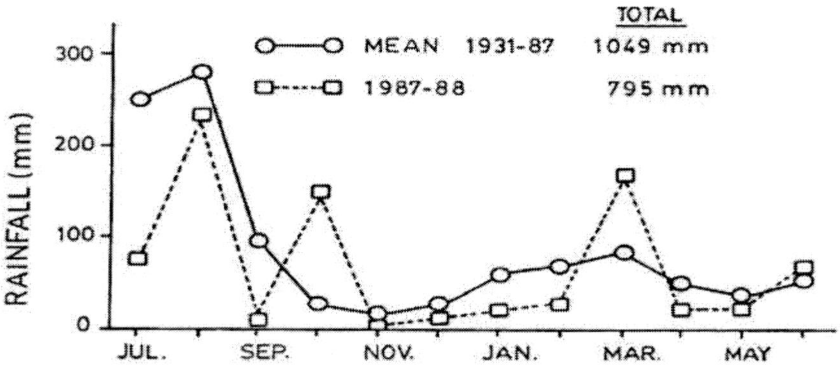

Fig. 22–5. Mean and seasonal rainfall in Potohar plateau of Pakistan—depicting bimodal and erratic patterns. Source: Rashid and Hussain (1994).

a crop failure in 3 to 5 yr out of 10, and even in 'good' rainfall years (2 to 3 out of 10) wheat yields are only 400 to 800 kg ha^{-1} (Rees et al., 1988). Thus, Pakistan's dryland agriculture is a high-risk, low-input enterprise for resource-poor farmers; most dryland holdings are small, and frequently fragmented, a factor which militates against progress. Consequently, crop yields are much below their demonstrated achievable potential. In addition, dryland cropping intensity is expected to increase in the future because of increased demand for food, feed, and fiber for a rapidly increasing population.

Fig. 22–6. Exploitation of overgrazed rangeland for fuel by the poverty-stricken inhabitants.

Table 22–2. Land capability classification of Pakistan's drylands.

Land class	Limitations	Mha
Class d II	Cropland with minor limitations	
d II c	Moisture shortage, irregular rainfall, and	
d II e	Sloping land, vulnerable to erosion	
Total		0.23
Class d III	Cropland with moderate potential	
d III e	Sloping soils	0.42
d III c	Inadequate rainfall	0.79
d III w	Flood watered land	0.86
Total		2.07
Class d IV	Poor dry farmed, economically marginal	
d IV c	Severe moisture shortage	1.65
d IV s	Sandy/gravelly or shallow soils	0.48
d IV e	Sloping soils	0.09
d IV x	Unstable soils	0.08
d IV w	Flood watered land	0.07
Total		2.37

This chapter is not exhaustive, but rather offers a perspective of major features of dryland agriculture in the country, summarizing prevalent situations, issues, and practices, listing suggestions for improvement, and highlighting efforts of development agencies to improve Pakistan's dryland sector.

RAINFED SOIL RESOURCES AND FARMING SYSTEMS

Soils

Pakistan is endowed with a rich soil resource base. The *U.S. Soil Taxonomy* soil orders prevalent in the drylands of Pakistan include Aridisol, Entisol, Inceptisol, and Alfisol (Rafiq, 1996). Though dryland soil productivity is less than in irrigated areas, it is primarily because of moisture constraints rather than poor soil resources per se. Land capability classification of Pakistan's drylands and their use potential is given in Table 22–2. Of the total dryland area, the highest proportion of the land has severe limitations, poorly farmed, and economically marginal (2.4 Mha). A slightly smaller fraction (2.07 Mha) has moderate development potential. However, only a tiny fraction (0.23 Mha) is considered as having only minor production limitations.

Rainfed Agriculture and Livestock

Non-irrigated farming systems in Pakistan are variable and are categorized as:

- *Barani*, which is entirely dependent upon rainfall, and mainly characteristic of the *Potohar* plateau and the northern mountains.
- *Sailaba*, which is based on residual moisture from summer floods and rains, and is practiced in riverine areas and active floodplains of the major rivers.

Fig. 22–7. Livestock feeding on scarce vegetation in a semi-arid ranglenad of Pakistan.

- Dry farming *Khushkaba*, which is based on diverting runoff from large, uncultivated fields, and is characteristic of the cooler, western highlands of Balochistan.

Livestock, mainly cattle (*Bos taurus*) and to a lesser extent sheep (*Ovine vignei*) and goats (*Caprine hircus*), are a major component of Pakistan's agriculture, including rainfed farming systems, and are closely integrated with crop production. As about 70% of dryland areas are rangelands, drylands support a major segment of total livestock population in the country. Though rangelands are generally overgrazed, a significant livestock population is sustained by drylands (Fig. 22–7). For example, in Punjab alone, 3.4 Mha of arid Cholistan Desert, and other rainfed areas amount to 7.3 Mha, of which 3.9 Mha are rangelands. This area supports 13 Mn head of cattle and poultry stock. The principal dryland agricultural activity in upland Balochistan (having hyper-arid to semi-arid continental climate, with 50–400 mm annual rainfall) is small-ruminant production (Buzdar et al., 1989; Rafiq, 1976), with rangeland grazing providing most of the animal feed.

ISSUES AND CHALLENGES OF RAINFED AGRICULTURE

Dryland crop production is confronted with both biophysical and socio-economic constraints. The foremost obstacle hampering dryland crop productivity is low, badly distributed, and erratic rainfall. While nothing can be done to influence rainfall, the only option is to manage its impact so as to increase its efficiency and reduce its potential degradation effects. Heavy monsoon showers, during hot months of July and August, are hardly beneficial for agriculture. Rather, these heavy downpours cause soil erosion and floods and experience heavy evaporation losses.

While rainfed crop productivity is uncertain and low, what is encouraging is that average farmers' yields are often 50 to 80% below the demonstrated achievable potential, thus indicating that improvement is possible. Because of low production efficiency and unreliable incomes, rainfed agriculture is considered a low priority enterprise. In the absence of reliable agro-based livelihoods, local manpower is constrained to migrate for alternate sources of income.

Potohar Plateau

A typical rainfed area (300–1500 mm, 65% monsoonal), located in northern Punjab, Pakistan, is the *Potohar* plateau, which consists mainly of a wide undulating plain, generally lying at 300 to 450 m above sea level (Beg et al., 1985). This area, which is badly gullied, because of severe water erosion, spans over 1.82 Mha, out of which about 0.82 Mha is cultivated. The main parent material in the area is loess, followed by piedmont alluvium, river alluvium, and tertiary rocks, and prevalent soil orders are Inceptisol, Alfisol, and Entisol. Major crops in the area are wheat, peanut, sorghum, maize, and rapeseed-mustard. Other soil-related constraints include: *shallow soils*, *low organic matter*, *nutrient depletion* (Table 22–3), and soil physical constraints such as *crusting*, *poor structure*, *coarse or gravely texture, hard pan,* and *less permeable/dense layer*.

Most dryland soils (more than 90%) are low in organic matter (<0.86%), and almost all are deficient in nitrogen (N), >90% are deficient in phosphorus (P), and 15 to 25% in potassium (K). Deficiencies of zinc (Zn), boron (B), and iron (Fe) are also widespread. The use of organic manures is minimal and is confined to village environs. Despite adequate moisture from rainfall being the dominant crop production constraint, fertilizer use of N, P, Zn, and B is essential for obtaining good crop yields. However, fertilizer use in dryland agriculture is minimal; for example, 42 kg nutrients ($N + P_2O_5$) ha^{-1} is used in the subhumid *Potohar* plateau compared to 117 kg nutrients in irrigated areas (Bajwa et al., 1991). Micronutrient use, though highly cost-effective, is rarely practiced in dryland agriculture (Rashid et al., 2002b). Consequently, macro- and micronutrient deficiencies are constraining potential yield realization. Faced with uncertain precipitation, the resource-poor dryland farmers are reluctant to invest in farm inputs, including fertilizers (Bajwa et al., 1991; Rashid and Hussain, 1994); without nutrient replenishment, this only promotes "nutrient mining". Thus, dryland soils are characterized by a serious fertility decline.

Causes of Low Productivity

The most obvious and major cause of low dryland productivity is not only the inadequate and/or erratic precipitation, but also its season-to-season and year-to-year variation. Other factors include poor moisture conservation, small and/or fragmented land holdings, scarce capital and labor, and primitive methods of cultivation (i.e., cultivation of low-yielding local varieties, inadequate fertilizer use, poor weeding, and minimal mechanization). Cropping intensity is low, more so in drier areas, with large areas left fallow for the whole year. Major challenges of rainfed agriculture involve reliable rainfall forecast, improved water harvesting, mois-

Table 22–3. Soil characteristics and nutrient deficiencies in typical rainfed and irrigated regions of Punjab province, Pakistan.

	Rainfed *Potohar* plateau†		Irrigated cotton–wheat‡	
Characteristics	Range	Mean	Range	Mean
Soil				
pH (1:1)	7.5–8.2	8.0	7.6–9.1	8.3
$CaCO_3$ (%)	0.8–21	6.4	1.0–10.8	6.9
Organic matter (%)	0.10–1.40	0.5	0.10–1.80	0.9
Texture	Loamy		Clay–Sandy	
Depth	Shallow–Deep		Deep	
Nutrients (mg kg^{-1})	Range	Deficient area %	Range	Deficient area %
Olsen-P	0.3–11	94	0.6–16	56
Extractable-K	22–332	25	26–380	18
AB-DTPA-Zn	0.12–4.12	72	0.1–2.6	46
Hot water-soluble-B	0.08–1.2	63	0.07–2.1	50

Source: †Rashid et al. (2002 a); ‡Rafique et al. (2002).

ture conservation and utilization, adoption of drought-tolerant, high-yielding varieties, efficient use of the expensive farm inputs, especially fertilizers, and better range management and cropping systems.

As irrigated agriculture is the driving force in Pakistan's economy, largely because of assured moisture supply and better management, including agrochemical inputs, its output per unit area of land is much higher than that of rainfed agriculture. It is thus of interest to compare soil properties between irrigated and rainfed conditions (Table 22–3), some differences being due to the influence of management and inputs while others are to some extent inherent. Therefore, irrigated cotton-wheat land, by comparison with the rainfed *Potohar* plateau, tended to have higher organic matter, and a wider range of available P, extractable K, AB-DTPA-Zn, and soluble B. Irrigated soils on average tended to have finer texture, as typical of lowland river valleys, and to be deeper, but to have higher pH values, which reflect salinity and possible sodic effects.

RAINFED AGRICULTURE EDUCATION, RESEARCH, AND DEVELOPMENT

Dryland agriculture in Pakistan had received comparatively less attention relative to irrigated agriculture, until the early 1960s, when numerous missions supported by the Food and Agriculture Organization (FAO) of the United Nations, World Bank, and Canadian International Development Agency (CIDA), evaluated *barani* (rainfed) agriculture in the country and considered means for improving its productivity. The *Barani* Commission, constituted in 1975 by the Punjab Government, took stock of the situation and emphasized developing appropriate technologies for realizing the potential of these areas. It was concluded that productivity of rainfed areas can be increased through intensive research in the following disciplines: (i) farming systems (i.e., crops, livestock, agroforestry, etc.); (ii) soil and water conservation; (iii) fertilizer practices; and (iv) practical and appropriate mechanization.

Some of the major donors and projects (mainly through the Pakistan Agricultural Research Council) are: *World Bank, Australian Center for Agricultural Research* (Dryland–biological nitrogen fixation); *International Fund for Agricultural Development*; *Canadian International Development Agency, Institut Mondial du Phosphat* (IMPHOS; applied P research); International Agricultural Research Centers, for example, *International Center for Agricultural Research in the Dry Areas* (ICARDA) in Syria, *International Center for Crops Research in Semi-arid Tropics* (ICRISAT) in India, and the *Centro International de Mejoriemento de Mais y Frigo* (CIMMYT) in Mexico, with a focus on germplasm development.

The number of organizations that deal with dryland agricultural research and development, either exclusively or in part, is large. Some operate at state level while others have a mandate to cover the entire country (Table 22–4). The major *Barani* (dryland) institutes are: Agency for *Barani* Area Development (Rawalpindi), *Barani* Agricultural Research Institute (Chakwal), *Barani* Agricultural Development Project (Peshawar), and *Barani* Agricultural Research and Development Project (Islamabad). Various institutes are operated under Pakistan's Agricultural Research Council, a major one being the Arid Zone Research Center (AZRC) in Quetta, which was established with the assistance of the main international dryland research center, ICARDA. Given the global mandate of this institution for dryland research and development, it is pertinent to elaborate on its current focus in Pakistan.

The *Barani* Village Development Project (1999–2005) is supported financially by the International Fund for Agricultural Development (IFAD) and involves various national institutions in collaboration with ICARDA, which provides scientific backstopping. The Project operates in a participatory mode, involving communities and farmers in all phases of the applied research and development. Broad objectives of the Project are to: (i) boost agricultural production and farmers' income, (ii) establish and institutionalize community organizations to provide the necessary services, (iii) improve the status of women in a culturally acceptable manner, (iv) expand the social infrastructure (water supply, roads, health centers), and (v) develop and expand off-farm employment.

The strategy of applied research was to implement technologies developed by the various national research institutes. The involvement of farmers and the tailoring of the applied research to location-specific agroclimatological and socioeconomic conditions are paramount. Broad areas of research include water and land use management, agroecological characterization and the use of integrated research sites, strengthening germplasm resources and developing new crop varieties, enhancing animal feed production and management, characterization of local communities, and the evaluation of economic and social land-use options.

In realization of the need for specialized training for rainfed agriculture, the country has established specialized educational institutions for addressing needs of dryland areas. Currently, there are a number of institutions that focus on rainfed agricultural education. These are: (1) University of Arid Agriculture, Rawalpindi, Punjab, established as a college in 1980 and upgraded to the status of a university in 1995; (2) College of Agriculture, Quetta, Balochistan, established in the early 1980s, (3) University College of Agriculture, Rawalakot, Azad Jammu and Kash-

Table 22–4. Research and development institutions in Pakistan dealing with rainfed agriculture.

Institution	Domain	Remarks
Agency for *Barani* Area Development (ABAD), Rawalpindi	Punjab	The major rainfed development agency in Punjab. Also executor of ICARDA-led, *Barani* Village Development Project
Soil Conservation Directorate, Rawalpindi	*Potohar* plateau, Punjab	Soil and water conservation development agency
Barani Agricultural Research Institute (BARI), Chakwal	*Potohar* plateau, Punjab	The main rainfed research institution in Punjab province
Soil & Water Conservation Research Institute (SAWCRI), Chakwal	*Potohar* plateau, Punjab	A soil and water conservation research organization
Cholistan Development Authority (CDA), Bahawalpur	Cholistan desert, Punjab	Development agency for the arid Cholistan desert
Barani Agricultural Development Project (BADP), Peshawar	NWFP	BADP's mandate in NWFP is similar to the ABAD' s in Punjab
Sindh Arid Zone Development Authority (SAZDA), Hyderabad	Sindh	SAZDA in Sindh is similar to ABAD and BADP in other provinces
Pakistan Agricultural Research Council (PARC), Islamabad	Pakistan	The main agricultural research organization in the country, addressing drylands throughout the country
(i) Arid Zone Research Centre (AZRC), Quetta	Pakistan	Dryland research throughout the country, with emphasis on Balochistan
(ii) Arid Zone Research Institute (AZRI), Umar Kot	Sindh	Dryland research in Sindh province, with emphasis on Thar Desert.
(iii) AZRI, Bahawalpur	Punjab	Dryland research in Cholistan Desert.
(iv) AZRI, DI Khan	NWFP	Dryland research in rainfed areas of NWFP
(v) National Agricultural Research Centre (NARC), Islamabad	Pakistan	Numerous dryland research activities; e.g., variety development, soil/water conservation, soil/nutrient management, technology developed for grain legumes, sorghum millet, etc.
(vi) *Barani* Agricultural Research & Development (BARD) Project, NARC, Islamabad	Pakistan	Dryland R&D throughout the country; project completed
(vii) Crop Maximization Project (CMP), NARC, Islamabad	Pakistan	Project completed
Pakistan Council for Research in Water Resources (PCRWR), Islamabad	Pakistan	Water resources R&D throughout the country
Water & Power Development Agency (WAPDA)	Pakistan	Has many dryland development activities, for example, watershed management

mir, established in the early 1990s; and (4) University College of Agriculture, Dera Ghazi Khan, Punjab, established in late 1990s. All the institutions offer BS and MS level training, while the University of Arid Agriculture also offers PhD level training in almost all disciplines.

STRATEGIES FOR DRYLAND AGRICULTURE

Dryland areas of Pakistan are diverse with regard to their soils, climate, resource base, production systems, socio-economics, and demographic features.

Thus, their development potential cannot be grouped broadly for strategy planning and its implementation. In fact, even within a defined agro-ecological zone, the strategies have to be tailored to suit local/regional situation. Rainfed research suggests the possibility of substantial potential for improving productivity per unit area. Remedial measures for enhancing dryland productivity include provision of adequate credit facilities; assured market demand; minimum support price for rainfed-oriented commodities like oilseeds, pulses, and peanut; effective rainwater harvesting, conservation and utilization; land consolidation; provision of good quality seed; effective nutrient management; and weed control. There has been a realization of the problems and potentials of dryland agriculture in Pakistan, which also has the advantage of having access to the relevant work/information elsewhere. However, the real job is to evaluate and adopt known and proven technologies.

In order to realize the potential of Pakistan's dryland agriculture sector, some needed actions in the area of research and development are:

- Development of site-specific technologies for water harvesting, moisture conservation, soil management, crop production, range management, agroforestry, and livestock.
- Effective dissemination of appropriate agricultural technologies to end-users.
- Arrangement of adequate agricultural credit, and provision of farm inputs (like good quality seed, fertilizers, etc.) and farm machinery.
- Introduction of effective support price mechanism for specific dryland crop produce.
- Land consolidation of fragmented holdings in dryland areas to help farmers manage their lands in a more effective manner.
- Development of agro-based, labor-intensive industries in the rainfed areas.
- Education and training of women in the relevant agro-based enterprises.
- A comprehensive program of large-scale groundwater recharge measures, including constructing of check dams, re-afforestation, and controlled grazing.

ACKNOWLEDGMENT

We are grateful to Dr. Muhammad Shafiq for reviewing the manuscript.

REFERENCES

Bajwa, M.I., M.S. Zia and P.A. Naim. 1991. Research findings in arid lands of Pakistan: A summary. Pakistan Agric. Res. Council, Islamabad, Pakistan.

Beg, A.R., M.S. Baig, Q. Ali, and C.M.A. Khan. 1985. Agro-ecological zonation of Potohar: An ecological basis for land-use planning and resource management. Natl. Agric. Res. Center, Islamabad, Pakistan.

Buzdar, N., G.F. Sabir, J.G. Nagy, and J.D.H. Keatinge. 1989. Animal raising in highlands: A socio-economic perspective. ICARDA, Quetta, Pakistan.

Government of Pakistan. 2002. Agricultural statistics of Pakistan, 2000–01. Economic Wing, Ministry of Food, Agriculture, and Livestock, Gov. of Pakistan, Islamabad. Pakistan.

Rafiq, M. 1976. Crop ecological zones of nine countries of the Near East region. Food and Agric. Org. of the United Nations, Rome, Italy.

Rafiq, M. 1996. Soil resources of Pakistan. p. 439–470. *In* A. Rashid and K.S. Memon (managing authors) Soil science. Natl. Book Found., Islamabad, Pakistan.

Rafique, E., A. Rashid, A.U. Bhatti, G. Rasool, and N. Bughio. 2002. Boron deficiency in cotton grown in calcareous soils of Pakistan. I. Distribution of B availability and comparison of soil testing methods. p. 349–356. *In* H.E. Goldbach et al. (ed.) Boron in plant and animal nutrition. Kluwer Academic/Plenum Publ., New York.

Rashid, A., and F. Hussain, 1994. Fertilizer efficiency in rainfed agriculture. p. 39–397. *In* N. Ahmad et al. (ed.) Efficient use of plant nutrients. Proc., 4th Natl. Congress of Soil Sci., Islamabad. 24–26 May 1992. Soil Sci. Soc. Pakistan, Islamabad, Pakistan.

Rashid, A., E. Rafique, S. Muhammed, and N. Bughio. 2002a. Boron deficiency in rainfed alkaline soils of Pakistan: Incidence and genotypic variation in rapeseed-mustard. p. 363–370. *In* H.E. Goldbach et al. (ed.) Boron in plant and animal nutrition. Kluwer Academic/Plenum Publ., New York.

Rashid, A., J. Ryan, and B. Amar. 2002b. Fertilizer use in Pakistan's dryland agriculture: Requirements, constraints, and options. Agron. Abstr. (online). ASA, Madison, WI.

Rees, D.J., M. Islam, A. Samiullah, F. Rehman, S.H. Raza, Z. Qureshi, and S. Mehmood. 1991. Rainfed crop production systems of upland Balochistan: Wheat (*Triticum aestivum*), barley (*Hordeum vulgare*) and forage legumes (*Vicia* species). Exp. Agric. 27:53–69.

Rees, D.J., J.G. Nagy, S.H. Raza, K. Mahmood, B.A. Chowdhry, and J.D.H. Keatinge. 1988. The dryland farming system of upland Balochistan: A case study. p. 81–98. *In* J.P. Srivastava et al. (ed.) Winter cereals and food legumes in mountainous areas. ICARDA, Aleppo, Syria.

23 Subsoil Constraints to Dryland Crop Production on the Low Rainfall Alkaline Soils of Southeastern Australia

James Nuttall and Roger Armstrong

Department of Primary Industries
Horsham, Victoria, Australia

Mark Imhof

Department of Primary Industries
Werribee, Victoria, Australia

Mohammad Abuzar

Department of Primary Industries
Tatura, Victoria, Australia

Robert Belford[1]

Department of Primary Industries
Horsham, Victoria, Australia

ABSTRACT

In the low rainfall cropping areas of southern Australia, particularly in northwestern Victoria and much of South Australia, soil constraints to crop production are widespread. These constraints include chemical problems of high pH, boron (B), sodicity, and salinity, which in turn are linked to physical problems caused by dense subsoils of limited water holding capacity and high resistance to root penetration. Soils are highly variable spatially (often at scales of <25 m), and have not been mapped at a scale consistent with the needs of land managers. In this environment, current varieties of pulse, cereal, and oilseed crops rarely reach their yield potential, and average yields are about 50% of the potential set by genetics and rainfall. This chapter describes soil properties and their spatial variation, and reviews the response of cereal crops to the soil constraints in this environment. The chapter concludes with a review of current and future options to overcome the limitations posed by these soils, in terms of both adaptation of crops, and amelioration of soils.

[1] Corresponding author (robert.belford@dpi.vic.gov.au)

INTRODUCTION

The Environment

The low rainfall southeastern cropping zone of Australia lies between 34° and 37° S lat, and 137° and 144° W long and occupies an area of approximately 185 000 km^2. The zone includes the eastern part of South Australia, and the northwestern Dune Fields and Plains geomorphic region of Victoria. It is an important area for crop and livestock production in southern Australia, with the farming system originally based on a wheat (*Triticum aestivum* L.)/pasture/fallow rotation, but now broadened to include a wide range of broad acre crops, with less pasture and fallow. The zone is characterized by low rainfall on calcareous soils, which are mainly derived from Quaternary windblown deposits and alluvial sediments. Rainfall is typically between 300 and 500 mm, with around 65% of this falling in the winter months (April–October). Diurnal temperature range in the summer months is between 13 and 30°C, with the winter temperature range between 4°C and 14°C. Frosts can occur in the winter months. In general terms, average temperatures are higher and average rainfall is lower in the northern parts of the zone, although there is wide variation from year to year and between locations. The combination of winter rainfall and hot dry summers gives the zone a Mediterranean climate, with potential evaporation ranging from 1100 to 1500 mm per annum.

Landform and Soils

Landforms consist of a combination of plains above flood level, sandstone ridges, east-west dunes and gentle to moderate hills, with a network of lakes, swamps, and rivers. Soils are all highly weathered, with the major soil types used for agricultural enterprises in the region being self-mulching Vertosols, Red Sodosols, and various Calcarosols (Isbell, 1996; Imhof et al., 2003).

Texture contrast or Duplex soils (Northcote, 1979) occur widely throughout Australian cropping zones, and are characterized by an abrupt change in soil texture between surface and subsoil horizons. In the southeastern zone of Australia, many of these are now described as Sodosols or Chromosols (Isbell, 1996), depending on the chemical nature of the subsoil. Sodosols have a strong texture contrast between surface (A) horizons and the sodic clay subsoil (B) horizons. They are widespread and frequently occur on the older alluvial plains in the northern part of the zone and on sedimentary hills and rises. There are some limited occurrences of Chromosols. Of the other major soil types, Calcarosols lack strong texture contrast between horizons and are calcareous throughout, and contain varying amounts of calcium carbonate as soft or hard fragments (Fig. 23–1). They are most common in the northeast of the zone. Vertosols are cracking clay soils that shrink and swell, and are widespread throughout the central and southern areas of the zone.

Land Use

The dominant agriculture, in terms of both land area and value of production, is broadacre cropping. More than 60% of the land used for agriculture is used for

Fig. 23–1. Calcarosol on lower slope of a rise near Birchip, southern Mallee, Victoria. Scale is in meters.

dryland cropping of cereals, pulses, oilseeds and pasture seed production. Cereals form the bulk of crop production, followed by pulses (chickpeas [*Cicer arietinum* L.], field pea [*Pisum sativum* L.], lentil [*Len culinaris* Medik.], faba bean [*Vicia faba* L.], and some lupin) and oilseeds (mostly canola [*Brassica* spp.]). The moderate rainfall and neutral to alkaline soil types have encouraged the growing of oilseed and pulse crops as an increasingly important segment of the region's overall agricultural production. These crops have also created the scope for local research and development initiatives, together with local processing opportunities. However, the move to a continuous cropping regime has increased the pressure on farmers to produce economic yields, and emphasized that the combination of low and unreliable rainfall and difficult soils requires a good awareness of underlying soil conditions and management options.

Crop Production and Yield

Historically, soils in the higher rainfall area ("Bordertown-Wimmera" agro-climatic zone [GRDC, 2003]) have been regarded as moderately fertile and productive. The soils in the more northerly "South Australian–Victorian Mallee" agro-climatic zone are generally less productive because of the lower and more variable rainfall and lighter soil types. Water supply is the major factor limiting dryland crop production in this environment, and crops must rely on root systems to access subsoil moisture, particularly after anthesis. The low yields, and low rate of yield improvement suggests that many crops cannot access sufficient water, particularly after flowering, that is, the performance of crops on many of the soils in the zone suffer from constraints over and above those induced by low seasonal rainfall. The economic impact of this low rate of yield improvement is exacerbated by the declining terms of trade which require growers to consistently increase production per unit input to maintain profitability.

Potential yield of wheat in this environment is often defined empirically as 20 kg grain ha^{-1} mm^{-1} of *effective* rainfall, which is usually defined as *growing season rainfall* less 100 mm for evaporation (French and Schultze, 1984). By this estimate, much of the zone should produce wheat yields of 4 to 5 t ha^{-1}. In practice, average wheat yields in the Bordertown–Wimmera zone are about 2 t ha^{-1}, and those in the South Australia–Victorian Mallee zone average <1.5 t ha^{-1}. Yields in both zones have increased at a slower rate over the same period than wheat yields in other parts of Australia with similar rainfall—particularly the central and eastern wheatbelt of Western Australia, and much of southern New South Wales (Stephens, 2002; Hamblin and Kyneur, 1993). Some of this low rate of yield improvement can be explained in terms of negative nitrogen (N) balance across the main cropping areas and a change to continuous cropping rotations (with no fallow period to store moisture or provide disease breaks). However, it has become clear that subsoil constraints in this zone are limiting access of crop roots to moisture and nutrients, thereby limiting crop growth and yield (Sadras et al., 2002; Nuttall et al., 2003b).

This chapter describes the soil characteristics and response of cereal and pulse crops in a region of this zone, the southern Mallee of Victoria. The chapter concludes with a discussion of options to improve the performance of crops in this difficult environment.

SOIL CONSTRAINTS AND CROP GROWTH

Soil Characteristics

These alkaline soils show a range of chemical toxicities and poor physical conditions. Since first used for agricultural purposes, the naturally low nutrient status has been corrected by fertilization and the inclusion of legume pastures in the rotation. However, these soils are now known to have dense subsoils with high pH, B, sodicity, and salinity, and have the potential to cause problems for root growth and function, particularly in the subsoil. The low conductivity of the subsoils, combined with low rainfall and high rates of transpiration and evaporation have

Table 23–1. Mean values of soil properties from 10 profiles on 15 Calcarosol sites in the southern Mallee of Victoria, Australia (after Nuttall et al., 2003a).

	Depth of sampling, m				
	0–0.10 m	0.10–0.20 m	0.20–0.40 m	0.40–0.60 m	0.60–1.00 m
Clay, %	33.7 (8.7)†	40.8 (7.7)	43.9 (7.1)	45.4 (5.9)	45.9 (5.1)
BD‡	1.24 (0.18)	1.37 (0.15)	1.34 (0.10)	1.37 (0.10)	1.42 (0.09)
$pH_{1:5}$	7.5 (0.6)	8.0 (0.4)	8.4 (0.3)	8.6 (0.3)	8.6 (0.2)
EC	2.0 (1.1)	2.4 (1.2)	4.1 (2.3)	6.3 (4.0)	8.4 (4.3)
ESP	3.1 (1.7)	6.3 (4.3)	10.9 (6.1)	15.4 (6.3)	19.1 (6.5)
B	2.4 (1.1)	5.5 (5.1)	13.3 (10.7)	21.1 (12.5)	24.7 (9.8)

† Values in parentheses are standard deviations.
‡ BD = bulk density, Mg m^{-3}; $pH_{1:5}$ measured in 0.01 *M* $CaCl_2$; EC = electrical conductivity, dS m^{-1}; ESP = exchangeable sodium percentage; B = boron, mg kg^{-1}.

caused the accumulation of salts within the root zone of annual crops (Rengasamy, 2002). This 'secondary' salinity (as distinct from that caused by rising watertables following clearance of native vegetation), has become an increasing problem in many areas within this low rainfall zone (National Land and Water Resource Audit, 2001).

The soil chemical factors of B, sodicity, and salinity are inter-related, and are also linked to physical problems. Clayey subsoils are typically sodic, dense, and poorly structured, and have low permeability to water and restricted porosity for root growth. In wetter years and sites, this low permeability can cause transient waterlogging at the interface between surface and subsoils, leading to loss of N by denitrification, and inhibited root growth. In drier years the subsoils may restrict root entry and access to water, effectively leaving only the surface soil for roots to explore. All soils have low levels of organic matter. A further difficulty is the large point to point variation in soil properties within a paddock, which means that selection of varieties and management inputs tends to be for the 'average' which further compromises optimum production per unit input.

The variation in soil properties across the area is well demonstrated by Nuttall et al. (2003a). This study covered an area of 3600 km^2 in the southern Mallee of Victoria, and sampled soil at five depths from each of 10 positions on 15 Calcarosol sites. Mean values for a range of soil parameters are shown in Table 23–1. Soil profiles generally showed increasing clay content and bulk density, an alkaline pH trend, and were calcareous throughout. Soil mean electrical conductivity (EC), exchangeable sodium percentage (ESP), and B all increased linearly with depth, to levels which are generally considered to be inhibitory to root growth. For salt tolerance in cereals, critical values are reported as >6.0 to 8.0 dS m^{-1} (Maas and Hoffman, 1977), and B toxicity is estimated to occur in cereals at tissue concentrations >15 mg kg^{-1} (Riley, 1987). From a physicochemical perspective, soils are considered sodic if the ESP exceeds 6.0% (Northcote and Skene, 1972). For the sites reported in Nuttall's study, these critical concentrations are exceeded below 0.40 m depth.

The variation between soil properties within a paddock can be of the same order as variation between sites (Nuttall et al., 2001; Imhof et al., 2003). In this landscape, much of this point to point variation can be related to the 'gilgai' microre-

Table 23–2. Variation in soil properties in relation to 'gilgai' microrelief on a Calcarosol, southern Mallee, Victoria, Australia (derived from Imhof et al., 2003).†

	Depth of sampling, m									
Depth	0–0.10 m		0.10–0.20 m		0.20–0.40 m		0.40–0.60 m		0.60–1.00 m	
Gilgai	M‡	D§	M	D	M	D	M	D	M	D
Clay, %	34.5	37.0	38.0	41.0	43.0	45.5	46.5	48.0	52.5	51.5
$pH_{1:5}$¶	8.60	8.50	9.48	8.75	9.78	9.08	9.92	9.30	9.82	9.52
EC	0.28	0.21	0.68	0.24	0.98	0.33	1.00	0.70	0.65	0.81
ESP	7.0	2.0	30.0	6.0	40.5	14.5	47.0	25.5	50.5	37.3
B	3.0	2.5	20.0	3.0	30.0	3.0	36.0	9.5	37.0	17.0

† Values are means derived from two adjacent soil profiles on 'M' and 'D', respectively.
‡ 'M' describes the mound of a Gilgai formation.
§ 'D' describes the depression.
¶ $pH_{1:5}$ measured in 0.01 *M*; EC = electrical conductivity, dS m^{-1}; ESP = exchangeable sodium percentage; B = boron, mg kg^{-1}

lief, which was originally formed through wetting and drying cycles which caused blocks of soil to move upwards and form mounds, typically about 25 m apart, with associated depressions intervening. Subsequent cultivation and land management has levelled these mounds somewhat, but the variation in soil properties across the surface and with depth remains pronounced (Table 23–2). The soils on former depressions in the gilgai landscape generally have slightly higher clay contents, and a slightly lower pH with depth; but markedly lower EC, ESP, and B levels compared to the mounds. Crop growth frequently reflects this mosaic of variation in soil properties.

Mapping Subsoil Constraints

Broad-scale and generalized soil maps are available for this landscape, but the remoteness of the area and the relatively low level of inputs and outputs has not justified intensive soil mapping. There are thus few detailed maps at farm or paddock level to help growers understand the variation that exists, and to manage this variation effectively. Traditional soil survey methods of examining profiles and describing variation through random or systematic core sampling (e.g., Imhof et al. 2003), are essential to provide some ground truthing of information, but are too expensive for routine use at a scale that matches the variability. Attention has therefore turned to using (i) inexpensive measures of soil characteristics as surrogates for other soil properties, which are expensive to measure but of significance for crop growth and (ii) remotely sensed data to derive maps of important soil properties at paddock level.

(i) The interrelationship of soil factors has been investigated by Nuttall et al. (2003a), who used regression techniques to explore the relations between a range of soil properties including clay content, bulk density, pH, carbonate, EC, ESP, extractable sodium (Na^+), and B. This showed that B concentrations were well correlated with pH values ($r^2 = 0.68$), and that soils with a pH of <8.1 were unlikely to have toxic levels of B. High correlations also existed between Na^+ and ESP and between ESP and EC. This work suggests that simple measurements of pH and Na^+

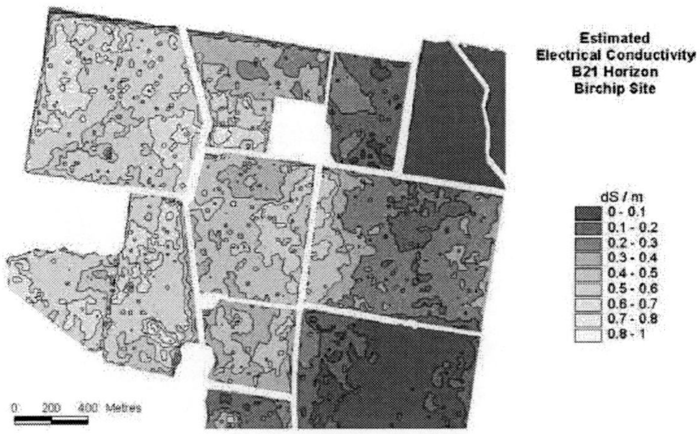

Fig. 23–2. Electrical conductivity (EC) in the subsoil of a Calcarosol estimated from remotely sensed data at Birchip, southern Mallee, Victoria, Australia. EC is defined by the equation:

$$EC = a + b_1\mathrm{relDEM} + b_2L^2_{hh} + b_3P^2_{vv} + b_4P^2_{hh}$$

where relDEM = Relative elevation; L = AirSAR L-band (25 cm wavelength); P = AirSar C-band (68 cm wavelength); hh = HH polarization; and vv = VV polarization. Coefficients for the equation are $a = -0.3429$; $b_1 = 0.0033$; $b_2 = 0.0006$; $b_3 = -0.003$; $b_4 = -0.00004$; $r^2 = 0.518$.

are likely to provide good indications of the important plant limiting factors of B, sodicity (as measured by ESP) and salinity (as measured by EC).

(ii) Remotely sensed data can be obtained from satellite imagery, airborne and ground based data collection. Satellite imagery provides reflectance signals across a wide range of wavebands and from several different satellite platforms. Airborne data includes multi-band radar, gamma radiometric emissions, digital elevation mapping (laser altimetry), and electrical conductivity. Similar information can also be collected in the paddock using single or multiple sensors on a vehicle. Preliminary work (M. Abuzar, personal communication, 2003) suggests that predictions of subsurface B and pH levels can be derived from airborne radar bands and digital elevation information. Predictions of sodicity and electrical conductivity can be derived from reflectance from a combination of information including radar and digital elevation monitoring. Preliminary paddock scale maps of predictions of subsoil EC and sodicity, derived from remotely sensed data and calibrated against information from soil cores are shown in Fig. 23–2 and 23–3.

Response of Crops to Subsoil Constraints

Soil measurement and mapping information confirms that subsoils are highly variable, physically difficult, and potentially toxic, which creates particular problems for growers in selecting crops and management regimes to minimize the impacts of subsoil constraints on crop growth.

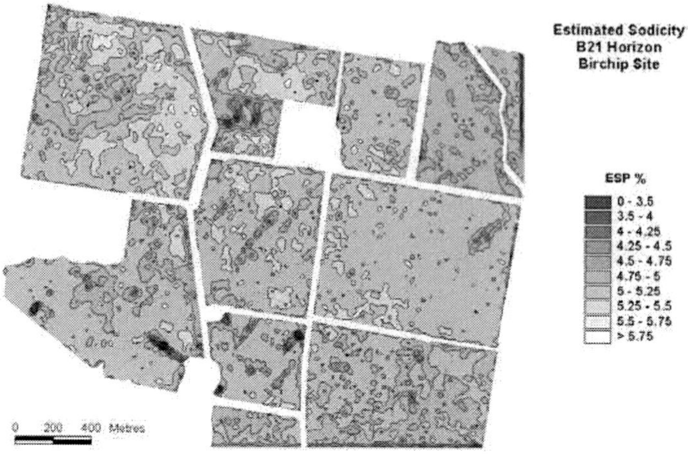

Fig. 23–3. Exchangeable sodium percentage (ESP) in the subsoil of a Calcarosol estimated from remotely sensed data at Birchip, southern Mallee, Victoria, Australia. ESP is defined by the equation:

$$\log \text{ESP} = a + b_1\text{relDEM} + b_2P_{hh} + b_3P_{vv} + b_4(P_{hh}/P_{vv})^2$$

where relDEM = Relative elevation; P = AirSAR C-band (68 cm wavelength), hh = HH polarization; vv = VV polarization. Coefficients for the equation are a = 3.611; $b_1 = 0.001$; $b_2 = -0.150$; $b_3 = 0.139$; $b_4 = -2.408$; $r^2 = 0.27$.

Cereals

At the landscape level, Nuttall et al.(2003a) followed up his study of variation in soil properties across 15 sites by an analysis of water use and yield of wheat at the same sampling positions (Nuttall et al., 2003b). A multivariate model which included rainfall around anthesis, available soil water and nitrate (NO_3^-) in the topsoil, and ESP from 0.60 to 1.00 m explained 51% of the variation in grain yield; of these factors, available soil water in the 0.10 to 0.40 m layer and ESP in the 0.60 to 1.00 m layer were the most important. For the year of the experiment, further analysis of water extraction and root density in the subsoil indicated that little water was extracted by wheat crops where ESP was >19% or where EC_e was >8 dS m^{-1}.

In this study, extractable B levels per se were generally not associated with water extraction and crop performance. This contrasts to other reports, in which both Sadras et al. (2002) and Cartwright et al. (1984, 1986) identified B as a constraint to crop growth. In these other studies, however, the separate effects of B and sodicity were not considered; it is also possible that differences in soils, seasons, species, and the varieties used confounds interpretation of the different results.

To further understand the nature of the constraints to growth of cereals, Nuttall et al. (2003c) used undisturbed soil columns of soil from the southern Mallee to separate the influence of different subsoil factors on yield of wheat and barley (*Hordeum vulgare* L.). The work examined the response of near-isogenic lines of wheat and barley (with and without tolerance to B) to rainfall, and ripping with gypsum. The soils had high levels of B (30mg kg^{-1}) below 0.60 m depth, and increasing levels of sodicity (ESP > 22%) and salinity (> 7 dS m^{-1}) with depth. The results showed that additional tolerance to B in both wheat and barley did not increase use of subsoil water, nor significantly increase grain yield. Simulated deep ripping (i.e., disturbance of the subsoil horizon at 0.25 m depth, and replacement of topsoil) with the addition of gypsum altered the pattern of water extraction but had no significant effect on grain yield. The study concluded that it was the increasing salinity and sodicity at depth, rather than B, that limited root growth and water extraction to 0.60 m.

In this low rainfall environment, the interaction of crop and season is extremely important. The extent to which particular species and varieties of crops extract available water depends on many factors, but particularly on the amount and distribution of rainfall during the year, and subsoil moisture content at critical stages of growth.

Pulses

The effect of B on growth of wheat is in marked contrast to that on pulses, and highlights the dangers in making generalized assumptions about the impact of subsoil constraints across a wide range of crops and species.

Pulse crops appear more sensitive to B, salinity, and sodicity than cereals. Although there is little information from field trials, soil solution B levels of 10 mg kg^{-1} (Chauhan and Power, 1978) and 4 mg kg^{-1} (Chauhan and Asthana, 1981) have produced B toxicity symptoms and growth reductions in field pea and lentil, respectively. However, there is a wide range of response to B in pea and lentil accessions (Hobson et al., 2003; Paull et al., 1992) which has yet to be incorporated into varieties.

The effect of salinity and sodicity on pulse species is poorly documented under field conditions in low rainfall environments; much of the literature is concerned with the impact of saline irrigation water on crop growth and yield. The effect of salinity seems to vary with stage of growth, and the impact of salinity on yield may be greater than on dry matter production (Lal, 1985). For sodicity, chickpea suffer delayed flower and pod formation and a 50% reduction in grain yield at ESP levels of around 17%, with similar values reported for field pea (20%) and lentil (14%), respectively; under the same conditions, the critical value of ESP for cereals was 40% (Gupta and Sharma, 1990).

OVERCOMING SOIL CONSTRAINTS

Fallowing has long been used in low rainfall environments as a way of managing risk, by increasing the storage of soil water for the subsequent crop, and as a disease break. Fallows can take many forms, but all are based on a period of bare ground during which a proportion of rainfall is stored in the soil and not lost by evaporation or transpiration. Accordingly, there is evidence that fallows increase groundwater recharge and may thus contribute to the increase in salinity in low rainfall areas (O'Connell et al., 1995). As a general principle, fallows do increase moisture storage (although storage efficiencies are low), and are usually followed by higher yields of wheat than after other crops or pastures (Ridge, 1996). However, the continued decline in terms of trade for Australian crop producers, the decline in wool prices and the number of sheep (*Ovis aries*), and the wider range of broad acre crops now available have meant that fallow practices are much less common than in the past. Crops rely on growing season rainfall and the highly variable storage of water over the preceding summer, allied to higher inputs of N and crop protection chemicals, to produce economic yields. Under these conditions, many producers are becoming increasingly aware of the impact of subsoil constraints, particularly in dry years. Overcoming the damaging effects of soil constraints can take two forms—avoidance and amelioration.

Avoidance Strategies

Avoidance implies either selection of crops with greater tolerance or resistance to the constraint, or development of varieties within species that show similar benefits. It follows that selection of appropriate crops/varieties for the season is extremely important—intolerant plants have little chance of yielding in dry seasons. In terms of species, pulse crops appear more sensitive to sodicity and salinity than do cereals, although there is considerable variation in germplasm of field pea, lentil, and chickpea. Many lines of lentil recently screened at the Department of Primary Industries have greater tolerance to both B toxicity and salinity than current Australian cultivars (Hobson et al., 2003; Maher et al., 2003).

For cereals, emphasis has been placed recently on developing wheat and barley with tolerance to B (Moody et al., 1993). However, the inability of wheat lines with B tolerance to increase use of subsoil water of grain yield (Nuttall et al., 2003c) suggests that B per se may not be the problem, and a broader approach to select for tolerance to salinity and sodicity may be more beneficial. In doing this, it will be important to recognize that the critical values accepted for toxicity of individual constraints may not apply when factors are considered in combination.

Amelioration Strategies

Amelioration of subsoil problems is difficult and expensive. Mechanical approaches (e.g., deep ripping, with or without application of lime or gypsum) to ameliorate subsoils for high value irrigated crops have had some success. The limited returns from broad acre cropping in Australia are likely to preclude expensive testing and manipulation of soils. However, some promising approaches have emerged from recent studies.

Agronomy and Nitrogen Management

Irrespective of the level of subsoil constraints, crops will perform closest to their potential if the basic agronomy is right. This includes an appropriate place in a sequence of crops, minimal tillage for crop establishment, sowing a crop variety with phenology matched to the time of sowing (and as early as possible), optimum seed rate, appropriate weed and pest control, and adequate levels of nutrients to meet the yield potential for crop and season. Of the nutrients, N represents the greatest single input cost on cereal and oilseed crops for many growers, but N response varies spatially within a paddock and with season. The development of variable rate application technology to match fertilizer addition and soil test information will help to fine tune fertilizer application, but the expense of acquiring subsoil information and lack of accurate seasonal rainfall forecasts are currently limiting widespread adoption of this technology. On these soils, growers currently incorporate a single application of N fertilizer at sowing; recent research by Norton et al. (2003) has shown that applying N between the rows, or using split top dressings of N gave positive yield responses in wheat. This was attributed to the benefits of separating N from the seed, and from salts in the subsoil, and provides a cost-effective way for growers to increase returns from similar inputs.

Gypsum

Gypsum is widely used in Australia to improve the structure of sodic clay soils. In principle, gypsum displaces sodium from the exchange complex, and replaces it with calcium; this can lower ESP, improve structural stability, and improve permeability to water. Application to the surface of clayey soils is relatively cheap and usually effective (e.g., Hamza and Anderson, 2002; Jarwal et al., 2001); application to subsoils through slotting, incorporation, or injection is much more expensive, and has produced inconsistent results on saline and sodic soils.

The effectiveness of gypsum can also be influenced by crop type and sequence. Work in northwestern Victoria, Australia, showed that 2.5 t ha^{-1} of gypsum applied to a chickpea and canola rotation lowered ESP to the same extent as 10 t ha^{-1} of gypsum applied to a wheat/safflower (*Carthamus tinctorius* L.) rotation. Furthermore, gypsum applied to chickpea/canola significantly lowered ESP to a depth of 0.25 m, whereas effects in the wheat crop were confined to the surface 0.10 m of soil. This suggests that (i) tap rooted crops may facilitate the movement of gypsum down the profile and (ii) the rhizosphere of some pulse crops may acidify the soil and thus dissolve $CaCO_3$, enhancing the displacement of sodium. This latter represents a form of biological remediation, which needs further study.

Raised Beds

Raised bed techniques have recently developed in high rainfall cropping zones in Australia to reduce the impact of waterlogging on crop yields. The technique creates a bed (usually 2–3m wide) by forming the surface soil, and leaving a furrow between beds for drainage of excess water. This system has several obvious advantages, irrespective of rainfall, in that it creates a deeper surface soil for root exploration and water and nutrient storage. It also minimizes the damaging effects of compaction by wheeled traffic once the beds are formed; and allows more accurate and cost effective application of crop management inputs (fertilizers, pesticides etc). There are clearly costs of establishment, maintenance, and machinery matching to be set against the advantages.

An experiment in northwest Victoria on two sites established a range of experimental treatments to raise yields on soils in the southern Mallee and Wimmera (Armstrong et al., 2000). The experiments aimed to tackle both physical and chemical limitations to crop growth, and included direct drilling, high levels of nutrients, pig (*Sus scrofa*) litter, gypsum, and raised beds as treatments, singly and in combination. In a year of below average rainfall on a site in the Victorian Wimmera (375 mm vs. 525 mm long-term average), single treatments of direct drilling and high

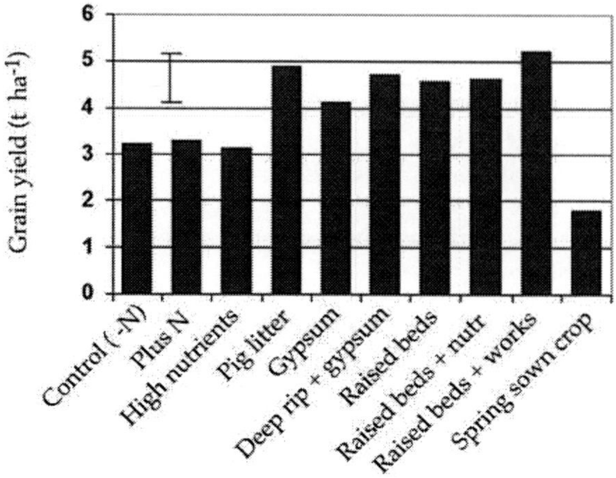

Fig. 23–4. Grain yield of wheat on a Vertosol in the Wimmera region of Victoria, Australia (1999). Vertical bar represents LSD at $P = 0.05$ (after Armstrong et al., 2000).

nutrients had no effect on yield; but single treatments of gypsum, pig litter, and raised beds raised yield by an average of 45% (Fig. 23– 4). All treatments that aimed to address the physical problems of highly sodic subsoils raised yields—treatments that included gypsum, pig litter (to add organic matter), and ripping raised yield by an average of 48%, and raised beds treatments increased yields by an average of 54%. Similar results, particularly in response to the raised beds, have been obtained on the same site in subsequent years, including a pulse crop (faba bean) in 2001. Gypsum, ripping, and high rates of use of bulky organic amendments are unlikely to be economically feasible in low rainfall, low-margin cropping systems. Raised beds, however, are a cheaper and easier option, and, provided the management difficulties can be overcome (particularly in relation to livestock), appear to have the potential to reduce the impact of subsoil constraints even in dry years.

Precision Agriculture Approaches

Although soil constraints vary widely within a paddock, it is usually possible to define two or three 'zones' within a production area in which soil properties are broadly similar. There are many possible techniques to define zones, including assessments from yield maps, classification based on remotely sensed imagery, and a range of more sophisticated geostatistical techniques to identify zones with stability in time and space (Whelan and McBratney, 2003). The remotely sensed data of greatest application is likely to include EC and topography (via digital elevation monitoring), as both relate to a number of fundamental soil properties, which influence crop growth. Once zones are defined, then specific soil sampling can be undertaken to get a better understanding of the drivers of crop growth and yield; and it follows that management can then be adjusted to best match crop type, variety, and input to the known characteristics of the zone.

CONCLUSIONS

In the low rainfall cropping areas of southern Australia, particularly in northwestern Victoria and much of South Australia, soil constraints are widespread and limit the ability of current varieties to reach their genetic yield potential for a given rainfall. Future progress in overcoming the damaging effects of subsoil constraints on crops in low rainfall environments requires some common sense, some new technologies, and a multi-disciplinary approach.

Common sense suggests that soil constraints are often present in the subsoil (usually below 60 cm depth), yet much routine soil testing does not assess either the physical or chemical properties of the subsoil. Also, soil testing often does not reflect the point to point variability that exists in surface and subsoil properties, leading to management recommendations for the average rather than for the broad zones or soil types that can usually be identified within a paddock.

New technologies include the development of surface and subsoil constraint maps at paddock level using surrogate measurements and/or remotely measured information. This can be used for decision support tools on application of N and other expensive inputs; 'precision agriculture' to identify and independently manage zones within a paddock, including the option to exclude unproductive zones from cropping; and better seasonal weather forecasts which will allow growers to make more robust decisions on choice of crops, varieties, and management inputs. There is also a need to investigate biological solutions to soil constraints, possibly by understanding and mimicking the behavior of native plants on these difficult soils. Hardware is likely to include development of systems for variable rate application of inputs, precision guidance mechanisms to help adoption of raised bed or controlled traffic systems, and real time measurements of crop and soil characteristics to help understand the causes of variation and to manage this variation.

The multi-disciplinary approach dictates that farmers, soil scientists, physiologists, agronomists, breeders, and water resource specialists all have an essential role to play to ensure that Australian broad acre croppers can deliver consistent high yields of good quality grains to export markets around the world.

ACKNOWLEDGMENTS

The Grains Research and Development Corporation and the Victorian Government's Science, Technology and Innovation Initiative have funded much of the research described in this chapter. We also thank Kristy Hobson and Laura Maher for information on pulse crops, and members of the Birchip Cropping Group in northwestern Victoria for their enthusiastic support of this work and generous donation of land and soil for experimental work.

REFERENCES

Armstrong, R.D., R. Flood, and J. Nuttall. 2000. Poor water use efficiency in north-western Victoria: Causes and possible solutions. p. 17–20. *In* The 2000 Shared solutions manual. Australian Grain. Greenmount Press, Toowoomba, Queensland.

Cartwright, B., B.A. Zarcinas, and A.H. Mayfield. 1984. Toxic concentrations of boron in a red-brown earth at Gladstone, South Australia. Aust. J. Soil Res. 22:261–272

Cartwright, B., B.A. Zarcinas, and L.R. Spouncer. 1986. Boron toxicity in South Australian barley crops. Aust. J. Agric. Res. 37:351–359.

Chauhan, R.P.S., and A.K. Asthana. 1981. Tolerance of lentil, barley and oats to boron in irrigation water. J. Agric. Sci. (Cambridge) 97:75–78.

Chauhan, R.P.S., and S.L. Powar. 1978. Tolerance of wheat and pea to boron in irrigation water. Plant Soil 50:145–149.

French, R.J., and J.E. Schultze. 1984. Water use of wheat in a Mediterranean-type environment. 1. The relation between yield, water use and climate. Aust. J. Agric. Res. 35:743–764

Grains Research and Development Corporation. 2003. BCA Agro-ecological zone maps. Available at www.grdc.com.au/researchers/bcazones.htm/2003 (verified 7 Jan. 2004).

Gupta, R.K., and S.K. Sharma. 1990. Response of crops to high exchangeable sodium percentage. Irrig. Sci. 11:173–179

Hamblin, A., and G. Kyneur. 1993. Trends in wheat yields and soil fertility in Australia. Bureau of Resource Sciences, Parkes, ACT.

Hamza, M.A., and W.K. Anderson. 2002. Improving soil physical fertility and crop yield on a clay soil in Western Australia. Aust. J. Agric. Res. 53:615–620.

Hobson, K.B., R.D. Armstrong, D.J. Connor, M.E. Nicolas, M.A. Materne. 2003. Genetic variation in response to high soil boron exists in lentil germplasm.. *In* M. Unkovich (ed.) Proc. of the 11th Australian Agronomy Conf., Geelong, Australia. 2–6 Feb. 2003. Conference Design Pty Ltd., Hobart. Available at www.regional.org.au/au/asa/2003 (verified 7 Jan. 2004).

Imhof, M., P. Rampant, S. Ryan, M. Abuzar, A. Murphy, T. Fay, and J. Martin. 2003. Soils of the Birchip Cropping Region. Dep. of Primary Industries, Victoria, Australia. Available at www.dpi.vic.gov.au/vro/2003 (verified 7 Jan. 2004).

Isbell, R.F. 1996. The Australian soil classification. CSIRO Publ., Melbourne.

Jarwal, S.D., R.D. Armstrong, and P. Rengasamy. 2001. Effect of gypsum and stubble retention on crop productivity in Western Victoria. *In* N. Mendham et al. (ed.) Proc. of the 10th Australian Agron. Conf., Hobart. 28 Jan.–1 Feb. 2001. Available at www.regional.org.au/au/asa/2001 (verified 7 Jan. 2004).

Lal, R.K. 1985. Effect of salinity applied at different stages of growth on seed yield and its constituents in field peas (*Pisum sativum* Linn *var.* Arvensis). Indian J. Agron. 30:296–299.

Maher, L., R.D. Armstrong, and D.J. Connor. 2003. Salt tolerant lentils—A possibility for the future? *In* M. Unkovich (ed.) Proc. of the 11th Australian Agron. Conf., Geelong, Australia. 2–6 Feb. 2003. Conference Design Pty. Ltd., Hobart. Available at www.regional.org.au/au/asa/2003 (verified 7 Jan. 2004).

Maas, E.V., and G.J. Hoffman. 1977. Crop salt tolerance—Current assessments. J. Irrig. Drain. Div. 103:115–134

Moody, D.B., A.J. Rathjen, and B. Cartwright. 1993. Yield evaluation of a gene for boron tolerance using backcross-derived lines. p. 363–366 *In* P.J. Randall et al. (ed.) Genetic aspects of plant mineral nutrition. Kluwer Academic Publ., The Netherlands.

National Land and Water Resources Audit. 2001. Australian dryland salinity assessment 2000: Extent, impacts, processes, monitoring and management options. Commonwealth of Australia, Canberra, Australia

Northcote, K.H. 1979. A factual key for the recognition of Australian soils. Rellim Technical Publ., Glenside, South Australia

Northcote, K.H., and Skene. 1972. Australian soils with saline and sodic properties. CSIRO Aust. Soil Publ. 27. CSIRO Publ., Melbourne.

Norton, R.M., J.F. Pedler, C.M. Walker, and J.F. Angus. 2003. Optimum management of N fertiliser for wheat growing on alkaline soils. *In* M. Unkovich (ed.) Proc. of the 11th Australian Agron. Conf., Geelong, Australia. 2–6 Feb. 2001.Conference Design Pty Ltd., Hobart. Available at www.regional.org.au/au/asa/2003 (verified 7 Jan. 2004).

Nuttall, J.G., R.D. Armstrong, D.J. Connor. 2001. Understanding subsoil water-use by cereals on southern Mallee soils. I. Spatial characteristics of sub-soil constraints. *In* N. Mendham et al. (ed.) Proc. of the 10th Australian Agron. Conf., Hobart. 28 Jan.–1 Feb. 2001. Available at www.regional.org.au/au/asa/2001 (verified 7 Jan. 2004).

Nuttall, J.G., R.D. Armstrong, D.J. Connor, and V.J. Matassa. 2003a. Interrelationships between edaphic factors potentially limiting cereal growth on alkaline soils in north-western Victoria. Aust. J. Soil Res. 41:277–292.

Nuttall, J.G., R.D. Armstrong, and D.J. Connor. 2003b. Evaluating physicochemical constraints of Calcarosols on wheat yield in the Victorian southern Mallee. Aust. J. Agric. Res. 54:487–497

Nuttall, J.G., R.D. Armstrong, and D.J. Connor. 2003c. Boron tolerance does not improve use of subsoil water or grain yield of wheat (*Triticum aestivum*) and barley (*Hordeum vulgare*) on a Calcarosol. Plant Soil (submitted)

O'Connell, M.G., G.J. O'Leary, and M. Incerti. 1995. Potential groundwater recharge from fallowing in north western Victoria, Australia. Agric. Water Manage. 29:37–52.

Paull, J.G., R.O. Nable, A.W.H Lake, M.A. Materne, and A.J. Rathjen. 1992. Response of annual medics (*Medicago* spp.) and field peas (*Pisum sativum*) to high concentration of boron: Genetic variation and the mechanisms of tolerance. Aust. J. Agric. Res. 43:203–213.

Rengasamy, P. 2002. Transient salinity and subsoil constraints to dryland farming in Australian sodic soils: An overview. Aust. J. Exp. Agric. 42:1–11.

Ridge, P.E. 1996. A review of long fallows for dryland wheat production in southern Australia. J. Aust. Inst. Agric. Sci. 52:37–44.

Riley, M.M. 1987. Boron toxicity in barley. J. Plant Nutrition 10:2109–2115

Ryan, S., M. Abuzar, M. Imhof, and P. Rampant, 2001. The use of multi-sourced remote sensing data for mapping soils and key soil properties in Victoria. p. 660–669. *In* Proc. Geospatial information and Agriculture. 5th Annual Symp., Precision Agriculture in Australasia, Sydney.

Sadras, V., D. Roget, and G. O'Leary. 2002. On farm assessment of environmental and management constraints to wheat yield and efficiency in the use of rainfall in the Mallee. Aust. J.Agric. Res. 53:587–598.

Stephens, D.J. 2002. National and regional assessments of crop yield trends and relative production efficiency. Final Report to the National Land and Water Resources Audit, Misc. Publ. 12/2002. Western Australian Dep. of Agric.

Whelan, B.M., and A. McBratney. 2003. Definition and interpretation of potential management zones in Australia. *In* M. Unkovich (ed.) Proc. of the 11th Australian Agron. Conf., Geelong, Australia. 2–6 Feb. 2003. Conference Design Pty Ltd., Hobart. Available at www.regional.org.au/au/asa/2003 (verified 7 Jan. 2004).

24 Impacts of Policies and Technologies in Dryland Agriculture: Evidence from Northern Ethiopia

John Pender[1]

International Food Policy Research Institute
Washington, DC

Berhanu Gebremedhin[1]

International Livestock Research Institute
Addis Ababa, Ethiopia

ABSTRACT

This chapter investigates the land management practices used in the highlands of Tigray, northern Ethiopia, the factors influencing them, and their implications for crop production. Several factors commonly hypothesized to have a major impact on land management and agricultural production—including population pressure, small landholdings, access to roads and irrigation, and extension and credit programs—are found to have limited direct impact on crop production, though most affect the intensity of production. The increase in farming intensity due to these factors has limited impact on total crop production due to low marginal product of labor in crop production, limited productivity impact of inputs such as fertilizer in the moisture-stressed environment of Tigray, and limited adoption of such inputs. We find that profitable opportunities exist to increase agricultural production and achieve more sustainable land management in the highlands of Tigray. These opportunities include improvement of crop production using low-external input investments and practices such as stone terraces, tree planting, manuring, reduced tillage, and reduced burning; and improved livestock management. The comparative advantage of people in the Tigray highlands is not in input-intensive cereal crop production but more in such low-input approaches and in alternative livelihood activities. As a result, greater emphasis on developing these alternatives in agricultural extension and other development programs may be fruitful. Food crop production should not be ignored in the development strategy, but less promotion of purchased inputs such as fertilizer and improved seeds and greater emphasis on profitable alternatives would be helpful.

[1] The authors are grateful to the Swiss Agency for Development and Cooperation, which provided financial support to this research, to Mekelle University for partnership and institutional support to the research, and especially to the many farmers who participated in the survey. Any errors or omissions are solely the responsibility of the authors.

INTRODUCTION

Low agricultural productivity, poverty, and land degradation are critical and closely related problems in the Ethiopian highlands. These problems are particularly severe in the highlands of Tigray in northern Ethiopia. Cereal yields average $<1\ t\ ha^{-1}$ in this region and over half of the area of the Tigray highlands has been characterized as severely degraded (Hurni, 1988). The average farm size is only 1 ha and most households subsist on incomes of less than one dollar per day.

In recognition of these problems, the regional government of Tigray has undertaken a massive program of investment and resource conservation since 1991. The regional development strategy of Conservation-based Agricultural Development-Led Industrialization has focused on promoting conservation of natural resources and improvement of agricultural productivity and welfare through a broad program of investment in infrastructure, agricultural extension, education, and other services. Empirical evidence of the impacts of these policies, and identification of specific areas where problems need to be addressed, is needed. Addressing this information need is the primary objective of this study.

This study is based upon a household and plot-level survey conducted in 100 villages in 50 tabias (the lowest administrative unit in Tigray) in the highlands of Tigray during 1999/2000.[2] It builds upon a prior study based upon tabia and village level surveys in the same communities in 1998/1999. This broad sample and the information collected at different levels enable investigation of the impacts of community level factors such as population density, investments in irrigation and roads, as well as household and plot-level factors such as household wealth, education, land tenure, and other factors on land management and the implications for agricultural productivity and land degradation.

CONCEPTUAL FRAMEWORK, RESEARCH HYPOTHESES, AND METHODOLOGY

There are many factors potentially affecting farmers' decisions about land management, and a complex set of linkages between government policies and these decisions. The impacts of government policies and programs on land management may vary greatly from one location to another, depending upon how such policies and programs are implemented in different locations, local agro-climatic conditions, the extent of local market development, farmers' endowments, and other local conditioning factors. Furthermore, the impacts of farmers' decisions may also vary substantially from place to place depending upon such conditioning factors. It is essential to account for this complexity and diversity if policy makers are to understand the implications of the decisions they make on land management at the local level. The conceptual framework guiding this research has been developed to address these challenges.

Conceptual Framework

[2] Highlands were defined to include areas at or above 1500 m above sea level.

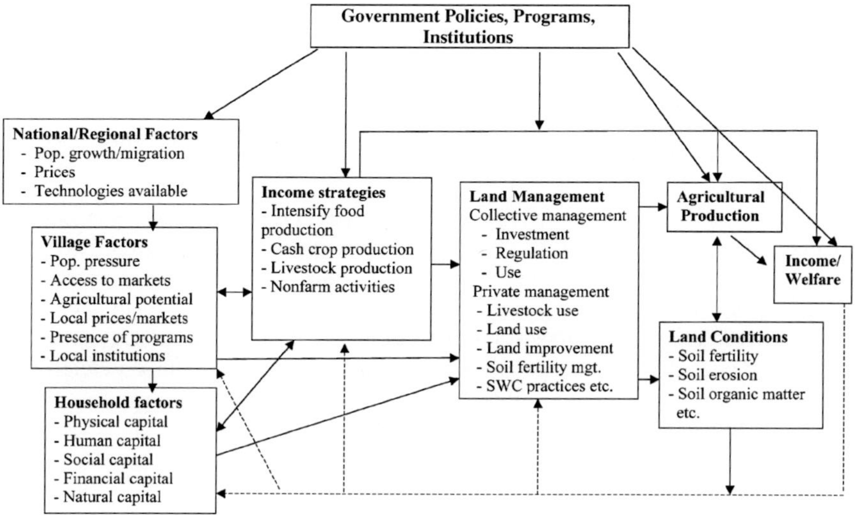

Fig. 24–1. Factors affecting income strategies, land management, and their implications.

The conceptual framework is illustrated in Fig. 24–1. Land management is determined by private decisions made at the farm household level, as well as by collective decisions made at the village or higher levels. For example, farm households choose what crops to plant and how to manage soil fertility or conserve soil and water on their own land; but these decisions may be affected by regulations on land use set by local councils. Communities may also regulate use of common lands, or may make collective investments in improving such resources, such as planting improved grasses or trees. This chapter focuses on determinants and impacts of farmers' private land management decisions.[3]

These household and collective decisions will determine current agricultural productivity and affect the condition of land resources (thus influencing future agricultural productivity), which in turn affect the level of farm income and rural poverty. It is important to emphasize that it is such outcomes (productivity, resource conditions, and household incomes), and not adoption of specific land management practices per se, that are of most concern to rural people and policy makers. It is thus critical to consider the ultimate impacts of any policy or technology on these outcomes, and the extent to which there may be trade-offs or complementarities among these objectives. This chapter focuses on impacts of land management on crop production and soil erosion.[4]

Land management decisions are determined by many factors operating at different scales (plot, household, village, region, nation, and international). Many of these factors influence land management directly; for example, the type of soil, topography of the land, and the climate will have a large impact on whether soil erosion is likely to be a problem and what options are feasible to address it. Demo-

[3] Gebremedhin et al. (2002, 2003) present results of research on collective action in natural resource management in Tigray.

[4] Due to space limitations, other outcomes, such as impacts on ousehold incomes or other resource conditions, are not presented in the chapter. More results are available in Pender et al. (2002).

graphic and socioeconomic factors—such as population density, access to markets, and the level of local prices—also influence land management. Some of these effects are direct; for example, access to markets and local prices determine the profitability of alternative practices. But some effects are indirect. For example, population pressure leads to smaller farm sizes and often to more fragmented holdings, which may reduce farmers' ability or incentive to fallow or to invest in land improvements.

The income strategies and land management practices used in a particular location may be influenced by many village level factors, such as agricultural potential, access to markets, population density, and presence of government programs and organizations.[5] These factors largely determine the comparative advantage of a location by determining the costs and risks of producing different commodities, the costs and constraints to marketing, and the opportunities and returns to alternative activities, such as farming vs. nonfarm employment. Household level factors such as households' endowments of physical capital (e.g., land, livestock), "human capital" (education, training, farming experience, household size, and composition), "social capital" (value of relationships, as reflected by participation in community organizations, leadership in community), "financial capital" (access to credit, savings), and "natural capital" (land quality, access to other resources) may also influence the livelihood strategy or land management practices chosen by particular households.

Government policies and programs may influence income strategies and land management and their implications at many levels. Macroeconomic and market liberalization policies will affect the relative prices of commodities and inputs in general. Agricultural research policies affect the types of technologies available to farmers in a particular agro-ecological region. Infrastructure development, agricultural extension, conservation technical assistance programs, land tenure policies, and rural credit and savings programs affect awareness, opportunities, or constraints at a village and household level. Programs may attempt to promote land management approaches directly, for example by promoting particular land management practices, or indirectly, by promoting particular livelihoods.

Research Hypotheses

A very large number of hypotheses could be considered concerning the linkages among the many causes and responses in the conceptual framework of Fig. 24–1. In this study we focus on the impacts of population pressure; access to roads and markets; irrigation, technical assistance, and credit programs; household endowments of physical, human, and social capital; and land tenure. We briefly discuss our hypotheses about these factors below.

Impacts of Population Pressure

Population pressure may induce expansion of production into more marginal and fragile lands, particularly where unsettled land is available. Where land is lim-

[5] Due to space limitations, this chapter does not include results of empirical work investigating determinants of income strategies. Such results are reported in Pender et al. (2002).

ited, intensification of labor per unit of land is expected to occur, including shortening of fallow cycles and adoption of more labor intensive products and practices (Boserup, 1965). Increased labor intensity may promote greater use of other inputs (such as fertilizer and seeds) if those are complementary to labor, but less use of inputs that are substitutes for labor (e.g., herbicides). Population growth may induce labor-intensive land improvements where land tenure is secure (Scherr and Hazell 1994; Tiffen et al., 1994). These changes may increase land productivity but, holding the state of technology and market development constant, may reduce labor productivity (Salehi-Isfahani, 1988) and per capita income (Pender, 1998).

Access to Roads and Markets

Increases in the profitability of agricultural products, whether resulting from infrastructure investment, market development, changes in market prices, technological innovation, or government policies affecting these factors, will promote expansion of agriculture into marginal areas if the costs of productive factors are unaffected by the change (Angelsen, 1999). If expansion of agricultural land is limited, increased profitability will cause intensification of labor and/or capital per unit of land, though the effects on capital relative to labor depend on whether capital and labor are complements and the nature of factor markets. Market or technology development may promote a shift to cash crops, and will tend to increase farm incomes (unless offset by falling farm prices). The implications for land management and degradation may be mixed. For example, changes in commodity prices have a theoretically ambiguous effect on soil conservation investments (LaFrance, 1992; Pagiola, 1996).

Irrigation

Irrigation is expected to increase the intensity of crop production by enabling production of multiple crops per year and increasing the return to use of inputs by relieving moisture constraints (Fitsum Hagos et al., 1999). This is expected to lead to higher yields, both by increased input intensity and increasing moisture availability. In areas with good access to markets, irrigation may enable production of high value perishable crops such as vegetables or fruits. Irrigation may be complementary to some forms of soil conservation investment, such as drainage ditches; but it may also displace some kinds of investment, such as stone terraces, if it reduces the need to conserve soil moisture. By increasing the value of land relative to labor, irrigation may promote greater effort to conserve land.

Technical Assistance and Credit Programs

Technical assistance and credit programs can have varied effects depending upon what technologies they promote and the terms and availability of credit. In Tigray, as elsewhere in the Ethiopian highlands, the agricultural extension program has strongly promoted increased use of inputs such as fertilizer and seeds, and has made credit available to help finance use of these inputs. Credit is provided in kind and repaid at harvest. We expect that these programs have contributed to increased use of fertilizer and seeds, and that this has contributed to increased crop production. Given the limited and uncertain rainfall and thin soils in Tigray, however, the yield response to these inputs may be limited. If such inputs are applied in combination with efforts to increase soil moisture availability, we expect the yield response to be greater. Thus, we expect farmers to be more likely to use these inputs where they have made investments in soil and water conservation measures such as stone terraces and soil bunds, or have access to irrigation.

Household Endowments of Physical, Human, and Social Capital

If markets for productive factors such as land and labor do not function fully efficiently, then there may be significant differences among households in their land management practices and agricultural productivity, as a result in differences in household endowments of these factors (de Janvry et al., 1991). For example, if hired labor is costly to monitor, households with a greater endowment of labor per unit of land may be able to farm the land more intensively and more able to conduct critical operations at the right time than other households, and thus may obtain higher yields than others. Similarly, households with more oxen may be able to use more oxen power and thus able to attain higher yields.

Farmers who have better access to land or other physical or financial assets may be more able to finance purchase of inputs such as fertilizer and seeds, using their savings or through better access to credit. However, to the extent that credit programs use collateral substitutes such as peer group monitoring (as used by the dominant credit program in Tigray), credit constraints may not be a problem.

Households with more education or other forms of human capital may have greater access to nonfarm income or be regarded as better credit risks, and thus be more able to purchase inputs. They may also be more aware of the benefits of modern technologies. On the other hand, more educated households may be less likely to invest in labor-intensive land investments and management practices, since they may be able to earn higher returns to their labor and capital if used in other activities.

Households with greater "social capital" (e.g., those with more social relationships, possibly through involvement in local organizations), may have better access to credit, and thus be more able to finance purchases of inputs or land investments. Involvement in agriculturally oriented organizations such as agricultural cadres may increase farmers' awareness of new technologies and management practices, and increase their productivity as a result. Social capital may also increase farmers' ability to earn high returns from marketing their products (e.g., membership in marketing cooperatives), which can affect land management in the same way as improvement in market access. On the other hand, as with human capital, some forms of social capital may also provide households higher returns to investing their labor and capital in nonfarm activities. For example, involvement of women in a women's group that promotes income diversification may reduce the intensity of the household's farming activities.

Land Tenure

The form of tenure on a plot of land can affect land management and productivity for several reasons. If there is insecurity of tenure, the household operating the plot may have less incentive to invest in land improvement (Feder et al., 1988). This is not necessarily the case, however, if the household can increase tenure security by investing (Besley, 1995; Otsuka and Place, 2001), in which case, there may be more investment on plots having insecure tenure. Tenure security and transferable land rights can increase farmers' access to credit, where land is used as collateral for credit (Feder et al., 1988; Pender and Kerr, 1999). In Ethiopia, however, land rights are not transferable by sale or mortgage.

The ability to transfer land through temporary arrangements such as cash rental, sharecropping, and borrowing (which are allowed in Tigray) can help households who own little or no land to overcome land constraints. To the extent such markets operate efficiently, land ownership will not affect the land management or yields, since land will be leased in or out to equalize productivity, as mentioned above. However, cash constraints or risk considerations may prevent tenants from being able or interested in using cash rental, and landowners are unlikely to lend land to people other than relatives or close friends. Thus sharecropping may be a dominant form of land tenancy. Since sharecroppers receive only a portion of the output, this reduces their incentive to provide inputs and can thus lead to lower input intensity and lower yields on sharecropped plots (Shaban, 1987; Otsuka and Hayami, 1988). However, if the landowner can readily monitor the tenant's use of inputs, or there is mutual trust between the landowner and tenant, the tenant may still provide the efficient level of inputs and achieve efficient production, if lease markets are not restricted by government policies (Otsuka and Hayami, 1988; Pender and Fafchamps, 2001). But the duration of land lease contracts is restricted to be no more than 2 yr in Tigray, and this may undermine the efficient functioning of the lease market (Gebremedhin et al., 2002).

Methods

Data Sources

This study is based on a survey of 500 households in 100 villages (kushets) in 50 communities (tabias) in the highlands of Tigray conducted in 1999 and 2000. Tabias below 1500 m.above sea level elevation were excluding from the sample frame. A random sample of tabias was used, stratified by distance to the woreda (district) town and whether an irrigation project was present in the tabia. Two villages were randomly selected within each sample tabia, and five households were randomly selected from each village. In addition to household-level information, information was collected on all plots owned or operated by the respondent households. The survey data were supplemented by data from tabia and village surveys on prices and other factors, secondary data from the 1994 Population Census on the population of each tabia, and maps of the boundaries of each tabia (used to calculate population density).

Econometric Approach

The dependent variables analyzed in this study include the use of inputs on each plot in 1998 (labor, draft animal power, seeds, fertilizer, manure, or compost), adoption of the most common land management practices in 1998 (burning to clear the plot, contour plowing, reduced tillage, intercropping, or mixed cropping), and the value of crop production on the plot. The econometric model used depends on the nature of the dependent variable. For use of labor, oxen-power and seeds on cultivated plots, and value of crop production, least squares regression was used. For explaining whether various land management practices were used, a probit model was used.

The explanatory variables include indicators of agricultural potential (average rainfall and altitude); population density; access to roads, transportation, and markets (walking time to nearest all-weather and seasonal road, bus service, woreda (district) town, and input supply shop); wealth (land and livestock owned); human capital (gender, age and education of household head, and household size); income strategy (primary and secondary income source); ownership of a radio (a determinant of access to information); savings; household social capital (membership in various organizations); use of formal or informal credit; contact of the household with the agricultural extension program; and various plot-level factors, including the land use (whether homestead, rainfed, or irrigated cultivated[6]), land tenure, presence of investments on the plot (stone terrace, soil bund, trees, fence, live barrier) and several indicators of different aspects of quality of the plot (size of plot, quality rank relative to the farmers' other plots, plot slope, position on slope, soil depth, color, texture, fertility, presence of gullies, and whether water logging was a problem). We controlled for these factors to reduce potential for omitted variable bias.

In the crop production regression and input use regressions, we used a logarithmic Cobb-Douglas specification. This leads to a theoretically consistent specification for output and input demands, and reduces problems due to outliers and nonnormality of the error term found when using a linear specification. In the structural crop production regression, we included the use of inputs and land management practices but excluded land tenure and household level factors, which are expected to influence production only via their impacts on land management. Since inputs and land management practices are endogenous choice variables, this may cause a bias in the results. We also ran a version of the models using predicted values of the inputs and management practices, since this avoids any endogeneity bias, and report the robustness of the results.[7] Finally, we estimated a reduced-form specification that excludes the use of inputs and management practices as explanatory variables. This specification eliminates the potential for endogeneity bias, and allows us to investigate the total effect of the explanatory variables on crop production, whether these are direct effects on production or via the impacts of the explanatory variables on input use and management practices. Examining all of these

[6] Pasture, woodlots, and fallow plots were excluded from the analysis.

[7] For all relevant regressions we also report robustness of our results to use of predicted choices of income strategies, membership in organizations, use of credit and participation in the extension program, or to excluding these in the reduced form regressions, since these may also be subject to endogeneity bias. The exogenous variables used to predict these choices include biophysical conditions (annual rainfall and altitude), population density, human capital (sex, age, and education of household head, household size), and land ownership. Unfortunately, it is difficult to identify instrumental variables that affect these choices that can be excluded as determinants of land management. Thus, the regressions with predicted values of these variables are not well identified, causing high multicollinearity (maximum variance inflation factor = 100, average vif = 10), and limiting the statistical confidence in these results.

specifications together provides a greater degree of confidence in the robustness of the results.

We included interaction terms between fertilizer use and presence of a stone terrace, a soil bund, or irrigation, to test whether there is complementarity between fertilizer use and these investments, due to the expected impact of these investments on soil moisture availability.

We tested the regression specifications for problems of multicollinearity, but found this not to be a serious problem in any of the specifications.[8] Various regression diagnostics were used to identify outliers and influential observations. Some data errors were discovered and corrected as a result of these tests.

AGRICULTURE AND LAND MANAGEMENT IN THE HIGHLANDS OF TIGRAY

Biophysical and Socioeconomic Conditions

The average annual rainfall is generally <1000 mm in the highlands of Tigray, and averages about 650 mm for all sample households (Table 24–1). Altitude in the highlands ranges from 1500 m above sea level to well over 3000 m above sea level, averaging 2174 m above sea level for the sample households.

The rural population is growing rapidly and population pressure is high in the Tigray highlands, with average population density of 137 persons per square kilometer in the sample communities. As a result, the average farm size in the Tigray highlands is only 1 ha (4 tsimad). One percent of the sample households were landless, and the maximum farm size was about 4 ha. Almost all households own livestock; with cattle (*Bos taurus*) most important (in value terms), followed by sheep (*Ovis aries*) and goats (*Capra hircus*).

Access to roads, transportation, and other services has improved substantially in Tigray since 1991. Nevertheless, most households are still far from roads, transportation services, and markets. In 1998, the average walking time to the nearest all weather road was more than 2 h, while other average walking times were 1 h to the nearest seasonal road, 4 h to the nearest bus service, and 3.5 h to the nearest woreda town.

Education has improved dramatically in Tigray since 1991 as a result of greatly increased number of schools and literacy campaigns. Still, only about 15% of household heads have had formal schooling (only 6% more than 2 yr), while 7% have participated in a literacy campaign.

The availability of agricultural extension and credit services has also greatly expanded. Nearly three-fifths of households had access to credit from formal sources, such as DEDEBIT, in 1998. Development agents of the extension service were involved in virtually every community, but only about 11% of households had direct contact with an extension agent.

[8] The maximum variance inflation factor was <5 in all cases (except when using predicted livelihoods, membership in organizations, use of credit and participation in extension, as noted in the previous footnote).

Table 24–1. Descriptive statistics of data.

Variable	Number of observations	Mean (standard error)
Secondary income source in 1998, % of households		
No secondary source	496	20.6 (2.2)
Cereals	496	2.9 (0.9)
Perishable annuals	496	3.3 (1.1)
Perennial crops	496	3.6 (1.2)
Cattle	496	35.0 (0.9)
Small ruminants	496	2.2 (0.9)
Beekeeping	496	0.4 (0.3)
Food for work	496	6.1 (1.1)
Salary employment	496	1.6 (0.6)
Farm employment	496	1.4 (0.6)
Trading	496	6.5 (1.3)
Food/other assistance	496	6.5 (1.3)
Other nonfarm	496	10.0 (1.7)
Land investments existing on plots, % of plots		
Stone terrace	1785	36.5 (1.9)
Soil bund	1785	8.3 (1.0)
Trees planted	1785	4.8 (0.9)
Constructed fence	1785	3.8 (0.7)
Live fence or barrier	1785	3.1 (0.6)
Use of inputs		
Labor, person-days ha^{-1}	1785	86.4 (9.5)
Oxen power, animal-days ha^{-1}	1785	25.3 (1.9)
Seed, kg ha^{-1}	1785	118.1 (7.7)
Improved seed, % of plots	1785	2.4 (0.5)
Fertilizer, % of plots	1785	27.0 (1.6)
Use of land management practices, % of plots		
Burning to prepare field	1785	11.0 (1.3)
Contour plowing	1785	87.5 (1.6)
Reduced tillage	1785	12.3 (1.2)
Intercropping/mixed cropping	1785	11.4 (1.2)
Manure or compost	1785	22.8 (1.3)
Value of crop production, EB† ha^{-1}	1593	1816 (176)
Annual rainfall, mm	480	652 (5)
Altitude, m above sea level	500	2174 (22)
Population density, km^{-2}	490	136.8 (4.4)
Female head of household	500	21.8 (2.2)
Age of household head, yr	500	46.0 (0.7)
Household size, no.	500	5.4 (0.1)
Education of household head, % of households		
1 to 2 yr	500	9.2 (1.6)
3+ yr	500	6.1 (1.3)
Literacy campaign	500	7.3 (1.5)
Walking time to nearest, h		
All weather road	496	2.33 (0.13)
Seasonal road	486	0.99 (0.12)
Bus service	497	4.11 (0.17)
Woreda town	497	3.54 (0.17)
Land owned, tsimad	477	4.20 (0.25)
Livestock owned, EB		
Total stock	500	1849 (86)
Cattle	500	1333 (62)
Sheep	500	154 (19)
Goats	500	141 (22)
Chickens	500	31.1 (1.8)
Beehives	500	58.7 (7.0)

(continued on next page)

Table 24–1. Continued.

Variable	Number of observations	Mean (standard error)
Membership in orgs.,% of households		
Farmers' association	500	64.7 (2.7)
Women's association	500	86.0 (1.9)
Youth association	500	32.3 (2.6)
Tabia baito	500	1.6 (0.7)
Kushet baito	500	2.0 (0.9)
Marketing coop.	500	6.4 (1.2)
Agricultural cadre	500	1.6 (0.7)
Use of credit, % of households		
Formal credit	500	57.7 (2.7)
Informal credit	500	18.9 (2.1)
Contact with extension, % of households	500	11.4 (1.7)
How plot acquired, % of plots		
Leased in	1785	13.7 (1.3)
Allocated by tabia	1785	84.0 (1.4)
Inherited	1785	1.4 (0.5)
Received as gift/other	1785	0.9 (0.3)
Plot slope, % of plots		
Flat	1779	57.8 (2.0)
Gentle	1779	32.3 (2.0)
Steep	1779	9.9 (1.4)
Position on slope, % of plots		
Top	1785	13.1 (1.4)
Middle	1785	21.1 (1.6)
Bottom	1785	28.0 (2.2)
Not on slope	1785	37.9 (2.2)
Relative plot quality rank, rank/total no. of plots	1754	0.571 (0.005)
Plot area, ha	1508	0.298 (0.013)
Soil depth, % of plots		
Deep	1767	21.9 (1.4)
Medium	1767	38.1 (1.7)
Shallow	1767	40.0 (1.8)
Soil color, % of plots		
Black	1767	27.8 (2.2)
Brown	1767	12.4 (1.1)
Grey	1767	22.9 (1.8)
Red	1767	36.8 (2.0)
Soil texture, % of plots		
Clay	1767	27.6 (2.2)
Loam	1767	31.7 (2.0)
Sand	1767	29.7 (1.8)
Silt	1767	11.1 (1.3)
Soil fertility, % of plots		
High fertility	1767	13.1 (1.1)
Moderate fertility	1767	61.1 (1.7)
Infertile	1767	25.8 (1.6)
Perceived erosion problem, % of plots		
None	1768	66.0 (2.0)
Moderate	1768	27.0 (1.9)
Severe	1768	7.0 (1.1)
Waterlogging problem, % of plots	1785	10.5 (1.1)
Gullies on plot, % of plots	1785	5.1 (0.7)
Land use, % of plots		
Homestead	1785	19.7 (1.0)
Rainfed cultivated	1785	73.7 (1.1)
Irrigated cultivated	1785	6.6 (1.0)

† EB = Ethiopian Birr.

Most households are involved in local farmers' associations and women's associations. About one-third of households also have members of a youth association. About 6% of households have members in a marketing cooperative, which are involved in marketing agricultural (mainly crop) outputs and providing inputs. About 2% of households have a member in an agricultural cadre, which focus on improving agricultural production. A similar small proportion of households are involved as community leaders in the local tabia or kushet (village) baito (council).

Income Strategies

The dominant source of income in the highlands of Tigray is cereal crop production, which is the primary source of income for 97% of households. Different income strategies are thus distinguished more by the secondary source of income. In about one-third of households, the secondary source of income is cattle production. One-fifth of households have no secondary source of income; cereal crop production is their sole income source. Production of other crops, including perishable annual crops or perennial crops, are the secondary income source of about 7% of households. Small ruminants are important for about 2% of households, while beekeeping for only about 0.4%. Farm wage labor is the secondary source for only 0.6% of households. Nonfarm activities are important for about one-fourth of households: trading activities are the secondary income source for 6.5% of households, food for work for 6%, salary employment for 1.6%, and other nonfarm activities (such as handicrafts, brewing beer, priest, local officials, etc.) for 10%. More than 6% of households had food aid, assistance from relatives, and other forms of assistance as their secondary source of income.

Land Management

The most common investments in land improvement in Tigray are stone terraces and soil bunds. Stone terraces existed on nearly 37% of cultivated plots in 1998, while soil bunds existed on about 8%. Other less common investments included planting trees or a live barrier, constructing a fence, and constructing irrigation canals and drainage ditches.

Preharvest labor use in crop production averaged 86 person-days per ha., most of this for plowing, planting, and weeding. Draft animal use (mainly oxen) averaged 25 animal days per ha. Seed use averages 118 kg. per ha. Fertilizer was used on 27% of plots and manure or compost used on about 20% of plots in 1998. Improved seeds were used on only about 2% of plots.

Several land management practices are commonly used in Tigray, including contour plowing, burning to prepare fields, reduced tillage, and intercropping or mixed cropping.[9] Contour plowing is very common, practiced on nearly 90% of plots.

[9] We investigated other land management practices such as mulching, use of green manures, cover crops, etc., but these were found to be rare.

Crop Production

The average estimated value of crop production on surveyed plots was 1816 Ethiopian Birr (EB) ha^{-1} in 1998.[10] The average value of production was significantly higher on plots where inorganic fertilizer was applied (2184 EB ha^{-1}) than where no fertilizer was applied (1684 EB/ha). The average value of production was substantially higher on irrigated field plots (6726 EB ha^{-1}) than on homestead plots (1838 EB ha^{-1}) or rainfed field plots (1428 EB ha^{-1}). These figures are the total value of production during the year, including multiple crops, which is why the irrigated production value was so much higher. These differences may be due to other factors besides fertilizer use or irrigation (such as differences in cropland quality); multivariate analysis is needed to control for such factors.

Land Quality

About 10% of the plots being cultivated are on "steep" slopes (in the perception of the farmer), and another third are on "gentle" slopes. Farmers consider about one-fourth of the plots being cultivated to be infertile, while about three-fifths are considered moderately fertile. Thirty percent of the plots have sandy soil, and 40% are considered by farmers to have shallow soils. Farmers consider soil erosion to be a severe problem on 7% of plots, and a moderate problem on 27% of plots. Gullies exist on about 5% of plots.

RESULTS OF ECONOMETRIC ANALYSIS

We focus our discussion on impacts of explanatory variables about which we present hypotheses, though other variables were also included in the regressions.

Input Use in Crop Production

Population pressure is associated with higher use of labor and animal draft power per hectare, and with a higher probability of use of fertilizer, manure, or compost (Table 24–2). We also find that households that own more land are less likely to apply fertilizer to a particular plot. These findings support the Boserup (1965) hypothesis that population pressure causes farmers to intensify use of labor and other inputs.

Access to roads and markets also affects the intensity of input use, though the effects are mixed. Households closer to an all-weather road use more labor per hectare and are more likely to use fertilizer. Households closer to a seasonal road use less draft animal power but more manure or compost. Households closer to a woreda town use more draft animal power per hectare but are less likely to use fer-

[10] The official exchange rate averaged about 7 EB per U.S. dollar in 1998. To compute the value of production, we used average prices in Tigray based upon the community and household level surveys. We were not able to compute value of production using local prices, due to limited number of observations for many crops. Thus, the data represent a weighted production index, where regional prices are used to weight production of different crops, and do not reflect local variation in prices.

tilizer. Households closer to an input supply shop are less likely to use manure or compost. Farmers are more likely to use fertilizer, manure, or compost closer to their residence, probably because of the difficulty of transporting bulky inputs (especially manure and compost) over long distances.

Table 24–2. Determinants of input use in crop production in 1998.

Variable	Labor ln (person-days ha^{-1})†	Oxen ln (animal-days ha–1)†	Seeds ln (kg ha^{-1})†	Fertilizer (whether used)‡	Manure or compost‡ (whether used)
ln (Annual rainfall), mm	–0.6237**	–0.4824**$^{--}$§	–2.4797**$^{-}$	–0.1715*	–0.3704**$^{-}$
ln(Altitude),					
m above sea level	–0.2691	–0.2751^{-}	1.7989**$^{+}$	–0.2818**	–0.0808
ln(Population density), km^{-2}	0.1453*	0.1656**	0.1438	0.0877*	0.0613*$^{+}$
Female head of household	–0.3672**	–0.1124^{-}	0.3300**	–0.0512	–0.1071**
ln(Age of household head), yr	0.2924**	0.0436	0.0421	0.0218	–0.0413
ln(Household size), no.	0.1612*	0.0795	0.1174	–0.0303	0.0179
Education of household head					
1 to 2 yr	0.0400	0.1189*	–0.2977*	–0.1222**	–0.0439
3^{+} yr	0.3743**	0.0251	0.2174	–0.0629	0.0140
Literacy campaign	–0.0095^{-}	0.1307	–0.0734	–0.0417	0.0334
Walking time to, h					
All weather road	–0.0617**	–0.0234	–0.0022	–0.0420**	–0.0083
Seasonal road	0.0074	0.1368**	0.1092	–0.0490^{-}	–0.0451*
Bus service	–0.0118	–0.0141^{-}	–0.0162	–0.0043	–0.0030
Woreda town	0.0038	–0.0658**	0.0084^{+}	0.0209*	0.0042^{-}
Input supply shop	0.0132	0.0155	–0.0545	–0.0180	0.0273*$^{++}$
Plot from residence	–0.1293	0.0003	0.0603	–0.1105*$^{--}$	–0.3413**$^{--}$
Ownership of assets					
Land, tsimad)	0.0030	–0.0051^{--}	–0.0079^{++}	–0.0118*	–0.0010^{-}
Oxen, no.	0.0694^{++}	0.0790*$^{++}$	0.0216	0.0012	0.0336*
Other cattle, no.	0.0120	0.0038	0.0322**$^{++}$	0.0148**$^{+}$	0.0051
Small ruminants, no.	–0.0093**$^{--}$	–0.0035	–0.0171**$^{--}$	–0.0026	0.0008
Pack animals, no.	0.0484	–0.0148	0.0573	–0.0170	–0.0042
Camels, no.	–0.0327	0.0074	–0.0240	–0.0229	–0.0178*$^{-}$
Radio, yes/no	0.0451	–0.0943^{-}	0.1193	0.0323	0.0008
Cash savings, yes/no	0.0996	–0.0428	0.1174^{+}	0.0832**$^{++}$	–0.0593*$^{-}$
Secondary income source					
Cereals	–0.0558	–0.3015	0.0884^{-}	0.0362	–0.0840
Perishable annuals	0.5282*	–0.0652	0.0907	0.0940	–0.0032
Perennial crops	0.1900	–0.0426	–0.0929	0.0979	–0.0585
Cattle	–0.2092**	–0.1537	0.1266	–0.0485	–0.0385^{+}
Small ruminants	–0.3367	–0.2233	0.5435	–0.0227^{-}	–0.0727
Beekeeping	–0.3331*$^{-}$	–0.2734*	–0.4115*	–0.1031*	–0.0812*
Food for work	–0.4375**	–0.3206**$^{-}$	0.3136*$^{++}$	–0.0209^{+}	–0.0797*
Salary employment	0.1385	–0.0334	0.2703*	–0.0292	–0.0753
Farm employment	–0.1909	–0.4684*	–0.0562	–0.1101	0.0605
Trading	–0.0257	–0.1156	–0.2177	0.0849	–0.0442
Food/other assistance	–0.2500	–0.2174*	0.4088*	–0.0914	–0.0622
Other nonfarm	0.0608	0.0058	0.1302	–0.0050	–0.0520
Membership in orgs.					
Farmers' association	–0.0145	0.0331	0.1927	0.0187	–0.0884*
Women's association	–0.0242	–0.1210*	–0.2367*	0.0247	–0.0037^{-}
Youth association	0.1443**	0.0923^{--}	0.0461	0.0092	–0.0030

(continued on next page)

Table 24–2. Continued.

Variable	Labor ln (person-days ha^{-1})†	Oxen ln (animal-days ha–1)†	Seeds ln (kg ha^{-1})†	Fertilizer (whether used)‡	Manure or compost‡ (whether used)
Tabia baito	–0.0849	–0.1286	–0.3164	–0.0472	–0.0030
Kushet baito	0.6429**$^{++}$	0.3412*	0.4652**	0.1821	0.3152**$^{++}$
Marketing coop.	–0.1071	–0.0155^{--}	0.2382^{+}	0.0198	–0.0093
Agricultural cadre	–0.3075**	0.1042	–0.6919**	0.2418	–0.0826^{--}
Use of credit					
Formal credit	0.0266	–0.0212	0.1825*	0.1924**	0.0064
Informal credit	0.0550	–0.0403	–0.0461	0.0370	–0.0307
Contact with extension	–0.1317	–0.1087^{--}	–0.0892	–0.0472	–0.0151
How plot acquired, cf. not owned					
Allocated by tabia	0.1293*$^{++}$	–0.0159	0.0932	0.0198	0.0175
Inherited	0.3268^{+}	0.1433	0.3980*	0.2892*	0.1315*$^{+}$
Received as gift/other	0.1245	–0.1981	–0.1737	–0.1624	0.3650*$^{+}$
ln(Plot area), ha	–0.5159**$^{--}$	–0.2308**$^{--}$	–0.4659**$^{--}$	0.0808**$^{++}$	0.0287*
Relative plot quality rank	–0.1862	–0.0962	–0.4235**$^{--}$	–0.1305*$^{-}$	–0.2558**$^{--}$
Plot slope, cf. flat					
Gentle	–0.1057	0.0439	–0.0280	–0.0147	–0.0374
Steep	–0.0905	0.0482	–0.0433	–0.0137	–0.0472
Soil depth, cf. deep					
Medium	0.1006	0.0681	0.1381	–0.0033	–0.0498
Shallow	0.1044	0.0194	0.2483*$^{+}$	0.0571	0.0317
Soil fertility, cf. high fertility					
Moderate fertility	–0.2107*$^{-}$	–0.1295^{-}	–0.0217	–0.0547	0.0304
Infertile	–0.3015**$^{--}$	–0.1212	0.0050	–0.0693	0.0508
Waterlogging problem, 1 = yes	–0.1292^{-}	–0.0330	0.1348^{+}	0.0616	0.0060
Gullies on plot, 1 = yes	–0.2942	–0.0637	–0.1258	0.0262	0.0220
Land use, cf. homestead plot					
Rainfed cultivated	–0.4140**$^{--}$	–0.1833**$^{--}$	–0.2096**$^{-}$	–0.0012	–0.3998**$^{--}$
Irrigated cultivated	0.4088*$^{+}$	0.1266	–0.2432	0.1881^{+}	–0.0921*
Initial investment on plot in 1998					
Stone terrace	0.0318	0.0129	0.0465	0.0761**$^{++}$	0.0326
Soil bund	0.0359	0.0770	0.0677	0.0647	–0.0104
Trees planted	0.2698*$^{++}$	–0.1390	–0.0647	–0.0399	–0.0607^{-}
Constructed fence	0.2961*	0.0217	0.0560	–0.0387	0.3111**$^{++}$
Live fence or barrier	0.1389^{+}	0.0426	0.1078	0.0581	0.0370
Intercept	7.6794**	7.4687**$^{++}$	4.7181	*	**
Number of observations	1368	1321	1401	1552	1552
Mean of dependent variable				0.270	0.245
Mean predicted probability of use	NA	NA	NA	0.269	0.246
R^2 or Pseudo R^2	0.553	0.375	0.516	0.235	0.482

*, **Mean coefficient statistically significant at 5% and 1% levels, respectively.

† Least squares regression. Coefficients and standard errors adjusted for sampling weights, clustering and stratification. Coefficients of plot position on slope, soil texture and color not reported to save space. (These results reported in Pender et al., 2002).

‡ Probit regression. Reported coefficients represent effect of a unit change in explanatory variable on probability of use of the mean of the data.

§ $^{+,++}$and $^{-,--}$Mean coefficient positive (negative) and statistically significant at 5% and 1% levels, respectively in regressions using predicted values of sources of income, participation in organizations, use of credit and contact with extension.

As expected, use of labor, oxen power, fertilizer, and manure or compost is much greater on irrigated plots than non-irrigated plots, due to producing multiple crops per year.

Ownership of livestock and other assets affects use of inputs. Households who own more oxen use more labor and oxen draft power per hectare (the effect on labor significant at 10% level), suggesting that imperfect markets for hiring oxen constrain households who own fewer oxen. Ownership of oxen is also associated with greater likelihood of applying manure or compost. Greater ownership of other types of cattle is associated with greater use of seeds and fertilizer, probably because income generated from cattle products helps farmers afford to buy these inputs. Consistent with this explanation, households with cash savings are more likely to use fertilizer, and less likely to apply manure or compost; suggesting that cash constraints limit use of fertilizer and cause farmers to use manure or compost as an alternative. By contrast, greater ownership of small ruminants is associated with less use of labor and seeds. This suggests that small ruminants producers focus less of their effort on crop production.

Human capital affects use of crop inputs. Female-headed households use significantly less labor but more seed per hectare, perhaps attempting to compensate for labor constraints by using more seeds. Female-headed households also are less likely to apply manure or compost, probably also due to labor constraints. Older and larger households use more labor, probably because of greater availability of family labor old enough to be involved in crop production. More educated farmers use more labor and oxen power, but use less seed and are less likely to use fertilizer. The positive impact of education on use of labor and oxen may be due to greater wealth of more educated households, enabling them to hire more labor and draft animal services.

Social capital also affects input use. Households having members of a women's association use less oxen power and seeds; probably because these households tend to rely more on income from other sources. Households with members of a youth association use more labor, probably because they have more labor of the appropriate age for crop production than other households. Households with members of a kushet baito (village council) appear to be very oriented towards intensive crop production, and use more labor, oxen power, and seeds per hectare, as well as being more likely to use manure or compost. By contrast, households with members of an agricultural cadre use less labor and seeds per hectare. These households appear to focus more on livestock production (Pender et al., 2002).

Not surprisingly, access to formal sector credit is strongly associated with greater use of seeds and fertilizer. This is because this credit is used primarily to purchase such crop inputs. Informal credit is not significantly associated with use of any crop inputs. Surprisingly, contact with the extension program is not associated with use of inputs. It appears that it is not the extension program per se that is leading to significant increases in use of fertilizer in Tigray, but rather availability of credit and other factors.

Labor inputs are greater on owner-operated plots (allocated by the tabia or inherited) than leased-in (mainly sharecropped) plots. We also find weak evidence (significant at the 10% level) that use of seeds, fertilizer, and manure or compost is greater on inherited than leased-in plots. Consistent with this, we find below that crop production is higher on inherited plots than leased-in plots. Use of manure or compost is also more likely on gift plots than leased-in plots. This evidence supports the hypothesis that share tenancy reduces the intensity of crop production and yields (Shaban, 1987).

Land Management Practices

Population pressure and smaller farm size are associated with greater use of intercropping/mixed cropping, and larger farms are less likely to adopt contour plowing but more likely to adopt reduced tillage (Table 24–3). These findings support the Boserup intensification hypothesis.

Better access to an all-weather road is associated with greater use of burning and contour plowing, while better access to a seasonal road is associated with greater use of reduced tillage. This is consistent with the finding noted earlier that use of draft animals is less closer to a road. Contour plowing is slightly more common further from a town. Use of manure and compost is more common further from an input supply shop, probably because of greater difficulty of obtaining fertilizer in such areas.

Greater ownership of oxen promotes contour plowing but reduces use of reduced tillage, as one would expect. Ownership of other cattle is positively associated with burning, while small ruminants are negatively associated with burning. The reason for the opposite effects of cattle vs. small ruminants is not clear. Ownership of pack animals is associated with less intercropping. Ownership of such animals may reflect lack of market access, and thus tend to reduce intensification.

Female-headed households are significantly less likely to adopt contour plowing, probably due to labor constraints as well as a cultural taboo against women plowing. Older households are more likely to use burning; this may be a reflection of greater adherence to traditional practices or labor constraints faced by older farmers. Larger households are not surprisingly less likely to use reduced tillage, since this is a less intensive use of their available family labor. Formal education has little impact on use of land management practices.

Burning was not practiced by any of the kushet baito members in the sample, perhaps reflecting greater awareness among these local leaders of the negative impacts of burning. All sample households having tabia baito or kushet baito members practiced contour plowing, again likely reflecting awareness of the importance of soil conservation. Interestingly, kushet baito members are more likely than other households to use intercropping, while none of the tabia baito members used this.

Access to credit or contact with the extension program have no significant impacts on adoption of these land management practices. These programs have focused more on promoting use of fertilizer and improved seed than adoption of other land management practices.

Differences in land tenure are associated with few differences in land management practices. Contour plowing is less common on owner-operated plots allocated by the tabia than on leased-in plots; but is practiced on all sample plots that were received as a gift. Perhaps recipients of gift plots, and sharecroppers to some extent, are expected to use such conservation measures in deference to the landowner who provided the land.

Table 24–3. Determinants of use of land management practices in crop production in 1998.†

Variable	Burn to prepare field	Contour plowing	Reduced tillage	Intercropping/ mixed cropping
ln(Annual rainfall), mm	–0.2585**—¶	–0.0795	–0.0353	–0.5033**—
ln(Altitude), m above sea level	–0.1407**—	0.1184	–0.0488	–0.0146^{-}
ln(Population density), km^{-2}	–0.0059	–0.0096	–0.0168	0.0587**$^{++}$
Female head of household	0.0304	–0.0917*	–0.0026	–0.0202
ln(Age of household head), yr	0.0435*	0.0260	–0.0331^{+}	0.0224
ln(Household size), no.	–0.0062	–0.0170	–0.0588*	0.0308^{+}
Education of household head				
1 to 2 yr	0.0107	0.0112	0.0663	–0.0144^{+}
3+ yr	0.0345^{++}	0.0248^{++}	0.0433^{+}	–0.0213
Literacy campaign	0.0265^{++}	–0.0051	–0.0048	–0.0042
Walking time to, h				
All weather road	–0.0131**—	–0.0160*	0.0118	0.0006^{-}
Seasonal road	0.0020	0.0107	–0.0631**$^{-}$	0.0113
Bus service	–0.0029	–0.0009	0.0003^{+}	0.0021
Woreda town	0.0043^{+}	0.0110*$^{+}$	–0.0000	–0.0106
Input supply shop	0.0021	0.0027	0.0120	–0.0064
Plot from residence	0.0063	–0.0239	0.0379	–0.0249
Ownership of assets				
Land, tsimad	0.00138^{-}	–0.0065*	0.0140**$^{++}$	–0.0028*
Oxen, no.	0.0077	0.0255^{*++}	–0.0434**—	0.0129^{+}
Other cattle, no.	0.0041*	0.0008	0.0014	0.0021
Small ruminants, no.	–0.0024*	0.0016^{++}	0.0015	–0.0011
Pack animals, no.	0.0008	0.0003	0.0137	–0.0254**$^{-}$
Camels, no.	–0.0090	+§	–0.0057	–0.0161
Radio, yes/no	–0.0107	–0.0359	0.0208	–0.0374
Cash savings, yes/no	0.0007	–0.0292	0.0135	–0.0022
Secondary income source				
Cereals	–0.0072	–0.0539^{+}	0.1336*$^{+}$	–0.0367
Perishable annuals	0.0032	–0.1269	0.1085	0.0513
Perennial crops	–0.0415*	0.0123	–0.0195	–‡
Cattle	–0.0680**	–0.0201	0.0197	–0.0382^{+}
Small ruminants	–0.0090	+§	–0.0127	–‡$^{+}$
Beekeeping	0.0589	–0.2355	0.2695**	–0.0483**
Food for work	–0.0393*	–0.0744	0.0962	0.0207
Salary employment	–0.0248^{-}	–0.0225	–0.0383	–0.0167
Farm employment	0.0511	–0.2801*	0.0476^{-}	–0.0390
Trading	–0.0289	0.0052	0.1297*	0.0040
Food/other assistance	–0.0425**	–0.0444	0.0140	–0.0453
Other nonfarm	–0.0085	–0.0364	–0.0323	0.0366
Membership in organizations				
Farmers' association	–0.0039	0.0582	0.0279	–0.0289
Women's association	0.0091	0.0079	0.0480*–	–0.0178^{-}
Youth association	–0.0020	0.0058^{-}	–0.0179	0.0169
Tabia baito	0.0387	+§	0.0573	–‡
Kushet baito	–‡—	+§	–0.0680*	0.2858*
Marketing coop.	–0.0250—	–0.0081	0.0280	0.0408—
Agricultural cadre	0.0426	0.0252	–0.0361	0.0069
Use of credit				
Formal credit	0.0153	0.0293	–0.0052	–0.0279
Informal credit	0.0203	0.0194	0.0256^{++}	–0.0154
Contact with extension	–0.0172—	0.0148^{-}	–0.0134	0.0266
How plot acquired, cf. not owned				
Allocated by tabia	0.0040	–0.0594**—	–0.0259	–0.0351
Inherited	0.0811	0.0407	–0.0425	–0.0447

(continued on next page)

Table 24–3. Continued.

Variable	Burn to prepare field	Contour plowing	Reduced tillage	Intercropping/ mixed cropping
Received as gift/other	0.1776	+§	–0.0284	–0.0207
ln(Plot area), ha	0.0194**	–0.0078	–0.0130	0.0349**$^{++}$
Relative plot quality rank	0.0156	–0.0426^{-}	–0.0104	–0.0258
Plot slope, cf. flat				
Gentle	0.0951**$^{++}$	0.0289	–0.0083	0.0075
Steep	0.1918**$^{++}$	–0.0238	–0.0505	0.0961**$^{+}$
Soil depth, cf. deep				
Medium	–0.0217	–0.0092	0.0362	0.0538*$^{+}$
Shallow	0.0062	–0.0483*$^{-}$	0.0666*$^{++}$	0.0479
Soil fertility, cf. high fertility				
Moderate fertility	–0.0291	0.0144	–0.0369	–0.0571
Infertile	–0.0511**$^{-}$	0.0202	–0.0464	–0.0249
Waterlogging problem, 1 = yes	–0.0197	0.0198	–0.0022	0.0264
Gullies on plot, 1 = yes	0.0061	–0.0202	–0.0311	0.0844*$^{+}$
Land use, cf. homestead plot				
Rainfed cultivated	–0.0045	–0.0402*	–0.0274	–0.0354*
Irrigated cultivated	0.0288	–0.0831	0.0749	–0.0483
Initial investment on plot in 1998				
Stone terrace	0.0142	0.0366*$^{+}$	0.0115	0.0095
Soil bund	0.0702**$^{++}$	0.0344	0.0045	0.0187
Trees planted	0.0175	–0.0167	0.0990*	–0.0372^{-}
Constructed fence	–0.0283	0.0235	–0.0189	–0.0183
Live fence or barrier	0.0309	0.0274	–0.0415	0.0686
Number of observations	1504	1387	1533	1457
Mean of dependent variable	0.112	0.877	0.119	0.130
Mean predicted probability of use	0.111	0.875	0.119	0.129
Pseudo R^2	0.301	0.326	0.222	0.291

*, ** Mean coefficient statistically significant at 5%, and 1% levels, respectively.

† Probit regressions. Reported coefficients represent effect of a unit change in explanatory variable on probability of use. Coefficients and standard errors adjusted for sampling weights, clustering, and stratification. Intercept not reported. Coefficients of plot position on slope, soil texture and color not reported to save space. (These results reported in Pender et al., 2002).

‡ No positive values of dependent variable for positive values of the explanatory variable. Observations with positive values of the explanatory variable dropped from the regression.

§ No nonpositive values of dependent variable for positive values of the explanatory variable. Observations dropped.

¶ $^{+, ++}$and $^{-, --}$Mean coefficient positive (negative) and statistically significant at 10%, 5% and 1% levels, respectively in regressions using predicted values of sources of income, participation in organizations, use of credit and contact with extension.

Crop Production

Several land investments and land management practices have a large and statistically significant impact on the value of crop production (Table 24–4). The predicted value of production is 17% higher on plots with stone terraces and 41% higher where trees are planted.[11] Use of burning to prepare the field is associated with 33% lower yields, contour plowing with 25% higher yields, and reduced tillage with 57% higher yields controlling for inputs and other factors. Use of fertilizer is associated with 13% higher yields, and manure or compost with 15% higher yields (both ef-

[11] For example, considering the logarithmic specification for the dependent variable, the predicted impact of stone terraces is exp(0.1584) = 1.172, or a 17% increase.

Table 24–4. Determinants of crop production and perceived land degradation in 1998.†

Variable	ln(Value of crop prod. ha^{-1})	
	Structural model	Reduced form
ln(Annual rainfall), mm	–0.5568*	–1.4001**—R‡§
ln(Altitude), m above sea level	–0.9527**—	–0.9936**–R
ln(Population density), km^{-2}		0.0244
Female head of household		–0.4213**R
ln(Age of household head), yr		–0.1710
ln(Household size), no.		–0.1003
Education of household head		
1 to 2 yr		–0.1239++
3+ yr		0.3513*R
Literacy campaign		0.0764
Walking time to, h		
All weather road		–0.0270
Seasonal road		0.0746
Bus service		0.0537**R
Woreda town		–0.0897**R
Input supply shop		–0.0131
Plot from residence		0.0739
Ownership of assets		
Land, tsimad		–0.0085
Oxen, number		–0.0435
Other cattle, no.		0.0528**++R
Small ruminants, no.		0.0038
Pack animals, no.		–0.0647–
Camels, number		–0.0409
Radio, yes/no		–0.0371
Cash savings, yes/no		0.0799
Secondary income source		
Cereals		–0.0070
Perishable annuals		–0.2394
Perennial crops		0.2542
Cattle		–0.1339
Small ruminants		0.0997
Beekeeping		0.1615
Food for work		0.1025
Salary employment		–0.1460—
Farm employment		0.2281+
Trading		0.1612
Food/other assistance		0.4697*
Other nonfarm		0.0921
Membership in orgs.		
Farmers' association		0.0983
Women's association		–0.2253
Youth associtíon		–0.1028
Tabia baito		0.1183++
Kushet baito		0.3269
Marketing coop.		0.4255**
Agricultural cadre		–0.5402*
Use of credit		
Formal credit		0.0221
Informal credit		–0.0845
Contact with extension		–0.0013

(continued on next page)

Table 24–4. Continued.

Variable	ln(Value of crop prod. ha^{-1}) Structural model	Reduced form
How plot acquired, cf. not owned		
Allocated by tabia		0.0470
Inherited		0.5087*$^{+R}$
Received as gift/other		0.5178*
ln(Plot area), ha	–0.2060**$^{--}$	–0.4769**$^{--R}$
Relative plot quality rank	–0.2201*	–0.3327**$^{--R}$
Plot slope, cf. flat		
Gentle	–0.1469	–0.1217
Steep	–0.1426	–0.0865
Soil depth, cf. deep		
Medium	–0.1186	–0.0662
Shallow	–0.2244^{--}	–0.1245
Soil fertility, cf. high fertility		
Moderate fertility	–0.0757	–0.1401
Infertile	–0.1089	–0.1673
Waterlogging problem, 1 = yes	0.0633	0.1050
Gullies on plot, 1 = yes	0.2791	0.0321
Land use, cf. homestead plot		
Rainfed cultivated	–0.0325	–0.386**$^{--R}$
Irrigated cultivated	–0.2803	–0.2933*$^{--}$
Initial investment on plot in 1998		
Stone terrace	0.1584*	0.1293*$^{++R}$
Soil bund	0.2132	–0.0897
Trees planted	0.3444*	0.1504
Constructed fence	–0.0024	0.0761
Live fence or barrier	–0.0152	–0.0306
Use of inputs		
Fertilizer, 1 = yes	0.1262	
Fertilizer × stone terrace	0.0205	
Fertilizer × soil bund	–0.4576**	
Fertilizer × irrigation	0.2302	
ln(Seed ha^{-1}), kg ha^{-1}	0.2797**$^{++}$	
Improved seed (1 = yes	0.1388^{++}	
ln(Labor ha^{-1}), d ha^{-1}	0.0232	
ln(Oxen labor ha^{-1}, d ha^{-1}	0.2671**$^{+}$	
Use of land management practices		
Burning to prepare field	–0.3904**	
Contour plowing	0.2243*	
Reduced tillage	0.4542**$^{+++}$	
Intercropping/mixed cropping	–0.0130	
Manure or compost	0.1395+	
Intercept	15.4853**$^{++R}$	24.4266**$^{++R}$
Number of observations	1224	1300
R^2	0.403	0.419

*, ** Mean coefficient statistically significant at 5% and 1% levels, respectively.

† Least squares regressions. Coefficients of plot position on slope, soil texture and color not reported to save space. (Results reported in Pender et al., 2002).

‡ $^{+, ++}$ and $^{-, --}$Mean coefficient positive (negative) and statistically significant at 5%, and 1% levels, respectively in regressions using predicted values of inputs and land management variables in the structural model, or predicted probabilities of secondary income source, participation in organizations, use of credit and contact with extension in the reduced form model.

§ ^{R} Means coefficient of the same sign and statistically significant at the 5% level in reduced form regression excluding sources of income, participation in organizations, use of credit and contact with extension.

fects statistically significant only at 10% level). The presence of a soil bund reduces the predicted return with fertilizer. This may be because of weed or pest problems caused by the combination of these technologies.

The amount of seed and oxen power used has relatively large and statistically significant positive impacts on production. By contrast, the impact of human labor is quantitatively small (elasticity = 0.02) and statistically insignificant. This suggests that surplus labor exists in crop production in Tigray, with additional labor yielding little positive impact. This is not surprising, given the very small farm sizes and marginal agricultural conditions in Tigray, and implies that population growth can have very negative consequences for human welfare, since the additional labor may not be productively used in agriculture (Lewis, 1954). Of course, as we have seen, population pressure and small farm sizes contribute to adoption of more intensive practices, which tend to increase yields, if not labor productivity. Thus, the negative consequences of population pressure can be mitigated to some extent by such Boserupian responses. We investigate the extent of this mitigation below.

Many of these impacts are robust to the regression specification. The impacts of seed and oxen labor, reduced tillage, and manure or compost are still statistically and quantitatively significant when using predicted values of inputs and practices. Some of the measures no longer have statistically significant impacts (i.e., stone terraces, trees, burning, contour plowing), though the predicted effects of these variables have the same sign and similar magnitudes as in the first regression. This suggests that endogeneity of these factors has not biased the results of the first regression, but attempting to correct for it has caused some of the coefficients to be statistically insignificant due to reduced statistical power.[12]

In the reduced form regressions, we include household characteristics and land tenure, but exclude endogenous input use and land management practices.[13] Stone terraces again have a significant positive impact on crop production in these specifications.

Population pressure and smaller farm sizes have an insignificant impact on crop production per hectare, even though we found above that these promote greater use of several inputs and labor-intensive practices. This does not support the Boserupian optimistic perspective about the responses of households to population pressure leading to increased yields, and suggests that food production per capita will not keep pace with increasing population as farm sizes decline. Unless households are able to depend on alternative livelihoods, food insecurity is thus likely to worsen as population continues to grow.

Households with better access to a woreda town had higher value of crop production, probably because of greater production of high value products. Surprisingly, households further from bus service attain higher value of crop production.

We do not find a statistically significant effect of irrigation on the value of crop production, controlling for other factors. However, irrigation does increase crop production indirectly by increasing the use of inputs, including labor, oxen, fertilizer, manure, and compost.

---

[12] We could not reject the null hypothesis of no endogeneity bias using the Hausman (1978) specification test.

[13] In one version of the reduced form regressions, secondary income source, predicted membership in organizations, use of credit and participation in the extension program was used, and in another version these were excluded, since these may be endogenous.

Use of credit (formal or informal) is not associated with significant increases in crop production, even though we found that formal credit promotes use of fertilizer. This is consistent with the fact that our evidence shows only limited impacts of fertilizer on crop production. Contact with the agricultural extension program also has insignificant impact on crop production.

Ownership of cattle other than oxen is associated with higher crop productivity. This may be related to the greater deposition of manure on plots operated by households owning more livestock, even if households are not explicitly collecting and applying manure or compost.

Female-headed households achieve yields averaging 33% lower than yields of male-headed households. More educated households (having 3 or more years of formal education) achieve 42% higher yields.

Membership in some organizations is also associated with differences in production. Households with members of a marketing cooperative attain substantially higher output value per hectare, probably because they focus on higher value crops. Members of agricultural cadres attain lower output value. As mentioned previously, these households appear to focus more on livestock production (Pender et al., 2002).

Land tenure also affects crop production in the reduced form regressions. Yields are higher on inherited plots and plots received as gifts (10% significance for gift plots) than on leased-in plots. This is consistent with the earlier findings that use of some inputs is lower on leased-in than on owner-operated plots, and supports the argument of inefficiency of sharecropping.[14]

KEY FINDINGS AND IMPLICATIONS

Here we summarize key findings with regard to our hypotheses, and their implications.

Population Pressure

Population pressure, as reflected by higher population density and smaller farms, was associated with more intensive use of labor, oxen power, fertilizer, manure, intercropping, and contour plowing, and less use of reduced tillage. These findings are consistent with the predictions of population-induced intensification, as hypothesized by Boserup (1965) and her followers. However, this increased farming intensity in more densely populated areas was not found to lead to significantly higher crop yields. This suggests that population growth and smaller farm sizes will lead to reduced food production per capita. It also suggests that pressure is causing land degradation.

Access to Roads, Transportation, and Markets

Access to markets, infrastructure, and transportation has mixed impacts on crop production. The value of crop production averages 5% lower 1 h closer to bus

[14] Ninety-five percent of the leased-in plots in the sample are sharecropped.

service, but 9% higher 1 h closer to a town. These differences are not due to different prices resulting from differences in market access, since we used constant prices to calculate the value of crop production. Probably the mix of crops includes higher value crops closer to a town. It is not clear why proximity to bus service would reduce production.

Fertilizer, Extension, and Credit

The agricultural extension and credit program has sought to boost productivity largely by promoting use of fertilizer and improved seeds. The evidence presented shows some impact of fertilizer use on crop production (though the impact is statistically weak and not robust), increasing predicted value of production by 240 EB ha^{-1} on average, controlling for other factors. This yield increase is insufficient to cover the average costs of fertilizer (about 280 EB ha^{-1} in 1998), indicating that fertilizer use was unprofitable on average and explaining why farmers' are reluctant to adopt it, despite substantial efforts to promote its use. In a semi-arid environment as in the highlands of Tigray, use of fertilizer can be risky as well as unprofitable if adequate soil moisture cannot be assured.

Land Investments and Conservation Practices

Our evidence shows that there are investments and management practices with potential to substantially increase crop yields. Stone terraces increase crop productivity by an estimated 17%. Since stone terraces help to conserve soil moisture, they also increase the benefit of using fertilizer, which is probably why we find more fertilizer adoption on plots that have stone terraces.[15] The estimated average rate of return to stone terraces is 34%, based on the predicted increase in annual crop income and our data on costs of constructing these terraces. This is somewhat lower than estimates of returns to stone terraces in south central Tigray by Gebremedhin et al. (1998), who estimated a 50% rate of return to stone terraces. Nevertheless, it is clear that investment in stone terraces is fairly profitable in Tigray.

Our evidence shows that adoption of several land management practices also could have major impacts on crop productivity. We estimate that application of manure/compost increases average yields 15%, contour plowing 25%, reduced burning 48%, and reduced tillage 57%.

Tree planting (on plot boundaries) was associated with 41% higher value of crop production (though result not robust). This may be due to beneficial effects of trees in reducing wind erosion and drying of the soil. Tree planting also may indirectly increase crop production, since it increases the likelihood of reduced tillage. The positive association of tree planting with crop production challenges common arguments about negative effects of trees on crops (Jagger and Pender, 2003) and the prohibition against tree planting in farmlands in Tigray.

[15] Even though the coefficient of the interaction term between presence of a stone terrace and fertilizer use is statistically insignificant, the separate impacts of terraces and fertilizer are multiplictive, given the logarithmic form of the production function. Thus a 17% increase in productivity due to stone terraces and a 13% increase in productivity due to fertilizer use increase total production by more than 32% ($1.17 \times 1.13 = 1.322$).

Irrigation

Irrigation increases the intensity of input use in crop production, especially labor, oxen labor, fertilizer, and manure or compost. By promoting increase in use of such inputs, irrigation contributes to increased crop production. The predicted average impact of irrigation, based upon the predicted impacts of irrigation on use of inputs, is a 26% increase in crop production relative to rainfed field plots. The impact of irrigation on the productivity of land management (i.e., controlling for use of inputs and practices) is statistically insignificant. Thus the main impacts of irrigation on crop production are by promoting increased intensity of farming, rather than increased productivity of farming practices.

Land Tenure and Factor Markets

As mentioned previously, the amount of land owned by a household has a quantitatively small and statistically insignificant impact on crop productivity, controlling for other factors. This is likely due in part to relatively equal distribution of land in Tigray as a result of prior land redistributions. Only 1% of the sample households are completely landless, the average landholding is only 1.0 ha, and the maximum landholding is <5 ha. In addition, sharecropping and other arrangements have helped to equalize households' access to land and other productive inputs. Households with limited land also make other adjustments in farming intensity to boost their crop production, such as being more likely to use fertilizer, contour plowing, intercropping, or mixed cropping.

Although land leasing helps to equalize households' crop income, lease markets do not operate perfectly in Tigray. We have found that labor use is higher on tabia-allocated plots than leased-in plots, and that use of labor, seeds, fertilizer, and manure tend to be higher on inherited than leased-in plots. As a result, the value of crop output per hectare is higher on inherited than leased-in plots, controlling for plot quality and other factors. This suggests that sharecropping (the dominant lease form) leads to some inefficiency by reducing farmers' incentives to provide inputs, as argued in some of the theoretical literature. This result contrasts with findings of a recent study of the impacts of land lease contracts in several villages in the Oromiya region, which found that sharecropping and other lease arrangements were relatively efficient in the villages studied (Pender and Fafchamps, 2001). One reason for the different impact in Tigray may be restrictions on the terms of land lease arrangements in Tigray, which limit such contracts to no more than 2 yr duration (Gebremedhin et al., 2002). Such restrictions may inhibit landowners' ability to lease to tenants who they know and trust, or whose input use they can monitor easily; thus contributing to the incentive problem of sharecropping.

The small and statistically insignificant impact of oxen ownership on crop productivity suggests that informal lease arrangements for oxen are similarly helping to equalize farmers' access to draft power. Indeed, nearly three-fourths of sample households used oxen belonging to other households in 1998, while three-fifths provided oxen to other households. These arrangements may not work perfectly, given that we have found that greater oxen ownership increases oxen and human labor use per hectare somewhat. Nevertheless, these arrangements, together with other adjustments made by households with fewer oxen (such as greater use of reduced tillage) appear to work well enough to limit differentials in crop production resulting from unequal oxen ownership.

Labor use per hectare is greater for larger households, suggesting that imperfections in labor markets such as imperfect substitutability and monitoring costs of nonfamily labor enable households with a larger labor endowment to use labor more intensively. This does not translate into higher crop production in larger households, however, because of the low marginal productivity of labor in crop production. Thus, imperfections in labor markets do not translate into significant losses in crop production.

Gender

Female-headed households attain 34% lower crop yields than those of male-headed households, probably due to labor constraints and cultural prohibitions against women plowing. As a result, their crop income and total income are substantially lower than those of male-headed households (Pender et al., 2002). Thus, poverty appears to be a greater concern for female-headed households.

Education

Better-educated household heads (3 or more years of formal education) achieved 42% higher crop yields than households with no formal education. Part of the difference is due to greater farming intensity by more educated households, who use 45% more labor per hectare.

Social Capital

Some forms of social capital, as measured by involvement in local organizations, have a significant impact on crop production. Members of a marketing cooperative attain more than 50% higher value of crop production per hectare, probably because they focus on producing higher value crops and/or have better access to inputs than other farmers. Members of an agricultural cadre obtain lower crop production, mainly because they use less labor and seeds, and appear to focus more on livestock production.

CONCLUSIONS

We have investigated the impacts of many factors commonly hypothesized to affect land management and agricultural productivity in the highlands of Tigray. Some of these factors—including population pressure, small landholdings, access to roads, irrigation, extension and credit programs—have weaker impacts on agricultural production than often thought. Most of these factors do affect the intensity of agricultural production and adoption of various land management practices. However, these impacts on intensity do not add up to much impact on total crop production, in part due to the low marginal product of labor in crop production and limited productivity impact of inputs such as fertilizer that have been promoted by some of these factors.

Some land management practices were found to substantially increase crop production, including construction of stone terraces, tree planting, contour plowing, reduced burning, reduced tillage, and application of manure or compost. These practices apparently contribute to productivity by helping to conserve soil moisture and organic matter. Greater ownership of cattle is also strongly associated with increased crop productivity, probably as a result of increased manure availability. Promotion of such conservation practices and exploitation of complementary livestock production show more promise to boost crop production than large application of modern inputs such as inorganic fertilizer and improved seeds. However, there appear to be opportunities to exploit complementarities between use of such inputs (especially fertilizer) and investment in stone terraces.

Overall, the findings of this study show that profitable opportunities exist to increase agricultural production and achieve more sustainable land management in the highlands of Tigray. These opportunities include improvement of crop production using low-external input investments and practices such as terraces, manuring, reduced tillage, and reduced burning; and improved livestock management. The comparative advantage of people in the Tigray highlands is not in input-intensive cereal crop production but more in low input technologies and alternative livelihood activities, such as raising dairy cows, poultry, beekeeping, and nonfarm activities (Pender et al., 2002). As a result, greater emphasis on developing these alternatives in agricultural extension and other development programs may be fruitful. Food crop production should not be ignored in the development strategy, but less promotion of purchased inputs such as fertilizer and improved seed and greater emphasis on low input sustainable land management practices may be helpful.

REFERENCES

Angelsen, A. 1999. Agricultural expansion and deforestation: Modelling the impact of population, market forces and property rights. J. Dev. Econ. 58:185–218.

Besley, T. 1995. Property rights and investment incentives: Theory and evidence from Ghana. J. Polit. Econ. 103(5):903–937.

Boserup, E. 1965. The conditions of agricultural growth. Aldine, New York.

Feder, G., T. Onchan, Y. Chalamwong, and Hongladaron. 1988. Land policies and farm productivity in Thailand. The Johns Hopkins Univ. Press, Baltimore.

Fitsum Hagos, J. Pender, and Nega Gebreselassie. 1999. Land degradation in the highlands of Tigray and strategies for sustainable land management. Socioeconomic and Policy Research Working Paper 25. Livestock Analysis Project, Int. Livestock Res. Inst., Addis Ababa, Ethiopia.

Gebremedhin, B., J. Pender, and G. Tesfaye. 2002. Collective action for grazing land management in mixed crop-livestock systems in the highlands of northern Ethiopia. Socio-economics and Policy Res. Workshop Paper 42. Int. Livestock Res. Inst., Addis Ababa, Ethopia.

Gebremedhin, B., J. Pender, and G. Tesfaye. 2003. Community resource managment: The case of woodlots in northern Ethiopia. Environ. Dev. Econ. 8:129–148.

Gebremedhin, B., S.M. Swinton, and Yibebe Tilahun. 1998. Effects of stone terraces on crop yields and farm profitability: Results of on-farm research in Tigray, northern Ethiopia. J. Soil Water Conserv. 54(3):568–573.

Hausman, J. 1978. Specification tests in econometrics. Econometrica 46:1251–1271.

Hurni, H. 1988. Degradation and conservation of the resources in the Ethiopian highlands. Mountain Res. Dev. 8(2/3):123–130.

Jagger, P., and J. Pender. 2003. The role of trees for sustainable management of less-favored lands: The case of eucalyptus in Ethiopia. For. Policy Econ. 3(1):83–95.

LaFrance, J.T. 1992. Do increased commodity prices lead to more or less soil degradation? Australian J. Agric. Econ. 36(1):57–82.

Lewis., W.A. 1954. Economic development with unlimited supplies of labour. Manchester School 22(May):139–191.

Otsuka, K., and Y. Hayami. 1988. Theories of share tenancy: A critical survey. Econ. Dev. Cultural Change 37(1):31–68.

Otsuka, K., and F. Place. 2001. Issues and theoretical framework. *In* K. Otsuka and F. Place (ed.) Land tenure and natural resource management: A comparative study of agrarian communities in Asia and Africa. Johns Hopkins Univ. Press, Baltimore.

Pagiola, S. 1996. Price policy and returns to soil conservation in semi-arid Kenya. Environ. Resource Econ. 8:251–271.

Pender, J. 1998. Population growth, agricultural intensification, induced innovation and natural resource sustainability: An application of neoclassical growth theory. Agric. Econ. 19:99–112.

Pender, J., and M. Fafchamps. 2001. Land lease markets and agricultural efficiency: Theory and evidence from Ethiopia. Environment and Production Technology Div. Discussion Paper 81. IFPRI, Washington, DC.

Pender, J., B. Gebremedhin, and M. Haile. 2002. Livelihood strategies and land management practices in the highlands of Tigray. Environment and Production Technol. Div., IFPRI, Washington, DC, Mimeo. (Summary of paper in: Benin, S., J. Pender, and S. Ehui. 2002a. Policies for sustainable land management in the East African highlands. Summary of papers and proceedings of the conference held at the United Nations Econ. Commission for Africa, Addis Ababa, Ethiopia, 24–26 April. Environment and Production Technology Div. Workshop Summary Paper 13, IFPRI, Washington, DC and ILRI Socioeconomics and Policy Working Paper 50.)

Pender, J., and J. Kerr. 1999. The effects of land sales restrictions: Evidence from south India. Agric. Econ. 21:279–294.

Scherr, S., and P. Hazell. 1994. Sustainable agricultural development strategies in fragile lands. Environment and Production Technol. Div. Discussion Paper 1. IFPRI, Washington, DC

Salehi-Isfahani, D. 1988. Technology and preferences in the Boserup model of agricultural growth. J. Develop. Econ. 28(2):175–191.

Shaban, R. 1987. Testing between competing models of sharecropping. J. Polit. Econ. 95(5):893–920.

Tiffen, M., M. Mortimore, and F. Gichuki. 1994. More people, less erosion: Environmental recovery in Kenya. John Wiley & Sons, New York.

25 Dryland Research at ICARDA: Achievements and Future Directions

Adel El-Beltagy, William Erskine, and John Ryan
ICARDA
Aleppo, Syria

ABSTRACT

The rationale for the establishment of the international agricultural research system was to assist the various national programs in the developing world to achieve food security and banish hunger and malnutrition from the world. Thus, the 16 research centers spanning the globe are organized under the umbrella of the Consultative Group on International Agricultural Research (CGIAR), which conducts strategic and applied research, with its products being international public goods; its research agenda is focused on problem-solving through interdisciplinary programs implemented by one or more of its international centers, in collaboration with a range of partners. Such programs concentrate on *increasing productivity*, *protecting the environment*, *saving biodiversity*, *improving policies*, and contributing to the *strengthening of agricultural* research in developing countries. While CGIAR centers represent various agroecological zones, the major one focusing on drylands, primarily Mediterranean-cropping environments, is the International Center for Agricultural Research in the Dry Areas (ICARDA). The Center serves the entire developing world for the improvement of lentil (*Lens culinaris* Medik.), barley (*Hordeum vulgare* L.), and faba bean (*Vicia faba* L.); all dry-area developing countries for the improvement of on-farm water-use efficiency, rangeland, and small-ruminant production; and the Central and West Asia and North Africa region for the improvement of bread and durum wheat (*Triticum* spp.), chickpea (*Cicer arietinum* L.), and farming systems. Its research provides global benefits of poverty alleviation through productivity improvements integrated with sustainable natural-resource management practices. The Center meets this challenge through research, training, and dissemination of information in partnership with the national agricultural research and development systems. This presentation gives a brief description of the evolution of ICARDA's research, and highlights its achievements, its mode of operation, and future direction.

INTRODUCTION

That the Western world today—and most developed countries—are characterized by abundant food supplies, and the capacity to continue to produce food sur-

 Challenges and Strategies for Dryland Agriculture. CSSA Special Publication no. 32.

pluses can largely be attributed to the application of technologies generated by agricultural research. It was the belief that research could achieve a similar transformation in the poorer or lesser-developed countries of the world that led to the creation in 1971 of the international network of research centers under the auspices of the Consultative Group on International Agricultural Research, or the CGIAR as it is known. This global network has undergone changes and restructuring since its inception. Today, the 16 worldwide centers (Fig. 26–1) represent the major developing country zones and address issues that range from specific commodities (International Rice Research Institute, International Potato Center, International Water Management Institute) to policies (International Food Policy Research Institute), genetic resources (International Plant Genetic Resources Institute), and livestock (International Livestock Research Institute). While most centers are based in tropical or subtropical regions of the world, only two centers embrace semi-arid or dryland areas.

As the International Crops Research Institute for the Semi-arid Tropics (ICRISAT) deals with semi-arid conditions in the Asian Subcontinent, its mandate climatic area is mainly monsoonal, with two rainy seasons and interspersed dry periods. However, the major center for rainfed dryland cropping is ICARDA, which largely deals with countries with a Mediterranean-type climate, moist cool winters, where cropping is possible, to long dry summers. While much has been written about such a Mediterranean climate (Kassam, 1981; De Pauw, 2004, this publication), the primary concern for the people who live in the lands that border the Mediterranean, especially in West Asia and North Africa (WANA), is the limitations that such a climate, merging into a continental one, imposes on crop production (Harris, 1995; Cooper et al., 1987). In addition to heat and drought, crops are subjected

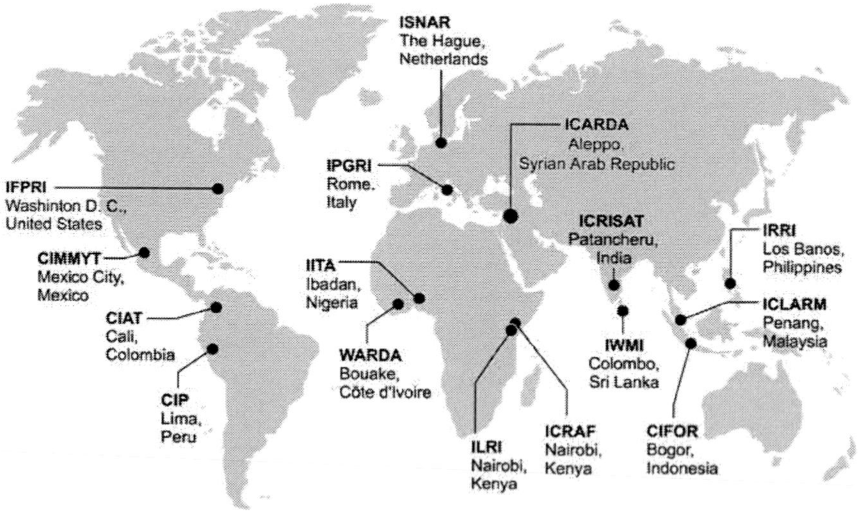

Fig. 26–1. International agricultural research centers under the auspices of the Consultative Group on International Agricultural Research.

to cold stress in high-elevation areas and salinity in low-lying areas, as well as a host of biotic stresses such as insect and fungal diseases and weeds.

The primary reason for the establishment of ICARDA in the WANA region is that it is a food-deficit one, the underlying concern being the rapid population growth rate, causing food demand to far exceed the supply. For instance, with most countries having growth rates in excess of 2.5%, the combined population of the region, including Central Asia, is estimated to be 1143 million people by 2020, up from only 656 million in 1995. As a consequence of the population-productive capacity imbalance, poverty is endemic for a large number of the region's people. In fact, of the total of one billion people in dry areas of the world 690 million are estimated to live on less than the equivalent of $2 per day. It is this underprivileged group that ICARDA's research addresses to alleviate poverty by sustainably increasing the quantity and quality of food while preserving or improving the resource base.

PRODUCTION CONSTRAINTS

Tackling the problems of dry areas is a daunting task. While rainfall is inevitably a constraint (Cooper et al., 1987), water scarcity is a problem that affects not only the agricultural sector but the urban one as well. Given the existing population growth rates, per capita water consumption will inevitably decline drastically within the next few decades. In 1990, only 8 of 23 WANA countries had per capita water availability of more than 1000 m^3, the threshold for water poverty level. Some countries such as Jordan have per capita water supplies of around 200 m^3, indicating severe scarcity. Water shortages in this region have already hampered the development in the Arabian Peninsula, Jordan, Palestinian Territories, Egypt, Tunisia, and Morocco. Other countries of the region such as Syria, Iraq, Algeria, and Lebanon are increasingly affected each year. Competition for the limited surface and groundwater will intensify, with urban growth diverting water away from the agricultural sector. The challenge is to use the available water more efficiently; the only growing source of water is urban and industrial wastewater. Therefore, better water management is a priority.

Attempts to produce more food must deal with soil issues. Soils in West Asia–North Africa, as elsewhere, show considerable variation in factors that influence crop growth (i.e., depth, texture, and mineralogy) and are related to a particular agroecology (Kassam, 1981). Again, as elsewhere, modern high crop yields are only possible with fertilizer input. Regardless of initial fertility, few soils can be cropped continuously, and produce acceptable yields without fertilizers (Ryan, 2004, this publication). The major soil problem is soil erosion by water and wind (Ryan, 1982; Sivakumar et al., 1998). Indeed, the eroded landscape of the Mediterranean region today is a result of abuse by man in the past (Carter, 1974). Of the 1.7 billion ha in the Central and West Asia and North Africa region (CWANA), soil degradation is a dominant feature with nearly half the total area of rainfed, irrigated, or rangelands being subjected to some degree of degradation. Drought exacerbates degradation and indeed hampers attempts at land rehabilitation in such an environment (De Pauw, 2004, this publication). While soil erosion has been termed the "silent earthquake", its impact on the productive capacity is loud and clear.

Drylands are often thought of as barren areas with little biodiversity. However, 17% of the global plant centers of diversity, 47% of the endemic bird areas, and 23% of the different terrestrial ecosystems are found in the drylands (White et al., 2002). The CWANA region contains three of the eight centers of plant genetic diversity identified by Vavilov, with many key crops and animals domesticated here. Small ruminants have been a part of farming systems since early times, perhaps as long ago as 8500 BC. But the rich biodiversity, which is key to future food security, is being lost through *habitat degradation*, *agricultural expansion,* and *intensification*, and *overgrazing*. A brief comment on rangelands of the CWANA region is pertinent, given the huge extent of these grazing areas.

Central Asia has evolved from a state of overexploitation to a state of underutilization. During the Soviet time, rangeland degradation due to high stocking rates was widespread, and millions of hectares of dry steppe were plowed, for example, northern Kazakhstan, to meet the grain production goals of the Soviet Union. After the breakup of the Soviet Union, lack of infrastructure maintenance and operating capital and loss of traditional markets for wool and karakul pelts rendered vast areas effectively ungrazeable or unattractive. As these areas begin to be re-utilized with increasing stock numbers, there is a need to develop grazing strategies that will avoid the degradation of previous times.

In contrast, overgrazing elsewhere in CWANA is a pervasive problem on rangelands, and severe soil erosion is commonplace. Studies in West Asia and North Africa clearly show that in most countries of the region the contribution of rangelands to livestock feeding is diminishing (Gintzburger et al., 2000). To meet the region's increasing demands for food and feed in WANA, there has been an expansion of rainfed crops, mainly barley (*Hordeum vulgare* L.) onto the best rangeland in the 150 to 250-mm zone, as well as an increase of irrigation activities wherever it is possible to establish a water source. The area available for grazing is becoming smaller, with more animals grazing fewer hectares. The perennial vegetation has been almost totally destroyed by overgrazing and firewood collection, leaving the soil exposed. The ever-decreasing biomass available for grazing encourages land users to plow and grow barley instead, thereby destroying the last vestiges of any dry-season soil cover.

It is against this background of relentless population growth, widespread and pervasive poverty, declining per capita food production in a vast area of the world beset with endemic production constraints that ICARDA was founded. By way of understanding how this international center is addressing dryland issues, particularly drought, a brief background is presented on the circumstances in which the Center came into being and how it evolved.

BACKGROUND

The historical context in which ICARDA was established has been recently described (Nour, 2002) on the occasion of the institution's 25th anniversary. In brief, the Center evolved from the Arid Lands Agricultural Development Program

(ALAD) and, after abortive attempts to establish a research center in Lebanon and in Iran, the site at Tel Hadya near Aleppo in Syria's northern dryland belt, was chosen (330 mm yr^{-1}). Subsequently, sub-stations with a range of rainfall and soil conditions (Ryan et al., 1997) to represent or simulate the range of agroecological conditions that exist throughout the entire dryland region south and east of the Mediterranean (Fig. 26–2) were established. Thus, on-site research could be conveniently conducted in rangelands, semi-arid barley-based systems, to the more favorable rainfall wheat-based system. The Center's brief was to work with scientists of the various national agricultural research programs in the countries of its West Asia–North Africa region – the area of Central Asia embracing the Republic of Uzbekistan, Kazakhstan, Turkmenistan, and Tajikistan was added to its mandate in 1996. While the socioeconomic/political circumstances in Central Asia differ markedly from the West Asia–North Africa, the biophysical conditions are relatively similar. The Center has extensive linkages with advanced research and development institutions and is thus both a conduit for and a catalyst of research and technology transfer from developed to developing institutions mainly through research networks (Ryan et al., 1995).

Following the expansion of the Center's mandate to Central Asia, there was a substantial shift in emphasis towards that region, especially in the area of soil and water (Karajeh et al., 2002). A more recent extension of activities involved reha-

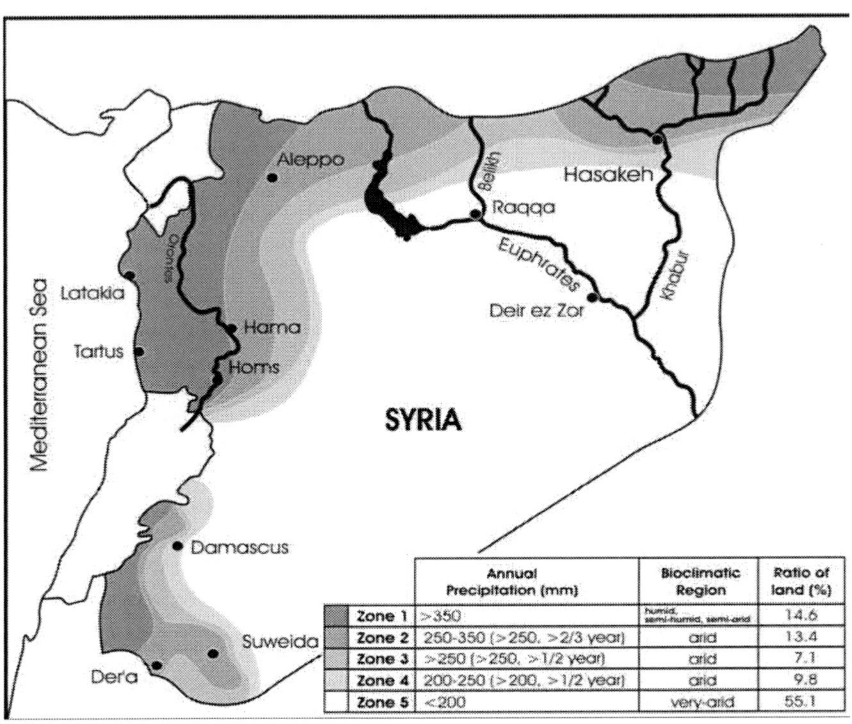

	Annual Precipitation (mm)	Bioclimatic Region	Ratio of land (%)
Zone 1	>350	humid, semi-humid, semi-arid	14.6
Zone 2	250-350 (>250, >2/3 year)	arid	13.4
Zone 3	>250 (>250, >1/2 year)	arid	7.1
Zone 4	200-250 (>200, >1/2 year)	arid	9.8
Zone 5	<200	very-arid	55.1

Fig. 26–2. Agricultural settlement zones in Syria.

bilitating a war-ravaged country on the fringes of its mandate area, Afghanistan, as part of the "Future Harvest Consortium to Rebuild Agriculture in Afghanistan" with funding from the U.S. Agency for International Development. ICARDA has been assisting with providing much-needed seed supplies to Afghan farmers, as well as developing policies for the country's seed industry. It has also facilitated efforts to identify and assess the countries agricultural development needs, in collaboration with other agencies and donors, and has provided technical support for restoration of the country's experiment stations. With sub-offices in the CWANA region and global collaborative programs, including other CGIAR centers, ICARDA can be described as a "center without walls." Its research objectives include:

- Serving as an international center for research into the improvement of barley, lentil, and faba bean—later the mandate was expanded to include water-use efficiency, rangeland, small ruminants, and natural resources management.
- Being a regional center, in cooperation with other centers, for research in bread wheat (*Triticum aestivum* L.) and durum wheat (*T. durum var turgidum*), with the International Wheat and Maize Center (CIMMYT) Mexico, and chickpea with ICRISAT, India.
- Conducting research into, and developing, promoting, and demonstrating improved systems of cropping, farming, and livestock husbandry.
- Collaborating with and fostering cooperation and communicable among national, regional, and international institutions.
- Supporting training and manpower development in the region.

In order to fulfill its mandate, the Center's research was organized into four major programs: *Cereal Improvement, Food-Legume Improvement, Pasture and Forage Improvement,* and *Farming Systems*. The cereal and legume programs were recently amalgamated into the Germplasm Program, while the other two programs were combined in the Natural Resources Management Program. The research activities were supported by smaller units—*Genetic Resources, Training and Communications*, and *Station Operations.* The focus of the Center's research was dryland rainfed agriculture across the semi-arid rainfall gradient from native pastures in the drier end of the spectrum (<200 mm yr^{-1}), through barley-based livestock systems in the 200 to 300 mm zone, and wheat-based systems in the more favorable rainfall zones 350 to 500 mm (Fig. 26–3). Purely irrigated conditions were excluded, as were horticultural crops in the high rainfall zone; however, supplemental irrigation in normally rainfed conditions developed to be a major research effort today.

Since its inception, the institution's research projects have changed and evolved as research goals were met and new challenges emerged. The evolving program brought with it concomitant changes in research disciplines and professional staffing. Various activities were reduced or phased out, while other areas of endeavor received increased research attention and support (Table 26–1). Even within specific projects there were shifts in emphasis; a major example of the former is the increased emphasis on water, biotechnology, and farmer participatory approaches, while examples of the latter include decentralized breeding, integrated pest management, in situ conservation, and adoption and impact assessment. As research programs adopted a project-based structure, 19 individual projects were arranged

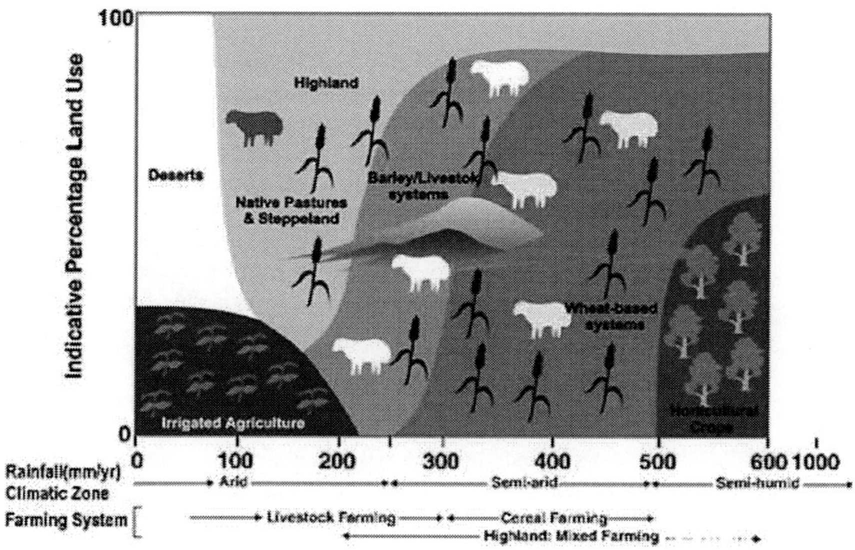

Fig. 26–3. The major farming systems in the ICARDA region.

under five themes: *germplasm enhancement*, *production systems management*, *natural resources management*, *socioeconomics and policy*, and *institutional strengthening,* with considerable interaction between projects.

Table 26–1. Increases and decreases in research agenda, and shifts of emphasis within projects.

Increase	Decrease	From	To
On-farm water management	Cereal with assured moisture,	Centralized breeding	Decentralized
Rangeland rehabilitation	In-house 'finished' cultivars,	Integrated pest management screening in isolation,	Systems,
Small ruminant nutrition,	Agronomy in high-rainfall zones	Center rotation trials	National programs
Agro-ecological characterization,	Wheat/medic systems,	Ex situ germplasm collecting,	Biodiversity in situ conservation,
Land and water,	Evaluation of Rhizobium strains,	Long-term, nondegree training,	Degree,
Resource economics	In-house software development,	Headquarter stress physiology,	Decentralized,
Prebreeding/ biotechnology	Technician training in crop improvement	Farming systems diagnostic survey, on-farm trial economic evaluation,	Adoption and impact assessment, rural poverty analysis,
Farmer participatory approaches, End-use quality/ added value, Faba bean		Technical seed production, that is, formal	Alternative seed production systems (informal sector) seed security, seed production economics

RESEARCH HIGHLIGHTS

During the past 27 yr of its existence, ICARDA scientists have produced a vast array of publications ranging from primary refereed articles that reflect the Center's contribution to global science, books that synthesize its research, conference proceedings that reflect ongoing issues and contributions from national programs, abstracts of international and regional meetings, and a variety of training and publicity materials, all centered primarily on dryland agriculture. In the following brief sections, reference is made to ongoing research activities, where significant progress has been or is likely to be achieved in a number of priority areas of research; anything approaching a complete review is obviously outside the scope of this article. Where appropriate, reference is made to milestone publications such as books and review articles.

Rotations, Long-Term Trials, and Soil Fertility

While annual trials repeated over 2 to 3 yr have value in answering various agronomic questions, sustainability of any particular practice or innovation can only be assessed by long-term trials that reflect actual cropping conditions (Jones, 1998). Consequently, most trials involved common rotations, including cereals and a range of feed and forage legumes compared to fallow or continuous cropping (Ryan and Abdel Monem, 1998). Of the 16 long-term trials initiated at Tel Hadya and its sub-stations, the "Cropping Systems" trial (Harris, 1995) clearly showed that vetch could replace fallow or continuous cereal cropping, though lentils may be more economical in some circumstances. The trial demonstrated the potential of vetch and medic in relation to increased soil organic matter and total N (Ryan, 1998). An ongoing barley-based grazing trial showed similar soil quality increases with vetch (Ryan et al., 2002). While most headquarters trials are now closed, collaborative trials are continuing in Egypt, including a dryland site. As many on-station trials have achieved their objectives, the rationale to apply such trials in farmers' field conditions was developed. Thus, Jones (2000) proposed the notion of anticipatory long-term research for sustainable productivity, including the socioeconomic dimension.

ICARDA's research on soil fertility reflected the importance of soil nutrients and crop nutrition in terms of enhancing crop yields, even within the limitations of rainfall in the Middle East region. As such, it pre-dated the establishment of long-term trials and was an intrinsic part of the concept of sustainability enshrined in those trials. The contributions of soil fertility research to the region and to science in general is reflected in books that emanated from international meetings (Ryan and Matar, 1990, 1992; Ryan, 1997), and comprehensive reviews (Matar et al., 1992). Through its soil test calibration program, the soil fertility program laid the foundation for rational use of fertilizers in the region's dryland cropping system. Indeed, as recently shown (Pala et al., 2004), adequate fertility and nutrition are crucial for improving national crop output.

Water: Technical and Institutional Factors

The supply of water for agriculture, whether from surface sources (streams, rivers, and lakes) or from underground, is limited and is likely to decrease if anything, given growing competition for urban uses—only wastewater is a source that will potentially increase. Since 1978, there has been increasing development of irrigation in normally rainfed areas (Perrier and Salkini, 1992). Using deficit or supplemental irrigation at less than full irrigation requirements, it is possible to substantially increase the crop yield per cubic meter of irrigation water used (Oweis et al., 1998). Similarly, there is a shift towards more efficient sprinkler and drip systems, and away from furrow irrigation. In addition to improved irrigation technology, substantial progress has been made in water harvesting in marginal areas, using technology such as geographic information systems to identify potential sites for harvesting as well as improved collection and storage practices (Oweis et al., 1999). Future research will inevitably focus on the use of wastewater, which is also a source of nutrients (Ryan et al., 1999).

Water must be more effectively managed in the region. *Water cost recovery* is one avenue. Although water is extremely valuable in the region, it is generally supplied free or at low subsidized cost, giving little incentive to farmers to increase water-use efficiency. There is a major campaign to adopt pricing schemes for water services based on total operational costs. The concept is seriously challenged in many countries of the region and innovative solutions are very much needed to put a real value on water for improving efficiency, but at the same time finding ways from within the local culture to protect the right of people to access water for their basic needs. We need to understand the incentives and disincentives for the careful use of water for agriculture.

Participation of all concerned in the management of scarce water resources is the key to successfully implementing more effective measures of water management. Users cannot, without appropriate policies, achieve the objectives of effective water management. It is widely agreed that lack of proper policies in this region is the main constraint to improved water use. With water being the main constraint to agriculture in the drylands, *technical options can only be adopted when there is an enabling institutional and policy environment*. One approach to tackle these aspects involves a focus on participatory approaches that involve all stakeholders. More work is needed, however, to integrate all these approaches in practical packages to achieve the largest return from the limited water available.

Other successful water-harvesting techniques that have for centuries sustained the population in the dry areas of WANA include the well-known cisterns and "qanats", traditional underground canal systems that guide groundwater by natural gravity flow to the surface. Many of these can be rehabilitated if the groundwater levels have not fallen beyond recovery levels. ICARDA has been working with communities to restore these systems that have fallen into neglect.

Water-Use Efficiency by Crops

Fundamental to dryland cropping is the efficient use of the limited water available from rainfall. ICARDA's early research (Cooper et al., 1987) clearly identifies the soil, crop, and environmental factors influencing water-use efficiency while fallow was traditionally the only way to effectively increase soil water for crop use, but its efficiency is low (Harris et al., 1991). However, having legumes in rotation with cereals by increasing total biomass production and reducing evaporation from bare soil leads to improved water-use efficiency (Harris, 1994).

To increase water-use efficiency, efforts need to be focused on improved soil, crop, and cropping system management. Water-use efficiency in rainfed agriculture depends on the maintenance of a permeable surface and the establishment of crop canopies that maximize productive transpiration relative to losses through weeds and surface evaporation. Well-timed tillage operations, in some cases including contour ridging or mulching, combined with skilled, locally adapted crop husbandry may double the proportion of water used productively. Therefore, various agronomic studies demonstrated the effectiveness of fertilizer, especially P (Gregory et al., 1986) and weed control (Matar et al., 1992) in improving the efficiency of water use by crops. While the basic agronomic aspects of water-use efficiency are now well established, efforts at breeding more efficient cereal cultivars are needed (Nachit et al., 2001). The use of techniques such as carbon (C) isotope discrimination in selection for water-use efficiency has been used by researchers for barley (Dakheel et al., 1994). There is an urgency to transfer what has been learned at ICARDA to the national agricultural systems for implementation at the farm level (Duivenboden et al., 2001).

Rangelands and Animal Fodder

Given the vast areas of range and grazing land in the region from Morocco to the Central Asian steppes, and the importance of animals, especially sheep, in the lives and farming systems of the region's inhabitants, the increased focus of ICARDA on such areas was understandable (Gibbon, 1981). Some of the approaches used—and the success stories—include development of use of fodder shrubs such as *Atriplex* as alley-cropping, plantation of spineless cactus, range pitting to hold water and increase seen germination, use of P fertilizer on marginal land, and production of feed-blocks from agro-industrial by-products as a supplemental animal feed. Much of the success of these endeavors stems from the community-action approach are described by Ngaido and El-Mourid (2004, this publication).

In addition to improving the available feed resources at critical periods of the year, attention is also given to better flock management via efforts to improve animal health and reproduction especially when animals are concentrated physically in pens or otherwise housed. The reproductive performance of small ruminants can be improved by use of hormone sponges to better synchronize mating and lambing dates with seasonal fluctuations in feed resources. Some publications are pivotal to our knowledge gained in the area of range and livestock; these include the proceedings of a meeting in Hammamet, Tunisia, which provides a comprehensive overview of issues related to fodder shrubs in semi-arid and arid drylands (Gintzburger et al., 2000), with research contributions from scientists throughout the Mediterranean zone. As livestock is central to Middle Eastern agriculture, a broad overview of livestock-related issues both in rainfed cropping areas and in rangelands is provided in a regional symposium on integrated crop-livestock systems (Haddad et al., 1997).

In the process of producing forage, food, or fibre crops in the Middle Eastern drylands, an issue of recent concerns has been soil organic matter as it relates to terrestrial-atmospheric dynamics, since elevated levels of carbon dioxide (CO_2) in the atmosphere are attributed to global warming. While burning of fossil fuel is the main contributor, agricultural production processes ranging from plowing and tillage to overgrazing and burning of biomass are also important factors. However, agricultural developments can reverse these trends. The increases in organic matter of about 1 t yr^{-1} with medic/vetch rotation noted by Ryan (1998) is an indication of the potential of rotations to sequester C; similarly, conservation tillage has been shown to conserve C in the surface soil layer (Pala et al., 2000). Given the large areas devoted to rangeland, any small increase in soil C would have a significant impact on C in the atmosphere. In a broad ranging review of the Middle East region, Lal (2002) suggested substantial amounts of C can be sequestered through improved nutrient and crop management, as well as controlled grazing. Current research using the Bowen Ratio is focused on examining CO_2 fluxes in rangelands of Central Asia, research that could have global implications.

Crop Improvement

Efforts to improve dryland crops in the Middle East region, mainly cereals and pulses, center around three basic approaches: (i) selection and subsequent development of the genetic potential of crops through incorporation of desirable traits; (ii) better agronomic and crop management; and (iii) improved inputs such as fertilizers, water, and pesticides. ICARDA's Germplasm Program, along with its associated Genetic Resources Unit, focuses primarily on the first and more fundamental aspect; the other more applied aspects fall within the concerns of the Natural Resources Management Program. As selection and breeding are based on existing populations or landraces, it is pertinent to briefly maintain ICARDA's efforts at identifying and characterizing the diversity of germplasm occurring naturally in the region.

Located in the center of origin of some of the major world food crops, ICARDA has been a focal point for international symposia on the history of crop domestication, species diversity in the agroecological zone of the Near East, approaches to conserving wild progenitors and evaluating their economic potential (Zakarieva et al., 201), and establishing genebanks and germplasm networks (Damania et al., 1998; Srivastava and Damania, 1990). Extensive collection missions have been made throughout the region, from deserts and steppes to cold mountainous terrain. Consequently, the Center's genebank has 128 000 accessions of mandate crops (cereals, food, and forage legumes).

Based on such collections, along with cultivated species, evaluation was made for biotic and abiotic stresses endemic in the region; specific examples include assessing food and forage legumes (Robertson et al., 1996; Hamdi and Erskine, 1997). Native and cultivated chickpea were shown to vary genetically in terms of tolerance to severe cold (Singh et al., 1995) and in terms of resistance to cereal cyst nematode (Di Vito et al., 1996). Similarly, Ethiopian lentil landraces were shown to vary in adaptation depending on specific environmental conditions (Bejiga et al., 1996a). Chickpea accessions at ICARDA were also used as a basis for comparing iron-chlorosis susceptibility of chickpea lines (Bejiga et al., 1996b). Much of what has been known and subsequently learned through research at ICARDA has been compiled in volumes such as breeding for stress tolerance in cool-season food legumes (Singh and Saxena, 1993), with lentil (Webb and Hawtin, 1981) and chickpea (Saxena and Singh, 1987) in particular.

While much success has been achieved through breeding for improvement in food legumes, one major breakthrough is worthy of some mention, as ascochyta blight is a major constraint to improving chickpea output, it devastates the crop if sown early in order to avail of the cooler conditions and soil moisture. Resistant varieties that could be grown in fall rather than late spring, and thus with a longer growing season, have significantly higher yields. The identification of ascochyta-resistant chickpea not only considerably out-yielded spring-sown cultivars (Pala and Mazid, 1992; Singh et al., 1997; Silim et al., 1991), but also increased water-use efficiency.

As cereals are the main crops of the Mediterranean dryland region, with most of the area under dryland cropping, considerable effort was made—and progress achieved—in the area of breeding and exploitation of desirable characteristics that contribute to yield and biotic and abiotic stress resistance in both wheat and barley. Approaches to achieving these goals range from conventional breeding, biotechnology, and the use of molecular markers, to farmer-assisted selection and participatory breeding.

In the case of barley, the most drought-tolerant cereal and one that is widely grown in all the agroecological zones of the Mediterranean and West Asia region, the many publications that emerged from barley research include: adaptation to harsh environments as reflected by morphological traits (van Oosterom and Acevedo, 1992), identification of genetic resources as a basis for crop improvement and sustainable agriculture (Ceccarelli et al., 1992), assessment of landrace diversity (Grando et al., 2000), selection of strategies for breeding in stress environments (Ceccarelli et al., 1998), and approaches to participatory breeding in target environments (Ceccarelli et al., 2000, 2001). Significant strides have also been made in screening for Barley Yellow Dwarf Virus resistance (Makkouk et al., 1994), common bunt pathotypes (Ismail et al., 1995), and heat tolerance (Tahir and Singh, 1993).

Similarly, while there has been progress in breeding for resistance to cereal insect pests, the achievements in combating the ravages of Hessian fly (*Mayetiola destructor*) are worthy of mention (El-Bouhssini et al., 2001).

Increasingly, indirect selection is used through identification and use of physiological and agronomic traits related to increased performance under dryland conditions (Araus et al., 1998). Selection for physiological traits via molecular markers linked to a particular trait is a major research thrust (Autrique et al., 1996; Eujayl et al., 1999; Nachit et al., 2001; Baum et al., 2003). The genetic diversity of durum wheat has been assessed by restricted fragment length polymorphism and morpho-physiological traits (Antrique et al., 1996).

While development of feed and forage legumes has been a major endeavor at ICARDA (Abd El-Moneim and Ryan, 2004, this publication), particular attention has been given to *Lathyrus* species in view of their potential adaptation in dry environment (Abd El-Moneim and Cocks, 1993); major emphasis has been given to selection by exploitation of somaclonal variation of cultivars low in B-ODAP a neurotoxic present in the seeds (Aletor et al., 1994; Abd El-Moneim and Ryan, 2000). As lathyrus seed is used extensively in the developing world (e.g., Ethiopia,

Bangladesh) as a source of human food, particularly in times of famine, making it safe to eat would be a major breakthrough.

CURRENT RESEARCH APPROACHES

The process of research, from the initial state of problem identification and subsequent resolution to implementation of the solution at the level where the problem occurs, normally under farm conditions, can involve varying philosophies at the various stages of the research continuum. With pressure on research institutes to deliver technologies that impact on farming communities, rural societies, and the entire national economy at times, the process has come under scrutiny to identify "bottlenecks" to effective technology application. A few approaches that are now in vogue and that have been adopted by ICARDA are worthy of mention.

The concept of *participatory action* involves the principle of the end-users of the intended research being involved in defining the program's goals and objectives as well as its direction during implementation and indeed participating in the evaluation of the program's effectiveness. Two examples of this principle will suffice. The Mashreq–Maghreb development project (Ngaido ad El-Mourid, 2002) in some countries of North Africa and West Asia involve the communities that are likely to be impacted by the Project. This innovative program successfully introduced technological packages to improve crop-livestock and pastoral production, stimulated the development of community institutions, and proposed options for policy and institutional reforms in regional communities. It is expected that technological options will require local refinement and adoption with due consideration of local or indigenous knowledge. The second example involved farmers in the field research, as in the case of varietal selection in Syria for barley breeding (Ceccarelli et al., 2000) as part of the philosophy of decentralized breeding; the approach was successfully adopted in other countries of the region (Ceccarelli et al., 2001).

Another approach or philosophy involves *integrated natural resource management*, a departure from tackling research problems from a single discipline perspective to adoption of a holistic view involving multi-disciplines. The approach attempts to integrate research on different types of natural resources into *stakeholder-driven processes of adaptive management* and innovation to improve livelihoods, agroecosystems resilience, agricultural productivity, and environmental services at the community, ecoregional, and global scales of intervention and impact. It takes into account the changing paradigms of agricultural research that include shifts from *agronomy to ecology*, *analytical research to system dynamics*, *top-down approaches to participatory action research*, *prescriptive processes to adaptive learning and management*, and *factor-orientated management to integrated natural resources management*. Such an approach also requires changes in the institutional environment towards more learning institutions that are willing to formulate and operate in multi-disciplinary teams to tackle the complexity of natural resource management and distill out those key driving variables that must be addressed to reverse degradation and at the same time build sustainable livelihoods for the rural poor who live in marginal drylands.

Traditional research is conducted under controlled on-station conditions or in more realistic conditions in farmers' fields. The integrated natural resources management approach takes this concept further with the adoption of "*integrated research sites*" in the broadest sense of the word. Such sites constitute a "living laboratory", central to which is the local farming community and the problems they face in their struggle to make a living from a harsh environment. In an effort to achieve greater integration and efficiency of research endeavors among CGIAR centers, as well as more effectively target communities, the concept of "challenge programs" was developed. A driving force behind this new initiative has been the change in funding for international centers whereby there has been a dramatic decrease in core funds and a parallel increase in donor-directed funds for specific projects such as the "*challenge programs*". The process of competing for funding under approved programs involves pre-screening of concept notes and then, if selected, advancement to the full proposal stage, and then final selection; all stages being scrutinized by an international panel of experts.

Some proposals are designated "fast tracks" for funding, while others are selected for the proposal stage. Currently, specific programs that have been approved include ones on *Water in Agriculture*, involving selected rivers basins around the world (e.g., Karkeh River Basin in Iran, and Yellow River in China), and *Genetic Resource and Bio-Fortification*. Other programs likely to be funded include *Climate Change* and its implication for *Desertification*, *Drought*, *Poverty*, and *Agriculture* (DDPA). The key question that this program asks is " how can poverty in resource-poor, desertification-prone areas be reduced and the poor achieve stable, secure livelihoods without undermining the ecosystem goods and services that they vitally depend on?". Multi-institutional and multi-disciplinary teams will focus on the main themes within this program: (i) understanding and coping with land degradation and drought risk; (ii) integrated ecosystem approach for the sustainable provision of agricultural and ecological goods and services; (iii) policy and institutional options; (iv) harnessing genetic resources; (v) income-increasing agricultural diversification to improve livelihoods and foster more sustainable land use; (vi) breaking technology and knowledge barriers; and (vii) increasing impact with a strategy for the use of information and communication technology for rural development. It is through collaborative efforts such as these that ICARDA is expanding its mandate and tackling the problems of dryland agriculture in an integrated holistic manner.

PERSPECTIVE

Because of their perceived low potential, as well as an inhospitable place to live in and produce crops, the arid and semi-arid areas of the world have been neglected in terms of research and funding for development in comparison with other climatic zones around the globe. However, because many of the world's poor live under semi-arid conditions and because such vast areas influence the global environment and are in turn influenced by climate, particularly perceived climate change, there has been a resurgence of interest in drylands in the past few decades. The interest of the world community in dryland areas is inflected in major international meetings on dryland agriculture (Unger et al., 1988), and on desert environments (Ryan, 2002), as well as international conventions such the United Nations Convention to Combat Desertification.

The establishment ICARDA in the draught-plagued Mediterranean—West Asia region was an act of faith by the global community that semi-arid drylands were important and that research in such an ecosystem could make a difference to the lives of the people who live there. Indeed, regional meetings, such as that in Amman, Jordan (Whitman et al., 1986), helped catalyze the growing research momentum. The symposium on "Challenges and Strategies of Dryland Agriculture into the New Millennium", sponsored by USDA-ARS, ICARDA, and the Crop Science Society of America, is a timely event and an opportune occasion to show the international scientific community what ICARDA scientists have contributed to dryland agriculture research and what an impact it has had on the mandate region of West and Central Asia and North Africa. The symposium also demonstrated the synergy that exists between dryland research scientists worldwide.

In the interest of world peace and the humanitarian need to relieve millions from the scourge of hunger, there is a moral imperative for wealthier nations and donor agencies to support this noble endeavor.

ACKNOWLEDGMENT

The authors wish to thank Dr. Richard Thomas for his input in an earlier version of this paper.

REFERENCES

Abd El-Moneim, A., and P.S. Cocks. 1993. Adaptation and yield stability of selected lines of Lathyrus spp. Under rainfed conditions in West Asia. Euphytica 66:89–97.

Abd El-Moneim, A.M., and J. Ryan. 2000. Nutritional influences on neurotoxin concentrations in grass-pea (*Lathyrus sativus*). p. 39. *In* Third Int. Crop Sci. Congr., Hamburg, Germany. 17–22 August. Abstracts.

Abd El-Moneim, A., and J. Ryan. 2004. Forage legumes for dryland West Asia and North Africa. p. 243–256 *In* S.C. Rao and J. Ryan (ed.) Challenges and strategies of dryland agriculture. CSSA Spec. Publ. 32. CSSA and ASA, Madison, WI.

Aletor, V.A., A. Abd El-Moneim, and A.V. Goodchild. 1994. Evaluation of seeds of selected lines of three Lathyrus spp for β-N-Oxalylamino-L-alanine (BOAA), tannins, trypsin inhibitor activity, and certain *in vitro* characteristics. J. Sci. Food Agric. 65:143–151.

Araus, J.L., T. Amaro, J. Voltas, H. Nakkoul, and M.M. Nachit. 1998. Chlorophyll fluorescence as a selection criterion for grain yield in durum wheat under Mediterranean conditions. Field Crops Res. 55:209–223.

Autrique, E., M. Nachit, P. Monneveaux, S.D. Tanksley, and M.E. Sorrells. 1996. Genetic diversity in durum wheat based on RFLPs, morph-physiological traits, and coefficient of parentage. Crop Sci. 36:735–742.

Baum M., S. Grando, G. Backes, A. Jahoor, A. Sabbagh, and S. Ceccarelli. 2003. QTLs for agronomic traits in the Mediterranean environment identified in recombinant inbred lines of the cross 'Arta' x H. spontaneum 41–1. Theor. Appli. Genet. 107:1215–1225.

Bejiga, G., K.B. Singh, and M.C. Saxena. 1996a. Evaluation of world collection of Kabuli chickpea for resistance to iron deficiency chlorosis. Genet. Res. Crop Evol. 43:257–259.

Bejiga, G., S. Tsegaye, A. Tullu, and W. Erskine. 1996b. Quantities evaluation of Ethiopian landraces of lentil (*Lens culinaris*). Genet. Res. Crop Evol. 43:293–301.

Carter, V.G. 1974. Topsoil and civilization. Univ. Oklahoma Press, Norman.

Ceccarelli, S., S. Grando, E. Bailey, A. Amri, M. El-Faleh, F. Nassif, S. Rezqui, and A. Yahyaoui. 2001. Farmer participation in barley breeding in Syria, Morocco, and Tunisia. Euphytica 122:521–536.

Ceccarelli, S., S. Grando, and A. Impiglia. 1998. Choice of selection strategy in breeding barley for stress environments. Euphytica 103:307–318.

Ceccarelli, S., S. Grando, R. Tutwiler, J. Baha, A.M. Martini, H. Salahieh, A. Goodchild, and M. Michael. 2000. A methodological study on participatory barley breeding I. Selection phase. Euphytica 111:91–104.

Ceccarelli, S., J. Valkoun, W. Erskine, S. Weigand, R. Miller, and J.A.G. Van Leur. 1992. Plant genetic resources and plant improvement as tools to develop sustainable agriculture. Exp. Agric. 28:89–98.

Cooper, P.J.M. P.J. Gregory, D. Tully, and H.C. Harris. 1987. Improving water use efficiency of annual crops on the rainfed farming systems of West Asia and North Africa. Exp. Agric. 23:113–158.

Dakkeel, A.J., J.M. Peacock, I. Auji, and E. Acevedo. 1994. Carbon isotope discrimination and water-use efficiency in barley under field conditions. Aspects Appl. Biol. 38:173–183.

Damania, A.B., J. Valkoun, G. Willcox, and C.O. Qualset (ed.) 1998. The origins of agriculture and crop domestication. The Harlan Symp. ICARDA, Aleppo, Syria.

De Pauw, E. 2004. Drought early warning systems for the Near East. p. 93–112. *In* S.C. Rao and J. Ryan (ed.) Challenges and strategies for dryland agriculture. CSSA Spec. Publ. 32. CSSA and ASA, Madison, WI.

Di Vito, M., K.B. Singh, N. Greco, and M.C. Saxena. 1996. Sources of resistance to cyst nematode in cultivated and wild *Cicer* species. Genet. Res. Crop Evol. 43:103–107.

El-Bouhssini, M., J.H. Hatchett, T.S. Cox, and G.E. Wilde. 2001, Genotypic interaction between resistance genes in wheat and virulence genes in the Hessian fly *Mayetiola destructor(Diptera: Cecidomyiidae)*. Bull. Entomol. Res. 91:327–331.

Eujayl, I., W. Erskine, M. Baum, and E. Pehu. 1999. Inheritance and linkage analysis of frost injury in lentil. Crop Sci. 39(3):639–642.

Gibbon, D. 1981. Rainfed systems in the Mediterranean region. Plant Soil 58:59–80.

Gintzburger, G., M. Bounejmate, and A. Nefzaoui (ed.). 2000. Fodder shrub development in arid and semi-arid zones. *In* Proc., Workshop on "Native and Exotic Fodder Shrubs in Arid and Semi-arid Zones", Hammamet, Tunisia. 27 Oct.–2 Nov. 1996. ICARDA, Aleppo, Syria.

Grando, S., S. Ceccarelli, and B. Tickle. 2000. Diversity of barley landraces from the Near East. Barley Genet. VIII:13–15.

Gregory, P.J., K.D. Shepherd, and P.J. Cooper. 1986. Effects of fertilizer on root growth and water use of barley in northern Syria. J. Agric. Sci. 103:429–438.

Haddad, N., R. Tutwiler, and E. Thomson (ed.) 1997. Improvement of crop-livestock integration systems in West Asia and North Africa. *In* Proc., Regional Symposium on "Integrated Crop- Livestock Systems in the Dry Areas of West Asia and North Africa", Amman, Jordan. 6–8 Nov. 1995. ICARDA, Aleppo, Syria.

Hamdi, A., and W. Erskine. 1997. Reaction of wild species of the genus *Lens* to drought. Euphytica 91:173–179.

Harmsen, K. 1984. Nitrogen fertilizer use in rainfed agriculture. Fert. Res. 5(4):371–382.

Harris, H.C. 1994. Water use efficiency of crop rotations in a Mediterranean environment. Aspects Appl. Biol. 38:165–172.

Harris, H. 1995. Long-term trials on soil and crop management at ICARDA. Adv. Soil Sci. 19:447–469.

Harris, H.C., P.J.M. Cooper, and M. Pala (ed.) 1991. Soil and crop management for improved water use efficiency in rainfed areas. *In* Proc., Int. Workshop, Ankara, Turkey. 15–19 May 1989. ICARDA, Aleppo, Syria.

Ismail, S.F., O.F. Mamluk, and M.F. Azmeh. 1995. New pathotypes of common bunt of wheat from Syria. Phytopathol. Medet. 34:1–6.

Jones, M.J. (ed.) 1998. The challenge of production sustainability: Long-term studies in agronomic research in dry areas. Abstracts and conclusions of a workshop held at ICARDA. 8–11 Dec. 1997. ICARDA, Aleppo, Syria.

Jones, M.J. 2000. Anticipatory long-term research for sustainable productivity. Exp. Agric. 36:137–150.

Karajeh, F., J. Ryan, and C. Studer (ed.) 2002. On-farm soil and water management in Central Asia. Proc., Int. Workshop, Tashkent, Uzbekistan. 17–19 May 1999. ICARDA, Aleppo, Syria.

Kassam, A.H. 1981. Climate, soil and land resources in the West Asia and North Africa. Plant Soil 58:1–28.

Lal, R. 2002. Carbon sequestration in dryland ecosystems of West Asia and North Africa. Land Degrad. Devel. 13:45–59.

Makkouk, K., A. Comeau, and C.A. St-Pierre. 1994. Screening of barley yellow dwarf luteovirus resistance in barley on the basis of virus movement. J. Phytopathol. 141:165–172.

Matar, A., J. Torrent, and J. Ryan. 1992. Soil and fertilizer phosphorus and crop responses in the dryland Mediterranean zone. Adv. Soil Sci. 18:82–146.

Nachit, M.M., I. Elouafi, M.A. Pagnotta, A. El-Saleh, E. Iacono, M. Labhilili, A. Asbati, M.Azrak, H. Hazzam, D. Benscher, M. Khairallah, J.M. Ribaut, O.A. Tanzarella, E.Porceddu, and M.E. Sorrells. 2001. Molecular linkage map for an intraspecific recombinant inbred population of durum wheat (*Triticum turgidum* L. var. durum). Theor. Appl. Genet. 102:177–186.

Ngaido, T., and M. El-Mourid. 2002. The Magreb-Mashreq project in West Asia-North Africa: A model for community-based action. *In* 2002 Annual meetings abstracts [CD-Rom]. ASA, CSSA, and SSSA, Madison, WI.

Nour, M. 2002. ICARDA 25: A promise of hope. ICARDA, Aleppo, Syria.

Oweis, T., A. Hachum, and J. Kijne. 1999. Water harvesting and supplemental irrigation for improved water-use efficiency in dry areas. System-Wide Initiative on Water Manage. (SWIM) Paper 7. IWMI, Colombo, Sri Lanka.

Oweis, T., M. Pala, and J. Ryan. 1998. Stabilizing rainfed wheat yields with supplemental irrigation in the Mediterranean region. Agron. J. 90:672–681.

Pala, M., H.C. Harris, J. Ryan, R. Makboul, and S. Dozom. 2000. Tillage systems and stubble management in a Mediterranean-type environment in relation to crop yield and soil moisture. Exp. Agric. 36:223–242.

Pala, M., and A. Mazid. 1992. On-farm assessment of improved crop production practices in northwest Syria: Chickpea. Exp. Agric. 28:175–184.

Pala, M., J. Ryan, A. Mazid, O. Abdallah, and M. Nachit. 2004. Wheat farming in Syria: An approach to economic transformation and sustainability. Renewable Agric. Food Syst. (in press).

Robertson, L.D., K.B. Singh, W. Erskine, and A.M. Abd El Moneim. 1996. Use of genetic diversity in germplasm collections to improve food and forage legumes for West Asia and North Africa. Genet. Res. Crop Evol. 43:447–460.

Ryan, J. 1982. Perspective on soil erosion and conservation. Publ. no. 69. Fac. Agric. and Food Sci., Am. Univ., Beirut, Lebanon.

Ryan, J. 1997. Accomplishments and future challenges in dryland soil fertility research in the Mediterranean area. Proc., Soil Fertility Workshop, Tel Hadya, Aleppo, Syria.19–23 Nov. 1995. ICARDA, Aleppo, Syria.

Ryan, J. 1998. Changes in organic carbon in long-term rotation and tillage trials in northern Syria. p. 285–295. *In* R. Lal et al. (ed.) Management of carbon sequestration in soil. Adv. Soil Sci. CRC, Boca Raton, FL.

Ryan, J. 2004. Soil fertility enhancement in Mediterranean-type dryland agriculture: A prerequisite for development. p. 275–290. *In* S.C. Rao and J. Ryan (ed.) Challenges and strategies of dryland agriculture. CSSA Spec. Publ. 32. CSSA and ASA, Madison, WI.

Ryan, J. (ed.) 2000. Plant nutrient management under pressurized irrigation systems in the Mediterranean region. Proc. Int. Workshop, Amman, Jordan. 25–27 Apr. 1999.

Ryan, J. (ed.) 2002. Desert and dryland development: Challenges and potential in the new millennium. Proc., Sixth Int. Conf. on the Development of Dry Lands, Cairo, Egypt. 22–27 Aug. 1999. ICARDA, Aleppo, Syria.

Ryan, J. (ed.) 2003. Desert and dryland development: Challenges and potential in the new millennium. *In* Proc., Int. Desert Development Conf., Cairo, Egypt. 22–27 Aug.1999. ICARDA, Aleppo, Syria.

Ryan, J., and M. Abdel Monem. 1998. Soil fertility for sustained production in West Asia-North Africa region: Need for long-term research. p. 155–174. *In* R. Lal (ed.) Soil quality and agricultural sustainability. Ann Arbor Press, Chelsea, MI.

Ryan, J., S. Masri, S. Garabet, J. Diekmann, and H. Habib. 1997. Soils of ICARDA's agricultural experimental stations and sites: Climate, classification, physical-chemical properties and land use. ICARDA Tech. Bull. ICARDA, Aleppo, Syria.

Ryan, J., S. Masri, and F. Karajeh, 1999. Use of untreated sewage water in Syria: A bane or blessing? p. 332–333. *In* Soils with a Mediterranean Type of Climate, 6th Int. Meet., Barcelona, Spain. 4–9 July. Extended abstracts.

Ryan, J., S. Masri, M. Pala, and M. Bounejmate. 2002. Barley—Based rotations in a typical Mediterranean agroecosystem: Crop production trends and soil quality. Options Medeterrenees, Ser. A No. 50:287–296.

Ryan, J., and A. Matar (ed.) 1990. Proc. Third Regional Soil Test Calibration Workshop. Amman, Jordan. 2–9 September. ICARDA, Aleppo, Syria.

Ryan, J., and A.E. Matar (ed.) 1992. Fertilizer use efficiency under rainfed agriculture. Proc. Fourth Regional Soil Test Calibration Workshop, Agadir, Morocco. 5–11 May 1991. ICARDA, Aleppo, Syria.

Ryan, J., L. Materon, and S. Christiansen. 1995. The networks for research collaboration in the dryland West Asia–North Africa region. J. Nat. Res. Life Sci. Edu. 24:155–[illegible].

Saxena, M.C., and K.B. Singh (ed.) 1984. Ascochyta blight and winter sowing of chickpeas. Martinus Nijhoff/Dr. W. Junk Publ. The Hague, The Netherlands.

Saxena, M.C., and K.B. Singh. 1987. The chickpea. CABI Int., Wallingford, UK.

Silim, S.N., M.C. Saxena, and W. Erskine. 1991. Effect of sowing date on the growth and yield of lentil in a rainfed Mediterranean environment. Exp. Agric. 27:145–154.

Singh, K.B., and M.C. Saxena. 1993. Breeding for stress tolerance in cool-season food legumes. John Wiley & Sons, Chichester, UK, and ICARDA, Aleppo, Syria.

Singh, K.B., R.S. Malhotra, and M.C. Saxena. 1995. Additional sources of tolerance to cold in cultivated and wild *Cicer* species. Crop Sci. 35:1491–1497.

Singh, K.B., R.S. Malhotra, M.C. Saxena, and G. Bejiga. 1997. Superiority of winter sowing over traditional spring sowing of chickpea in the Mediterranean region. Agron. J. 87: 692–698.

Sivakumar, M.U.K., M.A. Zöbish, S. Koala, and T. Maokonen. 1998. Wind erosion in Africa and West Asia: Problems and control strategies. ICARDA, Aleppo, Syria.

Srivastava, J.P., and A.B. Damania. 1990. Wheat genetic resources: meeting diverse needs. John Willey & Sons, Chichester, UK.

Tahir, M., and M. Singh. 1993. Assessment of screening techniques for heat tolerance in wheat. Crop Sci. 33:740–744.

Unger, P.W., T.V. Sneed, W.R. Jordan, and R. Jensen (ed.) 1988. Challenges in dryland agriculture: A global perspective. *In* Proc., Int. Conf. on Dryland Farming. 15–19 Aug. 1998. Texas Agric. Exp. Stn., College Station, Bushland.

Van Duivenboden, N., M. Pala, C. Studer, and C.L. Bielders. 1999. Efficient soil water use: The key to sustainable crop production in the dry areas of West Asia and North and Sub-Saharan Africa. ICARDA, Aleppo, Syria, and ICRISAT, Andhra Pradesh, India.

Van Oosterom, E.J., and E. Acevedo. 1992. Adaptation of barley (*Hordeum vulgare* L.) to harsh Mediterranean environments. I. Morphological trials. Euphytica. 62:1–14.

Webb, C., and G. Hawtin. 1981. Lentils. Commonwelt Agric. Bureaux, Slough, UK, and ICARDA, Aleppo, Syria.

White, R., D. Tunstall, and N. Henninger. 2002. An ecosystem approach to drylands: Building support for new development policies. World Resources Inst., Information Policy Brief no. 1, Feb. 2002.

Whitman, C.E., J.F. Parr, R.I. Papendick, and R. Meyer. 1989. Soil, water, and crop/livestock management systems for rainfed agriculture in the Near East region. Workshop Proc., 18–23 Jan. 1986. USAID, Washington, DC. and ICARDA, Aleppo, Syria.

Zaharieva, M., P. Monneveaux, M. Henry, R. Rivoal, J. Valkoun, and M.M. Nachit. 2001. Evaluation of a collection of wild wheat relative *Aegilops geniculata* Roth and identification of potential sources for useful traits. Euphytica 119:33–38.